Student Solutions Manual for

CHEMICAL PRINCIPLES

Kenton H. Whitmire and Charles Trapp

W. H. FREEMAN AND COMPANY
NEW YORK

ISBN 0-7167-3360-9

Printed in the United States of America

First printing 1999

PREFACE

This manual contains solutions and answers to the odd-numbered exercises in Atkins and Jones' *Chemical Principles.* The rules of significant figures have been adhered to in reporting the numerical answers for all exercises. In exercises with multiple parts, the properly rounded values were used for subsequent calculations.

Procedures and results that have been fully illustrated in the solutions to the exercises of an earlier chapter or in the introductory section, "Fundamentals," may not be repeated in the solutions to the exercises of later chapters. For example, the conversion between °C and K, the determination of molar mass from the formula of a compound, and other simple conversions such as mL to L or cm to m, are not usually worked out in the solutions.

Some abbreviations have been used to save space. These are usually defined at the start of a solution. Some that occur frequently are

n = neutron
p = proton
e = electron

I wish to thank Peter Atkins and Loretta Jones for their support and help during the production of this manual as well as for many stimulating exchanges about chemical principles and educational pedagogy. They, along with Julie Francis and Max Bishop, have reviewed the solutions and made many valuable suggestions. I would also like to acknowledge the staff at W. H. Freeman, especially Matthew Fitzpatrick, Jodi Isman, and Michelle Julet, who have been instrumental in getting this project to print.

K.H.W.

FUNDAMENTALS

A.1 (a) physical; (b) physical; (c) chemical

A.3 The temperature, humidity, and the evaporation of water are physical properties. The ripening of oranges is a chemical change.

A.5 (a) intensive; (b) intensive; (c) extensive; (d) extensive

A.7 $d = \frac{m}{V}$

$$= \left(\frac{3.60\ \text{g}}{9.8\ \text{mL} - 8.3\ \text{mL}}\right)\left(\frac{1\ \text{mL}}{1\ \text{cm}^3}\right)$$
$$= 2.4\ \text{g}\cdot\text{cm}^{-3}$$

A.9 $d = \frac{m}{V}$, rearranging gives $V = \frac{m}{d}$

$$= \left(\frac{0.300\ \text{carat}}{3.51\ \text{g}\cdot\text{cm}^{-3}}\right)\left(\frac{200\ \text{mg}}{1\ \text{carat}}\right)\left(\frac{1\ \text{g}}{1000\ \text{mg}}\right)$$
$$= 0.0171\ \text{cm}^3$$

A.11 $d = \frac{m}{V}$

$$= \left(\frac{2.0 \times 10^{-23}\ \text{g}}{\frac{4}{3}\pi(77\ \text{pm})^3}\right)\left(\frac{1\ \text{pm}}{1 \times 10^{-10}\ \text{cm}}\right)^3$$
$$= 10\ \text{g}\cdot\text{cm}^{-3}$$

Since the density of diamond is much less than the density of a carbon atom, the crystalline form of carbon must contain considerable empty space.

A.13 $E_K = \frac{1}{2}mv^2$

$$= \frac{1}{2}(4.2\ \text{kg})(14\ \text{km}\cdot\text{h}^{-1})^2\left(\frac{1\ \text{h}}{3600\ \text{s}}\right)^2\left(\frac{1000\ \text{m}}{1\ \text{km}}\right)^2$$
$$= 32\ \text{kg}\cdot\text{m}^2\cdot\text{s}^{-2}$$
$$= 32\ \text{J}$$

A.15 $E_K = \frac{1}{2}mv^2$

$$= \frac{1}{2}(2.5 \times 10^5\ \text{kg})(25\ \text{km}\cdot\text{s}^{-1})^2\left(\frac{1000\ \text{m}}{1\ \text{km}}\right)^2$$
$$= 7.8 \times 10^{13}\ \text{kg}\cdot\text{m}^2\cdot\text{s}^{-2}$$
$$= 7.8 \times 10^{13}\ \text{J}$$

A.17 $\mathcal{V} = mgh$

$$= (0.51\ \text{kg})(9.81\ \text{m}\cdot\text{s}^{-2})(3.0\ \text{m})$$
$$= 15\ \text{kg}\cdot\text{m}^2\cdot\text{s}^{-2}$$
$$= 15\ \text{J}$$

A.19 When the skier is at rest at the top of the slope, the kinetic energy is 0. The total energy is equal to the potential energy. At the bottom of the slope, the potential energy is 0; it has been converted into kinetic energy.

$$E = E_K + \mathcal{V}$$
$$= \mathcal{V}$$
$$= mgh$$
$$= (112\ \text{kg})(9.81\ \text{m}\cdot\text{s}^{-2})(140\ \text{m})$$
$$= 1.5 \times 10^5\ \text{kg}\cdot\text{m}^2\cdot\text{s}^{-2}$$
$$= 1.5 \times 10^5\ \text{J}$$

A.21 $F = ma$

$$= (9.1 \times 10^{-31}\ \text{kg})(1.2 \times 10^6\ \text{m}\cdot\text{s}^{-2})$$
$$= 1.1 \times 10^{-24}\ \text{kg}\cdot\text{m}\cdot\text{s}^{-2}$$
$$= 1.1 \times 10^{-24}\ \text{N}$$

B.1 number of beryllium atoms $= \dfrac{\text{mass of sample}}{\text{mass of one atom}}$

$$= \left(\frac{0.210\ \text{g}}{1.50 \times 10^{-26}\ \text{kg}\cdot\text{atom}^{-1}}\right)\left(\frac{1\ \text{kg}}{1000\ \text{g}}\right)$$
$$= 1.40 \times 10^{22}\ \text{atoms}$$

B.3 (a) Radiation may pass through a metal foil. (b) All light (electromagnetic radiation) travels at the same speed; the slower speed supports the particle model.

(c) This observation supports the radiation model. (d) This observation supports the particle model; radiation has no mass and no charge.

B.5 (a) A tritium atom has 1 proton, 2 neutrons, and 1 electron. (b) A ^{60}Co atom has 27 protons, 33 neutrons, and 27 electrons. (c) An ^{16}O atom has 8 protons, 8 neutrons, and 8 electrons. (d) A ^{204}Pb atom has 82 protons, 122 neutrons, and 82 electrons.

B.7 (a) ^{111}Cd; (b) ^{82}Kr; (c) ^{11}B

B.9 (a) ^{12}C, ^{13}C, and ^{14}C all have 6 protons and 6 electrons. (b) They have different numbers of neutrons: an atom of ^{12}C has 6 neutrons, an atom of ^{13}C has 7 neutrons, and an atom of ^{14}C has 8 neutrons.

B.11 (a) Atoms of ^{55}Mn, ^{56}Fe, and ^{58}Ni all have 30 neutrons. (b) They have different masses, numbers of protons, and numbers of electrons.

B.13 (a) In H_2O, the oxygen atom contributes 8 protons, 8 neutrons, and 8 electrons. Each hydrogen atom contributes 1 proton and 1 electron. There are 10 protons, 8 neutrons, and 10 electrons total.

(b) $\text{mass of protons} = (10)(1.673 \times 10^{-24}\text{ g}) = 1.673 \times 10^{-23}\text{ g}$

$\text{mass of neutrons} = (8)(1.675 \times 10^{-24}\text{ g}) = 1.340 \times 10^{-23}\text{ g}$

$\text{mass of electrons} = (10)(9.109 \times 10^{-28}\text{ g}) = 9.109 \times 10^{-27}\text{ g}$

(c) fraction of neutrons

$$= \frac{\text{mass of neutrons}}{\text{mass of water molecule}} \times 100\%$$

$$= \frac{1.340 \times 10^{-23}\text{ g}}{(1.673 \times 10^{-23}\text{ g} + 1.340 \times 10^{-23}\text{ g} + 9.109 \times 10^{-27}\text{ g})} \times 100\%$$

$$= 44.5\%$$

Note that the mass of an oxygen atom is actually not exactly the sum of its protons, neutrons and electrons due to the nuclear binding energy. See Chapter 17.

B.15 (a) Lithium is a Group 1 metal. (b) Gallium is a Group 13 metal. (c) Xenon is a Group 18 nonmetal. (d) Potassium is a Group 1 metal.

B.17 (a) Cd, metal; (b) Cr, metal; (c) Sb, metalloid

B.19 lithium, Li, 3; sodium, Na, 11; potassium, K, 19; rubidium, Rb, 37; cesium, Cs, 55. All the alkali metals react with water to produce hydrogen. Lithium reacts gently; the reaction becomes more vigorous down the periodic table. The melting points decrease down the periodic table.

B.21 (a) Metals are malleable, conduct electricity, and have luster. (b) Nonmetals are not malleable or ductile and do not conduct electricity.

C.1 (a) An ionic compound is made up of ions. Sodium chloride is an example of an ionic compound. (b) A molecular compound is made up of molecules. Sucrose (sugar) is an example of a molecular compound. Ionic compounds have higher melting points than solids made of molecules. Ionic compounds dissolve only in polar solvents if at all. Molecules are more likely to dissolve in nonpolar solvents. See Chapter 5.

C.3 (a) 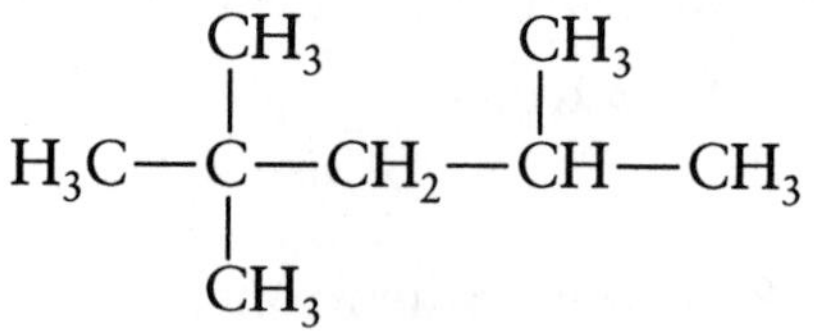

(b) 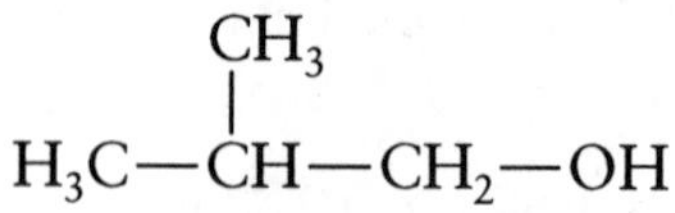

C.5 (a) Sulfur is a nonmetal in Group 16; it will form S^{2-} ions. (b) Potassium is a metal in Group 1 and will form K^+ ions. (c) Strontium is a Group 2 metal and will form Sr^{2+} ions. (d) Chlorine is a Group 17 nonmetal and will form Cl^- ions.

C.7 (a) $^{2}H^+$ has 1 proton, 1 neutron, and no electrons. (b) $^{9}Be^{2+}$ has 4 protons, 5 neutrons, and 2 electrons. (c) $^{80}Br^-$ has 35 protons, 45 neutrons, and 36 electrons. (d) $^{32}S^{2-}$ has 16 protons, 16 neutrons, and 18 electrons.

C.9 (a) $^{19}F^-$; (b) $^{24}Mg^{2+}$; (c) $^{128}Te^{2-}$; (d) $^{86}Rb^+$

C.11 (a) Aluminum forms Al^{3+} ions; sulfur forms S^{2-} ions. Two aluminum atoms produce a charge of $2 \times +3 = +6$. Three sulfur atoms produce a charge of $3 \times -2 = -6$. The formula for aluminum sulfide is Al_2S_3. (b) Sodium forms Na^+ ions and oxygen forms O^{2-} ions. Two sodium ions produce a charge of

$2 \times +1 = +2$, which is required to balance the charge on one O^{2-} ion. The formula for sodium oxide is Na_2O. (c) Magnesium forms +2 ions; sulfur forms −2 ions. The formula for magnesium sulfide is MgS. (d) Barium forms +2 ions and iodine forms −1 ions. Two iodide ions are required to balance the charge of one barium ion, so the formula is BaI_2.

D.1 (a) Magnesium forms +2 ions and oxygen forms −2 ions, so the formula is MgO. (b) Calcium forms +2 ions and the phosphate ion is PO_4^{3-}, so the formula is $Ca_3(PO_4)_2$. (c) Aluminum forms +3 ions and the sulfate ion is SO_4^{2-}, giving a formula of $Al_2(SO_4)_3$. (d) Calcium forms Ca^{2+} ions and the nitride ion is N^{3-}. The formula of calcium nitride is Ca_3N_2.

D.3 (a) potassium phosphate; (b) iron(II) iodide, ferrous iodide; (c) niobium(V) oxide; (d) copper(II) sulfate, cupric sulfate

D.5 (a) SeO_3; (b) CCl_4; (c) CS_2; (d) SF_6; (e) As_2S_3; (f) PCl_5; (g) N_2O; (h) ClF_3

D.7 (a) sulfur tetrafluoride; (b) dinitrogen pentoxide; (c) nitrogen triiodide; (d) xenon tetrafluoride; (e) arsenic tribromide; (f) chlorine dioxide

D.9 (a) hydrochloric acid; (b) sulfuric acid; (c) nitric acid; (d) acetic acid; (e) sulfurous acid; (f) phosphoric acid

D.11 (a) Na_2O; (b) K_2SO_4; (c) AgF; (d) $Zn(NO_3)_2$; (e) Al_2S_3

D.13 (a) silver sulfide; (b) zinc chloride; (c) chlorine pentafluoride; (d) magnesium hydroxide; (e) nickel(II) sulfate hexahydrate; (f) phosphorus pentachloride; (g) chromium(III) dihydrogen phosphate; (h) diarsenic trioxide; (i) manganese(II) chloride

D.15 (a) $(NH_4)_2SO_3$; (b) Fe_2O_3; (c) $Cu(BrO_3)_2$; (d) PH_3; (e) $Ca(HCO_3)_2$; (f) HCN; (g) $LiHSO_4$; (h) SeF_4; (i) $FeSO_4 \cdot 7H_2O$

D.17 (a) telluric acid; (b) sodium arsenate; (c) calcium selenite; (d) barium antimonate; (e) arsenic acid; (f) cobalt(III) tellurate

E.1 (a) $$\text{moles of people} = \frac{5.7 \times 10^{9}\ \text{people}}{6.022 \times 10^{23}\ \text{people}\cdot\text{mol}^{-1}}$$

$$= 9.5 \times 10^{-15}\ \text{mol}$$

(b) $$\text{time} = \frac{1\ \text{mol peas}}{9.5 \times 10^{-15}\ \text{mol}\cdot\text{s}^{-1}} = 1.1 \times 10^{14}\ \text{s}$$

$$(1.1 \times 10^{14}\ \text{s})\left(\frac{1\ \text{h}}{3600\ \text{s}}\right)\left(\frac{1\ \text{day}}{24\ \text{h}}\right)\left(\frac{1\ \text{yr}}{365\ \text{days}}\right) = 3.5 \times 10^{6}\ \text{years}$$

E.3 (a) mass of average Li atom

$$= \left(\frac{7.42}{100}\right)(9.988 \times 10^{-24}\ \text{g}) + \left(\frac{92.58}{100}\right)(1.165 \times 10^{-23}\ \text{g})$$

$$= 1.153 \times 10^{-23}\ \text{g}\cdot\text{atom}^{-1}$$

$$\text{molar mass} = (1.153 \times 10^{-23}\ \text{g}\cdot\text{atom}^{-1})(6.022 \times 10^{23}\ \text{atoms}\cdot\text{mol}^{-1})$$

$$= 6.94\ \text{g}\cdot\text{mol}^{-1}$$

(b) mass of average Li atom

$$= \left(\frac{5.67}{100}\right)(9.988 \times 10^{-24}\ \text{g}) + \left(\frac{100 - 5.67}{100}\right)(1.165 \times 10^{-23}\ \text{g})$$

$$= 1.1556 \times 10^{-23}\ \text{g}\cdot\text{atom}^{-1}$$

$$\text{molar mass} = (1.1556 \times 10^{-23}\ \text{g}\cdot\text{atom}^{-1})(6.022 \times 10^{23}\ \text{atoms}\cdot\text{mol}^{-1})$$

$$= 6.96\ \text{g}\cdot\text{mol}^{-1}$$

E.5 The percentage ^{10}B = 100 − percentage ^{11}B

$$\text{molar mass} = \left(\frac{\%\ ^{10}\text{B}}{100\%}\right)(\text{mass}\ ^{10}\text{B}) + \left(\frac{\%\ ^{11}\text{B}}{100\%}\right)(\text{mass}\ ^{11}\text{B})$$

$$= \left(\frac{100\% - \%\ ^{11}\text{B}}{100\%}\right)(\text{mass}\ ^{10}\text{B}) + \left(\frac{\%\ ^{11}\text{B}}{100\%}\right)(\text{mass}\ ^{11}\text{B})$$

Rearranging gives

$$\%\ ^{11}\text{B} = \frac{100\cdot\text{molar mass} - 100\cdot\text{mass}\ ^{10}\text{B}}{\text{mass}\ ^{11}\text{B} - \text{mass}\ ^{10}\text{B}}$$

$$= \frac{100(10.81\ \text{g}\cdot\text{mol}^{-1}) - 100(10.013\ \text{g}\cdot\text{mol}^{-1})}{11.093\ \text{g}\cdot\text{mol}^{-1} - 10.013\ \text{g}\cdot\text{mol}^{-1}}$$

$$= 73.8\%$$

$$\%\ ^{11}\text{B} = 26.2\%$$

E.7 (a) $$\frac{25\ \text{g}}{12.01\ \text{g}\cdot\text{mol}^{-1}} = 2.1\ \text{mol C}$$

$$\frac{35\ \text{g}}{28.09\ \text{g}\cdot\text{mol}^{-1}} = 1.3\ \text{mol Si}$$

25 g of carbon contains more moles of atoms than 35 g of silicon does

(b) $\dfrac{1.0\ \text{g}}{196.97\ \text{g}\cdot\text{mol}^{-1}} = 5.1 \times 10^{-3}$ mol Au

$\dfrac{1.0\ \text{g}}{200.59\ \text{g}\cdot\text{mol}^{-1}} = 5.0 \times 10^{-3}$ mol Hg

1.0 g of Au has slightly more atoms than 1.0 g of Hg does

(c) Because the two samples have the same number of atoms, they will have the same number of moles, which is given by $\dfrac{2.49 \times 10^{22}\ \text{atoms}}{6.022 \times 10^{23}\ \text{atoms}\cdot\text{mol}^{-1}} =$ 0.0413 mol

E.9 (a) $\left(\dfrac{12\ \text{g}}{12.01\ \text{g}\cdot\text{mol}^{-1}}\right)(6.022 \times 10^{23}\ \text{atoms}\cdot\text{mol}^{-1}) = 6.02 \times 10^{23}$ C atoms

$m_{\text{Ni}} = \left(\dfrac{6.02 \times 10^{23}\ \text{atoms}}{6.022 \times 10^{23}\ \text{atoms}\cdot\text{mol}^{-1}}\right)(58.71\ \text{g}\cdot\text{mol}^{-1}) = 59\ \text{g}$

(b) $\left(\dfrac{12\ \text{g}}{52.00\ \text{g}\cdot\text{mol}^{-1}}\right)(6.022 \times 10^{23}\ \text{atoms}\cdot\text{mol}^{-1}) = 1.4 \times 10^{23}$ Cr atoms

$m_{\text{Ni}} = \left(\dfrac{1.4 \times 10^{23}\ \text{atoms}}{6.022 \times 10^{23}\ \text{atoms}\cdot\text{mol}^{-1}}\right)(58.69\ \text{g}\cdot\text{mol}^{-1}) = 14\ \text{g}$

E.11 (a) molar mass of CCl_4 = 153.81 g·mol^{-1}

$n_{\text{CCl}_4} = \dfrac{10.0\ \text{g}}{153.81\ \text{g}\cdot\text{mol}^{-1}} = 0.0650\ \text{mol}$

$N_{\text{CCl}_4} = (0.0650\ \text{mol})(6.022 \times 10^{23}\ \text{atoms}\cdot\text{mol}^{-1}) = 3.92 \times 10^{22}$ molecules

(b) molar mass of HI = 127.91 g·mol^{-1}

$n_{\text{HI}} = \dfrac{1.65 \times 10^{-3}\ \text{g}}{127.91\ \text{g}\cdot\text{mol}^{-1}} = 1.29 \times 10^{-5}\ \text{mol}$

$N_{\text{HI}} = (1.29 \times 10^{-5}\ \text{mol})(6.022 \times 10^{23}\ \text{atoms}\cdot\text{mol}^{-1}) = 7.77 \times 10^{18}$ molecules

(c) molar mass of hydrazine = 32.05 g·mol^{-1}

$n_{\text{N}_2\text{H}_4} = \dfrac{3.77 \times 10^{-3}\ \text{g}}{32.05\ \text{g}\cdot\text{mol}^{-1}} = 1.18 \times 10^{-4}\ \text{mol}$

$N_{\text{N}_2\text{H}_4} = (1.18 \times 10^{-4}\ \text{mol})(6.022 \times 10^{23}\ \text{atoms}\cdot\text{mol}^{-1}) = 7.08 \times 10^{19}$ molecules

(d) molar mass of sucrose = 342.29 g·mol^{-1}

$n_{\text{sucrose}} = \dfrac{500\ \text{g}}{342.29\ \text{g}\cdot\text{mol}^{-1}} = 1.46\ \text{mol}$

$N_{\text{sucrose}} = (1.46\ \text{mol})(6.022 \times 10^{23}\ \text{atoms}\cdot\text{mol}^{-1}) = 8.79 \times 10^{23}$ molecules

(e) molar mass of O atoms = 16.00 g·mol^{-1}

$n_{\text{O}} = \dfrac{2.33\ \text{g}}{16.00\ \text{g}\cdot\text{mol}^{-1}} = 0.146\ \text{mol}$

$N_{\text{O}} = (0.146\ \text{mol})(6.022 \times 10^{23}\ \text{atoms}\cdot\text{mol}^{-1}) = 8.77 \times 10^{22}$ atoms molar mass of O_2 molecules $= 32.00\ g\cdot mol^{-1}$

$$n_{O_2} = \frac{2.33\ g}{32.00\ g\cdot mol^{-1}} = 0.0728\ mol$$

$$N_{O_2} = (0.0728\ mol)(6.022 \times 10^{23}\ atoms\cdot mol^{-1}) = 4.38 \times 10^{22}\ molecules$$

E.13 (a) molar mass of AgCl $= 143.32\ g\cdot mol^{-1}$

$$n_{AgCl} = \frac{2.00\ g}{143.32\ g\cdot mol^{-1}} = 0.0140\ mol$$

The number of moles of Ag^+ ions equals the number of moles of AgCl.

(b) molar mass of $UO_3 = 286.03\ g\cdot mol^{-1}$

$$n_{UO_3} = \frac{600\ g}{286.03\ g\cdot mol^{-1}} = 2.10\ mol$$

(c) molar mass of $FeCl_3 = 162.20$

$$n_{FeCl_3} = \left(\frac{4.19\ mg}{162.20\ g\cdot mol^{-1}}\right)\left(\frac{1\ g}{1000\ mg}\right) = 2.58 \times 10^{-5}\ mol$$

The number of moles of Cl^- ions equals 3 times the number of moles of $FeCl_3$.

$n_{Cl^-} = 7.75 \times 10^{-5}$ mol.

(d) molar mass of $AuCl_3\cdot 2H_2O = 339.35\ g\cdot mol^{-1}$

$$\left(\frac{2\ mol\ H_2O}{1\ mol\ AuCl_3\cdot 2H_2O}\right)(n_{AuCl_3\cdot 2H_2O}) = 2\left(\frac{1.00\ g}{339.35\ g\cdot mol^{-1}}\right)$$

$$= 5.89 \times 10^{-3}\ mol\ H_2O$$

E.15 (a) number of formula units $= (0.670\ mol)(6.022 \times 10^{23}\ formula\ units\cdot mol^{-1})$

$$= 4.03 \times 10^{23}\ formula\ units$$

(b) molar mass of $Rb_2SO_4 = 267.00\ g\cdot mol^{-1}$

$$\left(\frac{2.39 \times 10^{20}\ formula\ units}{6.022 \times 10^{23}\ formula\ units\cdot mol^{-1}}\right)(267.00\ g\cdot mol^{-1})\left(\frac{1000\ mg}{1\ g}\right) = 106\ mg$$

(c) molar mass of $NaHCO_2 = 68.01\ g\cdot mol^{-1}$

$$\left(\frac{6.66\ kg}{68.01\ g\cdot mol^{-1}}\right)\left(\frac{1000\ g}{1\ kg}\right)(6.022 \times 10^{23}\ formula\ units\cdot mol^{-1})$$

$$= 5.90 \times 10^{25}\ formula\ units$$

E.17 (a) molar mass of $H_2O = 18.02\ g\cdot mol^{-1}$

$$\left(\frac{18.02\ g\cdot mol^{-1}}{6.022 \times 10^{23}\ molecules\cdot mol^{-1}}\right) = 2.992 \times 10^{-23}\ g\cdot molecule^{-1}$$

(b) $N_{H_2O} = \left(\frac{1.00\ g}{18.02\ g\cdot mol^{-1}}\right)(6.022 \times 10^{23}\ molecules\cdot mol^{-1}) = 3.34 \times 10^{22}$ molecules

E.19 (a) molar mass of $CuBr_2 \cdot 4H_2O = 295.42\ g \cdot mol^{-1}$

$$\left(\frac{5.50\ g}{295.42\ g \cdot mol^{-1}}\right) = 1.86 \times 10^{-2}\ mol$$

(b) Because there are 2 mol Br^- per mole of compound, the number of moles will be twice the amount in part (a), 3.72×10^{-2} mol.

(c) $$\left(\frac{4\ mol\ H_2O}{1\ mol\ CuBr_2 \cdot 4H_2O}\right)(1.86 \times 10^{-2}\ mol)(6.022 \times 10^{23}\ molecules \cdot mol^{-1})$$

$$= 4.48 \times 10^{22}\ molecules\ H_2O$$

(d) $$\text{fraction of mass due to Cu} = \frac{63.54\ g}{295.42\ g \cdot mol^{-1}} = 0.2151$$

E.21 (a) S atoms required $= (0.683\ mol)(6.022 \times 10^{23}\ atoms \cdot mol^{-1}) = 4.11 \times 10^{23}$

(b) $$S_8 \text{ molecules required} = \frac{4.11 \times 10^{23}}{8} = 5.14 \times 10^{22}$$

(c) mass of sulfur needed $= (0.683\ mol)(32.06\ g \cdot mol^{-1}) = 21.9\ g$

E.23 (a) $MgSO_4 \cdot 7H_2O$ formula mass $= 246.48\ g \cdot mol^{-1}$

$$\text{atoms of Mg} = \left(\frac{2.00\ g}{246.48\ g \cdot mol^{-1}}\right)(6.022 \times 10^{23}\ atoms \cdot mol^{-1}) = 4.89 \times 10^{21}$$

(b) 4.89×10^{21} formula units

(c) $$\text{moles of } H_2O = 7\left(\frac{2.00\ g}{246.48\ g \cdot mol^{-1}}\right) = 5.68 \times 10^{-2}\ mol$$

F.1 molar mass of testosterone $= 288.41\ g \cdot mol^{-1}$

$$\%C = \frac{19 \times 12.01\ g \cdot mol^{-1}}{288.41\ g \cdot mol^{-1}} \times 100 = 79.12\%$$

$$\%H = \frac{28 \times 1.0079\ g \cdot mol^{-1}}{288.41\ g \cdot mol^{-1}} \times 100 = 9.78\%$$

$$\%O = \frac{2 \times 16.00\ g \cdot mol^{-1}}{288.41\ g \cdot mol^{-1}} \times 100 = 11.10\%$$

F.3 (a) For 100 g of compound,

$$\text{moles of Na} = \frac{32.79\ g}{22.99\ g \cdot mol^{-1}} = 1.426\ mol$$

$$\text{moles of Al} = \frac{13.02\ g}{26.98\ g \cdot mol^{-1}} = 0.4826\ mol$$

$$\text{moles of F} = \frac{54.19\ g}{19.00\ g \cdot mol^{-1}} = 2.852\ mol$$

Dividing each number by 0.4826 gives a ratio of 1 Al : 2.95 Na : 5.91 F. The formula is Na_3AlF_6.

(b) For 100 g of compound,

$$\text{moles of K} = \frac{31.91\ \text{g}}{39.10\ \text{g}\cdot\text{mol}^{-1}} = 0.8161\ \text{mol}$$

$$\text{moles of Cl} = \frac{28.93\ \text{g}}{35.45\ \text{g}\cdot\text{mol}^{-1}} = 0.8161\ \text{mol}$$

mass of O is obtained by difference:

$$\text{moles of O} = \frac{100\ \text{g} - 31.91\ \text{g} - 28.93\ \text{g}}{16.00\ \text{g}\cdot\text{mol}^{-1}} = 2.448\ \text{mol}$$

Dividing each number by 0.8161 gives a ratio of 1.00 K : 1 Cl : 3.00 O. The formula is $KClO_3$.

(c) For 100 g of compound,

$$\text{moles of N} = \frac{12.2\ \text{g}}{14.01\ \text{g}\cdot\text{mol}^{-1}} = 0.871\ \text{mol}$$

$$\text{moles of H} = \frac{5.26\ \text{g}}{1.0079\ \text{g}\cdot\text{mol}^{-1}} = 5.22\ \text{mol}$$

$$\text{moles of P} = \frac{26.9\ \text{g}}{30.97\ \text{g}\cdot\text{mol}^{-1}} = 0.869\ \text{mol}$$

$$\text{moles of O} = \frac{55.6\ \text{g}}{16.00\ \text{g}\cdot\text{mol}^{-1}} = 3.475\ \text{mol}$$

Dividing each number by 0.869 gives a ratio of 1.00 N : 6.01 H : 1.00 P : 4.00 O. The formula is NH_6PO_4 or $[NH_4][H_2PO_4]$, ammonium dihydrogen phosphate.

F.5 $$\text{moles of P} = \frac{4.14\ \text{g}}{30.97\ \text{g}\cdot\text{mol}^{-1}} = 0.134\ \text{mol}$$

$$\text{moles of Cl} = \frac{27.8\ \text{g} - 4.14\ \text{g}}{35.45\ \text{g}\cdot\text{mol}^{-1}} = 0.667\ \text{mol}$$

Dividing each number by 0.134 mol gives a ratio of 4.98 Cl : 1 P. The formula is PCl_5.

F.7 For 100 g of compound,

$$\text{moles of C} = \frac{54.82\ \text{g}}{12.01\ \text{g}\cdot\text{mol}^{-1}} = 4.565\ \text{mol}$$

$$\text{moles of H} = \frac{5.62\ \text{g}}{1.0079\ \text{g}\cdot\text{mol}^{-1}} = 5.58\ \text{mol}$$

$$\text{moles of N} = \frac{7.10\ \text{g}}{14.01\ \text{g}\cdot\text{mol}^{-1}} = 0.507\ \text{mol}$$

$$\text{moles of O} = \frac{32.46\ \text{g}}{16.00\ \text{g}\cdot\text{mol}^{-1}} = 2.029\ \text{mol}$$

Dividing each number by 0.507 mol gives a ratio of 9.00 C : 11.01 H : 1.00 N : 4.00 O. The formula is $C_9H_{11}NO_4$.

F.9 For 100 g of lindane,

$$\text{moles of C} = \frac{24.78\text{ g}}{12.01\text{ g}\cdot\text{mol}^{-1}} = 2.063\text{ mol}$$

$$\text{moles of H} = \frac{2.08\text{ g}}{1.0079\text{ g}\cdot\text{mol}^{-1}} = 2.064\text{ mol}$$

$$\text{moles of Cl} = \frac{73.14\text{ g}}{35.45\text{ g}\cdot\text{mol}^{-1}} = 2.063\text{ mol}$$

Dividing each number by 2.063 mol gives a ratio of 1.00 C : 1.00 H : 1.00 Cl. The formula is $(CHCl)_n$. The formula mass of CHCl is 48.47 $g\cdot mol^{-1}$. The molar mass is 290.81 $g\cdot mol^{-1}$ which is 6 times the formula mass, so $n = 6$ and the molecular formula is $C_6H_6Cl_6$.

F.11 For 100 g of caffeine,

$$\text{moles of C} = \frac{49.48\text{ g}}{12.01\text{ g}\cdot\text{mol}^{-1}} = 4.12\text{ mol}$$

$$\text{moles of H} = \frac{5.19\text{ g}}{1.0079\text{ g}\cdot\text{mol}^{-1}} = 5.15\text{ mol}$$

$$\text{moles of N} = \frac{28.85\text{ g}}{14.01\text{ g}\cdot\text{mol}^{-1}} = 2.059\text{ mol}$$

$$\text{moles of O} = \frac{16.48\text{ g}}{16.00\text{ g}\cdot\text{mol}^{-1}} = 1.03\text{ mol}$$

Dividing each number by 1.03 mol gives a ratio of 4.00 C : 5.00 H : 2.00 N : 1.00 O. The formula is $C_4H_5N_2O$ with a molar formula mass of 97.10 $g\cdot mol^{-1}$. Because the molecular molar mass is twice this value, the actual formula will be $C_8H_{10}N_4O_2$.

F.13 (a) For 100 g of compound,

$$\text{moles of C} = \frac{44.2\text{ g}}{12.01\text{ g}\cdot\text{mol}^{-1}} = 3.68\text{ mol}$$

$$\text{moles of Cl} = \frac{43.5\text{ g}}{35.45\text{ g}\cdot\text{mol}^{-1}} = 1.23\text{ mol}$$

$$\text{moles of O} = \frac{9.82\text{ g}}{16.00\text{ g}\cdot\text{mol}^{-1}} = 0.614\text{ mol}$$

$$\text{moles of H} = \frac{2.48\text{ g}}{1.0079\text{ g}\cdot\text{mol}^{-1}} = 2.46\text{ mol}$$

Dividing each number by 0.614 mol gives a ratio of 5.99 C : 2.00 Cl : 1.00 O : 4.01 H and an empirical formula of $C_6H_4Cl_2O$.

(b) The molar formula mass is 162.99 $g \cdot mol^{-1}$. Because the molecular molar mass is twice this value, the actual formula will be $C_{12}H_8Cl_4O_2$.

G.1 (a) solubility; (b) the abilities of the components to adsorb; (c) boiling points

G.3 (a) homogeneous, distillation; (b) heterogeneous, dissolving followed by filtration and distillation; (c) homogeneous, distillation

G.5 mass of $AgNO_3 = (0.155\ mol \cdot L^{-1})(0.025\ L)(169.88\ g \cdot mol^{-1}) = 0.658\ g$

G.7 (a) $$\text{molarity of } Ba(OH)_2 = \frac{\left(\dfrac{2.577\ g}{171.36\ g \cdot mol^{-1}}\right)}{0.250\ L} = 0.060\ 2\ mol \cdot L^{-1}$$

$$V = \frac{1.0 \times 10^{-3}\ mol}{0.060\ 2\ mol \cdot L^{-1}} = 1.7 \times 10^{-2}\ L \text{ or } 17\ mL$$

(b) $$V = \frac{3.5 \times 10^{-3}\ mol\ OH^-}{(0.060\ 2\ mol\ Ba(OH)_2 \cdot L^{-1})\left(\dfrac{2\ mol\ OH^-}{1\ mol\ Ba(OH)_2}\right)}$$

$$= 2.9 \times 10^{-2}\ L \text{ or } 29\ mL$$

(c) $$V = \frac{\left(\dfrac{50.0 \times 10^{-3}\ g}{171.36\ g \cdot mol^{-1}}\right)}{0.060\ 2\ mol \cdot L^{-1}} = 4.85 \times 10^{-3}\ L \text{ or } 4.85\ mL$$

G.9 (a) Weigh 1.6 g (0.010 mol, molar mass of $KMnO_4 = 158.04\ g \cdot mol^{-1}$) into a 1.0-L volumetric flask and add water to give a total volume of 1.0 L. Smaller (or larger) volumes could also be prepared by using a proportionally smaller (or larger) mass of $KMnO_4$.

(b) Starting with $0.050\ mol \cdot L^{-1}$ $KMnO_4$, you would need to add 4 volumes of water to 1 volume of starting solution, because the concentration desired is one-fifth of the starting solution. This relation can be derived from the expression

$$V_i \times \text{molarity}_i = V_f \times \text{molarity}_f$$

where i represents the initial solution and f the final solution. But $V_f = V_i + V_d$, where V_d represents the volume of solvent that must be added to dilute the initial solution. Rearranging the first equation gives

$$\frac{V_i}{V_f} = \frac{\text{molarity}_f}{\text{molarity}_i}$$

$$\frac{V_i}{V_i + V_d} = \frac{\text{molarity}_f}{\text{molarity}_i}$$

So if the ratio of final molarity to initial molarity is 1:5, we can write

$$\frac{V_i}{V_i + V_d} = \frac{1}{5}$$

$$5V_i = V_i + V_d$$

$$4V_i = V_d$$

For example, to prepare 50 mL of solution, you would add 40 mL of water to 10 mL of 0.050 $mol \cdot L^{-1}$ $KMnO_4$.

G.11 (a) $V(0.778\ mol \cdot L^{-1}) = (0.1500\ L)(0.0234\ mol \cdot L^{-1})$

$V = 4.51 \times 10^{-3}$ L or 4.51 mL

(b) The concentration desired is one-fifth of the starting NaOH solution, so the stockroom attendant will need to add 4 volumes of water to 1 volume of the 2.5 $mol \cdot L^{-1}$ solution. To prepare 60.0 mL of solution, divide 60.0 by 5; so 12.0 mL of 2.5 $mol \cdot L^{-1}$ NaOH solution are added to 48.0 mL of water. See the solution to G.9.

G.13 (a) mass of $CuSO_4 = (0.20\ mol \cdot L^{-1})(0.250\ L)(159.60\ g \cdot mol^{-1})$

$= 8.0$ g

(b) mass of $CuSO_4 \cdot 5H_2O = (0.20\ mol \cdot L^{-1})(0.250\ L)(249.68\ g \cdot mol^{-1})$

$= 13$ g

G.15 (a) moles of $(NH_4)_3PO_4 \cdot 3H_2O = \dfrac{2.50\ g}{203.14\ g \cdot mol^{-1}} = 0.0123$ mol

moles of $K_3PO_4 = \dfrac{1.50\ g}{212.27\ g \cdot mol^{-1}} = 0.00707$ mol

(b) total moles of $PO_4^{3-} = 0.0123\ mol + 0.00707\ mol = 0.0194$ mol

(c) mass of PO_4^{3-} ions $= (0.0194\ mol)(94.97\ g \cdot mol^{-1}) = 1.84$ g

(d) total mass of H_2O = mass of H_2O added as solvent + mass of H_2O present in $(NH_4)_3PO_4 \cdot 3H_2O$

total mass of H_2O

$$= 500\ g + \left(\frac{3\ mol\ H_2O}{1\ mol\ (NH_4)_3PO_4}\right)(0.0123\ mol\ (NH_4)_3PO_4)(18.02\ g \cdot mol^{-1})$$

$$= 500\ g + 0.665\ g = 501\ g$$

G.17 (a) mass of $NaCl = (0.10\ mol \cdot L^{-1})(1.0\ L)(58.44\ g \cdot mol^{-1}) = 5.8$ g

(b) mass of $CaCl_2 = (0.10\ mol \cdot L^{-1})(0.25\ L)(110.98\ g \cdot mol^{-1}) = 2.8$ g

(c) mass of $C_6H_{12}O_6 = (0.63\ mol \cdot L^{-1})(0.500\ L)(180.15\ g \cdot mol^{-1}) = 57$ g

G.19 (a) moles of water $= \dfrac{25.0\text{ g}}{18.02\text{ g}\cdot\text{mol}^{-1}} = 1.39\text{ mol}$

moles of ethanol $= \dfrac{50.0\text{ g}}{46.07\text{ g}\cdot\text{mol}^{-1}} = 1.09\text{ mol}$

$$X_{H_2O} = \frac{1.39\text{ mol}}{1.39\text{ mol} + 1.09\text{ mol}} = 0.560$$

$X_{ethanol} = 1 - X_{H_2O} = 0.440$

(b) moles of water $= \dfrac{25.0\text{ g}}{18.02\text{ g}\cdot\text{mol}^{-1}} = 1.39\text{ mol}$

moles of methanol $= \dfrac{50.0\text{ g}}{32.04\text{ g}\cdot\text{mol}^{-1}} = 1.56\text{ mol}$

$$X_{H_2O} = \frac{1.39\text{ mol}}{1.39\text{ mol} + 1.56\text{ mol}} = 0.471$$

$X_{methanol} = 1 - X_{H_2O} = 0.529$

(c) For 1 kg or 1000 g solvent, there are 0.10 mol $C_6H_{12}O_6$.

moles of water $= \dfrac{1000\text{ g}}{18.02\text{ g}\cdot\text{mol}^{-1}} = 55.49\text{ mol}$

$$X_{glucose} = \frac{0.10\text{ mol}}{55.49\text{ mol} + 0.10\text{ mol}} = 0.0018$$

$X_{H_2O} = 1 - X_{glucose} = 0.9982$

$X_{H_2O} \cong 1.00$

G.21 (a) $\dfrac{\left(\dfrac{10.0\text{ g}}{58.44\text{ g}\cdot\text{mol}^{-1}}\right)}{0.250\text{ kg}} = 0.684\ m\ \text{NaCl(aq)}$

(b) $\dfrac{0.48\text{ mol}}{0.050\text{ kg}} = 9.6\ m\ \text{KOH(aq)}$

(c) $\dfrac{\left(\dfrac{1.94\text{ g}}{60.06\text{ g}\cdot\text{mol}^{-1}}\right)}{0.200\text{ kg}} = 0.162\ m\ \text{CO(NH}_2\text{)}_2\text{(aq)}$

G.23 (a) moles of water $= \dfrac{150\text{ g}}{18.02\text{ g}\cdot\text{mol}^{-1}} = 8.32\text{ mol}$

moles of ethanol $= \dfrac{25.0\text{ g}}{46.07\text{ g}\cdot\text{mol}^{-1}} = 0.543\text{ mol}$

$$X_{H_2O} = \frac{8.32\text{ mol}}{8.32\text{ mol} + 0.543\text{ mol}} = 0.939$$

$X_{ethanol} = 1 - X_{H_2O} = 0.061$

(b) molality of ethanol $= \dfrac{0.543 \text{ mol}}{0.150 \text{ kg}} = 3.62\ m$

G.25 (a) For one liter of solution, the mass of the solution will be $(1000 \text{ mL})(1.027 \text{ g·mL}^{-1}) = 1027 \text{ g}$.

mass of $(NH_4)_2SO_4 = (0.35 \text{ mol})(132.14 \text{ g·mol}^{-1}) = 46 \text{ g}$

mass of $H_2O = 1027 \text{ g} - 46 \text{ g} = 981 \text{ g}$

molality of $(NH_4)_2SO_4(aq) = \dfrac{0.35 \text{ mol}}{0.981 \text{ kg}} = 0.36\ m$

(b) moles of $H_2O = \dfrac{981 \text{ g}}{18.02 \text{ g·mol}^{-1}} = 54.4 \text{ mol}$

mole fraction of $(NH_4)_2SO_4 = \dfrac{0.35 \text{ mol}}{0.35 \text{ mol} + 54.4 \text{ mol}} = 6.4 \times 10^{-3}$

G.27 (a) For 1.00 L of solution,

mass of solution $= (1.27 \text{ g·mL}^{-1})(1000 \text{ mL}) = 1270 \text{ g}$

$1 \text{ cm}^3 = 1 \text{ mL}$

mass of $MgCl_2 = (0.346)(1270 \text{ g}) = 439 \text{ g}$

mass of $H_2O = 1270 \text{ g} - 439 \text{ g} = 831 \text{ g}$

moles of $MgCl_2 = \dfrac{439 \text{ g}}{95.21 \text{ g·mol}^{-1}} = 4.61 \text{ mol}$

molality $= \dfrac{4.61 \text{ mol}}{0.831 \text{ kg}} = 5.55\ m\ MgCl_2(aq)$

(b) mole fraction of water $= \dfrac{(831 \text{ g})/(18.02 \text{ g·mol}^{-1})}{(831 \text{ g})/(18.02 \text{ g·mol}^{-1}) + 4.61 \text{ mol}} = 0.909$

(c) molarity of $MgCl_2 = \dfrac{4.61 \text{ mol}}{1.00 \text{ L}} = 4.61 \text{ M } MgCl_2$

H.1 (a) $P_4O_{10}(s) + 6\ H_2O(l) \longrightarrow 4\ H_3PO_4(l)$

(b) $Cd(NO_3)_2(aq) + Na_2S(aq) \longrightarrow CdS(s) + 2\ NaNO_3(aq)$

(c) $4\ KClO_3(s) \xrightarrow{\Delta} 3\ KClO_4(s) + KCl(s)$

(d) $2\ HCl(aq) + Ca(OH)_2(aq) \longrightarrow CaCl_2(aq) + 2\ H_2O(l)$

H.3 (a) $2\ Na(s) + 2\ H_2O(l) \longrightarrow H_2(g) + 2\ NaOH(aq)$

(b) $Na_2O(s) + H_2O(l) \longrightarrow 2\ NaOH(aq)$

(c) $6\ Li(s) + N_2(g) \longrightarrow 2\ Li_3N(s)$

(d) $Ca(s) + 2\ H_2O(l) \longrightarrow H_2(g) + Ca(OH)_2(aq)$

H.5 (I) $3\ Fe_2O_3(s) + CO(g) \longrightarrow 2\ Fe_3O_4(s) + CO_2(g)$
(II) $Fe_3O_4(s) + 4\ CO(g) \longrightarrow 3\ Fe(s) + 4\ CO_2(g)$

H.7 (I) $N_2(g) + O_2(g) \longrightarrow 2\ NO(g)$
(II) $2\ NO(g) + O_2(g) \longrightarrow 2\ NO_2(g)$

H.9 $4\ HF(aq) + SiO_2(s) \longrightarrow SiF_4(aq) + 2\ H_2O(l)$

H.11 $2\ C_8H_{18}(l) + 25\ O_2(g) \longrightarrow 16\ CO_2(g) + 18\ H_2O(g)$

H.13 $4\ C_{10}H_{15}N(s) + 55\ O_2(g) \longrightarrow 40\ CO_2(g) + 30\ H_2O(l) + 2\ N_2(g)$

H.15 (I) $H_2S(g) + 2\ NaOH(aq) \longrightarrow Na_2S(aq) + 2\ H_2O(l)$

(II) $4\ H_2S(g) + Na_2S(al) \xrightarrow{\text{alcohol}} Na_2S_5(al) + 4\ H_2(g)$
(III) $2\ Na_2S_5(al) + 9\ O_2(g) + 5\ H_2O(l) \longrightarrow 2\ Na_2S_2O_3 \cdot 5\ H_2O(s) + 6\ SO_2(g)$

I.1 (a) CH_3OH, nonelectrolyte; (b) $CaBr_2$, strong electrolyte; (c) KI, strong electrolyte

I.3 (a) soluble; (b) slightly soluble; (c) soluble; (d) soluble

I.5 (a) $Na^+(aq)$ and $I^-(aq)$; (b) $Ag^+(aq)$ and $CO_3^{2-}(aq)$, Ag_2CO_3 is insoluble. The very small amount that does go into solution will be present as Ag^+ and CO_3^{2-} ions. (c) $NH_4^+(aq)$ and $PO_4^{3-}(aq)$ (d) $Fe^{2+}(aq)$ and $SO_4^{2-}(aq)$

I.7 (a) $Fe(OH)_3$, precipitate
$Fe_2(SO_4)_3(aq) + 6\ NaOH(aq) \longrightarrow 2\ Fe(OH)_3(s) + 3\ Na_2SO_4(aq)$
(b) Ag_2CO_3, precipitate forms
$2\ AgNO_3(aq) + K_2CO_3(aq) \longrightarrow Ag_2CO_3(s) + 2\ KNO_3(aq)$
(c) No precipitate will form because all possible products are soluble in water.
$Pb(NO_3)_2(aq) + 2\ CH_3COONa(aq) \longrightarrow 2\ NaNO_3(aq) + Pb(CH_3COO)_2(aq)$

I.9 (a) $FeCl_3(aq) + 3\ NaOH(aq) \longrightarrow 3\ NaCl(aq) + Fe(OH)_3(s)$
net ionic equation: $Fe^{3+}(aq) + 3OH^-(aq) \longrightarrow Fe(OH)_3(s)$
spectator ions: Na^+, Cl^-
(b) $Ag(NO_3)(aq) + KI(aq) \longrightarrow AgI(s) + K^+(aq) + NO_3^-(aq)$
net ionic equation: $Ag^+(aq) + I^-(aq) \longrightarrow AgI(s)$
spectator ions: K^+, NO_3^-

(c) $Pb(NO_3)_2(aq) + K_2SO_4 \longrightarrow PbSO_4(s) + 2\ K^+(aq) + 2\ NO_3^-(aq)$

net ionic equation: $Pb^{2+}(aq) + SO_4^{2-}(aq) \longrightarrow PbSO_4(s)$

spectator ions: NO_3^-, K^+

(d) $2\ Na_2CrO_4(aq) + Pb(NO_3)_2(aq) \longrightarrow$

$2\ Na^+(aq) + 2\ NO_3^-(aq) + PbCrO_4(s)$

net ionic equation: $Pb^{2+}(aq) + CrO_4^{2-}(aq) \longrightarrow PbCrO_4(s)$

spectator ions: Na^+, NO_3^-

(e) $Hg_2(NO_3)_2(aq) + K_2SO_4(aq) \longrightarrow Hg_2SO_4(s) + 2\ K^+(aq) + 2\ NO_3^-(aq)$

net ionic equation: $Hg_2^{2+}(aq) + SO_4^{2-}(aq) \longrightarrow Hg_2SO_4(s)$

spectator ions: K^+, NO_3^-

I.11 (a) overall equation: $(NH_4)_2CrO_4(aq) + BaCl_2(aq) \longrightarrow$

$BaCrO_4(s) + 2\ NH_4Cl(aq)$

complete ionic equation:

$2\ NH_4^+(aq) + CrO_4^{2-}(aq) + Ba^{2+}(aq) + 2\ Cl^-(aq) \longrightarrow$

$BaCrO_4(s) + 2\ NH_4^+(aq) + 2\ Cl^-(aq)$

net ionic equation: $Ba^{2+}(aq) + CrO_4^{2-}(aq) \longrightarrow BaCrO_4(s)$

spectator ions: NH_4^+, Cl^-

(b) $CuSO_4(aq) + Na_2S(aq) \longrightarrow CuS(s) + Na_2SO_4(aq)$

complete ionic equation:

$Cu^{2+}(aq) + SO_4^{2-}(aq) + 2\ Na^+(aq) + S^{2-}(aq) \longrightarrow$

$CuS(s) + 2\ Na^+(aq) + SO_4^{2-}(aq)$

net ionic equation: $Cu^{2+}(aq) + S^{2-}(aq) \longrightarrow CuS(s)$

spectator ions: Na^+, SO_4^{2-}

(c) $3\ FeCl_2(aq) + 2\ (NH_4)_3PO_4(aq) \longrightarrow Fe_3(PO_4)_2(s) + 6\ NH_4^+(aq) + 6\ Cl^-(aq)$

complete ionic equation:

$3\ Fe^{2+}(aq) + 6\ Cl^-(aq) + 6\ NH_4^+(aq) + 2\ PO_4^{3-}(aq) \longrightarrow$

$Fe_3(PO_4)_2(s) + 6\ NH_4^+(aq) + 6\ Cl^-(aq)$

net ionic equation: $3\ Fe^{2+}(aq) + 2\ PO_4^{3-}(aq) \longrightarrow Fe_3(PO_4)_2(s)$

spectator ions: Cl^-, NH_4^+

(d) $K_2C_2O_4(aq) + Ca(NO_3)_2(aq) \longrightarrow CaC_2O_4(s) + 2\ KNO_3(aq)$

complete ionic equation:

$2\ K^+(aq) + C_2O_4^{2-}(aq) + Ca^{2+}(aq) + 2\ NO_3^-(aq) \longrightarrow$

$CaC_2O_4(s) + 2\ K^+(aq) + 2\ NO_3^-(aq)$

net ionic equation: $Ca^{2+}(aq) + C_2O_4^{2-}(aq) \longrightarrow CaC_2O_4(s)$

spectator ions: K^+, NO_3^-

(e) $NiSO_4(aq) + Ba(NO_3)_2(aq) \longrightarrow Ni(NO_3)_2(aq) + BaSO_4(s)$

complete ionic equation:

$Ni^{2+}(aq) + SO_4^{2-}(aq) + Ba^{2+}(aq) + 2\ NO_3^-(aq) \longrightarrow$
$Ni^{2+}(aq) + 2\ NO_3^-(aq) + BaSO_4(s)$
net ionic equation: $Ba^{2+}(aq) + SO_4^{2-}(aq) \longrightarrow BaSO_4(s)$
spectator ions: Ni^{2+}, NO_3^-

I.13 (a) $Pb^{2+}(aq) + 2\ ClO_4^-(aq) + 2\ Na^+(aq) + 2\ Br^-(aq) \longrightarrow$
$PbBr_2(s) + 2\ Na^+(aq) + 2\ ClO_4^-(aq)$
net ionic equation: $Pb^{2+}(aq) + 2\ Br^-(aq) \longrightarrow PbBr_2(s)$
(b) $Ag^+(aq) + NO_3^-(aq) + NH_4^+(aq) + Cl^-(aq) \longrightarrow$
$AgCl(s) + NH_4^+(aq) + NO_3^-(aq)$
net ionic equation: $Ag^+(aq) + Cl^-(aq) \longrightarrow AgCl(s)$
(c) $2\ Na^+(aq) + 2\ OH^-(aq) + Cu^{2+}(aq) + 2\ NO_3^-(aq) \longrightarrow$
$Cu(OH)_2(s) + 2\ Na^+(aq) + 2\ NO_3^-(aq)$
net ionic equation: $Cu^{2+}(aq) + 2\ OH^-(aq) \longrightarrow Cu(OH)_2(s)$

I.15 (a) $AgNO_3(aq)$ and $Na_2CrO_4(aq)$
(b) $CaCl_2(aq)$ and $Na_2CO_3(aq)$
(c) $Cd(ClO_4)_2(aq)$ and $(NH_4)_2S(aq)$

I.17 (a) $Pb^{2+}(aq) + SO_4^{2-}(aq) \longrightarrow PbSO_4(s)$
(b) $Cu^{2+}(aq) + S^{2-}(aq) \longrightarrow CuS(s)$
(c) $Co^{2+}(aq) + CO_3^{2-}(aq) \longrightarrow CoCO_3$
(d) $Pb(NO_3)_2$ and Na_2SO_4; Na^+, NO_3^-
$CuCl_2$ and K_2S; K^+, Cl^-
$CoSO_4$ and $(NH_4)_2CO_3$; NH_4^+, SO_4^{2-}

J.1 (a) base; (b) acid; (c) base; (d) acid; (e) base

J.3 (a) $HCl(aq) + NaOH(aq) \longrightarrow NaCl(aq) + H_2O(l)$
complete ionic equation:
$H_3O^+(aq) + Cl^-(aq) + Na^+(aq) + OH^-(aq) \longrightarrow Na^+(aq) + Cl^-(aq) + 2\ H_2O$
net ionic equation: $H_3O^+(aq) + OH^-(aq) \longrightarrow 2\ H_2O(l)$
(b) $NH_3(aq) + HNO_3(aq) \longrightarrow NH_4NO_3(aq)$
$NH_3(aq) + H_3O^+(aq) + NO_3^-(aq) \longrightarrow NH_4^+(aq) + NO_3^-(aq) + H_2O(l)$
$NH_3(aq) + H_3O^+(aq) \longrightarrow NH_4^+(aq) + H_2O(l)$
(c) $CH_3NH_2(aq) + HI(aq) \longrightarrow CH_3NH_3I(aq)$
$CH_3NH_2(aq) + H_3O^+(aq) + I^-(aq) \longrightarrow CH_3NH_3^+(aq) + H_2O(l) + I^-(aq)$
$CH_3NH_2(aq) + H_3O^+(aq) \longrightarrow CH_3NH_3^+(aq)$

J.5 The overall equations are:

(a) $HBr(aq) + KOH(aq) \longrightarrow KBr(aq) + H_2O(l)$

(b) $Zn(OH)_2(aq) + 2\ HNO_2(aq) \longrightarrow Zn(NO_2)_2(aq) + 2\ H_2O(l)$

(c) $Ca(OH)_2(aq) + 2\ HCN(aq) \longrightarrow Ca(CN)_2(aq) + 2\ H_2O(l)$

(d) $3\ KOH(aq) + H_3PO_4(aq) \longrightarrow K_3PO_4(aq) + 3\ H_2O(l)$

The net ionic reaction for all four reactions (a–d) is $H_3O^+(aq) + OH^-(aq) \longrightarrow 2\ H_2O(l)$.

J.7 (a) acid: $H_3O^+(aq)$; base: $CH_3NH_2(aq)$

(b) acid: $HCl(aq)$; base: $C_2H_5NH_2(aq)$

(c) acid: $HI(aq)$; base: $CaO(s)$

J.9 (a) basic solutions; (b) acidic solutions; (c) acidic solutions; (d) basic solutions

J.11 (a) $Ba(OH)_2$, base, and HNO_3, acid; (b) NaOH and H_2SO_4; (c) KOH and $HClO_4$; (d) $Ni(OH)_2$ and HCl

K.1 (a) $2\ P(s) + 3\ Br_2(l) \longrightarrow 2\ PBr_3(s)$

(b) $2\ Fe^{2+}(aq) + Sn^{4+}(aq) \longrightarrow 2\ Fe^{3+}(aq) + Sn^{2+}(aq)$

(c) $8\ H_2(g) + S_8(s) \longrightarrow 8\ H_2S(g)$

(d) $2\ NO(g) + O_2(g) \longrightarrow 2\ NO_2(g)$

K.3 (a) $Mg^0(s) + Cu^{2+}(aq) \longrightarrow Mg^{2+}(aq) + Cu^0(s)$

(b) $Fe^{2+}(aq) + Ce^{4+}(aq) \longrightarrow Fe^{3+}(aq) + Ce^{3+}(aq)$

(c) $H_2(g) + Cl_2(g) \longrightarrow 2\ HCl(g)$

(d) $4\ Fe(s) + 3\ O_2(g) \longrightarrow 2\ Fe_2O_3(s)$

K.5 (a) +4; (b) +3; (c) −3; (d) +3; (e) +6; (f) +3

K.7 (a) +5; (b) +6; (c) +4; (d) +7; (e) +3

K.9 (a) not a redox reaction

(b) Br in BrO_3^- is reduced, Br^- is oxidized

(c) F in F_2 is reduced, O in H_2O is oxidized

K.11 (a) Cl_2 will more easily be reduced and is therefore a stronger oxidizing agent than Cl^-. (b) N_2O_5 will be a stronger oxidizing agent because it will be readily reduced. N^{5+} will accept e^- more readily than N^+.

K.13 (a) oxidizing agent: H^+ in HCl(aq); reducing agent: Zn(s)
(b) oxidizing agent: $SO_2(g)$; reducing agent: $H_2S(g)$
(c) oxidizing agent: $B_2O_3(s)$; reducing agent: Mg(s)

K.15 (a) $ClO_3^- \longrightarrow ClO_2$, Cl goes from $+5 \longrightarrow +4$; reducing agent
(b) $SO_4^{2-} \longrightarrow S^{2-}$, S goes from $+6 \longrightarrow -2$; reducing agent
(c) $Mn^{2+} \longrightarrow MnO_2$, Mn goes from $+2 \longrightarrow +4$; oxidizing agent
(d) HCHO $\longrightarrow$ HCOOH, C goes from 0 to +2; oxidizing agent

K.17 (a) oxidizing agent: $WO_3(s)$; reducing agent: $H_2(g)$
(b) oxidizing agent: HCl(aq); reducing agent: Zn(s)
(c) oxidizing agent: $SnO_2(s)$; reducing agent: C(s)
(d) oxidizing agent: $N_2O_4(g)$; reducing agent: $N_2H_4(g)$

L.1 $2\ Na_2S_2O_3(aq) + AgBr(s) \longrightarrow NaBr(aq) + Na_3[Ag(S_2O_3)_2](aq)$
(a) moles of $Na_2S_2O_3$ needed to dissolve 1.0 mg AgBr

$$= 1.0\text{ mg AgBr}\left(\frac{1\text{ g AgBr}}{1000\text{ mg AgBr}}\right)\left(\frac{1\text{ mol AgBr}}{187.78\text{ g AgBr}}\right)\left(\frac{2\text{ mol Na}_2\text{S}_2\text{O}_3}{1\text{ mol AgBr}}\right)$$

$$= 1.1 \times 10^{-5}\text{ mol Na}_2\text{S}_2\text{O}_3$$

(b) mass of AgBr to produce 0.033 mol $Na_3[Ag(S_2O_3)_2]$

$$= 0.033\text{ mol Na}_3[\text{Ag(S}_2\text{O}_3)_2]\left(\frac{1\text{ mol AgBr}}{1\text{ mol Na}_3[\text{Ag(S}_2\text{O}_3)_2]}\right)\left(\frac{187.78\text{ g AgBr}}{1\text{ mol AgBr}}\right)$$

$$= 6.2\text{ g AgBr}$$

L.3 $6\ NH_4ClO_4(s) + 10\ Al(s) \longrightarrow 5\ Al_2O_3(s) + 3\ N_2(g) + 6\ HCl(g) + 9\ H_2O(g)$

(a) $$(1.5 \times 10^4\text{ kg NH}_4\text{ClO}_4)\left(\frac{1\text{ mol NH}_4\text{ClO}_4}{117.49\text{ g NH}_4\text{ClO}_4}\right)\left(\frac{1000\text{ g}}{1\text{ kg}}\right)$$

$$\left(\frac{10\text{ mol Al}}{6\text{ mol NH}_4\text{ClO}_4}\right)\left(\frac{26.98\text{ g Al}}{1\text{ mol Al}}\right) = 5.7 \times 10^6\text{ g Al or }5.7 \times 10^3\text{ kg Al}$$

(b) $$5000\text{ kg Al}\left(\frac{1000\text{ g Al}}{1\text{ kg Al}}\right)\left(\frac{1\text{ mol Al}}{26.98\text{ g Al}}\right)\left(\frac{5\text{ mol Al}_2\text{O}_3}{10\text{ mol Al}}\right)\left(\frac{101.96\text{ g Al}_2\text{O}_3}{1\text{ mol Al}_2\text{O}_3}\right)$$

$$= 9.448 \times 10^6\text{ g Al}_2\text{O}_3\text{ or }9.448 \times 10^3\text{ kg Al}_2\text{O}_3$$

L.5 $2\ C_{57}H_{110}O_6(s) + 163\ O_2(g) \longrightarrow 114\ CO_2(g) + 110\ H_2O(l)$

(a) $$(2.5\text{ kg fat})\left(\frac{1000\text{ g fat}}{1\text{ kg fat}}\right)\left(\frac{1\text{ mol fat}}{891.44\text{ g fat}}\right)\left(\frac{110\text{ mol H}_2\text{O}}{2\text{ mol fat}}\right)\left(\frac{18.02\text{ g H}_2\text{O}}{1\text{ mol H}_2\text{O}}\right)$$

$$= 2.8\text{ kg H}_2\text{O}$$

(b) $(2.5\text{ g fat})\left(\frac{1\text{ mol fat}}{891.44\text{ g}}\right)\left(\frac{163\text{ mol O}_2}{2\text{ mol fat}}\right)\left(\frac{32.00\text{ g O}_2}{1\text{ mol O}_2}\right) = 7.3\text{ g O}_2$

L.7 $2\,C_8H_{18}(l) + 25\,O_2(g) \longrightarrow 16\,CO_2(g) + 18\,H_2O(l)$

$d = 0.79\text{ g}\cdot\text{mL}^{-1}$, density of gasoline

$$(1\text{ L gas})\left(\frac{1000\text{ mL}}{1\text{ L}}\right)\left(\frac{0.79\text{ g gas}}{1\text{ mL}}\right)\left(\frac{1\text{ mol gas}}{114.22\text{ g gas}}\right)\left(\frac{18\text{ mol H}_2\text{O}}{2\text{ mol C}_8\text{H}_{18}}\right)$$
$$\left(\frac{18.02\text{ g H}_2\text{O}}{1\text{ mol H}_2\text{O}}\right) = 1.1 \times 10^3\text{ g H}_2\text{O}$$

L.9 $HCl + NaOH \longrightarrow NaCl + H_2O$

(a) $17.40\text{ mL}\left(\frac{0.234\text{ mol HCl}}{1000\text{ mL}}\right)\left(\frac{1\text{ mol NaOH}}{1\text{ mol HCl}}\right) = 0.00407\text{ mol}$

(b) concentration of NaOH $= \frac{0.00407\text{ mol}}{15.00 \times 10^{-3}\text{ L}} = 0.271\text{ M}$

$(0.271\text{ mol}\cdot\text{L}^{-1})(0.01500\text{ L})(40.00\text{ g}\cdot\text{mol}^{-1}) = 0.163\text{ g NaOH}$

L.11 (a) $Ba(OH)_2(aq) + 2\,HNO_3(aq) \longrightarrow Ba(NO_3)_2(aq) + 2\,H_2O(l)$

$$\text{molarity of HNO}_3 = \frac{\left(\frac{9.670\text{ g Ba(OH)}_2}{171.36\text{ g}\cdot\text{mol}^{-1}}\right)}{0.250\text{ L}} \times \frac{11.56\text{ mL}}{1000\text{ mL}\cdot\text{L}^{-1}} \times \frac{2\text{ mol HNO}_3}{1\text{ mol Ba(OH)}_2}$$
$$\div\ 0.0250\text{ L} = 0.209\text{ mol}\cdot\text{L}^{-1}$$

(b) mass of HNO_3 in solution:

$$\left(\frac{0.209\text{ mol}}{1000\text{ mL}}\right)(25.0\text{ mL})\left(\frac{63.02\text{ g HNO}_3}{1\text{ mol HNO}_3}\right) = 0.329\text{ g}$$

L.13 $HX(aq) + NaOH(aq) \longrightarrow NaX(aq) + H_2O(l)$

$(68.8\text{ mL})\left(\frac{0.750\text{ mol NaOH}}{1000\text{ mL NaOH}}\right) = 0.0516\text{ mol NaOH}$

3.25 g HX corresponds to 0.0516 mol NaOH used

$\frac{3.25\text{ g}}{0.0516\text{ mol}} = 63.0\text{ g}\cdot\text{mol}^{-1} = \text{molar mass of acid}$

L.15 (a) $Na_2CO_3(aq) + 2\,HCl(aq) \longrightarrow 2\,NaCl(aq) + H_2CO_3(aq)$

(b) $$(0.530\text{ g Na}_2\text{CO}_3)\left(\frac{1\text{ mol Na}_2\text{CO}_3}{105.99\text{ g Na}_2\text{CO}_3}\right)\left(\frac{0.025\text{ L}}{0.100\text{ L}}\right)\left(\frac{2\text{ mol HCl}}{1\text{ mol Na}_2\text{CO}_3}\right)$$
$$\left(\frac{1}{0.026\,50\text{ L}}\right) = 0.0943\text{ mol}\cdot\text{L}^{-1}\text{ HCl (diluted)}$$

The original HCl solution is 100 times more concentrated than the solution used for titration (diluted 10.00 mL to 1000 mL), so the original concentration of the HCl solution is 9.43 $mol\cdot L^{-1}$.

M.1 $CaCO_3(s) \longrightarrow CaO(s) + CO_2(g)$

theoretical yield:

$$(30.7\ g\ CaCO_3)\left(\frac{1\ mol\ CaCO_3}{100.09\ g\ CaCO_3}\right)\left(\frac{1\ mol\ CO_2}{1\ mol\ CaCO_3}\right)\left(\frac{44.01\ g\ CO_2}{1\ mol\ CO_2}\right)$$

$$= 13.5\ g\ CO_2$$

actual yield:

$$\frac{11.7\ g}{13.5\ g} \times 100\% = 86.7\%\ \text{yield}$$

M.3 $C_2H_5OH(aq) + O_2(g) \longrightarrow CH_3COOH(aq) + H_2O(l)$

$$(2.00\ g\ C_2H_5OH)\left(\frac{1\ mol\ ethanol}{46.07\ g\ C_2H_5OH}\right)\left(\frac{1\ mol\ O_2}{1\ mol\ ethanol}\right)\left(\frac{32.00\ g\ O_2}{1\ mol\ O_2}\right)$$

$$= 1.39\ g\ O_2$$

2.00 g ethanol can react with 1.39 g O_2, therefore oxygen is the limiting reagent.

M.5 (a) $P_4(s) + 3\ O_2(g) \longrightarrow P_4O_6(s)$

$P_4O_6(s) + 2\ O_2(g) \longrightarrow P_4O_{10}(s)$

In the first reaction, 5.77 g P_4 uses

$$(5.77\ g\ P_4)\left(\frac{1\ mol\ P_4}{123.88\ g\ P_4}\right)\left(\frac{3\ mol\ O_2}{1\ mol\ P_4}\right)\left(\frac{32.00\ g\ O_2}{1\ mol\ O_2}\right) = 4.47\ g\ O_2\ (g)$$

excess O_2 = 5.77 g − 4.47 g O_2 = 1.30 g O_2

In the second reaction, 5.77 g P_4 uses

$$\left(\frac{5.77\ g\ P_4}{123.88\ g\cdot mol^{-1}\ P_4}\right)\left(\frac{1\ mol\ P_4O_6}{1\ mol\ P_4}\right)\left(\frac{2\ mol\ O_2}{1\ mol\ P_4O_6}\right)\left(\frac{32.00\ g\ O_2}{1\ mol\ O_2}\right) = 2.98\ g\ O_2$$

limiting reagent: O_2

(b) $$\left(\frac{1.30\ g\ O_2}{32.00\ g\cdot mol^{-1}\ O_2}\right)\left(\frac{1\ mol\ P_4O_{10}}{2\ mol\ O_2}\right)\left(\frac{283.88\ g\ P_4O_{10}}{1\ mol\ P_4O_{10}}\right) = 5.77\ g\ P_4O_{10}$$

(c) $$\left(\frac{1.30\ g\ O_2}{32.00\ g\cdot mol^{-1}\ O_2}\right)\left(\frac{1\ mol\ P_4O_6}{2\ mol\ O_2}\right)\left(\frac{219.88\ g\ P_4O_6}{1\ mol\ P_4O_6}\right) = 4.47\ g\ P_4O_6\ \text{used}$$

In the first reaction, 5.77 g P_4 produces

$$\left(\frac{5.77\ g\ P_4}{123.88\ g\cdot mol^{-1}}\right)\left(\frac{219.88\ g\ P_4O_6}{1\ mol\ P_4O_6}\right)\left(\frac{1\ mol\ P_4O_6}{1\ mol\ P_4}\right) = 10.2\ g\ P_4O_6$$

excess reagent: 10.2 g − 4.47 g = 5.7 g P_4O_6

M.7 mass C = 0.714 g

mass N = 0.138 g

mass H = 0.992 g − (0.714 g + 0.138 g) = 0.070 g

$$\text{moles of C} = (0.714\text{ g C})\left(\frac{1\text{ mol C}}{12.01\text{ g C}}\right) = 0.0595\text{ mol}$$

$$\text{moles of N} = (0.138\text{ g N})\left(\frac{1\text{ mol N}}{14.01\text{ g N}}\right) = 0.009\ 85\text{ mol}$$

$$\text{moles of H} = (0.070\text{ g})\left(\frac{1\text{ mol H}}{1.0079\text{ g H}}\right) = 0.069\text{ mol}$$

Dividing each amount by 0.009 85 gives C:N:H ratios = 6.0:1:7.0. The empirical formula of aniline is C_6H_7N.

M.9 aspartame + $O_3(g) \longrightarrow CO_2(g) + N_2(g) + H_2O(g)$

$$(0.318\text{ g }CO_2)\left(\frac{1\text{ mol }CO_2}{44.01\text{ g }CO_2}\right)\left(\frac{1\text{ mol C}}{1\text{ mol }CO_2}\right) = 0.007\ 22\text{ mol C}$$

$$(0.007\ 22\text{ mol C})(12.01\text{ g}\cdot\text{mol}^{-1}\text{ C}) = 0.0868\text{ g C}$$

$$(0.084\text{ g }H_2O)\left(\frac{1\text{ mol }H_2O}{18.02\text{ g }H_2O}\right)\left(\frac{2\text{ mol H}}{1\text{ mol }H_2O}\right) = 0.0093\text{ mol H}$$

$$(0.0093\text{ mol H})(1.0079\text{ g}\cdot\text{mol}^{-1}\text{ H}) = 0.0094\text{ g H}$$

$$(0.0145\text{ g }N_2)\left(\frac{1\text{ mol }N_2}{28.02\text{ g }N_2}\right)\left(\frac{2\text{ mol N}}{1\text{ mol }N_2}\right) = 0.001\ 03\text{ mol N}$$

$$(0.001\ 03\text{ mol N})(14.01\text{ g}\cdot\text{mol}^{-1}\text{ N}) = 0.0145\text{ g N}$$

$$\text{mass of O} = 0.152 - (0.0868\text{ g} + 0.0094\text{ g} + 0.0145\text{ g}) = 0.041\text{ g O}$$

$$\left(\frac{0.041\text{ g O}}{16.00\text{ g}\cdot\text{mol}^{-1}\text{ O}}\right) = 0.0026\text{ mol O}$$

Dividing each amount by 0.001 03 gives C:H:N:O ratios of 7.01:9.1:1:2.5. Doubling these values will give a whole number of oxygen atoms. The empirical formula is $C_{14}H_{18}N_2O_5$. The empirical formula of aspartame, $C_{14}H_{18}N_2O_5$, gives a mass of 294.30 g·mol^{-1}. Because this value corresponds to the molar mass, the empirical formula and the molecular formula are the same.

M.11

$$(0.682\text{ g }CO_2)\left(\frac{1\text{ mol }CO_2}{44.01\text{ g }CO_2}\right)\left(\frac{1\text{ mol C}}{1\text{ mol }CO_2}\right) = 0.0155\text{ mol C}$$

$$(0.0155\text{ mol C})(12.01\text{ g}\cdot\text{mol}^{-1}\text{ C}) = 0.186\text{ g C}$$

$$(0.174\text{ g }H_2O)\left(\frac{1\text{ mol }H_2O}{18.02\text{ g }H_2O}\right)\left(\frac{2\text{ mol H}}{1\text{ mol }H_2O}\right) = 0.0193\text{ mol H}$$

$$(0.0193\text{ mol H})(1.0079\text{ g}\cdot\text{mol}^{-1}\text{ H}) = 0.0195\text{ g H}$$

$$(0.110\text{ g }N_2)\left(\frac{1\text{ mol }N_2}{28.02\text{ g }N_2}\right)\left(\frac{2\text{ mol N}}{1\text{ mol }N_2}\right) = 0.007\ 85\text{ mol N}$$

$$(0.007\ 85\text{ mol N})(14.01\text{ g}\cdot\text{mol}^{-1}\text{ N}) = 0.110\text{ g N}$$

mass of O = 0.376 g − (0.186 g + 0.0193 g + 0.110 g) = 0.061 g O

$$\frac{0.061\text{ g O}}{16.00\text{ g O}} = 0.0038\text{ mol O}$$

Dividing each amount by 0.0038 gives C:H:N:O ratios = 4.1:5.1:2.1:1. The empirical formula is $C_4H_5N_2O$.

The molecular mass of caffeine is 194 g·mol^{-1}. Its empirical mass is 97.10 g·mol^{-1}.

molecular formula = 2 × empirical formula = $C_8H_{10}N_4O_2$

$$2\ C_8H_{10}N_4O_2(s) + 19\ O_2(g) \longrightarrow 16\ CO_2(g) + 10\ H_2O(l) + 4\ N_2(g)$$

M.13 $3\ Ca(NO_3)_2(aq) + 2\ H_3PO_4(aq) \longrightarrow Ca_3(PO_4)_2(s) + 6\ HNO_3(aq)$

(a) The solid is calcium phosphate, $Ca_3(PO_4)_2$.

(b) $$(206\text{ g }Ca(NO_3)_2)\left(\frac{1\text{ mol }Ca(NO_3)_2}{164.10\text{ g }Ca(NO_3)_2}\right)\left(\frac{2\text{ mol }H_3PO_4}{3\text{ mol }Ca(NO_3)_2}\right)\left(\frac{97.99\text{ g }H_3PO_4}{1\text{ mol }H_3PO_4}\right) = 82.01\text{ g }H_3PO_4$$

Therefore $Ca(NO_3)_2$ is the limiting reagent.

$$(206\text{ g }Ca(NO_3)_2)\left(\frac{1\text{ mol }Ca(NO_3)_2}{164.10\text{ g }Ca(NO_3)_2}\right)\left(\frac{1\text{ mol }Ca_3(PO_4)_2}{3\text{ mol }Ca(NO_3)_2}\right)\left(\frac{310.18\text{ g }Ca_3(PO_4)_2}{1\text{ mol }Ca_3(PO_4)_2}\right) = 130\text{ g }Ca_3(PO_4)_2$$

CHAPTER 1
ATOMS: THE QUANTUM WORLD

1.1 ultraviolet radiation < visible light < infrared radiation < radio waves

1.3 microwaves < visible light < ultraviolet radiation < gamma rays ≈ x-rays

1.5 (a) If wavelength is known, the frequency can be obtained from the relation $c = \nu\lambda$:

$$2.997\,92 \times 10^{8}\ \text{m}\cdot\text{s}^{-1} = (\nu)(800 \times 10^{-9}\ \text{m})$$

$$\nu = \frac{2.997\,92 \times 10^{8}\ \text{m}\cdot\text{s}^{-1}}{800 \times 10^{-9}\ \text{m}}$$

$$= 3.75 \times 10^{14}\ \text{s}^{-1}$$

(b) $2.997\,92 \times 10^{8}\ \text{m}\cdot\text{s}^{-1} = (\nu)(3.0 \times 10^{-3}\ \text{m})$

$$\nu = \frac{2.997\,92 \times 10^{8}\ \text{m}\cdot\text{s}^{-1}}{3.0 \times 10^{-3}\ \text{m}}$$

$$= 1.0 \times 10^{11}\ \text{s}^{-1}$$

1.7 Wien's law states that $T\lambda_{max}$ = constant = 2.88×10^{-3} K·m.

If $T/\text{K} = 1540°\text{C} + 273°\text{C} = 1813$ K, then $\lambda_{max} = \dfrac{2.88 \times 10^{-3}\ \text{K}\cdot\text{m}}{1813\ \text{K}}$

$\lambda_{max} = 1.59 \times 10^{-6}$ m, or 1590 nm

1.9 (a) From $c = \nu\lambda$ and $E = h\nu$, we can write

$$E = hc\lambda^{-1}$$

$$= (6.626\,08 \times 10^{-34}\ \text{J}\cdot\text{s})(2.997\,92 \times 10^{8}\ \text{m}\cdot\text{s}^{-1})(589 \times 10^{-9}\ \text{m})^{-1}$$

$$= 3.37 \times 10^{-19}\ \text{J}$$

(b) 1.00 g of sodium atoms = 1.00 g/22.99 g·mol^{-1} = 0.0435 mol Na

0.0435 mol Na = 2.62×10^{22} atoms of Na

If one atom produces 3.37×10^{-19} J, then 2.62×10^{22} atoms will produce $(2.62 \times 10^{22})(3.37 \times 10^{-19}\ \text{J}) = 8830$ J or 8.83 kJ.

1.11 1 pm = 1×10^{-12} m; $c = \nu\lambda$

$$2.997\,92 \times 10^{8}\ \text{m}\cdot\text{s}^{-1} = \nu(86 \times 10^{-12}\ \text{m})$$

$$\nu = (2.997\,92 \times 10^{8}\ \text{m}\cdot\text{s}^{-1})/(86 \times 10^{-12}\ \text{m})$$
$$= 3.5 \times 10^{18}\ \text{s}^{-1}$$
$$E = h\nu = (6.626\,08 \times 10^{-34}\ \text{J}\cdot\text{s})(3.5 \times 10^{18}\ \text{s}^{-1}) = 2.3 \times 10^{-15}\ \text{J}$$

1.13 (a) false. The total intensity is proportional to T^4. (b) true; (c) false. Photons of radio-frequency radiation are lower in energy than photons of ultraviolet radiation.

1.15 To answer this question, we need to convert the quantities to a consistent set of units, in this case, SI units.

$$(5.15\ \text{ounce})(28.3\ \text{g}\cdot\text{ounce}^{-1})[1\ \text{kg}/1000\ \text{g}] = 0.146\ \text{kg}$$
$$(92\ \text{mi}\cdot\text{h}^{-1})[(3600\ \text{s}\cdot\text{h}^{-1})(0.6214\ \text{mi}\cdot\text{km}^{-1})(1\ \text{km}/1000\ \text{m})]^{-1} = 41\ \text{m}\cdot\text{s}^{-1}$$

Use the de Broglie relationship.

$$\lambda = hp^{-1} = h(mv)^{-1}$$
$$= h(mv)^{-1}$$
$$= \frac{6.626\,08 \times 10^{-34}\ \text{J}\cdot\text{s}}{(0.146\ \text{kg})(0.041\ \text{km}\cdot\text{s}^{-1})}$$
$$= \frac{6.626\,08 \times 10^{-34}\ \text{kg}\cdot\text{m}^2\cdot\text{s}^{-1}}{(0.146\ \text{kg})(41\ \text{m}\cdot\text{s}^{-1})}$$
$$= 1.1 \times 10^{-34}\ \text{m}$$

1.17 (a) Use the de Broglie relationship, $\lambda = hp^{-1} = h(mv)^{-1}$.

$$m_e = (9.109\,39 \times 10^{-28}\ \text{g})(1\ \text{kg}/1000\ \text{g}) = 9.109\,39 \times 10^{-31}\ \text{kg}$$
$$(2.2 \times 10^{3}\ \text{km}\cdot\text{s}^{-1})(1000\ \text{m}\cdot\text{km}^{-1}) = 2.2 \times 10^{6}\ \text{m}\cdot\text{s}^{-1}$$
$$\lambda = h(mv)^{-1}$$
$$= \frac{6.626\,08 \times 10^{-34}\ \text{J}\cdot\text{s}}{(9.109\,39 \times 10^{-31}\ \text{kg})(2.2 \times 10^{6}\ \text{m}\cdot\text{s}^{-1})}$$
$$= 3.3 \times 10^{-10}\ \text{m}$$

(b) $E = h\nu$
$$= (6.626\,08 \times 10^{-34}\ \text{J}\cdot\text{s})(1.00 \times 10^{15}\ \text{s}^{-1})$$
$$= 6.63 \times 10^{-19}\ \text{J}$$

(c) The photon needs to contain enough energy to eject the electron from the surface as well as to cause it to move at $2.2 \times 10^3\ \text{km}\cdot\text{s}^{-1}$. The energy involved is the kinetic energy of the electron, which equals $\frac{1}{2}mv^2$.

$$E_{\text{photon}} = 6.63 \times 10^{-19}\ \text{J} + \frac{1}{2}mv^2$$
$$= 6.63 \times 10^{-19}\ \text{J} + \frac{1}{2}(9.109\,39 \times 10^{-31}\ \text{kg})(2.2 \times 10^{6}\ \text{m}\cdot\text{s}^{-1})^2$$
$$= 6.63 \times 10^{-19}\ \text{J} + 2.2 \times 10^{-18}\ \text{J}$$
$$= 2.9 \times 10^{-19}\ \text{J}$$

But we are asked for the wavelength of the photon, which we can get from $E = h\nu$ and $c = \nu\lambda$ or $E = hc\lambda^{-1}$.

$2.9 \times 10^{-19}\ \text{kg}\cdot\text{m}^2\cdot\text{s}^{-2} = (6.626\,08 \times 10^{-34}\ \text{kg}\cdot\text{m}^2\cdot\text{s}^{-1})(2.997\,92\cdot 10^{8}\ \text{m}\cdot\text{s}^{-1})\lambda^{-1}$

$\lambda = 6.8 \times 10^{-8}\ \text{m}$

$= 68\ \text{nm}$

(d) 68 nm is in the region lying in the far uv, almost to the x-ray region.

1.19 The uncertainty principle states that the uncertainty in momentum times the uncertainty in position must be greater than a constant $h/2\pi$:

$$\Delta p \Delta x \geq \frac{1}{2}\hbar, \text{ where } \hbar = \frac{h}{2\pi} = \frac{6.626\,08 \times 10^{-34}\ \text{J}\cdot\text{s}}{2\pi} = 1.054\,57 \times 10^{-34}\ \text{J}\cdot\text{s}$$

$$\Delta p \Delta x \geq \frac{1}{2}(1.054\,57 \times 10^{-34}\ \text{J}\cdot\text{s}) = 5.272\,85 \times 10^{-35}\ \text{J}\cdot\text{s}$$

The momentum is given by $p = mv$, so we will attribute the uncertainty in the momentum to the uncertainty in speed. Because the error in speed is given as a $\pm$ value, the overall uncertainty will be twice that value $= 2(0.05\ \text{m}\cdot\text{s}^{-1})(8.00\ \text{kg})$, or $0.8\ \text{kg}\cdot\text{m}\cdot\text{s}^{-1}$. The minimum uncertainty will be given by $\Delta p \Delta x = 5.272\,85 \times 10^{-35}\ \text{J}\cdot\text{s}$

$(0.8\ \text{kg}\cdot\text{m}\cdot\text{s}^{-1})\Delta x = 5.272\,85 \times 10^{-35}\ \text{J}\cdot\text{s} = 5.272\,85 \times 10^{-35}\ \text{kg}\cdot\text{m}^2\cdot\text{s}^{-1}$

$\Delta x = 7 \times 10^{-35}\ \text{m}$

1.21 (a) This problem is the same as that solved in Example 1.5, but the electron is moving between different energy levels. For movement between energy levels separated by a difference of 1 in principal quantum number, the expression is

$$\Delta E = E_{n+1} - E_n = \frac{(n+1)^2h^2}{8mL^2} - \frac{n^2h^2}{8mL^2} = \frac{(2n+1)h^2}{8mL^2}$$

For $n = 2$ and $n + 1 = 3$, $\Delta E = \dfrac{5h^2}{8mL^2}$

Then $\lambda_{3,2} = \dfrac{hc}{E} = \dfrac{8mhcL^2}{5h^2} = \dfrac{8mcL^2}{5h}$

For an electron in a 150-pm box, the expression becomes

$$\lambda_{3,2} = \frac{8(9.109\,39 \times 10^{-31}\ \text{kg})(2.997\,92 \times 10^{8}\text{m}\cdot\text{s}^{-1})(150 \times 10^{-12}\ \text{m})^2}{5(6.626\,08 \times 10^{-34}\ \text{J}\cdot\text{s})}$$

$= 1.48 \times 10^{-8}\ \text{m}$

(b) We need to remember that the equation for ΔE was originally determined for energy separations between successive energy levels, so the expression needs to be altered to make it general for energy levels two units apart:

$$\Delta E = E_{n+2} - E_n = \frac{(n+2)^2h^2}{8mL^2} - \frac{n^2h^2}{8mL^2} = \frac{(n^2 + 4n + 4 - n^2)h^2}{8mL^2} = \frac{(4n+4)h^2}{8mL^2}$$

$$\lambda = \frac{hc}{\Delta E} = \frac{hc(8mL^2)}{h[4n+4]} = \frac{8mcL^2}{h[4n+4]}$$

For $n = 2$, the expression becomes

$$\lambda = \frac{8mcL^2}{h[(4 \times 2) + 4]} = \frac{8mcL^2}{9h}$$

$$= \frac{8(9.109\ 39 \times 10^{-31}\ \text{kg})(2.997\ 92 \times 10^{8}\ \text{m}\cdot\text{s}^{-1})(150 \times 10^{-12}\ \text{m})^2}{12(6.626\ 08 \times 10^{-34}\ \text{kg}\cdot\text{m}^2\cdot\text{s}^{-1})}$$

$$= 6.18 \times 10^{-9}\ \text{m}$$

1.23 (a) According to the postulates of quantum mechanics—in particular, the Heisenberg uncertainty principle—the energy of a very small particle cannot be exactly 0 because that would require knowing its position and momentum exactly. Therefore, the zero point energy is the minimum amount of energy a particle will have.
(b) The zero point energy, which is very small to begin with, is not observed for macroscopic systems because m and L are large. The value of the zero point energy in such systems is extremely small and lies well within the error range of values measured for macroscopic systems.

1.25 In each of these series, the principal quantum number for the lower energy level involved is the same for each absorption line. Thus, for the Lyman series, the lower energy level is $n = 1$; for the Balmer series, $n = 2$; for the Paschen series, $n = 3$; and for the Brackett series, $n = 4$.

1.27 $E = h\nu = hc\lambda^{-1} = (6.626\ 08 \times 10^{-34}\ \text{J}\cdot\text{s})(2.997\ 92 \times 10^{8}\ \text{m}\cdot\text{s}^{-1})(589 \times 10^{-9}\ \text{m})^{-1}$
$= 3.37 \times 10^{-19}\ \text{J}$

1.29 The Rydberg equation gives ν when $\mathcal{R} = 3.29 \times 10^{15}\ \text{s}^{-1}$, from which one can calculate λ from the relationship $c = \nu\lambda$.

$$\nu = \mathcal{R}\left(\frac{1}{n_2^{\ 2} - n_1^{\ 2}}\right)$$

and $c = \nu\lambda = 2.997\ 92 \times 10^{8}\ \text{m}\cdot\text{s}^{-1}$

$$c = \mathcal{R}\left(\frac{1}{n_2^{\ 2} - n_1^{\ 2}}\right)\lambda$$

$$2.997\ 92 \times 10^{8}\ \text{m}\cdot\text{s}^{-1} = (3.29 \times 10^{15}\ \text{s}^{-1})\left(\frac{1}{4} - \frac{1}{36}\right)\lambda$$

$\lambda = 4.10 \times 10^{-7}\ \text{m} = 410\ \text{nm}$

(b) violet

1.31 The equation derived in Example 1.7 can be used:

$$\frac{\psi^2(r = 0.25a_0,\theta,\phi)}{\psi^2(0,\theta,\phi)} = \frac{\dfrac{e^{-2(0.25a_0)/a_0}}{\pi a_0{}^3}}{\left(\dfrac{1}{\pi a_0{}^3}\right)} = 0.61$$

1.33 (a) For $n = 2$, there are two subshells because l can be 0 or 1 (s- or p-orbitals).
(b) For $n = 3$, there are three possibilities because $l = 0, 1, 2$ (s-, p-, or d-orbitals).
(c) When $n = 3$, l can be 0, 1, 2.

1.35 (a) 1 orbital; (b) 5 orbitals; (c) 3 orbitals; (d) 7 orbitals

1.37 (a) $n = 3, l = 1$; (b) $n = 5, l = 2$; (c) $n = 4, l = 3$; (d) $n = 6, l = 0$

1.39 (a) $-1, 0, 1$; (b) $-2, -1, 0, 1, 2$; (c) $-3, -2, -1, 0, 1, 2, 3$; (d) 0

1.41 (a) 6 electrons; (b) 10 electrons; (c) 2 electrons; (d) 14 electrons

1.43 (a) $5d$, five; (b) $1s$, one; (c) $6f$, seven; (d) $2p$, three

1.45 (a) $n = 2, l = 1$; (b) $n = 5, l = 3$; (c) $n = 3, l = 0$; (d) $n = 4, l = 2$

1.47 (a) six; (b) two; (c) eight; (d) two

1.49 (a) cannot exist; (b) exists; (c) cannot exist; (d) exists

1.51 (a) The total Coulomb potential energy $V(r)$ is the sum of the individual coulombic attractions and repulsions. There will be one attraction between the nucleus and each electron plus a repulsive term to represent the interaction between each pair of electrons. For lithium, there are three protons in the nucleus and three electrons. Each attractive Coulomb potential will be equal to

$$\frac{(-e)(+3e)}{4\pi\epsilon_0 r} = \frac{-3e^2}{4\pi\epsilon_0 r}$$

where $-e$ is the charge on the electron and $+3e$ is the charge on the nucleus, ϵ_0 is the vacuum permittivity, and r is the distance from the electron to the nucleus. The total attractive potential will thus be

$$\left(\frac{-3e^2}{4\pi\epsilon_0 r_1}\right) + \left(\frac{-3e^2}{4\pi\epsilon_0 r_2}\right) + \left(\frac{-3e^2}{4\pi\epsilon_0 r_3}\right) = \left(\frac{-3e^2}{4\pi\epsilon_0}\right)\left(\frac{1}{r_1} + \frac{1}{r_2} + \frac{1}{r_3}\right)$$

The repulsive terms will have the form

$$\frac{(-e)(-e)}{4\pi\epsilon_0 r_{ab}} = \frac{e^2}{4\pi\epsilon_0 r_{ab}}$$

where r_{ab} represents the distance between two electrons a and b. The total repulsive term will thus be

$$\frac{e^2}{4\pi\epsilon_0 r_{12}} + \frac{e^2}{4\pi\epsilon_0 r_{13}} + \frac{e^2}{4\pi\epsilon_0 r_{23}} = \frac{e^2}{4\pi\epsilon_0}\left(\frac{1}{r_{12}} + \frac{1}{r_{13}} + \frac{1}{r_{23}}\right)$$

This gives

$$V(r) = \left(\frac{-3e^2}{4\pi\epsilon_0}\right)\left(\frac{1}{r_1} + \frac{1}{r_2} + \frac{1}{r_3}\right) + \frac{e^2}{4\pi\epsilon_0}\left(\frac{1}{r_{12}} + \frac{1}{r_{13}} + \frac{1}{r_{23}}\right)$$

(b) The first term represents the coulombic attractions between the nucleus and each electron, and the second term represents the coulombic repulsions between each pair of electrons.

1.53 (a) false. Z_{eff} is considerably affected by the total number of electrons present in the atom because the electrons in the lower energy orbitals will "shield" the electrons in the higher energy orbitals from the nucleus. This effect arises because the *e-e* repulsions tend to offset the attraction of the electron to the nucleus. (b) true; (c) false. The electrons are increasingly less able to penetrate to the nucleus as l increases. (d) true.

1.55 Only (d) is the configuration expected for a ground-state atom; the others all represent excited-state configurations.

1.57 (a) This configuration is possible. (b) This configuration is not possible because $l = 0$ here, so m_l must also equal 0. (c) This configuration is not possible because the maximum value l can have is $n - 1$; $n = 4$, so $l_{max} = 3$.

1.59

Ga	$[Ar]3d^{10}4s^24p^1$	one unpaired electron
Ge	$[Ar]3d^{10}4s^24p^2$	two unpaired electrons
As	$[Ar]3d^{10}4s^24p^3$	three unpaired electrons
Se	$[Ar]3d^{10}4s^24p^4$	two unpaired electrons
Br	$[Ar]3d^{10}4s^24p^5$	one unpaired electron

1.61 (a) Rh, rhodium; (b) P, phosphorus; (c) Ca, calcium

1.63 (a) positive: +5. This oxidation number is determined by the number of valence electrons present. (b) negative: -3. This oxidation number is determined by the number of empty spaces in the valence shell available to accommodate additional electrons.

1.65 (a) $4p$; (b) $4s$; (c) $6s$; (d) $6s$

1.67 (a) $+1$; (b) -2; (c) -1

1.69 (a) $+8$; (b) $+6$; (c) $+3$; (d) $+7$

1.71 (a) 5; (b) 3; (c) 6; (d) 8

1.73 (a) 2; (b) 2; (c) 4; (d) 4

1.75 (a) ns^1; (b) ns^2np^1; (c) $(n-1)d^7ns^2$; (d) $(n-1)d^{10}ns^1$

1.77 (a) phosphorus; (b) cadmium; (c) calcium

1.79 The atomic radii (in pm) are

Sc	164	Fe	124
Ti	147	Co	125
V	135	Ni	125
Cr	129	Cu	128
Mn	137	Zn	137

The major trend is for decreasing radius as the nuclear charge increases, with the exception that Cu and Zn begin to show the effects of electron-electron repulsions and become larger as the *d*-subshell becomes filled. Mn is also an exception as found for other properties, and this may be attributed to having the *d*-shell half-filled.

1.81 $O^{2-} < S^{2-} < Se^{2-} < Te^{2-}$

1.83 (d) chlorine

1.85 From $c = \nu\lambda$ and $E = h\nu$: $E = hc\lambda^{-1}$.

$$E \text{ (for one atom)} = (6.626\,08 \times 10^{-34}\ \text{J·s})(2.997\,92 \times 10^{8}\ \text{m·s}^{-1})(434 \times 10^{-9}\ \text{m})^{-1}$$

$$= 4.58 \times 10^{-19}\ \text{J per photon}$$

$$E \text{ (for } 9.8 \times 10^{20} \text{ photons)} = (9.8 \times 10^{20}\ \text{photons})(4.58 \times 10^{-19}\ \text{J·photon}^{-1})$$

$$= 4.5 \times 10^{2}\ \text{J}$$

1.87 (a) $\frac{\nu}{c} = 2143\ \text{cm}^{-1}$

$\nu = c(2143\ \text{cm}^{-1})$

$\nu = (2.997\ 92 \times 10^{8}\ \text{m}\cdot\text{s}^{-1})(2143\ \text{cm}^{-1})$

$\nu = (2.997\ 92 \times 10^{10}\ \text{cm}\cdot\text{s}^{-1})(2143\ \text{cm}^{-1})$

$\nu = 6.424 \times 10^{13}\ \text{s}^{-1}$

(b) From $E = h\nu$: $E = (6.626\ 08 \times 10^{-34}\ \text{J}\cdot\text{s})(6.424 \times 10^{13}\ \text{s}^{-1})$
$= 4.257 \times 10^{-20}\ \text{J}$.

(c) 1.00 mol of molecules $= 6.022 \times 10^{23}$ molecules, so the energy absorbed by 1.00 mol will be $(4.25 \times 10^{-20}\ \text{J}\cdot\text{molecule}^{-1})(6.022 \times 10^{23}\ \text{molecules}\cdot\text{mol}^{-1})$
$= 2.564 \times 10^{4}\ \text{J}\cdot\text{mol}^{-1}$ or $25.64\ \text{kJ}\cdot\text{mol}^{-1}$.

1.89 (a) See Figures 1.24, 1.27, and 1.28. (b) A node is a region in space where the wavefunction ψ passes through 0. (c) The simplest *s*-orbital has 0 nodes, the simplest *p*-orbital has 1 nodal plane, and the simplest *d*-orbital has 2 nodal planes. (d) Given the increase in number of nodes, an *f*-orbital would be expected to have 3 nodal planes.

1.91 (a) Bk, berkelium; (b) Au, gold; (c) Co, cobalt; (d) S, sulfur

1.93 (a) 6 values: 0, 1, 2, 3, 4, 5; (b) 5 values; $-2, -1, 0, 1, 2$; (c) 3 values: $-1, 0, 1$; (d) 4 subshells: 4*s*, 4*p*, 4*d*, and 4*f*

1.95 A diagonal relationship is a similarity in chemical properties between an element in the periodic table and one lying one period lower and one group to the right. It is caused by the similarity in size of the ions. The lower-right element in the pair would generally be larger because it lies in a higher period, but it also will have a higher oxidation state, which will cause the ion to be smaller. For example, Al^{3+} and Ge^{4+} compounds show the diagonal relationship, as do Li^{+} and Mg^{2+}.

1.97 This trend is attributed to the inert-pair effect, which states that the *s*-electrons are less available for bonding in the heavier elements. Thus, there is an increasing trend as we descend the periodic table for the preferred oxidation number to be 2 units lower than the maximum one.

1.99 (a) metal; (b) nonmetal; (c) metal; (d) metalloid

1.101 The ionization energies of the *s*-block metals are considerably lower, thus making it easier for them to lose electrons in chemical reactions.

1.103 From $E = hc\lambda^{-1}$, we can calculate the wavelength of light that matches the 348 $kJ\cdot mol^{-1}$ necessary to break the C—C bond. First we calculate the energy needed to break one bond and use that value in the equation above to calculate λ:

$$E = \frac{3.48 \times 10^5\ J\cdot mol^{-1}}{6.022 \times 10^{23}\ molecules\cdot mol^{-1}} = 5.78 \times 10^{-19}\ J$$

$$= hc\lambda^{-1} = \frac{(6.626 \times 10^{-34}\ J\cdot s^{-1})(2.997\ 92 \times 10^8\ m\cdot s^{-1})}{\lambda}$$

$\lambda = 3.44 \times 10^{-7}$ m or 344 nm

334-nm light is in the ultraviolet region, so visible light will not have enough energy to break the C—C bond. UV radiation of 344 nm would be suitable, however.

1.105 (a) The relation is derived as follows: the energy of the photon entering, E_{total}, must be equal to the energy to eject the electron, $E_{ejection}$, plus the energy that ends up as kinetic energy, $E_{kinetic}$, in the movement of the electron:

$$E_{total} = E_{ejection} + E_{kinetic}$$

But E_{total} for the photon $= h\nu$ and $E_{kinetic} = (\frac{1}{2})mv^2$, where m is the mass of the object and v is its velocity. $E_{ejection}$ corresponds to the ionization energy, I, so we arrive at the final relationship desired.

(b) $E_{total} = h\nu = hc\lambda^{-1}$

$$= (6.626\ 08 \times 10^{-34}\ J\cdot s)(2.997\ 92 \times 10^8\ m\cdot s^{-1})(58.4 \times 10^{-9}\ m)^{-1}$$

$$= 3.40 \times 10^{-18}\ J$$

$$= E_{ejection} + E_{kinetic}$$

$$E_{kinetic} = (\tfrac{1}{2})mv^2 = (\tfrac{1}{2})(9.109\ 39 \times 10^{-28}\ g)(2450\ km\cdot s^{-1})^2$$

$$= (\tfrac{1}{2})(9.109\ 39 \times 10^{-31}\ kg)(2.450 \times 10^6\ m\cdot s^{-1})^2$$

$$= 2.73 \times 10^{-18}\ kg\cdot m^2\cdot s^{-2} = 2.73 \times 10^{-18}\ J$$

$$3.40 \times 10^{-18}\ J = E_{ejection} + 2.73 \times 10^{-18}\ J$$

$$E_{ejection} = 6.7 \times 10^{-19}\ J$$

1.107 By the time we get to the lanthanides and actinides—the two series of *f*-orbital filling elements—the energy levels become very close together and minor changes in environment cause the different types of orbitals to switch in energy-level ordering. For the elements mentioned, the electronic configurations are

La	$[Xe]6s^25d^1$	Lu	$[Xe]6s^24f^{14}5d^1$
Ac	$[Rn]7s^26d^1$	Lr	$[Rn]7s^25f^{14}6d^1$

As can be seen, all these elements have one electron in a *d*-orbital, so placement in the third column of the periodic table could be considered appropriate for either, depending on what aspects of the chemistry of these elements we are comparing. The choice is not without argument, and it is discussed by W. B. Jensen, *J. Chem. Ed.* 59, 634 (1982).

1.109 (a) We can use the hydrogen 2*s* wavefunction found in Table 1.2. Remember that the probability of locating an electron at a small region in space is proportional to ψ^2, not ψ.

$$\psi_{2s} = \frac{1}{4}\left(\frac{1}{2\pi a_0^3}\right)^{1/2}\left(2 - \frac{r}{a_0}\right)e^{-r/2a_0}$$

$$\psi_{2s}^2 = \left(\frac{1}{32\pi a_0^3}\right)\left(2 - \frac{r}{a_0}\right)^2 e^{-r/a_0}$$

$$\frac{\psi_{2s}^2(r,\theta,\phi)}{\psi_{2s}^2(0,\theta,\phi)} = \frac{\left(\frac{1}{32\pi a_0^3}\right)\left(2 - \frac{a_0}{a_0}\right)^2 e^{-a_0/a_0}}{\left(\frac{1}{32\pi a_0^3}\right)(2)^2 e^{-0/a_0}} = \frac{e^{-1}}{4} = 0.092$$

(b) By examining the 2*s* wavefunctions, we see that the value of ψ and also ψ^2 will be 0 when $2 - \frac{r}{a_0} = 0$ or $r = 2a_0$.

CHAPTER 2
CHEMICAL BONDS

2.1 The coulombic attraction is inversely proportional to the distance between the two oppositely charged ions (Equation 1) so the ions with the shorter radii will give the greater coulombic attraction. The answer is therefore (a) Na^+, F^-.

2.3 The Mg^{2+} ion is smaller than the Ba^{2+} ion (72 vs 136 pm). Because the lattice energy is related to the coulombic attraction between the ions, it will be inversely proportional to the distance between the ions (see Equation 2). Hence the larger barium ion will have the lower lattice energy for a given anion.

2.5 (a) 3; (b) 2; (c) 6; (d) 4

2.7 (a) $[Kr]4d^{10}5s^2$; (b) $[Kr]4d^{10}5s^2$; (c) $[Xe]4f^{14}5d^4$; (d) [Ar]

2.9 (a) $[Ar]3d^4$; (b) $[Ar]3d^2$; (c) $[Kr]4d^2$; (d) $[Kr]4d^8$

2.11 (a) Ni^{2+}; (b) Sc^{2+}; (c) Tc^{2+}; (d) Nb^{2+}

2.13 (a) 4*s*; (b) 3*p*; (c) 3*p*; (d) 4*s*

2.15 (a) +1; (b) −2; (c) +2; (d) +3

2.17 (a) 2; (b) 8; (c) 7; (d) 5

2.19 (a) +7; (b) −1; (c) [Kr] for +7, [Xe] for −1; (d) electrons are lost or added to give noble gas configuration

2.21 (a) $[Xe]4f^{14}5d^{10}6s^2$; (b) $[Kr]4d^{10}$; (c) [Ar]; (d) [Kr]; (e) [Ar]; (f) $[Ar]3d^8$

2.23 (a) 3*d*; (b) 5*s*; (c) 5*p*; (d) 4*d*

2.25 (a) Mg_3As_2; (b) In_2S_3. In_2S is also possible but not as common as In_2S_3; (c) AlH_3; (d) H_2Te; (e) BiF_3

2.27 (a) GaAs; (b) MgO; (c) Al_2Te_3; (d) RuO_2; (e) V_2O_5

2.29

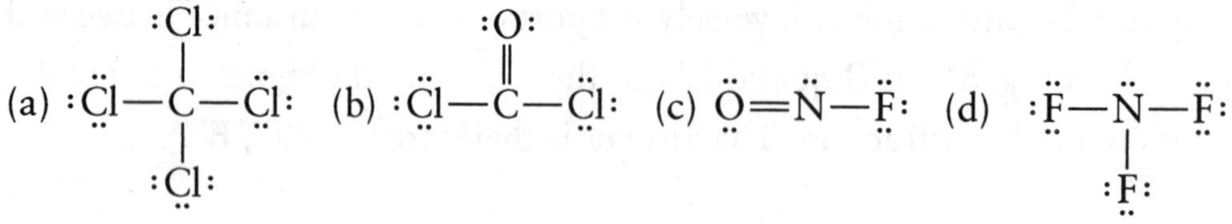

2.31

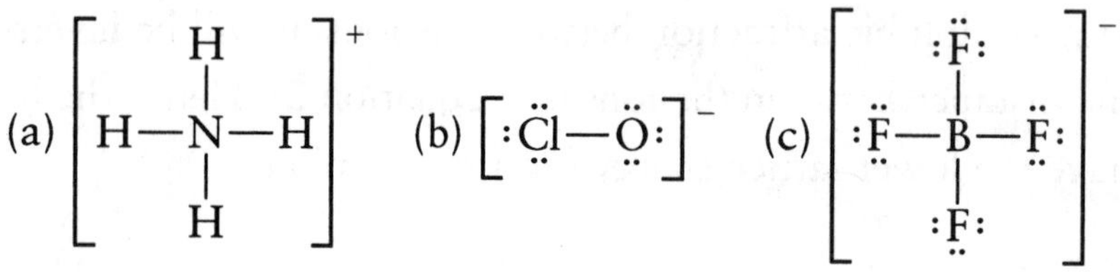

2.33

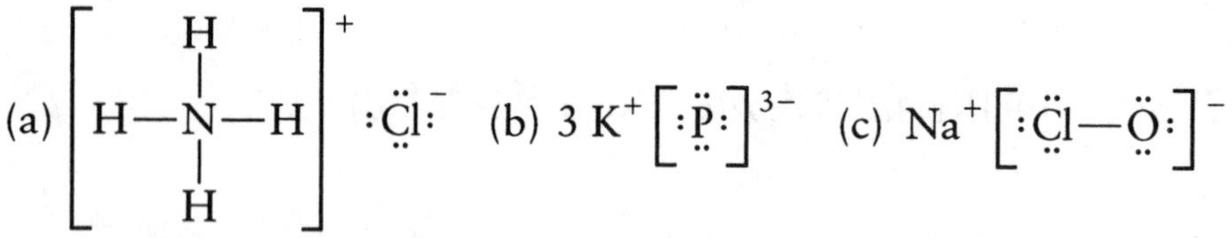

2.35

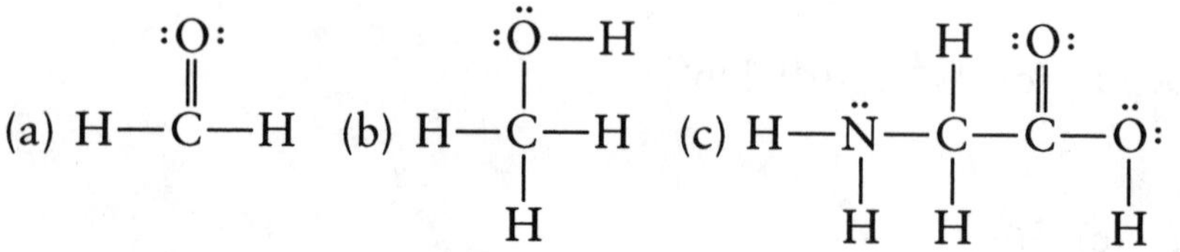

2.37

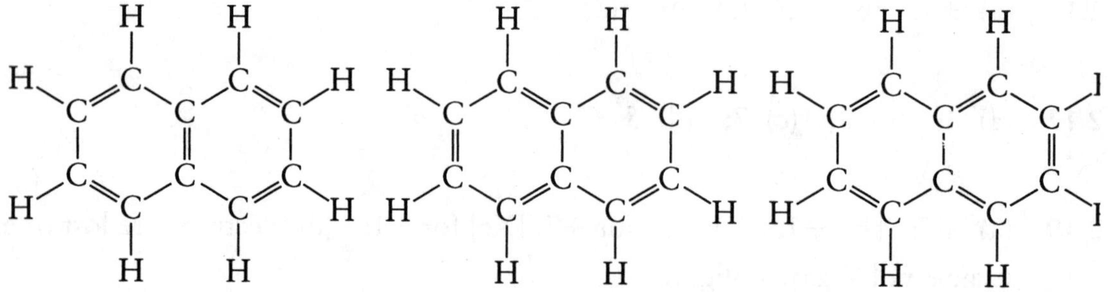

2.39

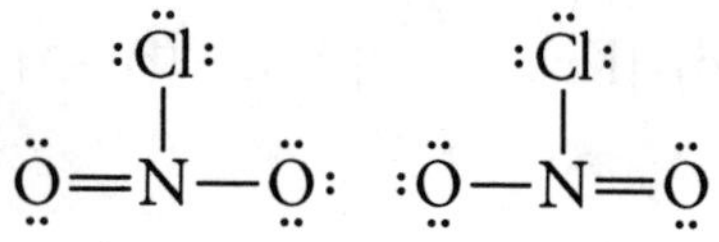

2.41 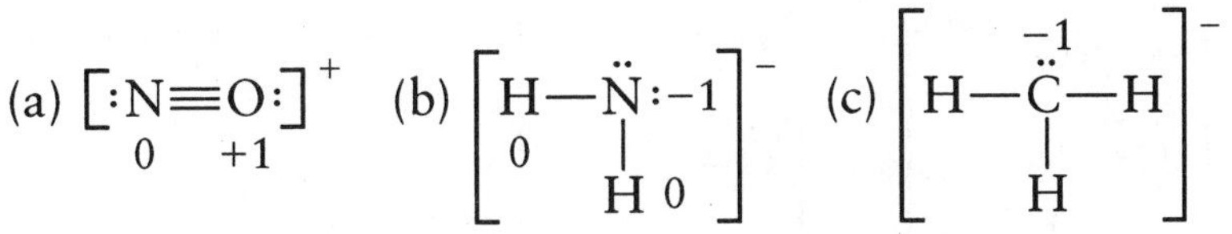

all H's are 0

2.43 (a) The first structure has a formal charge of 0 at S, 0 at two of the oxygen atoms and -1 at the third. The S atom in the second structure has a formal charge of $+1$, 0 at one of the oxygen atoms and -1 at the other two. The first structure would be expected to be lower in energy. (b) The formal charge at S in the first structure is $+1$, at two O atoms it is 0, and at the other two it is -1. In the second structure, the formal charge at S is 0, at three O atoms it is also 0, and at the fourth O atom it is -1. The second structure would be expected to be lower in energy.

2.45 (a) The formal charges at Xe and F are 0 in the first structure, whereas in the second structure Xe is -1, one F is 0, and one F is $+1$. The first structure is favored based on formal charges. (b) In the first structure all of the atoms have formal charges of 0 whereas in the second, one oxygen atom has a formal charge of $+1$ and one has a formal charge of -1. The first structure is thus preferred.

2.47 (a) The sulfite ion has one Lewis structure that obeys the octet rule. The formal charge of $+1$ at S can be reduced to 0 by including 1 double bond to oxygen, a form with three resonance structures.

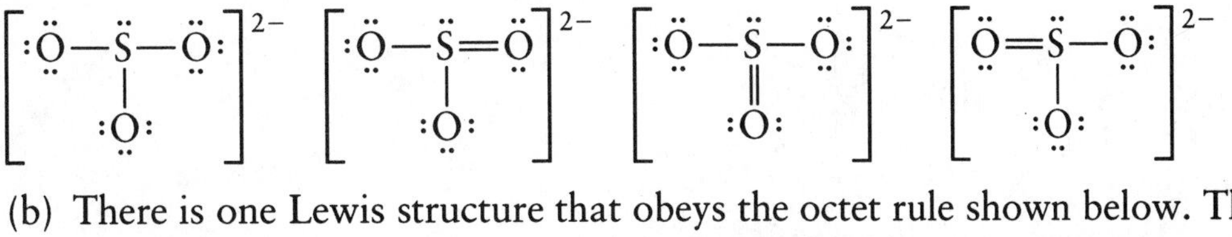

(b) There is one Lewis structure that obeys the octet rule shown below. The formal charge at sulfur can be reduced to 0 by including one double bond contribution. This gives rise to two expanded octet structures. Notice that, unlike the case of the sulfite ion where there are three resonance forms, the presence of the hydrogen ion restricts the electrons on the oxygen atom to which it is attached. Because H is electropositive, its placement near an oxygen atom makes it less likely for that oxygen to donate a lone pair to an adjacent atom.

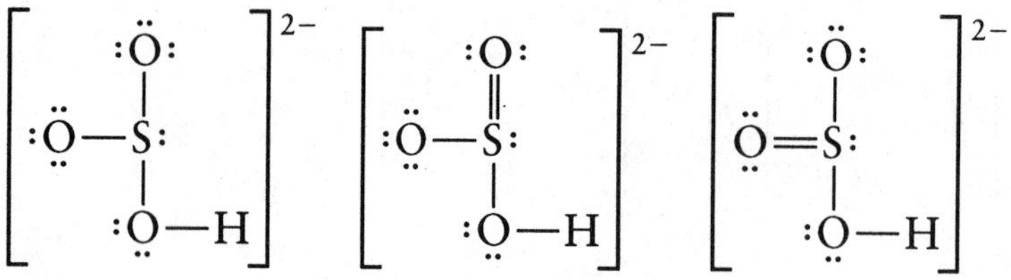

(c) There is one Lewis structure for the perchlorate ion that obeys the octet rule. The formal charge at Cl can be reduced to 0 by including three double bond contributions, giving rise to four resonance forms.

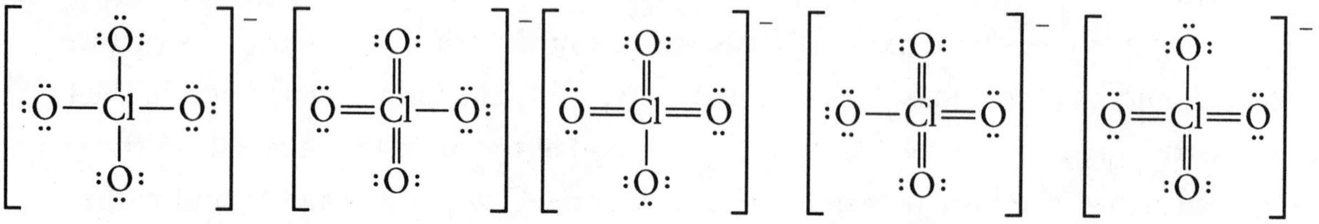

(d) For the nitrite ion, there are two resonance forms, which both obey the octet rule.

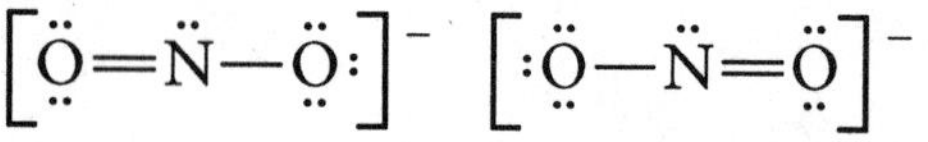

2.49 The Lewis structures are

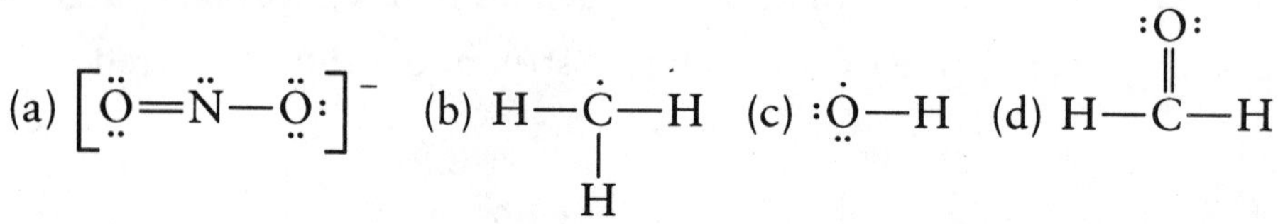

Radicals are species with an unpaired electron, therefore (b) and (c) are radicals.

2.51

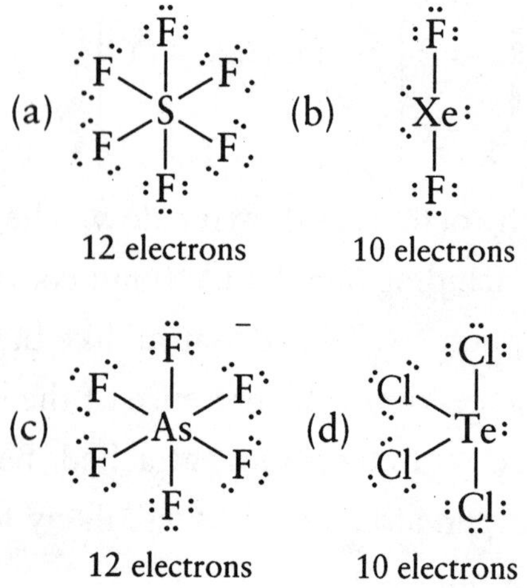

2.53 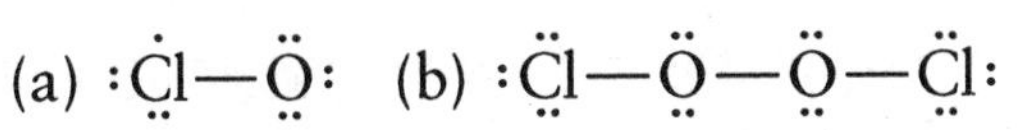

radical — not a radical

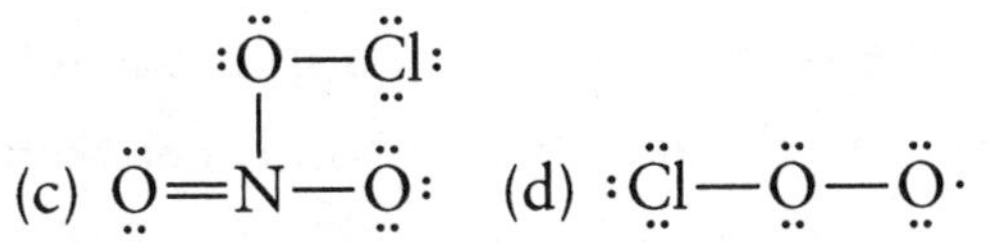

not a radical — radical

2.55 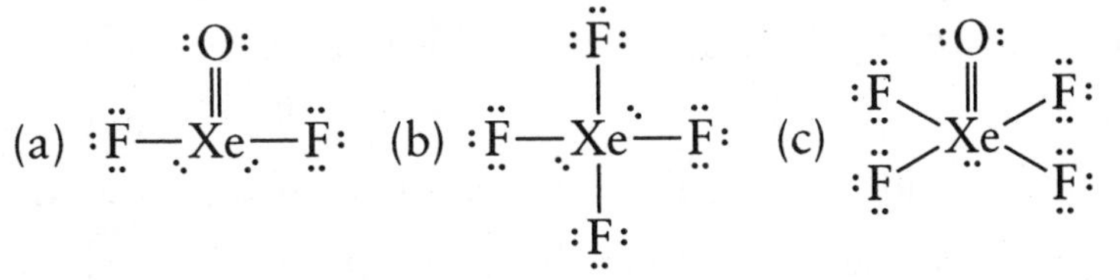

12 electrons, 2 lone pairs — 12 electrons, 2 lone pairs — 14 electrons, 1 lone pair

2.57 (a) Lewis base; (b) Lewis acid; (c) Lewis acid; (d) Lewis base

2.59 (a) PF_5 is the Lewis acid, F^- is the Lewis base, the product is PF_6^-.
(b) SO_2 is the Lewis acid, Cl^- is the Lewis base, the product is SO_2Cl^-.
(c) SO_3 is the Lewis acid, OH^- is the Lewis base, the product is HSO_4^-.

2.61 $Rb^+ < Sr^{2+} < Be^{2+}$, the smaller, more highly charged cations will have the greater polarizing power. The ionic radii are 149, 116, 27 pm, respectively.

2.63 $O^{2-} < N^{3-} < Cl^- < Br^-$; the polarizability increases as the ion gets larger and less electronegative. The ionic radii for these species are 140, 171, 181, 196, respectively.

2.65 (a) Iodide is more polarizable than Cl^-, so HI would be more covalent than HCl. (b) The bonds in CF_4 would be more ionic. The electronegativity difference is greater between C and F than between C and H, making the C—F bonds more ionic. (c) C and S have nearly identical electronegativities, so the C—S bonds would be expected to be almost completely covalent, whereas the C—O bonds would be more ionic.

2.67 (a) CO_3^{2-} will have the longest C—O bond length. In CO there is a triple bond and in CO_2 the C—O bonds are double bonds. In carbonate, the bond is an average of three Lewis structures in which the bond is double in one form and single in two of the forms. We would thus expect the bond order to be approximately 1.3. Because the bond length is inversely related to the number of bonds between the atoms, we expect the bond length to be longest in carbonate. (b) Similar arguments can be used for these molecules as in part (a). In SO_2 and SO_3, the Lewis structures with the lowest formal charge at S have double bonds between S and each O. In the sulfite ion, however, there are three Lewis structures that have a 0 formal charge at S. Each has one S—O double bond and two S—O single bonds. Because these S—O bonds would have a substantial amount of single bond character, they would be expected to be longer than those in SO_2 or SO_3. This is consistent with the experimental data that show the S—O bond lengths in SO_2 and SO_3 to be 143 pm, whereas those in SO_3^{2-} range from about 145 to 152 pm depending on the compound. (c) The C—N bond in HCN is a triple bond, in CH_2NH it is a double bond, and in CH_3NH_2 it is a single bond. The C—N bond in the last molecule would, therefore, be expected to be the longest.

2.69 (a) The covalent radius of N is 75 pm, so the N—N single bond in hydrazine would be expected to be ca. 150 pm. The experimental value is 145 pm. (b) The C—O bonds in carbon dioxide are double bonds. The covalent radius for doubly bonded carbon is 67 pm and that of O is 60 pm. Thus we predict the C=O in CO_2 to be ca. 127 pm. The experimental bond length is 116.3 pm. (c) The C—O bond is a double bond so it would be expected to be the same as in (b). This is the experimentally found value. The C—N bonds are single bonds and so one might expect the bond distance to be the sum of the single bond C radius and the single bond N radius (77 plus 75 pm) which is 152 pm. However, because the C atom is involved in a multiple bond, its radius is actually smaller. The sum of that radius (67 pm) and the N single bond radius gives 132 pm, which is close to the experimental value of 133 pm. (d) The N—N bond is a double bond so we expect the bond distance to be two times the double bond covalent radius of N, which is $2 \times (60 \text{ pm})$ or 120 pm. The experimental value is 123.0 pm.

2.71 Sodium fluoride, because the lattice energy is inversely proportional to the distance between the ions. The fluoride ion will be smaller than chloride ion.

2.73 (a) 4; (b) 3; (c) 4; (d) 5

2.75 (a) [Kr]; (b) [Xe]$4f^{14}5d^{10}6s^2$; (c) [Ar]$3d^{10}$; (d) [Xe]$4f^{14}5d^{10}$

2.77 (a) Ni^{3+}; (b) Fe^{3+}; (c) Ru^{3+}; (d) Mo^{3+}

2.79 (a) V_2O_5; (b) PbO_2; (c) Tl_2O_3

2.81 (a) 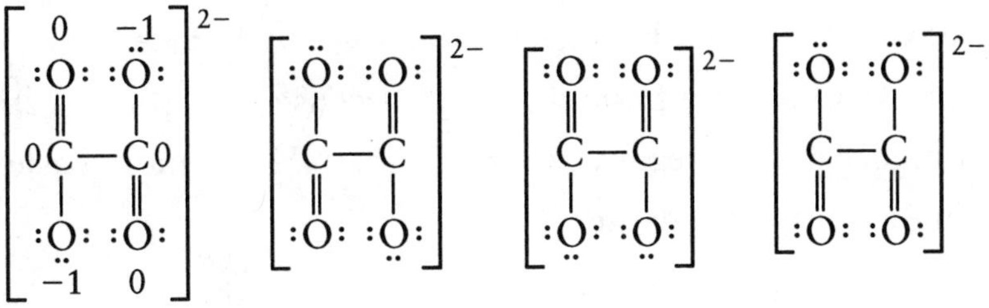

(b) 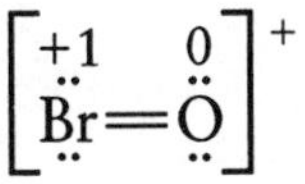

(c) 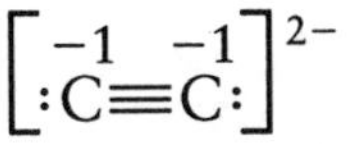

2.83 (a) 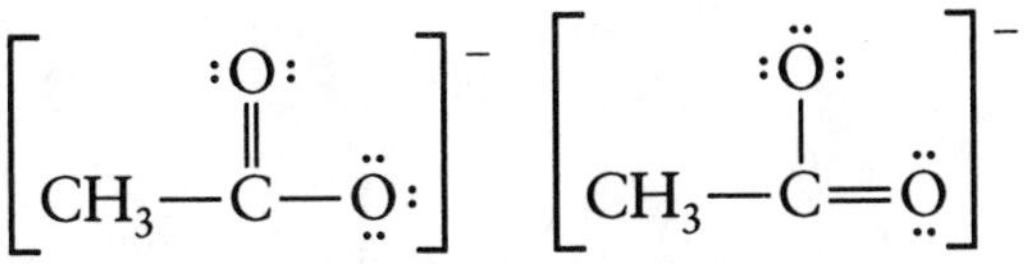

(b) 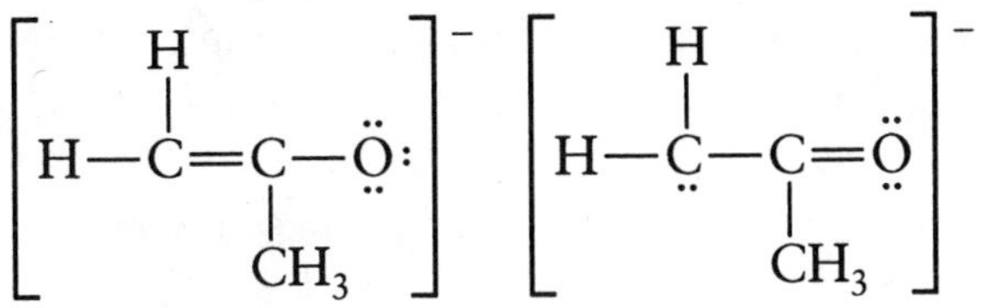

(c) 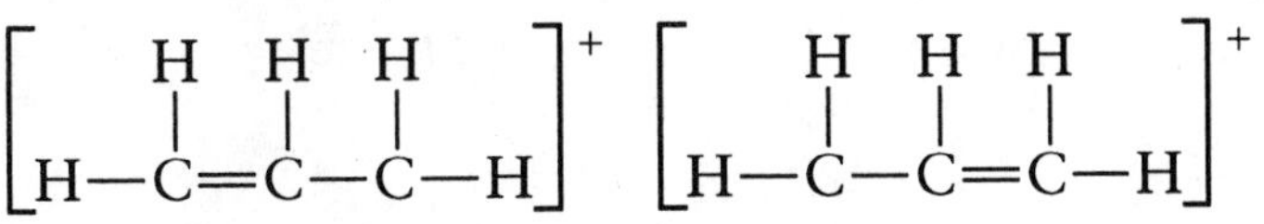

(d) 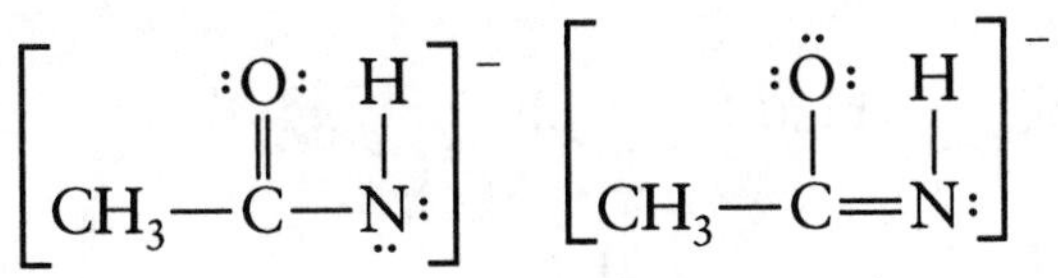

2.85 (a) 77 pm + 72 pm = 149 pm (b) 111 pm + 72 pm = 183 pm

(c) 141 pm + 72 pm = 213 pm; the bond distance increases as you go down the periodic table because the atoms are getting larger.

2.87 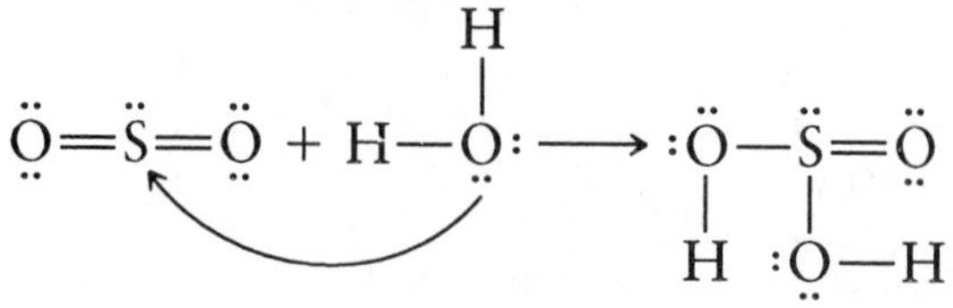

2.89 (a) CH_3^- is more basic than CH_4 because it has a lone pair of electrons, whereas CH_4 does not. (b) H_2O is more basic than H_2S because O is more electronegative than S and has a higher partial negative charge. (c) NH_2^- is more basic than NH_3 due to the higher negative charge.

2.91 The chemical formula for phosgene is $COCl_2$ as determined in section F-2 and F-3. The Lewis structure is

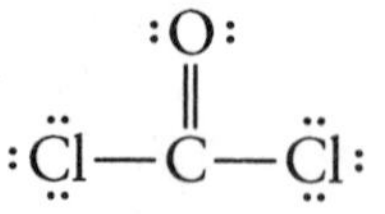 molar mass = 98.91 g·mol^{-1}

2.93 P and S are larger atoms which are less able to form multiple bonds to themselves. Unlike the small N and O atoms. All bonds in P_4 and S_8 are single bonds, whereas N_2 has a triple bond and O_2 a double bond.

2.95 (a) H—C≡C—H H—C≡Si—H H—Si≡Si—H
H—C≡N: :N≡N:

(b) 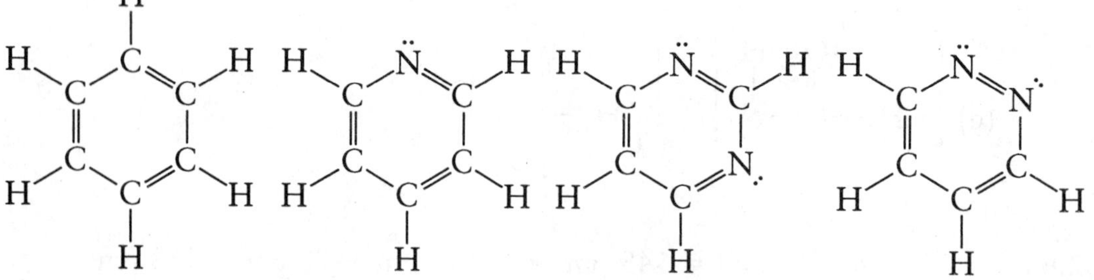

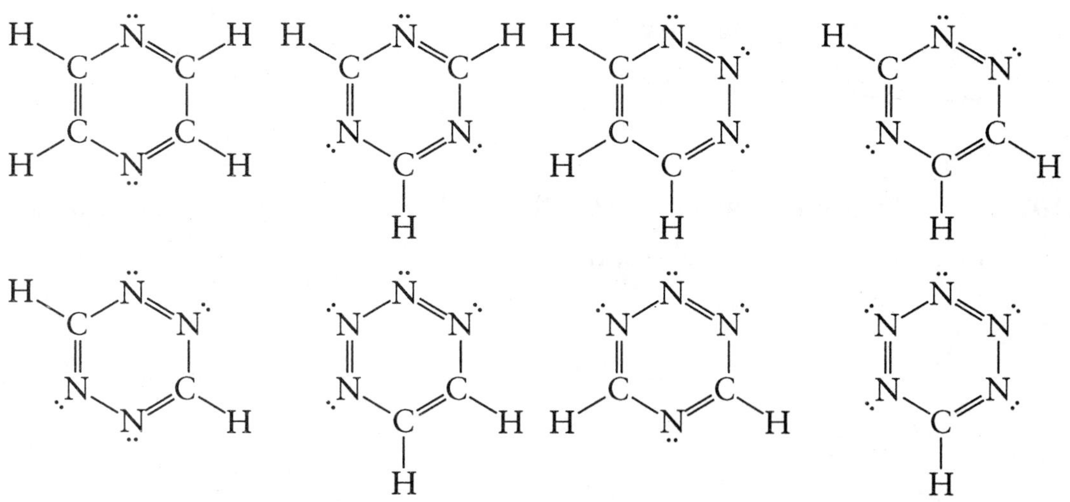

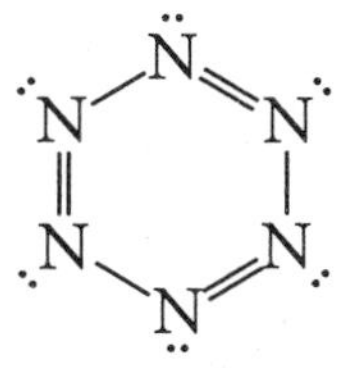

$^+$suitable resonance forms

Note: the ability to draw the structures suggests that the compounds are feasible but it does not indicate stability. For example, the N_6 ring gives a reasonable Lewis structure but would be very unstable.

2.97 The Lewis structures for NO and NO_2 are

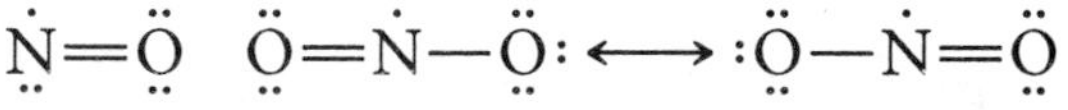

Both compounds are radicals.

(a) NO has a double bond, but NO_2 has an N—O bond that is the average of a single bond and a double bond. Thus NO would be expected to have a shorter, stronger bond. This is indicated by the bond energies.

(b) The fact that the two N—O bonds in NO_2 are equal is a result of the two available resonance forms.

2.99 The resonance forms are achieved by moving the location of the radical from one carbon atom to the next as shown.

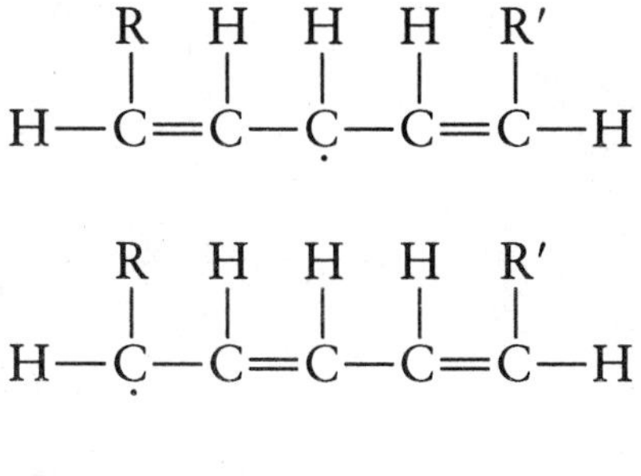

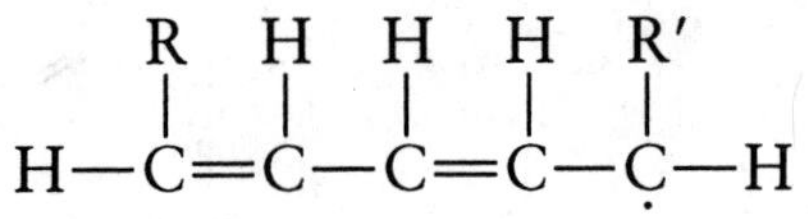

2.101 The alkyne group will suffer C—H bond because a large force constant k results in a higher frequency absorption.

CHAPTER 3
MOLECULAR SHAPE AND STRUCTURE

3.1 (a) 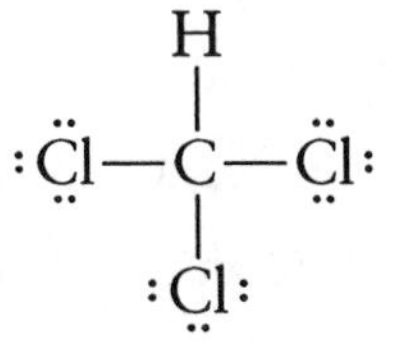

(a) tetrahedral; (b) one; (c) 109.5°

3.3 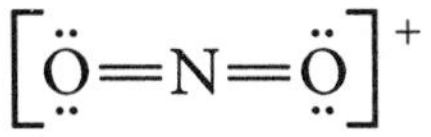

(a) linear; (b) 180°

3.5 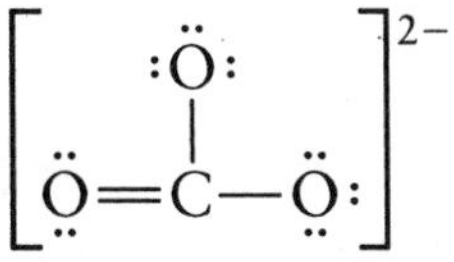 + 2 resonance structures

(a) trigonal planar; (b) all O—C—O bond angles are equivalent (c) 120°

3.7 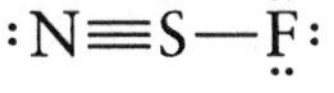

(a) angular or bent; (b) slightly less than 120°

3.9 The Lewis structures are

(a) 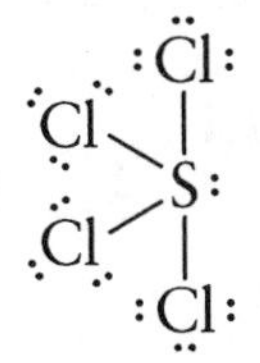 (b) 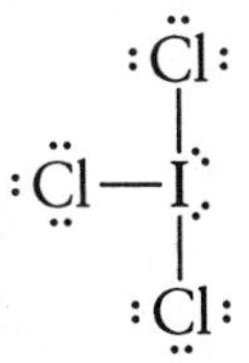 (c) 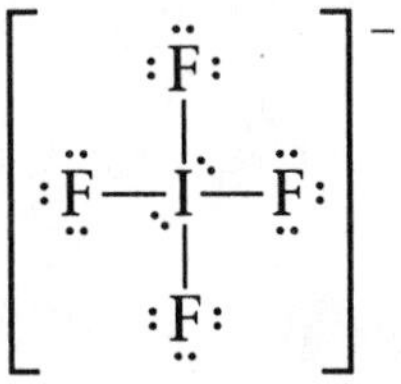 (d) 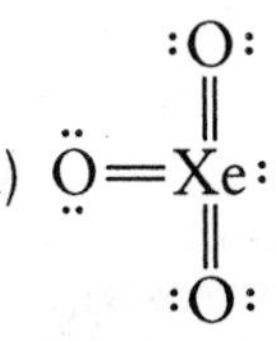

(a) The sulfur atom will have five pairs of electrons about it: one nonbonding pair and four bonding pairs to chlorine atoms. The arrangement of electron pairs will be trigonal bipyramidal; the nonbonding pair of electrons will prefer to lie in an equatorial position, because in that location the e-e repulsions will be lowest. The actual structure is described as a seesaw. AX_4E

(b) Like the sulfur atom in (a), the iodine in iodine trichloride has five pairs of electrons about it, but here there are two lone pairs and three bonding pairs. The arrangement of electron pairs will be the same as in (a), and again the lone pairs will occupy the equatorial positions. Because the name of the molecule ignores the lone pairs, it will be classified as T-shaped. AX_3E_2

(c) There are six pairs of electrons about the central iodine atom in IF_4^-. Of these, two are lone pairs and four are bonding pairs. The pairs will be placed about the central atom in an octahedral arrangement with the lone pairs opposite each other. This will minimize repulsions between them. The name given to the structure is square planar. AX_4E_2

(d) In determining the shape of a molecule, double bonds count the same as single bonds. The XeO_3 structure has four "objects" about the central Xe atom: three bonds and one lone pair. These will be placed in a tetrahedral arrangement. Because the lone pair is ignored in naming the molecule, it will be classified as pyramidal. AX_3E

3.11 The Lewis structures for the ions and molecules are

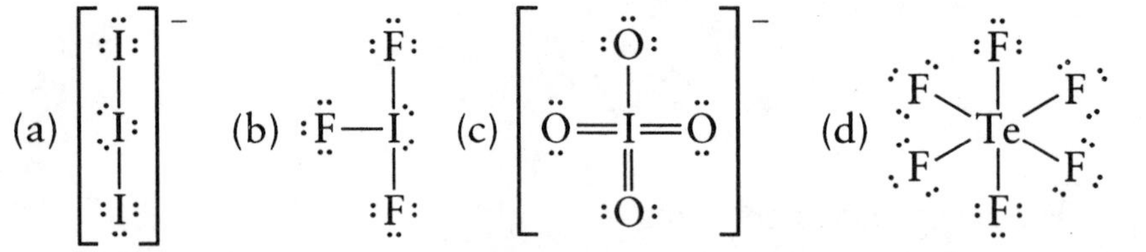

(a) The I_3^- molecule is predicted to be linear, so the < I—I—I should equal 180°. AX_2E_3

(b) The IF_3 molecule is T-shaped. There should be two F—I—F angles of 90° and one of 180°. AX_3E_2

(c) The structure of IO_4^- will be tetrahedral, so the O—I—O bond angles should be 109.5°. AX_4

(d) The structure of TeF_6 is octahedral, so the F—Te—F angles should be either 90° or 180°. AX_6

3.13 The Lewis structures are

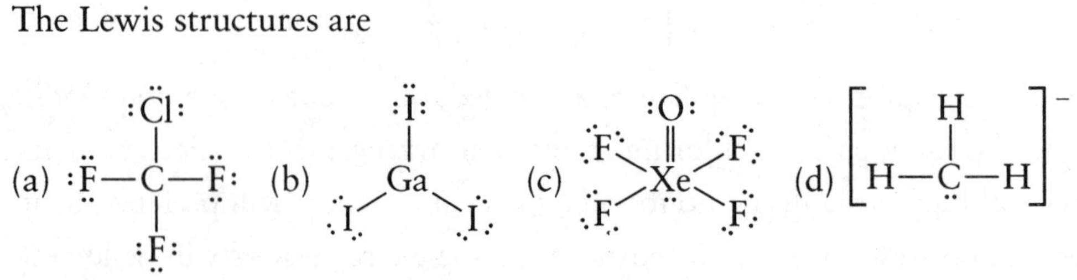

(a) The shape of CF_3Cl is tetrahedral; all halogen—C—halogen angles should be approximately 109.5°. AX_4; (b) GaI_3 molecules will be trigonal planar with

I—Ga—I bond angles of approximately 120°. AX_3; (c) $XeOF_4$ molecules will be square pyramidal with O—Xe—F angles of 90°, and F—Xe—F angles of 90° and 180°. AX_5E; (d) CH_3^- ions will be pyramidal with H—C—H angles of slightly less than 109.5°. AX_3E

3.15 Angles a and b are both expected to be approximately 120°.

3.17 (a) slightly less than 120°; (b) 180°; (c) 180°; (d) slightly less than 109.5°

3.19 The Lewis structures are

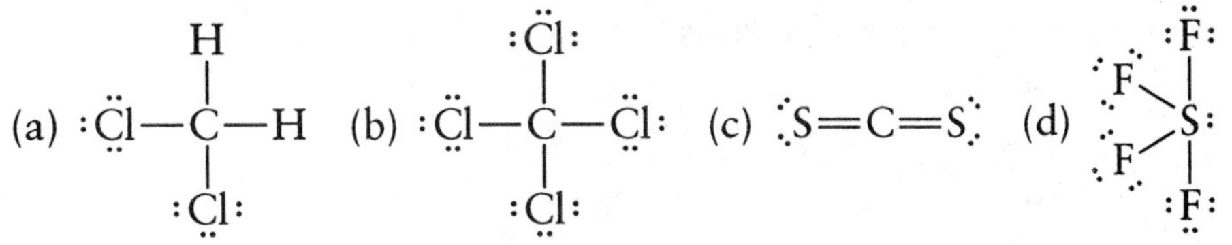

Molecules (a) and (d) will be polar; (b) and (c) will be nonpolar.

3.21 (a)

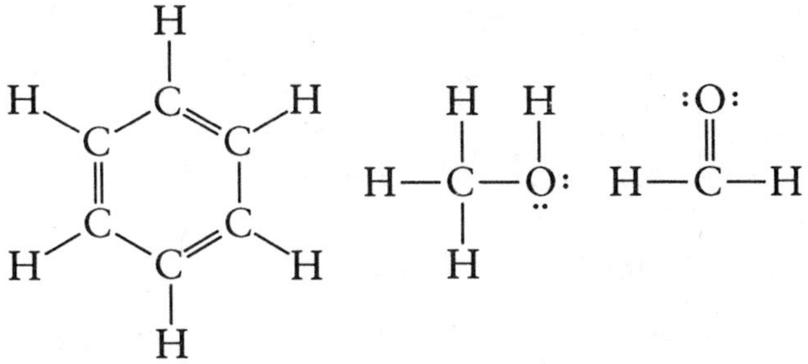

benzene methanol formaldehyde

(a) nonpolar; (b) polar; (c) polar

3.23 Of the three forms, only **3** is nonpolar. This is because the C—Cl bond dipoles are pointing in exactly opposite directions in **3**. The dipole moment for **1** would be the largest because the C—Cl bond vectors are pointing most nearly in the same direction in **1** (60° apart) whereas in **2** the C—Cl vectors point more away from each other, giving a larger cancellation of dipole (120°).

3.25 (a) sp^3, orbitals oriented toward corners of a tetrahedron (109.5° apart); (b) *sp*, orbitals oriented directly opposite to each other (180° apart); (c) sp^3d^2, orbitals oriented toward the corners of an octahedron (interorbital angles of 90° and 180°); (d) sp^2, orbitals oriented toward the corners of an equilateral triangle trigonal planar array (angles = 120°); trigonal planar.

3.27 (a) sp^3d; (b) sp^2; (c) sp^3; (d) sp

3.29 (a) sp^3; (b) sp; (c) sp^3d; (d) sp^3

3.31 (a) Li_2 BO = $\frac{1}{2}(2 + 2 - 2) = 1$

— 2sσ*

2s ↑ ↑ 2s

↑↓ 2sσ

↑↓ 1sσ*

1s ↑↓ ↑↓ 1s

↑↓ 1sσ

diamagnetic, no unpaired electrons

(b) Li_2^+ BO = $\frac{1}{2}(2 + 2 - 2 - 1) = \frac{1}{2}$

— 2sσ*

2s ↑ — 2s

↑ 2sσ

↑↓ 1sσ*

1s ↑↓ ↑↓ 1s

↑↓ 1sσ

paramagnetic, one unpaired electron

(c) Li_2^- BO = $\frac{1}{2}(2 + 2 - 2 - 1) = \frac{1}{2}$

↑ 2sσ*

2s ↑ ↑↓ 2s

↑↓ 2sσ

↑↓ 1sσ*

1s ↑↓ ↑↓ 1s

↑↓ 1sσ

paramagnetic, one unpaired electron

3.33 (a) (1) $(2s\sigma)^2(2s\sigma^*)^2(2p\sigma)^2(2p\pi_x)^2(2p\pi_y)^2(2p\pi_x{}^*)^2(2p\pi_y{}^*)^1$
(2) $(2s\sigma)^2(2s\sigma^*)^2(2p\sigma)^2(2p\pi_x)^2(2p\pi_y)^2(2p\pi^*)^1$
(3) $(2s\sigma)^2(2s\sigma^*)^2(2p\sigma)^2(2p\pi_x)^2(2p\pi_y)^2(2p\pi_x{}^*)^2(2p\pi_y{}^*)^2$

(b) (1) 1.5; (2) 2.5; (3) 1

(c) (1) and (2) are paramagnetic with one unpaired electron each

(d) π in all three cases

3.35 An N—N double bond is composed of a σ and a π bond, whereas the N—N single bond is a σ-type bond. Because the types of orbitals that compose the σ and π

bonds are different, it is reasonable to expect that the resulting molecular orbitals would have different energies.

3.37 (a) See Figure 3.35 for the energy level diagram for N_2. (b) The nitrogen atom is more electronegative, which will make its orbitals lower in energy than those of C. The revised energy-level diagram is shown below. This will make all of the bonding orbitals closer to N than to C in energy and will make all the antibonding orbitals closer to C than to N in energy.

Energy level diagram for CN^-

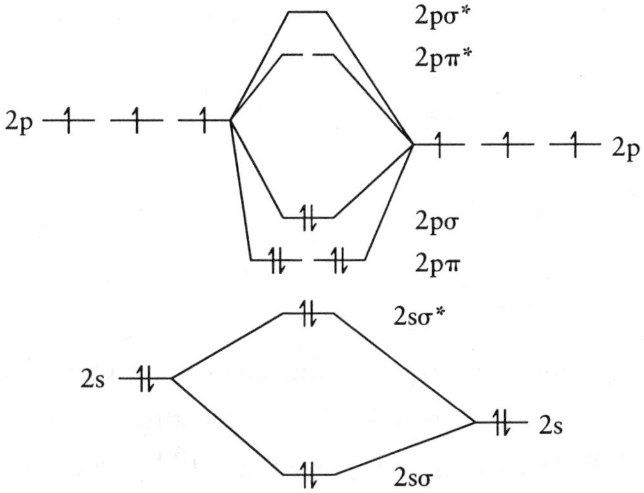

energy levels on C energy levels on N

(c) The electrons in the bonding orbitals will have a higher probability of being at N because it is the more electronegative atom and its orbitals are lower in energy.

3.39 The conductivity of a semiconductor increases with temperature as increasing numbers of electrons are promoted into the conduction band, whereas the conductivity of a metal will decrease as the motion of the atoms will slow down the migration of electrons.

Note: The electron due to the charge has arbitrarily been placed in a 2p orbital on C on the left-hand side of the diagram.

3.41 (a) In and Ga; (b) P and Sb

3.43

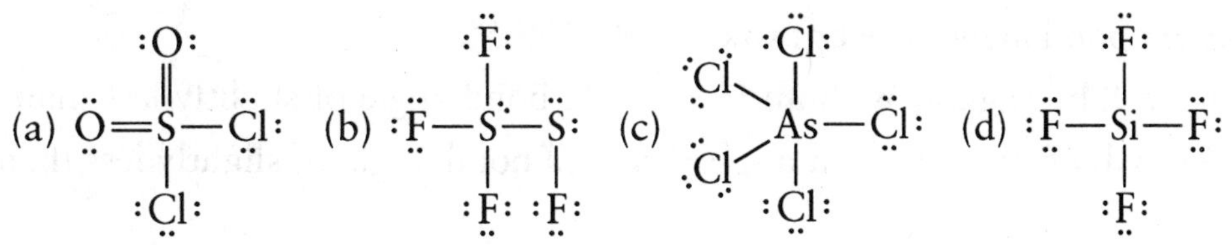

(a) tetrahedral, polar; (b) tetrahedral at the SF_3 group, bent at the S attached to SF_3 and one F atom, polar; (c) trigonal bipyramidal, nonpolar; (d) tetrahedral, nonpolar

3.45 (a) sp^2, trigonal planar, 120°; (b) sp^3, pyramidal, slightly less than 109.5°; (c) sp^3d, linear, 180°; (d) sp^3d^2, octahedral, 90° and 180°.

3.47 (a) PF_3; (b) SF_4; (c) IF_5

3.49

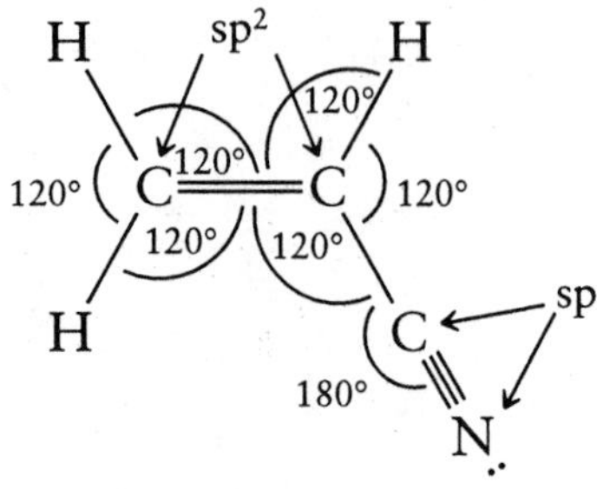

3.51 The elemental composition gives an empirical formula of CH_4O, which agrees with the molar mass. There is only one reasonable Lewis structure; this corresponds to the compound methanol. Both carbon and oxygen are sp^3 hybridized. All of the bond angles at carbon should be 109.5°. The bond angles about oxygen should be close to 109.5° but will be somewhat less, due to the repulsions by the lone pairs. The molecule is polar.

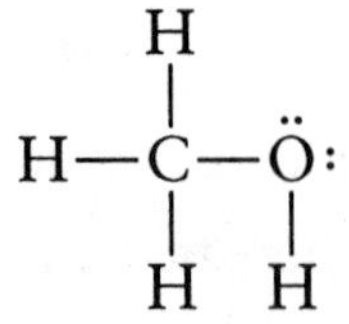

3.53 (a) 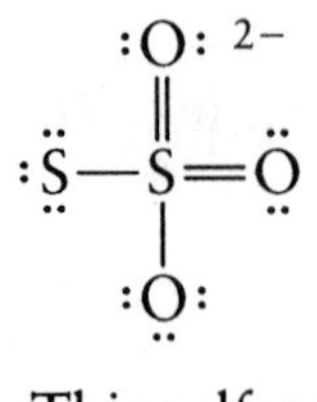 (b) H_3C—Be—CH_3 (c) 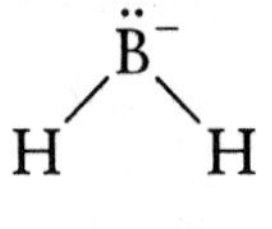 (d) 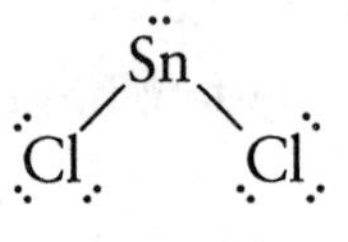

(a) Thiosulfate is tetrahedral with bond angles approximately equal to 109.5°.
(b) Dimethyl beryllium is linear with a C—Be—C bond angle of 180°; the H—C—H bond angles are approximately 109.5°.
(c) BH_2^- will be angular with an H—B—H bond angle of slightly less than 120°.
(d) $SnCl_2$ will be angular with a Cl—Sn—Cl bond angle of slightly less than 120°.

3.55 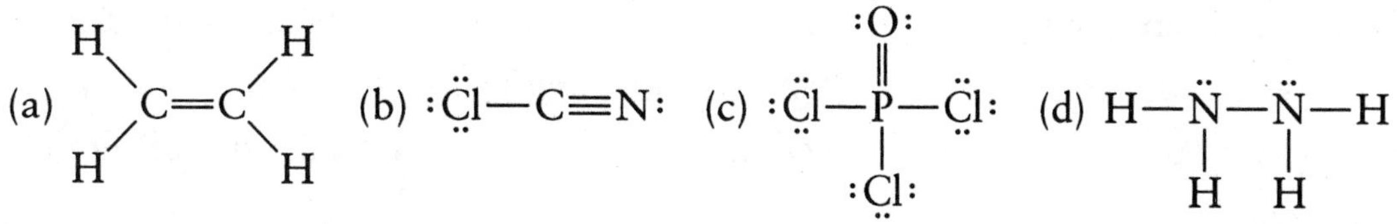

(a) bond angles are all ca. 120°; (b) bond angle = 180°; (c) bond angles = 109.5°; (d) bond angles about both nitrogen atoms are slightly less than 109.5°

3.57 (a)

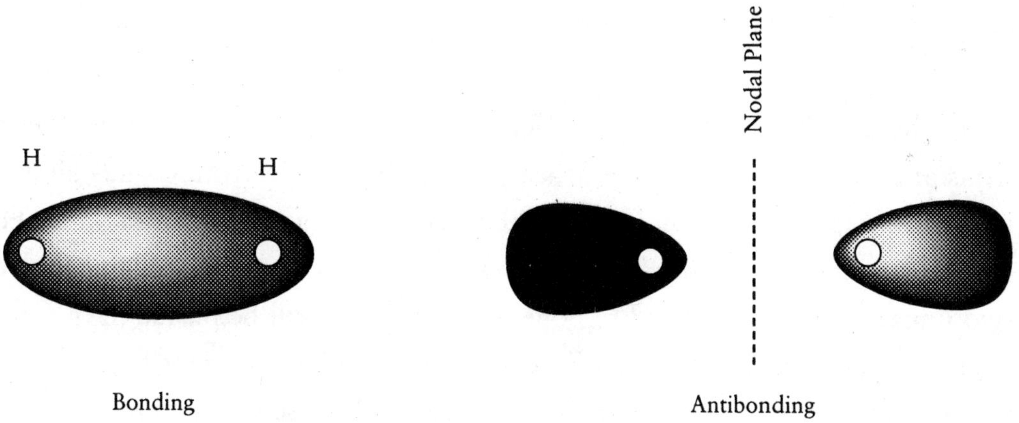

(b)

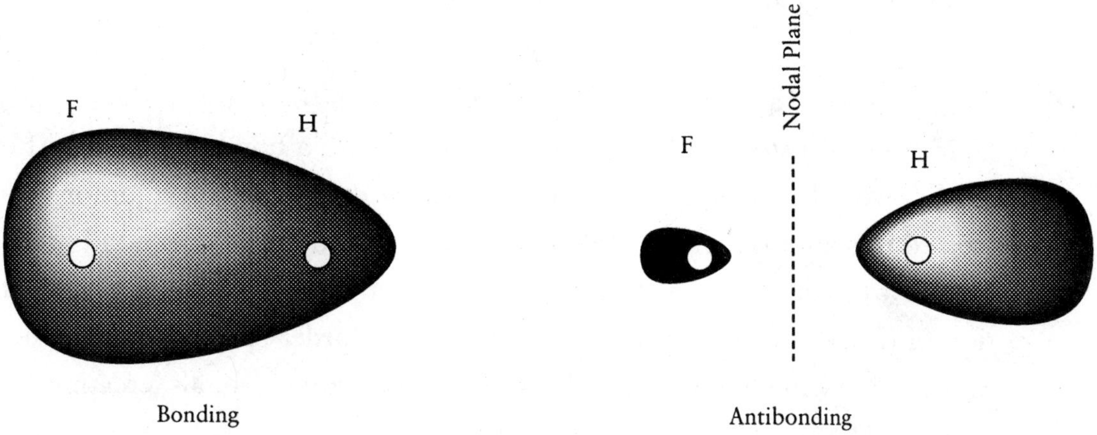

3.59 (a) Yes. This molecule has multiple bonds in the six-membered benzene-like rings that can participate in delocalization. The amide linkages —C(=O)NH— can also become involved in this delocalization, due to their resonance form as shown:

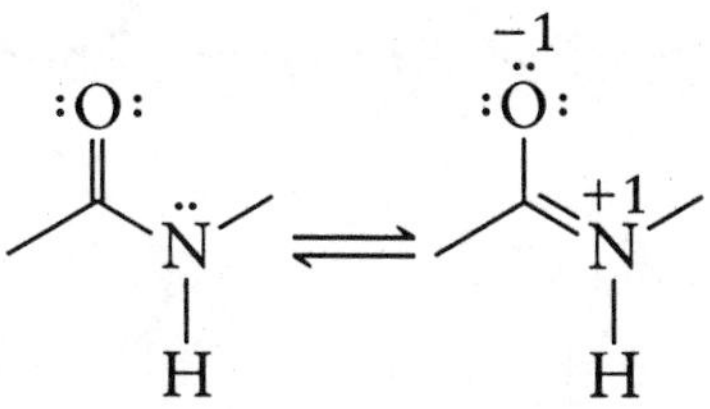

(b) Suramin would be polar. There are a number of polar groups attached to the molecule: in particular the sulfonate groups ($—SO_3^-$), which are highly polar. These groups cannot be arranged in the molecule so that their dipole moments will cancel.

3.61 (a) F_2 (14 valence electrons): $(2s\sigma)^2\,(2s\sigma^*)^2\,(2p\sigma)^2(2p\pi_x)^2(2p\pi_y)^2(2p\pi_x^*)^2(2p\pi_y^*)^2$; bond order = 1. (b) Be_2 (4 valence electrons): $(2s\sigma)^2\,(2s\sigma^*)^2$, bond order = 0. (c) Ne_2 (16 valence electrons): $(2s\sigma)^2\,(2s\sigma^*)^2\,(2p\sigma)^2(2p\pi_x)^2(2p\pi_y)^2(2p\pi_x^*)^2$ $(2p\pi_y^*)^2(2p\sigma^*)^2$; bond order = 0.

3.63 All of these molecules are paramagnetic. O_2^- and O_2^+ have an odd number of electrons and must, therefore, have at least one unpaired electron. O_2 has an even number of electrons, but in its molecular orbital energy level diagram, the HOMO is a degenerate set of orbitals that are each singly occupied, giving this molecule two unpaired electrons. For O_2^-, one more electron will be placed in this degenerate set of orbitals, causing one of the original unpaired electrons to now be paired. O_2^- will therefore have one unpaired electron. Likewise O_2^+ will have one less electron than O_2; thus one of the originally unpaired electrons will be removed, leaving one unpaired electron in this molecule.

3.65 (a) F_2 with 14 valence electrons has a valence electron configuration of $(2s\sigma)^2$ $(2s\sigma^*)^2\,(2p\sigma)^2(2p\pi_x)^2(2p\pi_y)^2(2p\pi_x^*)^2(2p\pi_y^*)^2$ with a bond order of 1. After forming F_2^- from F_2, an electron is added into a $2p\sigma^*$ orbital. The addition of an electron to this antibonding orbital will result in a reduction of the bond order to 1/2 (See 51). F_2 will have the stronger bond. (b) B_2 will have an electron configuration of $(2s\sigma)^2(2s\sigma^*)^2(2p\pi_x)^1(2p\pi_y)^1$ with a bond order of 1. Removing one electron to form B_2^+ will eliminate one electron in the bonding orbitals, creating a bond order of 1/2. B_2 will have the stronger bond.

3.67 CF (11 valence electrons) has the following configuration: $(2s\sigma)^2(2s\sigma^*)^2(2p\pi_x)^2(2p\pi_y)^2(2p\sigma)^2(2p\pi_x^*)^1$, bond order = 2.5. CF^+ (10 valence electrons) has the following configuration: $(2s\sigma)^2(2s\sigma^*)^2(2p\pi_x)^2(2p\pi_y)^2(2p\sigma)^2$, bond order = 3; the bond in CF^+ has the greater bond order and therefore is shorter than the bond in CF.

3.69 The Lewis structure of borazine is nearly identical to that of benzene. It is obtained by replacing alternating C atoms in the benzene structure with B and N, as shown. The orbitals at each B and N atom will be sp^2 hybridized.

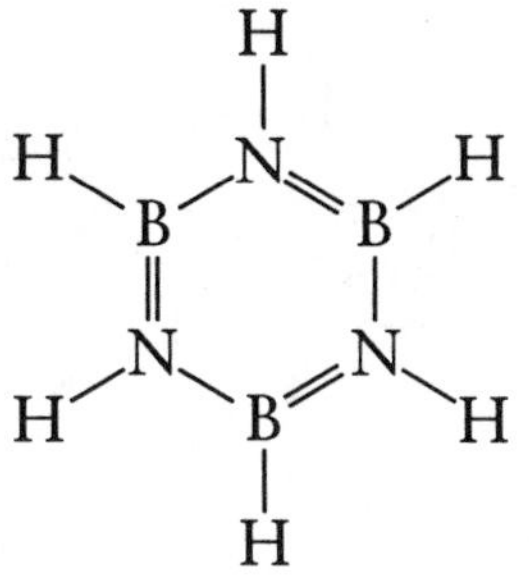

3.71 The predicted bond angles in each species based upon the Lewis structure and VSEPR theory will be

CH_3^+	AX_3	trigonal planar	120°
CH_4	AX_4	tetrahedral	109.5°
CH_3^-	AX_3E	pyramidal	slightly less than 109.5°
CH_2	AX_2E	angular	slightly less than 120°
CH_2^{2+}	AX_2	linear	180°
CH_2^{2-}	AX_2E_2	angular	less than 109.5°, more so than CH_3^- due to the presence of two lone pairs

The order of increasing H—C—H bond angle will be:

$CH_2^{2-} < CH_3^- < CH_4 < CH_2 < CH_3^+ < CH_2^{2+}$

3.73 $H—C \equiv C—H \longrightarrow$ 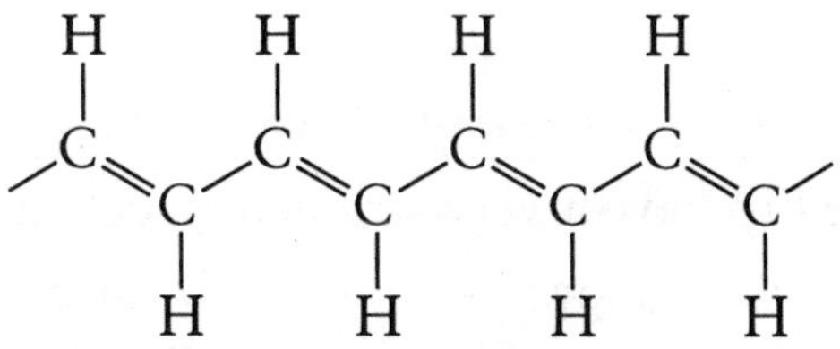

Polyacetylene retains multiple bonds along the chain. It is through the series of orbitals that electrons can be conducted. A resonance form of the Lewis structure can be drawn showing that the electrons may be delocalized along the polyacetylene chain. No such resonance form is possible for polyethylene.

Note: The dark color of the material results from the formation of a large number of molecular orbitals that are not very different in energy. These molecular orbitals are made up of combinations of the p-orbitals on the carbons that make up the double bonds. Because the orbitals are closely spaced in energy, electrons in them can readily absorb visible light to be promoted to a higher energy orbital. See section 3.13.

3.75

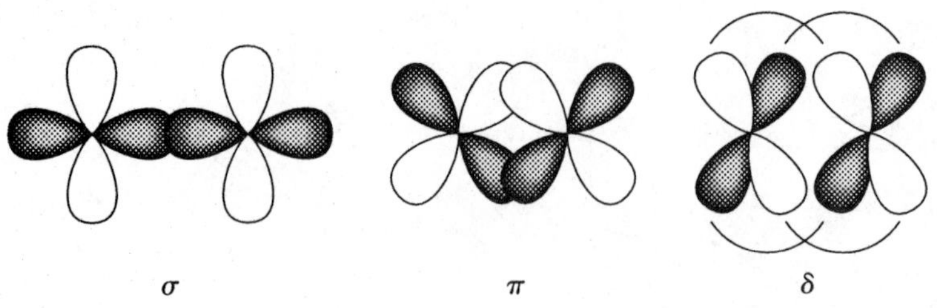

3.77 The effect of both changes (a) and (b) will be similar. The overall bond order will change. In the first case, electrons will be removed from π orbitals so that the net π-bond order will drop from three to two. The same thing will happen in (b), but because two electrons are added to antibonding orbitals so that a net total of one π-bond will be broken. Based on this simple model, the ions formed should be paramagnetic because the electrons are added to or taken from doubly degenerate orbitals.

3.79 (a) The Lewis structure of benzyne is

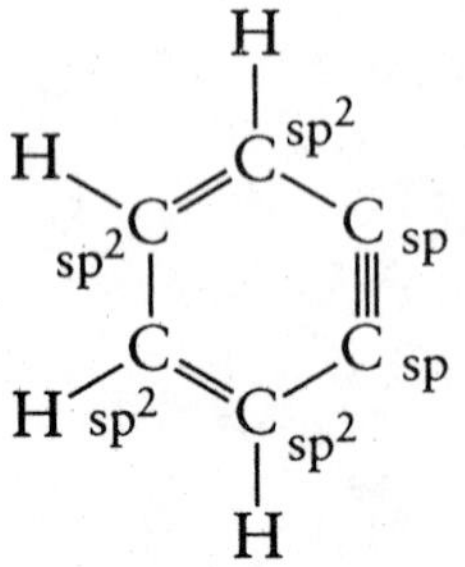

(b) Benzyne would be highly reactive because the two carbons that are sp hybridized are constrained to have a very strained structure compared to what their hybridization would like to adopt—namely a linear arrangement. Instead of 100° angles at these carbon atoms, the angles by necessity of being in a six-membered ring are constrained to be close to 120°. A possibility that allows the carbons to adopt more reasonable angles is the formation of a diradical:

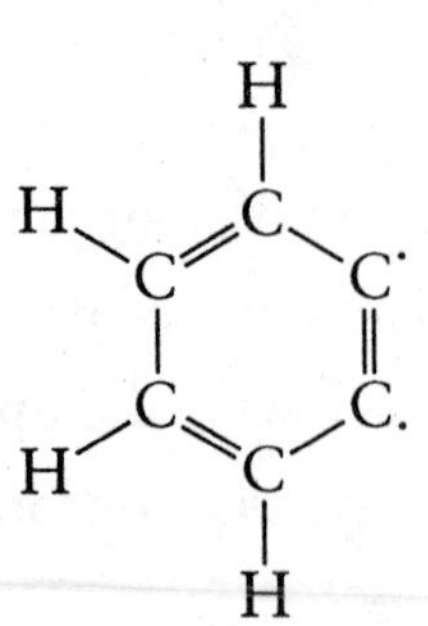

Radicals are also highly reactive in general because orbitals in which the unpaired electrons reside can easily accommodate another electron. This process can result in bond formation that is energetically a favored process.

3.81 (a) and (d) have the possibility of n-to-π^* transitions because these molecules possess both an atom with a lone pair of electrons (on O in HCOOH and on N in HCN) and a π-bond to that atom. The other molecules have either a lone pair or a π-bond, but not both.

CHAPTER 4
THE PROPERTIES OF GASES

4.1 (a) 8×10^9 Pa; (b) 80 kilobars

4.3 (a) 2.556 atm; (b) 1943 Torr

4.5 $d_1h_1 = d_2h_2$

$$73.5\ \text{cm} \times \frac{13.6\ \text{g}\cdot\text{cm}^{-3}}{1.1\ \text{g}\cdot\text{cm}^{-3}} = 909\ \text{cm or } 9.09\ \text{m}$$

4.7 (a) from $P_1V_1 = P_2V_2$, we have $(2.0 \times 10^5\ \text{kPa})\ (1.0\ \text{mL}) = (P_2)\ (1000\ \text{mL})$; solving for P_2 we get 2.0×10^2 kPa; (b) similar to (a), $P_1V_1 = P_2V_2$ or $(600\ \text{Torr})(30.0\ \text{cm}^3) = (P_2)(5.0\ \text{cm}^3)$, $P_2 = 3.6 \times 10^3$ Torr

4.9 (a) Using $\dfrac{P_1}{T_1} = \dfrac{P_2}{T_2}$ and expressing T in Kelvins, $\dfrac{1.10\ \text{atm}}{298\ \text{K}} = \dfrac{P_2}{898\ \text{K}} \cdot P_2 = 3.31$ atm.

(b) Since P is a constant, we can use $\dfrac{V_1}{T_1} = \dfrac{V_2}{T_2}$, which gives $\dfrac{22.5\ \text{L}}{291\ \text{K}} = \dfrac{V_2}{258\ \text{K}}$.

$V_2 = 19.9$ L.

4.11 Since we want P and T to remain constant, we can use the relation $\dfrac{V_1}{n_1} = \dfrac{V_2}{n_2}$. Substituting the numbers that we have, we get $\dfrac{V_1}{0.100\ \text{mol}} = \dfrac{V_2}{0.110\ \text{mol}}$. Solving for V_2 in terms of V_1, we obtain $V_2 = \dfrac{n_2V_1}{n_1} = \dfrac{0.110V_1}{0.10} = 1.10V_1$. Thus, the volume must be increased by 10% to keep P and T constant.

4.13 (a) Since P, V, and T all change, we use the relation $\dfrac{P_1V_1}{T_1} = \dfrac{P_2V_2}{T_2}$. Substituting for the appropriate values we get

$$\frac{(0.480\ \text{atm})(20.0\ \text{mL})}{258\ \text{K}} = \frac{(1.00\ \text{atm})(V_2)}{298\ \text{K}} \cdot V_2 = 11.1\ \text{mL}.$$

(b) The same relation holds as in (a) but here the final temperature and volume are known: $\dfrac{(0.480\text{ atm})(20.0\text{ mL})}{258\text{ K}} = \dfrac{(P_2)(12.0\text{ mL})}{293\text{ K}} \cdot P_2 = 0.909\text{ atm}.$ (c) Similarly, we can use the same expression, with P and V known and T wanted.

$$\frac{(0.480\text{ atm})(20.0\text{ mL})}{258\text{ K}} = \frac{\left(\dfrac{(500\text{ Torr})}{760\text{ Torr}\cdot\text{atm}^{-1}}\right)(12.0\text{ mL})}{T_2}.\ T_2 = 212\text{ K}$$

4.15 (a) Using the ideal gas law with the gas constant R expressed in kPa:

$P(0.3500\text{ L}) = (0.1500\text{ mol})(8.314\ 51\text{ L}\cdot\text{kPa}\cdot\text{K}^{-1}\cdot\text{mol}^{-1})(297\text{ K})$;

$P = 1.06 \times 10^3\text{ kPa}$

(b) BrF_3 has a molar mass of $136.91\text{ g}\cdot\text{mol}^{-1}$. We then substitute into the ideal gas equation:

$$\left(\frac{10.0\text{ Torr}}{760\text{ Torr}\cdot\text{atm}^{-1}}\right)V = \left(\frac{23.9 \times 10^{-3}\text{g}}{136.91\text{ g}\cdot\text{mol}^{-1}}\right)(0.082\ 06\text{ L}\cdot\text{atm}\cdot\text{K}^{-1}\cdot\text{mol}^{-1})(373\text{ K})$$

$V = 4.06 \times 10^2\text{ mL}$

(c) $(0.77\text{ atm})(0.1000\text{ L}) = \left(\dfrac{m}{64.06\text{ g}\cdot\text{mol}^{-1}}\right)(0.082\ 06\text{ L}\cdot\text{atm}\cdot\text{K}^{-1}\cdot\text{mol}^{-1})$

(303 K)

$m = 0.20\text{ g}$

(d) $(129\text{ kPa})(6.00 \times 10^3\text{ m}^3)\left(\dfrac{1 \times 10^6\text{ cm}^3}{\text{m}^3}\right)\left(\dfrac{1\text{ L}}{1000\text{ cm}^3}\right) =$

$n(8.314\ 51\text{ L}\cdot\text{kPa}\cdot\text{K}^{-1}\cdot\text{mol}^{-1})(287\text{ K})$

$n = 3.24 \times 10^5\text{ mol } CH_4$

(e) The number of He atoms is Avogadro's constant N_A multiplied by the number of moles. The number of moles is obtained from the ideal gas equation:

$PV = nRT$

$n = \dfrac{PV}{RT}$; $n = \dfrac{N}{N_A}$

so the number of atoms N will be given by

$$N = N_A\left(\frac{PV}{RT}\right)$$

$$= (6.022 \times 10^{23}\text{ atoms}\cdot\text{mol}^{-1})\left(\frac{(2.00\text{ kPa})(1.0 \times 10^{-6}\text{ L})}{(8.314\ 51\text{ L}\cdot\text{kPa}\cdot\text{K}^{-1}\cdot\text{mol}^{-1})(158\text{ K})}\right)$$

$$= 9.2 \times 10^{14}\text{ atoms}$$

4.17 The pressure of the Ar sample will be given by

$$P_{\text{Ar}} = \frac{nRT}{V} = \frac{\left(\dfrac{2.00 \times 10^{-3}\text{ g}}{39.95\text{ g}\cdot\text{mol}^{-1}}\right)(0.082\ 06\text{ L}\cdot\text{atm}\cdot\text{K}^{-1}\cdot\text{mol}^{-1})(293\text{ K})}{0.050\ 0\text{ L}}$$

$$P_{\mathrm{Kr}} = \frac{\left(\dfrac{2.00 \times 10^{-3}\ \mathrm{g}}{83.80\ \mathrm{g \cdot mol^{-1}}}\right)(0.082\ 06\ \mathrm{L \cdot atm \cdot K^{-1} \cdot mol^{-1}})(\mathrm{T_2})}{0.050\ 0\ \mathrm{L}}$$

Because we want the pressure to be the same, we can set these two equal to each other. Because volume, mass of the gases, and the gas constant R are the same on both sides of the equation, they will cancel.

$$\left(\frac{1}{83.80\ \mathrm{g \cdot mol^{-1}}}\right)(T_2) = \left(\frac{1}{39.95\ \mathrm{g \cdot mol^{-1}}}\right)(293\ \mathrm{K})$$

Solving for T_2, we obtain temperature = 615 K or 342°C.

4.19 For all problems, the number of molecules will be given by the number of moles n multiplied by Avogadro's constant N_A.

$$N_A \cdot n = N_A\left(\frac{PV}{RT}\right)$$

$$= (6.022 \times 10^{23}\ \mathrm{molecules \cdot mol^{-1}})\left(\frac{(1.00\ \mathrm{atm})V}{(0.082\ 06\ \mathrm{L \cdot atm \cdot K^{-1} \cdot mol^{-1}})(298\ \mathrm{K})}\right)$$

(a) $$N = (6.022 \times 10^{23}\ \mathrm{molecules \cdot mol^{-1}})\left(\frac{(1.00\ \mathrm{atm})(6.99 \times 10^{-9}\ \mathrm{L})}{(0.082\ 06\ \mathrm{L \cdot atm \cdot K^{-1} \cdot mol^{-1}})(298\ \mathrm{K})}\right)$$

$$= 1.72 \times 10^{14}\ \text{molecules}$$

(b) $$N = (6.022 \times 10^{23}\ \mathrm{molecules \cdot mol^{-1}})\left(\frac{(1.00\ \mathrm{atm})\left(\dfrac{\frac{4}{3}\pi(5.00\ \mathrm{cm})^3}{1000\ \mathrm{cm^3 \cdot L^{-1}}}\right)}{(0.082\ 06\ \mathrm{L \cdot atm \cdot K^{-1} \cdot mol^{-1}})(298\ \mathrm{K})}\right)$$

$$= 1.29 \times 10^{22}\ \text{molecules}$$

(c) $$N = (6.022 \times 10^{23}\ \mathrm{molecules \cdot mol^{-1}})\left(\frac{(1.00\ \mathrm{atm})\left(\dfrac{(\pi(0.070\ \mathrm{cm})^2(1.6\ \mathrm{cm})}{1000\ \mathrm{cm^3 \cdot L^{-1}}}\right)}{(0.082\ 06\ \mathrm{L \cdot atm \cdot K^{-1} \cdot mol^{-1}})(298\ \mathrm{K})}\right)$$

$$= 6.1 \times 10^{17}\ \text{molecules}$$

4.21 (a) $$V = \frac{nRT}{P} = \frac{(1\ \mathrm{mol})(0.082\ 06\ \mathrm{L \cdot atm \cdot K^{-1} \cdot mol^{-1}})(773\ \mathrm{K})}{1\ \mathrm{atm}} = 63.4\ \mathrm{L}$$

(b) $$V = \frac{nRT}{P} = \frac{(1\ \mathrm{mol})(0.082\ 06\ \mathrm{L \cdot atm \cdot K^{-1} \cdot mol^{-1}})(77\ \mathrm{K})}{1\ \mathrm{atm}} = 6.32\ \mathrm{L}$$

4.23 Density is proportional to the molar mass of the gas as seen from the ideal gas law:

$$PV = nRT$$

$$PV = \frac{m}{M}RT$$

$$\text{density} = \text{mass per unit volume} = \frac{m}{V} = \frac{MP}{RT}$$

The molar masses of the gases in question are 28.01 $g \cdot mol^{-1}$ for CO(g), 44.01 $g \cdot mol^{-1}$ for $CO_2(g)$, and 34.01 $g \cdot mol^{-1}$ for $H_2S(g)$. The most dense will be the one with the highest molar mass, which in this case is CO_2.

4.25 (a) Density is proportional to the molar mass of the gas as seen from the ideal gas law. See Exercise 4.23.

$$d = \frac{(119.37\ g \cdot mol^{-1})\left(\dfrac{200\ Torr}{760\ Torr \cdot atm^{-1}}\right)}{(0.080\ 26\ L \cdot atm \cdot K^{-1} \cdot mol^{-1})(298\ K)} = 1.28\ g \cdot L^{-1}$$

(b) $$d = \frac{(119.37\ g \cdot mol^{-1})(1.00\ atm)}{(0.080\ 26\ L \cdot atm \cdot K^{-1} \cdot mol^{-1})(373\ K)} = 3.90\ g \cdot L^{-1}$$

4.27 (a) $$M = \frac{dRT}{P} = \frac{(8.0\ g \cdot L^{-1})(0.082\ 06\ L \cdot atm \cdot K^{-1} \cdot mol^{-1})(300\ K)}{2.81\ atm} = 70\ g \cdot mol^{-1}$$

(b) The compound is most likely CHF_3 for which $M = 70\ g \cdot mol^{-1}$. It might also be $C_2H_4F_2$ for which $M = 66\ g \cdot mol^{-1}$.

(c) You can use the relation in (a) to calculate the new density, or you can apply the proportionality changes expected from the change in pressure and temperature to the original density:

$$d_2 = (8.0\ g \cdot L^{-1})\left(\frac{1.00\ atm}{2.81\ atm}\right)\left(\frac{300\ K}{298\ K}\right) = 2.9\ g \cdot L^{-1}$$

4.29 From the analytical data, an empirical formula of CHCl is calculated. The empirical formula mass is 48.47 $g \cdot mol^{-1}$. The problem may be solved using the ideal gas law:

$$PV = nRT$$

$$PV = \frac{m}{M}RT$$

$$M = \frac{mRT}{PV}$$

$$M = \frac{(3.557\ g)(0.082\ 06\ L \cdot atm \cdot K^{-1} \cdot mol^{-1})(273\ K)}{(1.10\ atm)(0.755\ L)} = 95.9\ g \cdot mol^{-1}$$

The value of n in the formula $(CHCl)_n$ is, therefore, equal to
2(95.9 $g \cdot mol^{-1} \div 48.47 = 1.98$). The formula is $C_2H_2Cl_2$.

4.31 (a) The number of moles of H_2 needed will be 1.5 times the amount of NH_3 produced, as seen from the balanced equation:

$$\tfrac{1}{2}\ N_2(g) + \tfrac{3}{2}\ H_2(g) \longrightarrow NH_3(g)$$

or

$N_2(g) + 3\,H_2(g) \longrightarrow 2\,NH_3(g)$

Once the number of moles is known, the volume can be obtained from the ideal gas law.

$$V = \frac{n_{H_2}RT}{P} = \frac{(\frac{3}{2}n_{NH_3})RT}{P} = \frac{\left(\left(\frac{3\text{ mol H}_2}{2\text{ mol NH}_3}\right)\frac{(10^3\text{ kg})(10^3\text{ g}\cdot\text{kg}^{-1})}{17.03\text{ g}\cdot\text{mol}^{-1}}\right)RT}{P}$$

$$= \frac{\left(\frac{3}{2}\right)\left(\frac{10^6\text{ g}}{17.03\text{ g}\cdot\text{mol}^{-1}}\right)(0.082\,06\text{ L}\cdot\text{atm}\cdot\text{K}^{-1}\cdot\text{mol}^{-1})(298\text{ K})}{1.00\text{ atm}}$$

$$= 2.15 \times 10^6\text{ L}$$

(b) The ideal gas equation

$$\frac{P_1V_1}{n_1RT_1} = \frac{P_2V_2}{n_2RT_2} \quad \text{simplifies to} \quad \frac{P_1V_1}{T_1} = \frac{P_2V_2}{T_2}$$

because R and n are constant for this problem.

$$\frac{(1.00\text{ atm})(2.15 \times 10^6\text{ L})}{298\text{ K}} = \frac{(200\text{ atm})V_2}{(673\text{ K})}$$

$$V_2 = 2.43 \times 10^4\text{ L}$$

4.33 To answer this, we need to know the number of moles of $CH_4(g)$ present in each case. Because the combustion reaction is the same in both cases, as are the temperature and pressure, the larger number of moles of $CH_4(g)$ should produce the larger volume of $CO_2(g)$. We will use the ideal gas equation to solve for n in the first case:

$$n = \frac{PV}{RT} = \frac{(1.00\text{ atm})(2.00\text{ L})}{(0.082\,06\text{ L}\cdot\text{atm}\cdot\text{K}^{-1}\cdot\text{mol}^{-1})(348\text{ K})} = 0.0700\text{ mol CH}_4$$

2.00 g of CH_4 will be $\dfrac{2.00\text{ g}}{16.04\text{ g}\cdot\text{mol}^{-1}} = 0.124$ mol

The latter case will have the greater number of moles of CH_4 and should produce the larger amount of $CO_2(g)$.

4.35 (a) We find the pressure of $SO_2(g)$ originally present by difference. The initial data gives us the total number of moles present, whereas the data for the gas sample after being passed over $CaSO_3(s)$ represents the number of moles of $N_2(g)$.

$$PV = nRT$$

$$n_{total} = \frac{PV}{RT} = \frac{(1.09\text{ atm})(0.500\text{ L})}{(0.082\,06\text{ L}\cdot\text{atm}\cdot\text{K}^{-1}\cdot\text{mol}^{-1})(298\text{ K})} = 0.0223\text{ mol}$$

$$n_{N_2} = \frac{PV}{RT} = \frac{(1.09\text{ atm})(0.150\text{ L})}{(0.082\,06\text{ L}\cdot\text{atm}\cdot\text{K}^{-1}\cdot\text{mol}^{-1})(323\text{ K})} = 0.006\,17\text{ mol}$$

The number of moles of SO_2 gas is 0.0223 − 0.006 17 mol = 0.0161 mol. The partial pressure will be given by the mole fraction multiplied by the total pressure. The mole fraction of $SO_2(g)$ will be 0.0161 mol ÷ 0.0223 = 0.722. The pressure due to SO_3 in the original mixture is (0.722)(1.09 atm) = 0.787 atm.

(b) The mass of SO_2 will be obtained by multiplying the number of moles of SO_2 by the molar mass of SO_2:

$m_{SO_2} = (0.0161 \text{ mol})(64.06 \text{ g}\cdot\text{mol}^{-1}) = 1.03 \text{ g}$

4.37 (a) Of the 756.7 Torr measured, 17.54 Torr will be due to water vapor. The pressure due to $H_2(g)$ will, therefore, be 756.7 Torr − 17.54 Torr = 739.2 Torr.

(b) $H_2O(l) \longrightarrow H_2(g) + \frac{1}{2} O_2(g)$; (c) To answer this question, we must determine the number of moles of H_2 produced in the reaction. Using the partial pressure of H_2 calculated in part (a) and the ideal gas equation, we can set up the following:

$$\left(\frac{739.2 \text{ Torr}}{760 \text{ Torr}\cdot\text{atm}^{-1}}\right)(0.220 \text{ L}) = n\,(0.082\ 06 \text{ L}\cdot\text{atm}\cdot\text{K}^{-1}\cdot\text{mol}^{-1})(293 \text{ K})$$

Solving for n, we obtain n = 0.008 90 mol. According to the stoichiometry of the reaction, half as much oxygen as hydrogen should be produced, so the number of moles of O_2 = 0.004 45 mol. The mass of O_2 will be given by

$(0.004\ 45 \text{ mol})(32.00 \text{ g}\cdot\text{mol}^{-1}) = 0.142 \text{ g}$.

4.39 Dalton's law states that the total pressure of a mixture is equal to the sum of the partial pressure of each of its components. In the kinetic model, pressure arises from the collisions of the molecules against the wall of the vessel in which it is contained. The result is that the pressure will be a simple function of the number of particles and will not depend upon the nature of the particles.

4.41 Graham's law of effusion states that the rate of effusion of a gas is inversely proportional to the square root of its molar mass:

$$\text{rate of effusion} = \frac{1}{\sqrt{M}}$$

If we have two different gases whose rates of effusion are measured under identical conditions, we can take the ratio

$$\frac{\text{rate}_1}{\text{rate}_2} = \frac{\frac{1}{\sqrt{M_1}}}{\frac{1}{\sqrt{M_2}}} = \sqrt{\frac{M_2}{M_1}}$$

If a compound takes 1.24 times as long to effuse as Kr gas, the rate of effusion of

Kr is 1.24 times that of the unknown. We can now use the expression to calculate the molar mass of the unknown, given the mass of Kr:

$$\frac{1.24}{1} = \sqrt{\frac{M_2}{83.80 \text{ g·mol}^{-1}}}$$

$M_2 = 129 \text{ g·mol}^{-1}$

A mass of 129 g·mol^{-1} corresponds to a molecular formula of $C_{10}H_{10}$.

4.43 (a) The average kinetic energy is obtained from the expression: average kinetic energy $= \frac{3}{2}RT$. The value is independent of the nature of the monatomic ideal gas. The numerical values are:

(a) $4103.2 \text{ J·mol}^{-1}$; (b) $4090.7 \text{ J·mol}^{-1}$;

(c) $4103.2 \text{ J·mol}^{-1} - 4090.7 \text{ J·mol}^{-1} = 12.5 \text{ J·mol}^{-1}$

4.45 Hydrogen bonding is important in HF. At low temperatures, this hydrogen bonding causes the molecules of HF to be attracted to each other more strongly, thus lowering the pressure. As the temperature is increased, the hydrogen bonds are broken and the pressure rises more quickly than for an ideal gas. Dimers (2 HF molecules bonded to each other) and chains of HF molecules are known to form.

4.47 The pressures are calculated very simply from the ideal gas law:

$$P = \frac{nRT}{V} = \frac{(1.00 \text{ mol})(0.082\ 06 \text{ L·atm·K}^{-1}\text{·mol}^{-1})(298 \text{ K})}{V}$$

Calculating for the volumes requested, we obtain $P =$ (a) 1.63 atm; (b) 48.9 atm; (c) 489 atm. The calculations can now be repeated using the van der Waals equation:

$$\left(P + \frac{an^2}{V^2}\right)(V - nb) = nRT$$

We can rearrange this to solve for P:

$$P = \left(\frac{nRT}{V - nb}\right) - \left(\frac{an^2}{V^2}\right)$$

$$= \left(\frac{(1.00 \text{ mol})(0.082\ 06 \text{ L·atm·K}^{-1}\text{·mol}^{-1})(298 \text{ K})}{V - (1.00 \text{ mol})(0.04267 \text{ L·mol}^{-1})}\right) - \left(\frac{(3.640 \text{ L}^2\text{·atm·mol}^{-2})(1.00^2)}{V^2}\right)$$

Using the three values for V, we calculate for $P =$ (a) 1.62; (b) 38.9; (c) 1.88×10^3 atm. Note that at low pressures, the ideal gas law gives essentially the same values as the van der Waals equation, but at high pressures there is a very significant difference.

4.49 $(20\ \text{in})(10\ \text{in})(14.7\ \text{lb}\cdot\text{in}^2) = 2.9 \times 10^3\ \text{lb}$

4.51 Because P, V, and T are state functions, the intermediate conditions are irrelevant to the final states. We can simply use the ideal gas law in the form

$$\frac{P_1V_1}{T_1} = \frac{P_2V_2}{T_2}$$

$$\frac{\left(\dfrac{759\ \text{Torr}}{760\ \text{Torr}\cdot\text{atm}^{-1}}\right)(1.00\ \text{L})}{253\ \text{K}} = \frac{\left(\dfrac{128\ \text{Torr}}{760\ \text{Torr}\cdot\text{atm}^{-1}}\right)(V_2)}{1523\ \text{K}}$$

$$V_2 = 35.7\ \text{L}$$

4.53 Since T is constant we can use

$$P_1V_1 = P_2V_2$$

$$(1.00\ \text{atm})(1.00\ \text{L}) = P_2(0.100\ \text{L})$$

$$P_2 = 10.0\ \text{atm}$$

4.55 $PV = nRT$

$$\left(\frac{24.5\ \text{kPa}}{101.325\ \text{kPa}\cdot\text{atm}^{-1}}\right)(0.2500\ \text{L}) = n(0.082\ 06\ \text{L}\cdot\text{atm}\cdot\text{K}^{-1}\cdot\text{mol}^{-1})(292.7\ \text{K})$$

$$n = 2.52 \times 10^{-3}\ \text{mol}$$

4.57 (a) $\dfrac{P_1V_1}{T_1} = \dfrac{P_2V_2}{T_2}$

$$\frac{(104\ \text{kPa})(2.0\ \text{m}^3)}{294.3\ \text{K}} = \frac{(52\ \text{kPa})V_2}{268.2\ \text{K}}$$

$$V_2 = 3.6\ \text{m}^3$$

(b) $\dfrac{P_1V_1}{T_1} = \dfrac{P_2V_2}{T_2}$

$$\frac{(104\ \text{kPa})(2.0\ \text{m}^3)}{294.3\ \text{K}} = \frac{(0.880\ \text{kPa})V_2}{221.2\ \text{K}}$$

$$V_2 = 1.8 \times 10^2\ \text{m}^3$$

4.59 Density is proportional to the molar mass of the gas as seen from the ideal gas law:

$$PV = nRT$$

$$PV = \frac{m}{M}RT$$

$$\text{density} = \text{mass per unit volume} = \frac{m}{V} = \frac{MP}{RT}$$

$$1.23\ \mathrm{g \cdot L^{-1}} = \frac{M\left(\dfrac{25.5\ \mathrm{kPa}}{101.325\ \mathrm{kPa \cdot atm^{-1}}}\right)}{(0.082\ 06\ \mathrm{L \cdot atm \cdot K^{-1} \cdot mol^{-1}})(330\ \mathrm{K})}$$

$$M = 132\ \mathrm{g \cdot mol^{-1}}$$

4.61 (a) The molar volume of an ideal gas is 22.4 L at 273.15 K. 1.0 mol of ideal gas will exert a pressure of 1.0 atm under those conditions. The partial pressure of N_2(g) will be 1.0 atm. Because there are 2.0 mol of H_2(g), the partial pressure of H_2(g) will be 2.0 atm. (b) The total pressure will be 1.0 atm + 2.0 atm = 3.0 atm.

4.63 From $P_1V_1 = P_2V_2$

$(764\ \mathrm{Torr})(545\ \mathrm{cm^3}) = (P_2)(58.8\ \mathrm{cm^3})$

$P_2 = 7.08 \times 10^3$ Torr, or 9.32 atm

4.65 From $P_1V_1 = P_2V_2$ we can calculate the *total pressure* of the final sample:

$(765\ \mathrm{Torr})(555\ \mathrm{mL}) = (P_2)(125\ \mathrm{mL})$

$P_2 = 3.40 \times 10^3$ Torr

3.40×10^3 Torr should be the final pressure. The additional pressure needed will be 3.40×10^3 Torr $-$ 765 Torr $= 2.63 \times 10^3$ Torr.

4.67 From the ideal gas equation we can solve for the mass of gas collected, using the average molar mass:

$$PV = nRT$$

$$PV = \frac{m}{M}RT$$

$$m = \frac{PVM}{RT} = \frac{(0.84\ \mathrm{atm})(22.1\ \mathrm{L})(33.8\ \mathrm{g \cdot mol^{-1}})}{(0.082\ 06\ \mathrm{L \cdot atm \cdot K^{-1} \cdot mol^{-1}})(299\ \mathrm{K})} = 26\ \mathrm{g}$$

4.69 (a) First, we need to determine the number of moles of TiO_2 formed, which can be obtained from the number of moles of $TiCl_4$ that reacted. The moles of $TiCl_4$ can be calculated from the ideal gas law:

$$n = \frac{PV}{RT} = \frac{\left(\dfrac{15.5\ \mathrm{kPa}}{101.325\ \mathrm{kPa \cdot atm^{-1}}}\right)(215\ \mathrm{L})}{(0.082\ 06\ \mathrm{L \cdot atm \cdot K^{-1} \cdot mol^{-1}})(308\ \mathrm{K})} = 1.30\ \mathrm{mol}$$

1.30 mol of $TiCl_4$ will produce 1.30 mol TiO_2(s). The molar mass of TiO_2(s) is $79.88\ \mathrm{g \cdot mol^{-1}}$.

mass of TiO_2 produced $= 1.30 \text{ mol} \times 79.88 \text{ g}\cdot\text{mol}^{-1} = 104 \text{ g}$

(b) The number of moles of HCl produced will be 4 times the number of moles of $TiCl_4$ reacted, or $4 \times 1.30 \text{ mol} = 5.20 \text{ mol}$. The volume of HCl produced can then be determined from the ideal gas law:

$PV = nRT$

$(1.00 \text{ atm})(V) = (5.20 \text{ mol})(0.082\,06 \text{ L}\cdot\text{atm}\cdot\text{K}^{-1}\cdot\text{mol}^{-1})(298 \text{ K})$

$V = 127 \text{ L}$

4.71 The molar mass of glucose is $180.15 \text{ g}\cdot\text{mol}^{-1}$. From this, we can calculate the number of moles of glucose formed and, using the reaction stoichiometry, determine the number of moles of CO_2 needed. With that information and the other information provided in the problem, we can use the ideal gas law to calculate the volume of air that is needed:

$PV = nRT$

$$V = \frac{\left[\left(\dfrac{10.0 \text{ g glucose}}{180.15 \text{ g glucose}\cdot\text{mol}^{-1} \text{ glucose}}\right)\left(\dfrac{6 \text{ mol } CO_2}{1 \text{ mol glucose}}\right)\right](0.082\,06 \text{ L}\cdot\text{atm}\cdot\text{K}^{-1}\cdot\text{mol}^{-1})(298 \text{ K})}{\left(\dfrac{0.26 \text{ Torr}}{760 \text{ Torr}\cdot\text{atm}^{-1}}\right)}$$

$= 2.4 \times 10^4 \text{ L}$

4.73 The elemental analyses yield an empirical formula of NH_2. The formula unit has a mass of $16.02 \text{ g}\cdot\text{mol}^{-1}$. The mass, volume, pressure and temperature data will allow us to calculate the molar mass using the ideal gas equation:

$PV = nRT$

$$PV = \frac{m}{M}RT$$

$$M = \frac{mRT}{PV} = \frac{(0.473 \text{ g})(0.082\,06 \text{ L}\cdot\text{atm}\cdot\text{K}^{-1}\cdot\text{mol}^{-1})(298 \text{ K})}{(1.81 \text{ atm})(0.200 \text{ L})} = 31.9 \text{ g}\cdot\text{mol}^{-1}$$

The molar mass divided by the mass of the empirical formula mass will give the value of n in the formula $(NH_2)_n$. $31.9 \text{ g}\cdot\text{mol}^{-1} \div 16.02 \text{ g}\cdot\text{mol}^{-1} = 1.99$, so the molecular formula is N_2H_4, which corresponds to the molecule known as hydrazine.

4.75 The rate of effusion is inversely proportional to the square root of the molar mass.

Using a ratio as follows allows us to calculate the time of effusion without knowing the exact conditions of pressure and temperature:

$$\frac{\text{rate}_1}{\text{rate}_2} = \frac{\dfrac{1}{\sqrt{M_1}}}{\dfrac{1}{\sqrt{M_2}}} = \sqrt{\frac{M_2}{M_1}}$$

The rate will be equal to the number of molecules N which effuse in a given time interval. For the conditions given, N will be the same for argon and for the second gas chosen.

$$\frac{\dfrac{N}{\text{time}}}{\dfrac{N}{147\text{ s}}} = \frac{\dfrac{1}{\text{time}}}{\dfrac{1}{147\text{ s}}} = \sqrt{\frac{39.95\text{ g}\cdot\text{mol}^{-1}}{M_1}}$$

In order to calculate the time of effusion, we need to know only the molar mass of the gases.

(a) For CO_2 with a molar mass of 44.01 $\text{g}\cdot\text{mol}^{-1}$: $\dfrac{\dfrac{1}{\text{time}_{CO_2}}}{\dfrac{1}{147\text{ s}}} = \sqrt{\dfrac{39.95\text{ g}\cdot\text{mol}^{-1}}{44.01\text{ g}\cdot\text{mol}^{-1}}}$

$$\text{time} = 154\text{ s}$$

(b) For C_2H_4 with a molar mass of 28.05 $\text{g}\cdot\text{mol}^{-1}$: $\dfrac{\dfrac{1}{\text{time}_{C_2H_4}}}{\dfrac{1}{147\text{ s}}} = \sqrt{\dfrac{39.95\text{ g}\cdot\text{mol}^{-1}}{28.05\text{ g}\cdot\text{mol}^{-1}}}$

$$\text{time} = 123\text{ s}$$

(c) For H_2 with a molar mass of 2.01 $\text{g}\cdot\text{mol}^{-1}$: $\dfrac{\dfrac{1}{\text{time}_{CO_2}}}{\dfrac{1}{147\text{ s}}} = \sqrt{\dfrac{39.95\text{ g}\cdot\text{mol}^{-1}}{2.01\text{ g}\cdot\text{mol}^{-1}}}$

$$\text{time} = 33.0\text{ s}$$

(d) For SO_2 with a molar mass of 64.06 $\text{g}\cdot\text{mol}^{-1}$: $\dfrac{\dfrac{1}{\text{time}_{CO_2}}}{\dfrac{1}{147\text{ s}}} = \sqrt{\dfrac{39.95\text{ g}\cdot\text{mol}^{-1}}{64.06\text{ g}\cdot\text{mol}^{-1}}}$

$$\text{time} = 186\text{ s}$$

4.77 The formula mass of C_2H_3 is 27.04 $\text{g}\cdot\text{mol}^{-1}$. From the effusion data, we can calculate the molecular weight of the sample.

$$\frac{\text{rate}_1}{\text{rate}_2} = \frac{\dfrac{1}{\sqrt{M_1}}}{\dfrac{1}{\sqrt{M_2}}} = \sqrt{\frac{M_2}{M_1}}$$

Because time is inversely proportional to rate, we can write alternatively

$$\frac{\dfrac{1}{349\text{ s}}}{\dfrac{1}{210\text{ s}}} = \sqrt{\frac{39.95\text{ g}\cdot\text{mol}^{-1}}{M_1}}$$

$$\frac{210}{349} = \sqrt{\frac{39.95\text{ g}\cdot\text{mol}^{-1}}{M_1}}$$

$$M_1 = 110\text{ g}\cdot\text{mol}^{-1}$$

The molar mass is 4.8 times that of the empirical formula mass, so the molecular formula is C_8H_{12}.

4.79 The values for the pressure of gas with varying numbers of moles of CO_2 present are calculated as follows:

The ideal gas law values are calculated from:

$$P = \frac{nRT}{V}$$

while those for the van der Waals equation can be obtained by rearranging the equation:

$$\left(P + \frac{an^2}{V^2}\right)(V - nb) = nRT$$

$$P = \left(\frac{nRT}{V - nb}\right) - \left(\frac{an^2}{V^2}\right)$$

$$= \left(\frac{(n)(0.082\ 06\text{ L}\cdot\text{atm}\cdot\text{K}^{-1}\cdot\text{mol}^{-1})(300\text{ K})}{1.00\text{ L} - (n)(0.042\ 67\text{ L}\cdot\text{mol}^{-1})}\right) - \left(\frac{(3.640\text{ L}^2\cdot\text{atm}\cdot\text{mol}^{-2})(n^2)}{(1.00\text{ L})^2}\right)$$

The resulting values are

n	P_{ideal}	$P_{\text{van der Waals}}$	% deviation*
0.100	2.46	2.44	0.8
0.200	4.92	4.82	2.1
0.300	7.38	7.15	3.2
0.400	9.85	9.44	4.3
0.500	12.31	11.67	5.5

*% deviation $= \dfrac{P\text{ ideal} - P\text{ van der Waals}}{P\text{ van der Waals}} \times 100$

Consider one point, for example the case for $n = 0.400$ mol. The term $V\text{-}nb$ will increase the ideal value $\frac{nRT}{V}$ by 1.7% of the ideal value (10.02 atm versus 9.85 atm) while the correction from $\frac{an^2}{V^2}$ will decrease the value by 0.58 atm, a change of 5.9% over the ideal gas value. The second effect, which is due to the intermolecular attractions, dominates in this case.

4.81 The overly high pressure obtained when S_8 is vaporized is an indication that S_8 units have decomposed to a greater number of units, S_6, S_4, S_2, etc. This dissociation increases with temperature. In other words, the number of moles of gas increased upon vaporization.

4.83 The molar mass of quinine can be obtained from the ideal gas law:

$$PV = nRT$$

$$PV = \frac{m}{M}RT$$

$$M = \frac{mRT}{PV} = \frac{(27.20 \times 10^{-3}\ \text{g})(0.082\ 06\ \text{L}\cdot\text{atm}\cdot\text{K}^{-1}\cdot\text{mol}^{-1})(453.1\ \text{K})}{\left(\dfrac{9.48\ \text{Torr}}{760\ \text{Torr}\cdot\text{atm}^{-1}}\right)(0.2500\ \text{L})}$$

$$= 324\ \text{g}\cdot\text{mol}^{-1}$$

The combustion data will give the elemental composition.

$$n_{CO_2} = \frac{47.0 \times 10^{-3}\ \text{g}}{44.01\ \text{g}\cdot\text{mol}^{-1}} = 1.07 \times 10^{-3}\ \text{mol}$$

$$n_{H_2O} = \frac{11.5 \times 10^{-3}\ \text{g}}{18.01\ \text{g}\cdot\text{mol}^{-1}} = 6.38 \times 10^{-4}\ \text{mol}$$

$$n_{N_2} = \frac{1.51 \times 10^{-3}\ \text{g}}{28.01\ \text{g}\cdot\text{mol}^{-1}} = 5.39 \times 10^{-5}\ \text{mol}$$

We need to check the mass balance to make sure all the composition is accounted for:

m of CO_2	$= (1.07 \times 10^{-3}\ \text{mol})(12.01\ \text{g}\cdot\text{mol}^{-1})$	$= 0.0128$ g
m of H in H_2O	$= 2(6.38 \times 10^{-4})(1.0079\ \text{g}\cdot\text{mol}^{-1})$	$= 0.00129$ g
m of N in N_2	$= 2(5.39 \times 10^{-5}\ \text{mol})(14.01\ \text{g}\cdot\text{mol}^{-1})$	$= 0.00151$ g
Total mass		$= 0.0156$ g

The amount of quinine burned was 17.2 mg but only 15.6 mg ended up in the products. The difference must be due to the presence of oxygen as a component of quinine: $C_aH_bN_cO_d$. To balance the mass, there must have been 1.6 mg of O in the original sample. From the combustion analysis, we can see that original sample had the following numbers of moles of the consistent elements and their relative ratios:

C	1.07×10^{-3} mol	10.7
H	1.28×10^{-3} mol	12.8
N	1.07×10^{-4} mol	1.07
O	1.0×10^{-4} mol	1

The numbers for C and H are intermediate suggesting that the analytical data may have some error. We expect then, that the values of a, b, c, and d in the empirical formula will be 10–11, 12–13, 1 and 1, respectively. The mass of $C_{10}H_{12}NO$ is 162.2 $g \cdot mol^{-1}$, which is right at exactly half of the experimentally determined molar mass. We can conclude then that the molecular formula of quinine is $C_{20}H_{24}N_2O_2$. As a check, we can calculate the expected amounts of CO_2, H_2O, and N_2 that would arise from combustion of 17.2 mg of quinine:

Theoretical values: $C_{20}H_{24}N_2O_2 + 25\ O_2(g) \longrightarrow 20\ CO_2(g) + 12\ H_2O(l) + N_2(g)$

17.2 mg of quinine is $(17.2 \times 10^{-3}$ g quinine $\div$ 324 $g \cdot mol^{-1}) = 5.31 \times 10^{-5}$ mol quinine

We would thus expect:

$20(5.31 \times 10^{-5}\ mol)\ CO_2 = 1.06 \times 10^{3}\ mol\ CO_2 = 4.67 \times 10^{-2}$ g or 46.7 mg

$12(5.31 \times 10^{-5}\ mol)\ H_2O = 6.37 \times 10^{-4}\ mol\ H_2O = 1.15 \times 10^{-2}$ g or 11.5 mg

$1(5.31 \times 10^{-5}\ mol)\ N_2 = 5.31 \times 10^{-5}\ mol\ N_2 = 1.48 \times 10^{-3}$ g or 1.48 mg

Note that the nitrogen analysis is the one that is furthest in error, and this may be expected since it is the one that has the least abundance in the sample.

4.85 The molar mass calculation follows from the ideal gas law:

$$PV = nRT$$

$$PV = \frac{m}{M}RT$$

$$M = \frac{mRT}{PV} = \frac{(1.509\ g)(0.082\ 06\ L \cdot atm \cdot K^{-1} \cdot mol^{-1})(473\ K)}{\left(\frac{745\ Torr}{760\ Torr \cdot atm^{-1}}\right)(0.235\ L)} = 254\ g \cdot mol^{-1}$$

If the molecular formula is OsO_x, then the molar mass will be given by:

$190.2\ g \cdot mol^{-1} + x(16.00\ g \cdot mol^{-1}) = 254\ g \cdot mol^{-1}$

$x = 3.99$

The formula is OsO_4.

4.87 (a) This is a limiting reactant problem. Our first task is to determine the number of moles of NH_3 and HCl that are present to start with. This can be done from the ideal gas equation:

$PV = nRT$

$$n_{NH_3} = \frac{PV}{RT} = \frac{\left(\dfrac{100\ \text{Torr}}{760\ \text{Torr}\cdot\text{atm}^{-1}}\right)(0.0150\ \text{L})}{(0.082\ 06\ \text{L}\cdot\text{atm}\cdot\text{K}^{-1}\cdot\text{mol}^{-1})(303\ \text{K})} = 7.94 \times 10^{-5}\ \text{mol}$$

$$n_{HCl} = \frac{PV}{RT} = \frac{\left(\dfrac{150\ \text{Torr}}{760\ \text{Torr}\cdot\text{atm}^{-1}}\right)(0.0250\ \text{L})}{(0.082\ 06\ \text{L}\cdot\text{atm}\cdot\text{K}^{-1}\cdot\text{mol}^{-1})(298\ \text{K})} = 2.02 \times 10^{-4}\ \text{mol}$$

The ammonia is the limiting reactant. The number of moles of $NH_4Cl(s)$ that form will be equal to the number of moles of NH_3 that react. From the molar mass of NH_4Cl ($53.49\ \text{g}\cdot\text{mol}^{-1}$) and the number of moles we can calculate the mass of NH_4Cl that forms:

$(7.94 \times 10^{-5}\ \text{mol}\ NH_4Cl(s))(53.49\ \text{g}\cdot\text{mol}^{-1}) = 4.25 \times 10^{-3}\ \text{g}$

(b) There will be $(2.02 \times 10^{-4}\ \text{mol} - 7.94 \times 10^{-5}\ \text{mol}) = 1.23 \times 10^{-4}$ mol HCl left after the reaction. This quantity will exist in a total volume after mixing of 40.0 mL or 0.0400 L. Again, we use the ideal gas law to determine the final pressure:

$PV = nRT$

$$P = \frac{(1.23 \times 10^{-4}\ \text{mol})(0.082\ 06\ \text{L}\cdot\text{atm}\cdot\text{K}^{-1}\cdot\text{mol}^{-1})(300\ \text{K})}{0.0400\ \text{L}} = 0.0757\ \text{atm}$$

4.89 In this problem, the volume, pressure and molar mass of the substance stay constant. In order to calculate the new mass with the same conditions we can resort to using the ideal gas equation rearranged to group the constant terms on one side of the equation:

$PV = nRT$

$$PV = \frac{m}{M}RT$$

but M, P and V are constants, so we can write

$$\frac{MPV}{R} = mT$$

Now we have two conditions, 1 and 2 for which these conditions are constant so we can set them equal:

$m_1T_1 = m_2T_2$

$(32.5\ \text{g})(295\ \text{K}) = (m_2)(485\ \text{K})$; therefore $m_2 = 19.8\ \text{g}$

The mass of gas released must therefore be $32.5\ \text{g} - 19.8\ \text{g} = 12.7\ \text{g}$.

4.91 (a) The most probable speed is the one that corresponds to the maximum on the distribution curve. (b) The percentage of molecules having the most probable speed decreases as the temperature is raised (the distribution spreads out).

CHAPTER 5
LIQUIDS AND SOLIDS

5.1 (a) London forces; (b) dipole-dipole, London forces; (c) London forces; (d) dipole-dipole, London forces

5.3 Only (b) CH_3Cl, (c) CH_2Cl_2 and (d) $CHCl_3$ will have dipole-dipole interactions. The molecules CH_4 and CCl_4 do not have dipole moments.

5.5 The interaction energies can be ordered based on the relation the energy has to the distance separating the interacting species. Thus ion-ion interactions are the strongest and are directly proportional to the distance separating the two interacting species. Ion-dipole energies are inversely proportional to d^2, whereas dipole-dipole for constrained molecules (i.e., solid state) is inversely proportional to d^3. Dipole-dipole interactions where the molecules are free to rotate become comparable to induced dipole-induced dipole interactions, which are both inversely related to d^6. The order thus derived is: (b) dipole-induced dipole $\cong$ (c) dipole-dipole in the gas phase $<$ (e) dipole-dipole in the solid phase $<$ (a) ion-dipole $<$ (d) ion-ion.

5.7 Only molecules with H attached to the electronegative atoms F, N, and O can hydrogen bond. Additionally, there must be lone pairs available for the H's to bond to. This is true of (a) HF, (c) NH_3, and (d) CH_3OH.

5.9 (a) NaCl (801°C vs. −114.8°C) because it is an ionic compound as opposed to a molecular compound; (b) butanol (−90°C vs. −116°C) due to hydrogen bonding in butanol which is not possible in diethyl ether; (c) HF because it exhibits strong hydrogen bonding whereas HCl does not; (d) H_2O because the number of possible hydrogen bonds is greater than the number in methanol.

5.11 (a) PF_3 and PCl_3 are both pyramidal and should have similar intermolecular forces, but PCl_3 has the greater number of electrons and should have the higher boiling point. The boiling point of PF_3 is −101.5°C and that of PCl_3 is 75.5°C. (b) SO_2 is bent and has a dipole moment whereas CO_2 is linear and will be nonpolar. SO_2 should have the higher boiling point. SO_2 boils at −10°C whereas CO_2 sublimes at −78°C. (c) BF_3 and BCl_3 are both trigonal planar, so the choice of higher boiling point depends on the difference in total number of electrons. BCl_3 should have the higher boiling point (12.5°C vs. −99.9°C). (d) AsF_3 is pyramidal and has a dipole moment, whereas AsF_5 is a trigonal bipyramid and is nonpolar. AsF_5, on the other hand, has more electrons, so the two effects oppose each other. The actual boiling points of AsF_3 and AsF_5 are 63°C and −53°C, respectively, so the effect of the polarity is greater than the effect of the increased mass in this case.

5.13 The ionic radius of Al^{3+} is 53 pm and that of Be^{2+} is 27 pm. The ratio of energies will be given by

$$V \propto \frac{-|z|\mu}{d^2}$$

$$V_{Al^{3+}} \propto \frac{-|z|\mu}{d^2} = \frac{-|3|\mu}{(53)^2}$$

$$V_{Be^{2+}} \propto \frac{-|z|\mu}{d^2} = \frac{-|2|\mu}{(27)^2}$$

The electric dipole moment of the water molecule (μ) will cancel:

$$\text{ratio}\left(\frac{V_{Al^{3+}}}{Be^{2+}}\right) = \frac{-|3|\mu/(53)^2}{-|2|\mu/(27)^2} = \frac{3(27)^2}{2(53)^2} = 0.39$$

The attraction of the Be^{2+} ion will be greater than that of the Al^{3+} ion. Even though it has a lower charge, its radius is much smaller than that of Al^{3+}, making the attraction greater.

5.15 (a) *cis*-Dichloroethene is polar, whereas *trans*-dichloroethene, whose individual bond dipole moments cancel, is nonpolar. Therefore, *cis*-dichloroethene has the greater intermolecular forces and the greater surface tension. (b) Surface tension of liquids decreases with increasing temperature as a result of thermal motion as temperature rises. Increased thermal motion allows the molecules to more easily break away from each other, which manifests itself as decreased surface tension.

5.17 At 50°C all three compounds are liquids. C_6H_6 (nonpolar) < C_6H_5SH (polar, but no hydrogen bonding) < C_6H_5OH (polar and with hydrogen bonding). The viscosity will show the same ordering as the boiling points, which are 80°C for C_6H_6, 169°C for C_6H_5SH, 182° for C_6H_5OH.

5.19 Using $h = \dfrac{2\gamma}{gdr}$ we can calculate the height. For water:

$$r = \frac{1}{2}\,diameter = \frac{1}{2}(0.15\text{ mm})\left(\frac{1\text{ m}}{1000\text{ mm}}\right) = 7.5 \times 10^{-5}\text{ m}$$

$$d = 1.00\ \text{g}\cdot\text{cm}^{-3}\left(\frac{1\ \text{kg}}{1000\ \text{g}}\right)\left(\frac{10^6\ \text{cm}^3}{\text{m}^3}\right) = 1.00 \times 10^3\ \text{kg}\cdot\text{m}^{-3}$$

$$h = \frac{2(72.75 \times 10^{-3}\ \text{N}\cdot\text{m}^{-1})}{(9.81\ \text{m}\cdot\text{s}^{-1})(1.00 \times 10^3\ \text{kg}\cdot\text{m}^{-3})(7.5 \times 10^{-5}\ \text{m})} = 0.20\ \text{m or } 200\ \text{mm}$$

Remember that $1\ \text{N} = 1\ \text{kg}\cdot\text{m}^{-1}\cdot\text{s}^{-2}$.

For ethanol:

$$d = 0.79\ \text{g}\cdot\text{cm}^{-3}\left(\frac{1\ \text{kg}}{1000\ \text{g}}\right)\left(\frac{10^6\ \text{cm}^3}{\text{m}^3}\right) = 7.9 \times 10^2\ \text{kg}\cdot\text{m}^{-3}$$

$$h = \frac{2(22.8 \times 10^{-3}\ \text{N}\cdot\text{m}^{-1})}{(9.81\ \text{m}\cdot\text{s}^{-1})(7.9 \times 10^2\ \text{kg}\cdot\text{m}^{-3})(7.5 \times 10^{-5}\ \text{m})} = 0.078\ \text{m or } 78\ \text{mm}$$

Water will rise to a higher level than hexane. There are two opposing effects to consider. While the greater density of water as compared to ethanol acts against it rising as high, it has a much higher surface tension.

5.21 (X) molecular (no conductivity, moderate m.p.); (Y) network (very high m.p.—compare to SiO_2's m.p. of about 1730°C, no conductivity, insolubility); (Z) ionic (conductivity in solution, solubility in water)

5.23 (a) At center: 1 center × 1 atom·center^{-1} = 1 atom; at 8 corners, 8 corners × $\frac{1}{8}$ atom·corner^{-1} = 1 atom; total = 2 atoms; (b) There are 8 nearest neighbors, hence a coordination number of 8; (c) The direction along which atoms touch each other is the body diagonal of the unit cell. This body diagonal will be composed of 4 times the radius of the atom. In terms of the unit cell edge length a, the body diagonal will be $\sqrt{3}a$. The unit cell edge length will, therefore, be given by

$$4r = \sqrt{3}a \text{ or } a = \frac{4r}{\sqrt{3}} = \frac{4\cdot(235\ \text{pm})}{\sqrt{3}} = 543\ \text{pm}$$

5.25 (a) a = length of side for a unit cell; for an fcc unit cell, $a = \sqrt{8}\,r$ or $2\sqrt{2}\,r$ = 393 pm. $V = a^3 = (393\ \text{pm} \times 10^{-12}\ \text{m}\cdot\text{pm}^{-1})^3 = 6.07 \times 10^{-29}\ \text{m}^3 = 6.07 \times 10^{-23}\ \text{cm}^3$. Because for a fcc unit cell there are 4 atoms per unit cell, we have

$$\text{mass(g)} = 4\ \text{Pt atoms} \times \frac{1\ \text{mol Pt atoms}}{6.022 \times 10^{23}\ \text{atoms}\cdot\text{mol}^{-1}} \times \frac{195.09\ \text{g}}{\text{mol Pt atoms}}$$

$$= 1.30 \times 10^{-21}\ \text{g}$$

$$d = \frac{1.30 \times 10^{-21}\ \text{g}}{6.07 \times 10^{-23}\ \text{cm}^3} = 21.4\ \text{g}\cdot\text{cm}^{-3}$$

(b) $$a = \frac{4r}{\sqrt{3}} = \frac{4 \times 272\ \text{pm}}{\sqrt{3}} = 628\ \text{pm}$$

$$V = (628 \times 10^{-12}\ \text{m})^3 = 2.48 \times 10^{-28}\ \text{m}^3 = 2.48 \times 10^{-22}\ \text{cm}^3$$

There are 2 atoms per bcc unit cell:

$$\text{mass(g)} = 2 \text{ Cs atoms} \times \frac{1 \text{ mol Cs atoms}}{6.022 \times 10^{23} \text{ atoms}\cdot\text{mol}^{-1}} \times \frac{132.91 \text{ g}}{\text{mol Cs atoms}}$$

$$= 4.41 \times 10^{-22} \text{ g}$$

$$d = \frac{4.41 \times 10^{-22} \text{ g}}{2.48 \times 10^{-22} \text{ cm}^3} = 1.78 \text{ g}\cdot\text{cm}^{-3}$$

5.27 a = length of unit cell edge

$$V = \frac{\text{mass of unit cell}}{d}$$

(a) $$V = a^3 = \frac{(1 \text{ unit cell})\left(\frac{192.2 \text{ g Ir}}{\text{mol Ir}}\right)\left(\frac{1 \text{ mol Ir}}{6.022 \times 10^{23} \text{ atoms Ir}}\right)\left(\frac{4 \text{ atoms}}{1 \text{ unit cell}}\right)}{22.6 \text{ g}\cdot\text{cm}^3}$$

$a = 3.84 \times 10^{-8}$ cm

Because for a fcc cell, a $= \sqrt{8}\, r$, $r = \frac{\sqrt{2}\, a}{4} = \frac{\sqrt{2}\,(3.84 \times 10^{-8} \text{ cm})}{4}$

$$= 1.36 \times 10^{-8} \text{ cm} = 136 \text{ pm}$$

(b) $V = a^3$

$$= \frac{(1 \text{ unit cell})\left(\frac{95.94 \text{ g Mo}}{1 \text{ mol Mo}}\right)\left(\frac{1 \text{ mol Mo}}{6.022 \times 10^{23} \text{ atoms Mo}}\right)\left(\frac{2 \text{ atoms}}{1 \text{ unit cell}}\right)}{10.2 \text{ g}\cdot\text{cm}^3}$$

$$= 3.12 \times 10^{-23} \text{ cm}^3$$

$a = 3.15 \times 10^{-8}$ cm

$$r = \frac{\sqrt{3}\, a}{4} = \frac{\sqrt{3}(3.15 \times 10^{-8} \text{ cm})}{4} = 1.36 \times 10^{-8} \text{ cm} = 136 \text{ pm}$$

5.29 (a) The alloy is undoubtedly interstitial, because the atomic radius of nitrogen is much smaller (74 pm vs. 124 pm) than that of iron. The rule of thumb is that the solute atom be less than 60% the solvent atom in radius, in order for an interstitial alloy to form. That criterion is met here. (b) We expect that nitriding will make iron harder and stronger, with a lower electrical conductivity.

5.31 Graphite is a metallic conductor parallel to the planes; the electrons are quite free to move within them. Between planes, however, there is an energy barrier to conduction, though this barrier can be partially overcome by raising the temperature. Thus, graphite is a semiconductor perpendicular to the planes and a conductor parallel to the planes.

5.33 (a) There are eight chloride ions at the eight corners, giving a total of 8 corners $\times \frac{1}{8}$ atom$\cdot$corner^{-1} = 1 Cl^- ion

There is one Cs^+ which lies at the center of the unit cell. All of this ion belongs to the unit cell. The ratio is thus 1:1 for an empirical formula of CsCl with one formula unit per unit cell.

(b) The titanium atoms lie at the corners of the unit cell and at the body center:

$8 \text{ corners} \times \frac{1}{8} \text{ atom} \cdot \text{corner}^{-1} + 1 \text{ at body center} = 2 \text{ atoms per unit cell}$

Four oxygen atoms lie on the faces of the unit cell and two lie completely within the unit cell giving:

$4 \text{ atoms in faces} \times \frac{1}{2} \text{ atom} \cdot \text{face}^{-1} + 2 \text{ atoms wholely within cell} = 4 \text{ atoms}$

The ratio is thus two Ti per four O, or an empirical formula of TiO_2 with two formula units per unit cell. The Ti atoms are 6-coordinate and the O atoms are 3-coordinate.

5.35 Y: $8 \text{ atoms} \times \frac{1}{8} \text{ atom} \cdot \text{corner}^{-1} = 1 \text{ Y atom}$

Ba: $8 \text{ atoms} \times \frac{1}{4} \text{ atom} \cdot \text{edge}^{-1} = 2 \text{ Ba atoms}$

Cu: 3 Cu atoms completely inside unit cell = 3 Cu atoms

O: $10 \text{ atoms on faces} \times \frac{1}{2} \text{ atom} \cdot \text{face}^{-1} + 2 \text{ atoms completely inside unit cell} = 7 \text{ O atoms}$

Formula = $YBa_2Cu_3O_7$

5.37 (a) $\text{ratio} = \dfrac{149 \text{ pm}}{133 \text{ pm}} = 1.12$, predict cesium-chloride structure with (8,8) coordination; however, rubidium fluoride actually adopts the rock-salt structure

(b) $\text{ratio} = \dfrac{72 \text{ pm}}{140 \text{ pm}} = 0.51$, predict rock-salt structure with (6,6) coordination

(c) $\text{ratio} = \dfrac{102 \text{ pm}}{196 \text{ pm}} = 0.520$, predict rock-salt structure with (6,6) coordination

5.39 (a) In the rock-salt structure, the unit cell edge length is equal to 2 times the radius of the cation plus two times the radius of the anion. Thus for CaO, $a = 2(100 \text{ pm}) + 2(140 \text{ pm}) = 480 \text{ pm}$. The volume of the unit cell will be given by (converting to cm^3 because density is normally given in terms of $g \cdot cm^{-3}$)

$$V = \left(480 \text{ pm} \times \frac{10^{-12} \text{ m}}{\text{pm}} \times \frac{100 \text{ cm}}{\text{m}}\right)^3 = 1.11 \times 10^{-22} \text{ cm}^3$$

There are four formula units in the unit cell so the mass in the unit cell will be given by

$$\text{mass in unit cell} = \frac{\left(\dfrac{4 \text{ formula units}}{1 \text{ unit cell}}\right) \times \left(\dfrac{56.08 \text{ g CaO}}{1 \text{ mol CaO}}\right)}{6.022 \times 10^{23} \text{ molecules} \cdot \text{mol}^{-1}} = 3.725 \times 10^{-22} \text{ g}$$

The density will be given by the mass in the unit cell divided by the volume of the unit cell:

$$d = \frac{3.725 \times 10^{-22}\ \text{g}}{1.11 \times 10^{-22}\ \text{cm}^3} = 3.37\ \text{g}\cdot\text{cm}^{-3}$$

(b) For a cesium chloride-like structure, it is the body diagonal that represents two-times the radius of the cation and plus two times the radius of the anion. Thus the body diagonal for CsBr is equal to 2(170 pm) + 2(196) = 732 pm.
For a cubic cell, the body diagonal = $\sqrt{3}\, a$ = 732 pm

$$a = 423\ \text{pm}$$

$$a^3 = V = \left(423\ \text{pm} \times \frac{10^{-12}\ \text{m}}{\text{pm}} \times \frac{100\ \text{cm}}{\text{m}}\right)^3 = 7.57 \times 10^{-23}\ \text{cm}^3$$

There is one formula unit of CsBr in the unit cell, so the mass in the unit cell will be given by

$$\text{mass in unit cell} = \frac{\left(\frac{1\ \text{formula units}}{1\ \text{unit cell}}\right)\left(\frac{212.82\ \text{g CsBr}}{1\ \text{mol CsBr}}\right)}{6.022 \times 10^{23}\ \text{molecules}\cdot\text{mol}^{-1}} = 3.534 \times 10^{-22}\ \text{g}$$

$$d = \frac{3.534 \times 10^{-22}\ \text{g}}{7.57 \times 10^{-23}\ \text{cm}^3} = 4.67\ \text{g}\cdot\text{cm}^3$$

5.41 There is a total of 10 atoms in the unit cell.

5.43 Glucose will be held in the solid by London forces, dipole-dipole interactions, and hydrogen bonds; benzophenone will be held in the solid by dipole-dipole interactions and London forces; methane will be held together by London forces only. London forces are strongest in benzophenone, but glucose can experience hydrogen bonding, which is a strong interaction and dominates intermolecular forces. Methane has few electrons so experiences only weak London forces. We would expect the melting points to increase in the order CH_4 (m.p. = −182°C) < benzophenone (m.p. = 48°C) < glucose (m.p. = 148 − 155°C).

5.45 one form of boron nitride, silicon dioxide, plus many others

5.47 There are too many ways that these molecules can rotate and twist so that they do not remain rod-like. The molecular backbone when the molecule is stretched out is rod-like, but the molecules tend to curl up on themselves, destroying any possibility of long range order with neighboring molecules. This is partly due to the fact that the molecules have only single bonds which allow rotation about the bonds, so that each molecule can adopt many configurations. If multiple bonds are present, the bonding is more rigid.

5.49 Use of a nonpolar solvent such as hexane or benzene (etc.) in place of water should give rise to the formation of inverse micelles.

5.51 (a) Xenon is larger, with more electrons, giving rise to larger London forces which increase the melting point. (b) Hydrogen bonding in water causes the molecules to be held together more tightly than in diethyl ether. (c) Both molecules have the same molar mass, but pentane is a linear molecule compared to dimethylpropane, which is a compact, spherical molecule. The compactness of the dimethyl propane gives it a lower surface area. That means that the intermolecular attractive forces, which are of the same type (London forces) for both molecules, will have a larger effect for pentane.

5.53 (a) These problems are most easily solved by assuming 100 g of substance. In 100 g of Ni-Cu alloy there will be 25 g Ni and 75 g Cu, corresponding to 0.43 mol Ni and 1.2 mol Cu. The atom ratio will be the same as the mole ratio:

$$\frac{1.2 \text{ mol Cu}}{0.43 \text{ mo Ni}} = 2.8 \text{ Cu per Ni}$$

(b) Pewter, which is 7% Sb, 3% Cu, and 90% Sn, will contain 7 g Sb, 3 g Cu, and 90 g Sn per 100 g of alloy. This will correspond to 0.06 mol Sb, 0.05 Cu, and 0.76 mol Sn per 100 g. The atom ratio will be 15 Sn : 1.2 Sb : 1 Cu.

(c) The residual composition of Wood's metal will be Bi; this will constitute 51% of the alloy by mass. The number of moles per 100 g of alloy will be: 0.105 mol Sn, 0.111 mol Cd, 0.11 mol Pb, and 0.24 mol Bi. The atom ratio will be 2.3 Bi : 1.1 Cd : 1.1 Pb : 1 Sn.

5.55 (a) To answer this question we will need to have the dimensions in comparable units. The edge length of the unit cell will be

$$a = 562.8 \text{ pm} \times \frac{1 \text{ m}}{10^{12} \text{ pm}} = 5.628 \times 10^{-10} \text{ m}$$

$$V_{\text{unit cell}} = a^3 = (5.628 \times 10^{-10} \text{ m})^3 = 1.783 = 10^{-28} \text{ m}^3 \cdot \text{unit cell}^{-1}$$

The volume of the crystal will be given by

$$V_{\text{crystal}} = \left(1.00 \text{ mm} \times \frac{1 \text{ m}}{10^3 \text{ m}}\right)^3 = 1.00 \times 10^{-9} \text{ m}^3$$

The number of units cells in the crystal will then be given by $V_{\text{crystal}} \div V_{\text{unit cell}}$:

$$\# \text{ of unit cells in crystal} = \frac{1.00 \times 10^{-9} \text{ m}^3 \cdot \text{crystal}^{-1}}{1.783 \times 10^{-28} \text{ m}^3 \cdot \text{unit cell}^{-1}}$$

$$= 5.61 \times 10^{18} \text{ unit cells} \cdot \text{crystal}^{-1}$$

(b) Because the unit cell is fcc, there will be four formula units of NaCl per unit cell so the number of NaCl units will be given by $4(5.61 \times 10^{18}$ unit cells$) = 2.24 \times 10^{19}$ formula units. To get the number of moles of NaCl, we divide this number by Avogadro's number: $(2.24 \times 10^{19}) \div (6.022 \times 10^{23}) = 3.72 \times 10^{-5}$ mol.

5.57 (a) There are three types of cubic unit cells that are possible choices. These include the simple cubic cell, the body-centered cubic cell, and the face-centered cubic cell. They differ in that the simple cubic cell has one atom per unit cell, the body-centered has a total of two atoms per unit cell, and the fcc cell has four atoms. The mass of an Al atom will be

$26.98\ \text{g}\cdot\text{mol}^{-1} \div (6.022 \times 10^{23})\ \text{atoms}\cdot\text{mol}^{-1} = 4.480 \times 10^{-23}\ \text{g}\cdot\text{atom}^{-1}$

The volume of the unit cell will equal

$$V_{\text{unit cell}} = \left(404\ \text{pm} \times \frac{10^{-12}\ \text{m}}{\text{pm}} \times \frac{100\ \text{cm}}{\text{m}}\right)^3 = 6.594 \times 10^{-23}\ \text{cm}^3\cdot\text{unit cell}^{-1}$$

The density will be given by

$$d = \frac{\text{mass in unit cell}}{\text{volume of unit cell}} = \frac{Z\cdot(4.480 \times 10^{-23}\ \text{g}\cdot\text{unit cell}^{-1})}{6.594 \times 10^{-23}\ \text{cm}^3\cdot\text{unit cell}^{-1}} = Z\cdot 0.6794\ \text{g}\cdot\text{cm}^3$$

where Z is the number of atoms in the unit cell.

Because the observed density is $2.70\ \text{g}\cdot\text{cm}^3$, we can calculate Z

$$Z\cdot 0.6794\ \text{g}\cdot\text{cm}^3 = 2.70\ \text{g}\cdot\text{cm}^3$$

$$Z = \frac{2.70\ \text{g}\cdot\text{cm}^3}{0.6794\ \text{g}\cdot\text{cm}^3} = 3.97$$

Because Z is 4, the lattice type must be fcc.

(b) For a metal packing in fcc unit cell (cubic close-packing), the coordination number will be 12.

5.59 The unit cell for a cubic close-packed lattice is the fcc unit cell. For this cell, the relation between the radius of the atom r and the unit cell edge length a is

$$4r = \sqrt{2}\,a$$

$$a = \frac{4r}{\sqrt{2}}$$

The volume of the unit cell is given by

$$V = a^3 = \left(\frac{4r}{\sqrt{2}}\right)^3$$

If r is given in pm, then a conversion factor to cm is required:

$$V = a^3 = \left(\frac{4r}{\sqrt{2}} \times \frac{10^{-12}\ \text{m}}{\text{pm}} \times \frac{100\ \text{cm}}{\text{m}}\right)^3$$

Because there are four atoms per fcc unit cell, the mass in the unit cell will be given by

$$\text{mass} = \left(\frac{4\ \text{atoms}}{\text{unit cell}}\right)\left(\frac{M}{6.022 \times 10^{23}\ \text{atoms}\cdot\text{mol}^{-1}}\right)$$

The density will be given by

$$d = \frac{\text{mass of unit cell}}{\text{volume of unit cell}} = \frac{\left(\dfrac{4\ \text{atoms}}{\text{unit cell}}\right) \times \left(\dfrac{M}{6.022 \times 10^{23}\ \text{atoms}\cdot\text{mol}^{-1}}\right)}{\left(\dfrac{4\,r}{\sqrt{2}} \times \dfrac{10^{-12}\ \text{m}}{\text{pm}} \times \dfrac{100\ \text{cm}}{\text{m}}\right)^3}$$

$$= \frac{(2.936 \times 10^5)M}{r^3}$$

or

$$r = \sqrt[3]{\frac{(2.936 \times 10^5)M}{d}}$$

where M is the atomic mass in $\text{g}\cdot\text{mol}^{-1}$ and r is the radius in pm.

For the different gases we calculate the results given in the following table:

Gas	Density ($\text{g}\cdot\text{cm}^3$)	Molar mass ($\text{g}\cdot\text{mol}^{-1}$)	Radius (pm)
Neon	1.20	20.18	170
Argon	1.40	39.95	203
Krypton	2.16	83.80	225
Xenon	2.83	131.30	239
Radon	4.4	222	246

5.61 There are two approaches to this problem. The information given does not specify the radius of the tungsten atom. This value can be looked up in Appendix 2D. We can calculate an answer, however, based simply upon the fact that the density is $19.3\ \text{g}\cdot\text{cm}^{-3}$ for the bcc cell and taking the ratio between the expected densities based upon the assumption that the atomic radius of tungsten will be the same for both. The unit cell for a cubic close-packed lattice is the fcc unit cell. For this cell, the relation between the radius of the atom r and the unit cell edge length a is

$$4\,r = \sqrt{2}\,a$$

$$a = \frac{4\,r}{\sqrt{2}}$$

The volume of the unit cell is given by

$$V = a^3 = \left(\frac{4\,r}{\sqrt{2}}\right)^3$$

If r is given in pm, then a conversion factor to cm is required:

$$V = a^3 = \left(\frac{4\,r}{\sqrt{2}} \times \frac{10^{-12}\ \text{m}}{\text{pm}} \times \frac{100\ \text{cm}}{\text{m}}\right)^3$$

Because there are four atoms per fcc unit cell, the mass in the unit cell will be given by

$$\text{mass} = \left(\frac{4\ \text{atoms}}{\text{unit cell}}\right)\left(\frac{M}{6.022 \times 10^{23}\ \text{atoms}\cdot\text{mol}^{-1}}\right)$$

The density will be given by

$$d = \frac{\text{mass of unit cell}}{\text{volume of unit cell}} = \frac{\left(\frac{4\text{ atoms}}{\text{unit cell}}\right)\left(\frac{M}{6.022 \times 10^{23}\text{ atoms}\cdot\text{mol}^{-1}}\right)}{\left(\frac{4\,r}{\sqrt{2}} \times \frac{10^{-12}\text{ m}}{\text{pm}} \times \frac{100\text{ cm}}{\text{m}}\right)^3}$$

$$= \frac{(2.936 \times 10^5)M}{r^3}$$

or

$$r = \sqrt[3]{\frac{(2.936 \times 10^5)M}{d}}$$

where M is the atomic mass in $\text{g}\cdot\text{mol}^{-1}$ and r is the radius in pm.

Likewise, for a body-centered cubic lattice there will be two atoms per unit cell. For this cell, the relation between the radius of the atom r and the unit cell edge length a is derived from the body diagonal of the cell, which is equal to 4 times the radius of the atom. The body diagonal is found from the Pythagorean theorum to be equal to the $\sqrt{3}\,a$.

$$4\,r = \sqrt{3}\,a$$

$$a = \frac{4\,r}{\sqrt{3}}$$

The volume of the unit cell is given by

$$V = a^3 = \left(\frac{4\,r}{\sqrt{3}}\right)^3$$

If r is given in pm, then a conversion factor to cm is required:

$$V = a^3 = \left(\frac{4\,r}{\sqrt{3}} \times \frac{10^{-12}\text{ m}}{\text{pm}} \times \frac{100\text{ cm}}{\text{m}}\right)^3$$

Because there are two atoms per bcc unit cell, the mass in the unit cell will be given by

$$\text{mass} = \left(\frac{2\text{ atoms}}{\text{unit cell}}\right) \times \left(\frac{M}{6.022 \times 10^{23}\text{ atoms}\cdot\text{mol}^{-1}}\right)$$

The density will be given by

$$d = \frac{\text{mass of unit cell}}{\text{volume of unit cell}} = \frac{\left(\frac{2\text{ atoms}}{\text{unit cell}}\right)\left(\frac{M}{6.022 \times 10^{23}\text{ atoms}\cdot\text{mol}^{-1}}\right)}{\left(\frac{4\,r}{\sqrt{3}} \times \frac{10^{-12}\text{ m}}{\text{pm}} \times \frac{100\text{ cm}}{\text{m}}\right)^3}$$

$$= \frac{\left(\frac{2\text{ atoms}}{\text{unit cell}}\right)\left(\frac{M}{6.022 \times 10^{23}\text{ atoms}\cdot\text{mol}^{-1}}\right)}{(2.309 \times 10^{-10}\,r)^3}$$

$$= \frac{(2.698 \times 10^5)M}{r^3}$$

or

$$r = \sqrt[3]{\frac{(2.698 \times 10^5)M}{d}}$$

Setting these two equal and cubing both sides we obtain

$$\frac{(2.936 \times 10^5)M}{d_{fcc}} = \frac{(2.698 \times 10^5)M}{d_{bcc}}$$

The molar mass of tungsten M is the same for both ratios and will cancel from the equation.

$$\frac{(2.936 \times 10^5)}{d_{fcc}} = \frac{(2.698 \times 10^5)}{d_{bcc}}$$

Rearranging, we get

$$d_{fcc} = \frac{(2.936 \times 10^5)}{(2.698 \times 10^5)} d_{bcc}$$

$$= 1.088\ d_{bcc}$$

For W, $d_{fcc} = 1.088 \times 19.6\ g{\cdot}cm^3$

$$= 21.3\ g{\cdot}cm^{-3}$$

5.63 (a) The oxidation state of the titanium atoms must balance the charge on the oxide ions, O^{2-}. The presence of 1.18 O^{2-} ions will mean that the Ti present must have a charge to compensate the -2.36 charge on the oxide ions. The average oxidation state of Ti will thus be $+2.36$. (b) This is most easily solved by setting up a set of two equations in two unknowns. We know that the total charge on the titanium atoms present must equal 2.36, so if we multiply the charge on each type of titanium by the fraction of titanium present in that oxidation state and sum the values, we should get 2.36:

let x = fraction of Ti^{2+}, y = fraction of Ti^{3+}, then

$2x + 3y = 2.36$

Also, because we are assuming all the titanium is either $+2$ or $+3$, the fractions of each present must add up to 1:

$x + y = 1$

Solving these two equations simultaneously, we obtain y = 0.36, x = 0.64.

5.65 (a) true. If this is not the case, the unit cell will not match with other unit cells of the same type when stacked to form the entire lattice.

(b) false. Unit cells do not have to have atoms at the corners.

(c) true. In order for the unit cell to repeat properly, opposite faces must have the same composition.

(d) false. If one face is centered, the opposing face must be centered, but the other faces do not necessarily have to be centered.

5.67 The density is calculated as for cubic structures but the volume = a^2c for a tetragonal unit cell.

$V = a^2c = (459\ \text{pm})^2(296\ \text{pm}) = 6.24 \times 10^7\ \text{pm}^3$ or $6.24 \times 10^{-23}\ \text{cm}^3$

Density is then calculated as before:

density =

$$\frac{(2\text{TiO}_2\ \text{formula units}\cdot\text{unit cell}^{-1})(79.88\ \text{g}\cdot\text{mol}^{-1}\ \text{TiO}_2)\left(\dfrac{1\ \text{mol}}{6.022 \times 10^{23}\ \text{formula units}}\right)}{6.24 \times 10^{-23}\ \text{cm}^3}$$

$= 4.25\ \text{g}\cdot\text{cm}^{-3}$

5.69 There are several ways to draw unit cells that will repeat to generate the entire lattice. Some examples are shown below. The choice of unit cell is determined by conventions that are beyond the scope of this text (the smallest unit cell that indicates all of the symmetry present in the lattice is typically the one of choice).

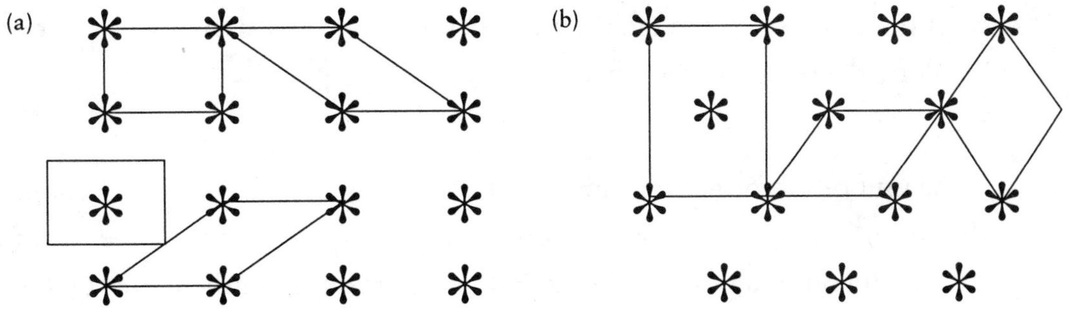

5.71 In a salt such as sodium chloride or sodium bromide, all of the interactions will be ionic. In sodium acetate or sodium methoxide, however, the organic anions CH_3COO^- and CH_3O^- will have the negative charges localized largely on the oxygen atoms, which are the most electronegative atoms. The carbon part of the molecule will interact with other parts of the crystal lattice via London forces rather than ionic interactions. In general, solid organic salts will not be as hard as purely ionic substances and will also have lower melting points.

5.73 Fused silica, also known as fused quartz, is predominantly SiO_2 with very few impurities. This glass is the most refractory, which means it can be used at the highest temperatures. Quartz vessels are routinely used for reactions that must be carried out at temperatures up to 1000°C. Vycor, which is 96% SiO_2, can be used up to ca. 900°C, and normal borosilicate glasses up to approximately 550°C. This is due to the fact that the glasses melt at lower temperatures as the amount of materials other than SiO_2 increases. The borosilicate glasses are commonly used

(Pyrex or Kimax) because the lower softening points allow them to be more easily molded and shaped into different types of glass objects, such as reaction flasks, beakers, and other types of laboratory and technical glassware.

5.75 For M, there will be a total of one atom in the unit cell from the corners

cations: 8 corners $\times \frac{1}{8}$ atom·corner^{-1} + 6 facts + $\frac{1}{2}$ atom·face^{-1} = 4 atoms

anions: 8 tetrahedral holes $\times$ 1 atom·tetrahedral hole^{-1} = 8 atoms

The cation to anion ratio is thus 4:8 or 1:2; the empirical formula is MA_2.

5.77 Because the bisulfate anions occupy spaces between the planes of the graphite layers, they will separate the layers more than they are separated in pure graphite. The conduction between layers will be reduced. The conduction within the layers, however, is enhanced because the partial oxidation produces some empty spaces in the delocalized molecular orbitals which extend over the graphite sheet. The presence of these empty spaces makes it easier for the electrons to migrate through the solid along the planar carbon networks.

5.79 (a) The unit cell information shows that calcium fluoride exists in a fcc unit cell, with the Ca^{2+} ions occupying the corners and face centers of the cube, and the fluoride ions found in all of the tetrahedral holes produced in the lattice of Ca^{2+} ions. This arrangement will contain 4 formula units of CaF_2 in the unit cell. From this information and the density, we can calculate the unit cell edge length a:

a = length of unit cell edge

$$V = \frac{\text{mass of unit cell}}{\text{density}}$$

$$V = a^3 = \frac{(1\ \text{unit cell})\left(\dfrac{4\ \text{atoms}}{1\ \text{unit cell}}\right)\left(\dfrac{78.08\ \text{g CaF}_2}{\text{mol CaF}_2}\right)\left(\dfrac{1\ \text{mol CaF}_2}{6.022 \times 10^{23}\ \text{molecules CaF}_2}\right)}{3.180\ \text{g}\cdot\text{cm}^3}$$

$a = 5.46 \times 10^{-8}$ cm or 546 pm

(b) There are several approaches that one can take to analyzing the geometry in order to calculate the Ca-F distance. Two will be presented here. By looking at the structure, one can see that the fluoride ion sits in the center of a regular tetrahedron of Ca^{2+} ions. The distance between the Ca^{2+} ions will be given by half the length of the face diagonal of the unit cell. The face diagonal of the fcc unit cell is equal to $a\sqrt{2}$ = 772 pm. The Ca^{2+}-Ca^{2+} separation is, therefore, 386 pm. Because the F^- sits at the center of a regular tetrahedron, the Ca-F-Ca angle will be 109.47°. From these two pieces of information we can construct the triangle shown below. A right triangle can be constructed by dropping a line from the F^- ion to the midpoint of the line connecting the two Ca^{2+} ions.

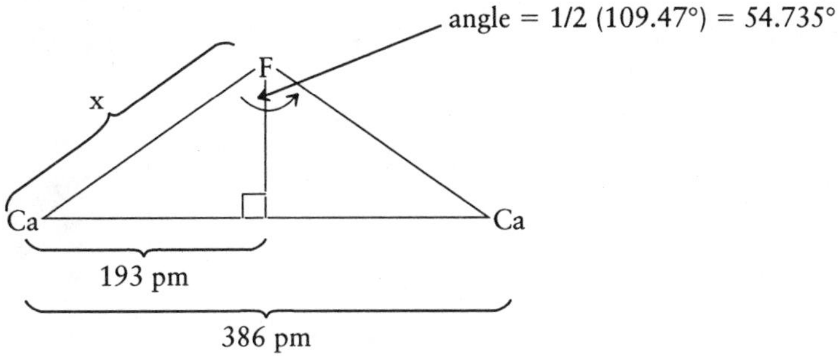

The Ca-F distance (labeled x) can be calculated using simple trigonometry:

$$\sin 54.735° = \frac{193 \text{ pm}}{x}$$

$$x = 236 \text{ pm}$$

An alternative approach is to realize that the Ca-F separation is one-fourth the body diagonal. Because the body diagonal is a $\sqrt{3}$, the Ca-F separation will be 236 pm.

(c) This is in good agreement with the sum of the ionic radii found in the Appendix, which gives $r(Ca^{2+}) + r(F^-) = 100 \text{ pm} + 133 \text{ pm} = 233 \text{ pm}$.

5.81 In order to work this problem, we must first realize that the unit cell dimension does not correspond to the direction of closest packing. The close-packed layers in a fcc cell will lie perpendicular to the body diagonal, which will be given by $a\sqrt{3} = (408 \text{ pm})\sqrt{3} = 707 \text{ pm}$. In examining the unit cell along the body diagonal, it is apparent that there are four layers of close-packing that lie perpendicular to the body diagonal within one unit cell; the separation between these layers will be $(707 \text{ pm}) \div 3$. Each layer will add 236 pm. In a deposit that is 0.125 mm we will then expect to find

$$\text{number of layers} = \frac{0.125 \text{ mm} \times \dfrac{1 \text{ m}}{10^3 \text{ mm}}}{236 \text{ pm} \times \dfrac{1 \text{ m}}{10^{12} \text{ pm}}} = 5.30 \times 10^5 \text{ layers}$$

5.83 The answer to this problem is obtained from Bragg's law: $\lambda = 2\,d \sin \Theta$ where λ is the wavelength of radiation, d is the interplanar spacing, and Θ is the angle of incidence of the x-ray beam. Here $d = 401.8$ pm and $\lambda = 71.037$ pm.

$71.07 \text{ pm} = 2(401.8 \text{ pm})\sin \Theta$

$\Theta = 5.074°$

CHAPTER 6
THERMODYNAMICS: THE FIRST LAW

6.1 (a) isolated; (b) closed; (c) isolated

6.3 The change in internal energy ΔU is given simply by summing the two energy terms involved in this process. We must be careful, however, that the signs on the energy changes are appropriate. In this case, internal energy will be added to the gas sample both by heating and by compression, so both are positive numbers.
$\Delta U = 375\text{ kJ} + 645\text{ kJ} = 1020\text{ kJ}$

6.5 To calculate this we use the relationship $\Delta U = q + w$ which arises from the first law of thermodynamics. Because the system absorbs heat, q will be a positive number as will ΔU; as it is an increase in internal energy.

$$120\text{ kJ} = 150\text{ kJ} + w$$
$$w = -30\text{ kJ}$$

Because work is negative, the system will do work on its surroundings.

6.7 To get the entire internal energy change, we must sum the changes due to heat and work. In this problem, $q = +7000$ kJ. Work will be given by $w = -P_{\text{ext}}\Delta V$ because it is an expansion against a constant opposing pressure:

$$w = -\left(\frac{750\text{ Torr}}{760\text{ Torr}\cdot\text{atm}^{-1}}\right)\left(\frac{1450\text{ mL} - 700\text{ mL}}{1000\text{ mL}\cdot\text{L}^{-1}}\right) = -0.740\text{ L}\cdot\text{atm}$$

To convert to J we use the equivalency of the ideal gas constants:

$$w = -(0.740\text{ L}\cdot\text{atm})\left(\frac{8.134\text{ J}\cdot\text{K}^{-1}\cdot\text{mol}^{-1}}{0.082\,06\text{ L}\cdot\text{atm}\cdot\text{K}^{-1}\cdot\text{mol}^{-1}}\right) = -75.0\text{ J}$$

$$\Delta U = q + w = 7000\text{ kJ} - 0.075\text{ kJ} = 7000\text{ kJ}$$

The energy change due to the work term turns out to be negligible in this problem.

6.9 Using $\Delta U = q + w$ where $\Delta U = -2573$ kJ and $q = -947$ kJ

$$-2573\text{ kJ} = -947\text{ kJ} + w$$
$$w = -1626\text{ kJ}$$

1626 kJ of work can be done by the system on its surroundings.

6.11 From $\Delta H = \Delta U + P\Delta V$ at constant pressure, or $\Delta U = \Delta H - P\Delta V$. Because $w = -P\Delta V = +22$ kJ, we get -15 kJ $+ 22$ kJ $= \Delta U = +7$ kJ.

6.13 (a) The irreversible work of expansion against a constant opposing pressure is given by

$$w = -P_{ex}\Delta V$$

$$w = -(1.00\ \text{atm})(6.52\ \text{L} - 4.29\ \text{L})$$
$$= -2.23\ \text{L·atm}$$
$$= -2.23\ \text{L·atm} \times 101.325\ \text{J·L}^{-1}\text{·atm}^{-1} = -226\ \text{J}$$

(b) An isothermal expansion will be given by

$$w = -nRT\frac{V_2}{V_1}$$

n is calculated from the idea gas law:

$$n = \frac{PV}{RT} = \frac{(1.79\ \text{atm})(4.29\ \text{L})}{(0.082\ 06\ \text{L·atm·K}^{-1}\text{·mol}^{-1})(305\ \text{K})} = 0.307\ \text{mol}$$

$$w = -(0.307\ \text{mol})(8.314\ \text{J·K}^{-1}\text{·mol}^{-1})(305\ \text{K})\ln\frac{6.52}{4.29}$$
$$= -326\ \text{J}$$

Note that the work done is greater when the process is carried out reversibly.

6.15 (a) $q = (150.0\ \text{g})(16.6°\text{C} - 50.0°\text{C})(2.42\ \text{J·(°C)}^{-1}\text{·g}^{-1}) = -12.1\ \text{kJ}$
(12.1 kJ of heat removed)
(b) $q = (\text{mass})(205°\text{C} - 15°\text{C})(0.38\ \text{J·(°C)}^{-1}\text{·g}^{-1}) = 425 \times 10^3\ \text{J}$
mass = 5.9 kg

6.17 (a) The heat change will be made up of two terms: one term to raise the temperature of the copper and the other to raise the temperature of the water:

$$q = (450.0\ \text{g})(4.18\ \text{J·(°C)}^{-1}\text{·g}^{-1})(100.0°\text{C} - 25.0°\text{C})$$
$$+ (500.0\ \text{g})(0.38\ \text{J·(°C)}^{-1}\text{·g}^{-1})(100°\text{C} - 25°\text{C})$$
$$= 141\ \text{kJ} + 14\ \text{kJ} = 155\ \text{kJ}$$

(b) The percentage of heat attributable to raising the temperature of water will be

$$\left(\frac{141\ \text{kJ}}{155\ \text{kJ}}\right)(100) = 91.1\%$$

6.19 heat lost by metal = −heat gained by water

$$(20.0\ \text{g})(T_{\text{final}} - 100.0°\text{C})(0.38\ \text{J·(°C)}^{-1}\text{·g}^{-1})$$
$$= -(50.7\ \text{g})(4.18\ \text{J·(°C)}^{-1}\text{·g}^{-1})(T_{\text{final}} - 22.0°\text{C})$$
$$(T_{\text{final}} - 100.0°\text{C})(7.6\ \text{J·(°C)}^{-1}) = -(212\ \text{J·(°C)}^{-1})(T_{\text{final}} - 22.0°\text{C})$$
$$T_{\text{final}} - 100.0°\text{C} = -28(T_{\text{final}} - 22.0°\text{C})$$

$T_{\text{final}} + 28\ T_{\text{final}} = 100.0°\text{C} + 616°\text{C}$

$29\ T_{\text{final}} = 716°\text{C}$

$T_{\text{final}} = 25°\text{C}$

6.21 $C_{\text{cal}} = \dfrac{22.5\ \text{kJ}}{23.97°\text{C} - 22.45°\text{C}} = 14.8\ \text{kJ}\cdot(°\text{C})^{-1}$

6.23 (a) The molar heat capacity of a monatomic ideal gas at constant volume is given by $C_{V,\text{m}} = \frac{3}{2}R = 12.5\ \text{J}\cdot\text{mol}^{-1}\cdot\text{K}^{-1}$, which is derived from the kinetic molecular theory of gases.

$\Delta U = (0.325\ \text{mol})\ (12.5\ \text{J}\cdot\text{mol}^{-1}\cdot(°\text{C})^{-1})\ (50°\text{C} - (-25°\text{C})) = 305\ \text{J}$

(b) The molar heat capacity of a monatomic ideal gas at constant pressure is given by $C_{P,\text{m}} = \frac{5}{2}R = 20.8\ \text{J}\cdot\text{mol}^{-1}\cdot K^{-1}$

$\Delta H = (0.325\ \text{mol})(20.8\ \text{J}\cdot\text{mol}^{-1}\cdot(°\text{C})^{-1})\ (50°\text{C} - (-25°\text{C})) = 507\ \text{J}$

Note that the heat capacities can be expressed in terms of °C as well as K because a Celsius degree is the same size as a kelvin degree.

(c) At constant pressure, the gas is expanding and thus does work on its surroundings. Some of the extra heat thus spills over into the surroundings in the form of expansion work.

6.25 NO_2. The heat capacity increases with molecular complexity—as more atoms are present in the molecule there are more possible bond vibrations that can absorb added energy.

6.27 (a) The molar heat capacity of a monatomic ideal gas at constant pressure is $C_{P,\text{m}} = \frac{5}{2}R$. The heat released will be given by

$$q = \left(\frac{1.078\ \text{g}}{83.80\ \text{g}\cdot\text{mol}^{-1}}\right)(25.0°\text{C} - 97.6°\text{C})(20.8\ \text{J}\cdot\text{mol}^{-1}\cdot(°\text{C})^{-1}) = -19.4\ \text{J}$$

(b) Similarly, the molar heat capacity of a monatomic ideal gas at constant volume is $C_{V,\text{m}} = \frac{3}{2}R$. The heat released will be given by

$$q = \left(\frac{1.078\ \text{g}}{83.80\ \text{g}\cdot\text{mol}^{-1}}\right)(25.0°\text{C} - 97.6°\text{C})(12.5\ \text{J}\cdot\text{mol}^{-1}\cdot(°\text{C})^{-1}) = -11.7\ \text{J}$$

6.29 (a) $\Delta H_{\text{vap}} = \dfrac{1.93\ \text{kJ}}{0.235\ \text{mol}} = 8.21\ \text{kJ}\cdot\text{mol}^{-1}$

(b) $\Delta H_{\text{vap}} = \dfrac{21.2\ \text{kJ}}{\left(\dfrac{22.45\ \text{g}}{46.07\ \text{g}\cdot\text{mol}^{-1}}\right)} = 43.5\ \text{kJ}\cdot\text{mol}^{-1}$

6.31 (a) $\Delta H = \left(\dfrac{100.0 \text{ g } H_2O}{18.02 \text{ g}\cdot\text{mol}^{-1} \text{ } H_2O}\right)(40.7 \text{ kJ}\cdot\text{mol}^{-1}) = 226 \text{ kJ}$

(b) $\Delta H = \left(\dfrac{612 \text{ g } NH_3}{17.03 \text{ g}\cdot\text{mol}^{-1} \text{ } NH_3}\right)(5.65 \text{ kJ}\cdot\text{mol}^{-1}) = 203 \text{ kJ}$

6.33 This process is composed of two steps: melting the ice at 0°C and then raising the temperature of the liquid water from 0°C to 25°C:

Step 1: $\Delta H = \left(\dfrac{50.0 \text{ g}}{18.02 \text{ g}\cdot\text{mol}^{-1}}\right)(6.01 \text{ kJ}\cdot\text{mol}^{-1}) = 16.7 \text{ kJ}$

Step 2: $\Delta H = (50.0 \text{ g})(4.18 \text{ J}\cdot(°\text{C})^{-1}\cdot\text{g}^{-1})(25°\text{C} - 0°\text{C}) = 5.2 \text{ kJ}$

Total heat required = 16.7 kJ + 5.2 kJ = 21.9 kJ

6.35 (a) $\Delta H = (0.20 \text{ mol})(+358.8 \text{ kJ}\cdot\text{mol}^{-1}) = 72 \text{ kJ}$

(b) $\Delta H = \left(\dfrac{20.0 \text{ g C}}{12.011 \text{ g}\cdot\text{mol}^{-1} \text{ C}}\right)\left(\dfrac{358.8 \text{ kJ}}{4 \text{ mol C}}\right) = 149 \text{ kJ}$

(c) $\Delta H = 217 \text{ kJ} = (n_{CS_2})\left(\dfrac{353.8 \text{ kJ}\cdot\text{mol}^{-1})}{4 \text{ mol } CS_2}\right)$

$n_{CS_2} = 2.42 \text{ mol } CS_2$ or $(2.42 \text{ mol})(76.13 \text{ g}\cdot\text{mol}^{-1}) = 184 \text{ g } CS_2$

6.37 (a) mass of octane $= \left(\dfrac{-12.0 \times 10^3 \text{ kJ}}{-10\,942 \text{ kJ}}\right)(2 \text{ mol octane})(114.22 \text{ g}\cdot\text{mol}^{-1})$

$= 2.5 \times 10^2 \text{ g}$

(b) $\Delta H = \left(\dfrac{(1.0 \text{ gal})(3.785 \times 10^3 \text{ mL}\cdot\text{gal}^{-1})(0.70 \text{ g}\cdot\text{mL}^{-1})}{114.22 \text{ g}\cdot\text{mol}^{-1}}\right)\left(\dfrac{-10\,942 \text{ kJ}}{2 \text{ mol octane}}\right)$

$= -1.3 \times 10^5 \text{ kJ}$

6.39 $\dfrac{50.0 \text{ g } Fe_2O_3}{159.7 \text{ g}\cdot\text{mol}^{-1} \text{ } Fe_2O_3} = 0.313 \text{ mol Fe}; \dfrac{25.0 \text{ g Al}}{26.98 \text{ g}\cdot\text{mol}^{-1}} = 0.927 \text{ mol Al}$

The limiting reactant is Fe_2O_3 because 0.313 mol will react with no more than 0.626 mol Al. The amount of heat produced will be $(0.313 \text{ mol } Fe_2O_3)(-851.5 \text{ kJ}\cdot\text{mol}^{-1} \text{ } Fe_2O_3) = -267 \text{ kJ}$.

6.41 (a) For $H_2O(l)$, we want to find the enthalpy of the reaction

$H_2(g) + \frac{1}{2} O_2(g) \longrightarrow H_2O(l)$

The enthalpy change can be estimated from bond enthalpies. We will need to put in $(1 \text{ mol})(436 \text{ kJ}\cdot\text{mol}^{-1})$ to break the H—H bonds in 1 mol H_2, $(\frac{1}{2} \text{ mol})(496 \text{ kJ}\cdot\text{mol}^{-1})$ to break the O—O bonds in $\frac{1}{2}$ mol $O_2(g)$; we will get back $(2 \text{ mol})(463 \text{ kJ}\cdot\text{mol}^{-1})$ for the formation of 2 mol O—H bonds. This will give $\Delta H = -242 \text{ kJ}\cdot\text{mol}^{-1}$. This value, however, will be to produce water in the gas phase. In order to get the value for the liquid, we will need to take into account the amount of heat given off when the gaseous water condenses to the liquid phase. This is 44.0 kJ·mol^{-1} at 298 K:

$$\Delta H°_{f,\ water(l)} = \Delta H°_{f,\ water(g)} - \Delta H°_{vap} = -242\ kJ - (1\ mol)(44.0\ kJ\cdot mol^{-1})$$
$$= -286\ kJ\cdot mol^{-1}$$

(b) The calculation for methanol is done similarly:

$C(gr) + 2\ H_2(g) + \frac{1}{2}\ O_2(g) \longrightarrow CH_3OH(l)$

ΔH for individual bond contributions:

atomize 1 mol C(gr)	$(1\ mol)(717\ kJ\cdot mol^{-1})$
break 2 mol H—H bonds	$(2\ mol)(436\ kJ\cdot mol^{-1})$
break $\frac{1}{2}$ mol O_2 bonds	$(\frac{1}{2}\ mol)(496\ kJ\cdot mol^{-1})$
form 3 mol C—H bonds	$-(3\ mol)(412\ kJ\cdot mol^{-1})$
form 1 mol C—O bonds	$-(1\ mol)(360\ kJ\cdot mol^{-1})$
form 1 mol O—H bonds	$-(1\ mol)(463\ kJ\cdot mol^{-1})$
Total	−222 kJ

$$\Delta H°_{f,\ methanol(l)} = \Delta H°_{f,\ methanol(g)} - \Delta H°_{vap} = -222\ kJ - (1\ mol)(35.3\ kJ\cdot mol^{-1})$$
$$= -257\ kJ$$

(c) $6\ C(gr) + 3\ H_2(g) \rightarrow C_6H_6(l)$

Without resonance, we do the calculation considering benzene to have 3 double and 3 single C—C bonds:

atomize 6 mol C(gr)	$(6\ mol)(717\ kJ\cdot mol^{-1})$
break 3 mol H—H bonds	$(3\ mol)(436\ kJ\cdot mol^{-1})$
form 3 mol C=C bonds	$-(3\ mol)(612\ kJ\cdot mol^{-1})$
form 3 mol C—C bonds	$-(3\ mol)(348\ kJ\cdot mol^{-1})$
form 6 mol C—H bonds	$-(6\ mol)(412\ kJ\cdot mol^{-1})$
Total	+258 kJ

$$\Delta H°_{f,\ benzene(l)} = \Delta H°_{f,\ benzene(g)} - \Delta H°_{vap} = +258\ kJ - (1\ mol)(30.8\ kJ\cdot mol^{-1})$$
$$= +227\ kJ$$

(d) $6\ C(gr) + 3\ H_2(g) \longrightarrow C_6H_6(l)$

With resonance, we repeat the calculation considering benzene to have 6 resonance-stabilized C—C bonds:

atomize 6 mol C(gr)	$(6\ mol)(717\ kJ\cdot mol^{-1})$
break 3 mol H—H bonds	$(3\ mol)(436\ kJ\cdot mol^{-1})$
form 6 mol C—C bonds, resonance	$-(6\ mol)(518\ kJ\cdot mol^{-1})$
form 6 mol C—H bonds	$-(6\ mol)(412\ kJ\cdot mol^{-1})$
Total	+30 kJ

$$\Delta H°_{f,\ benzene(l)} = \Delta H°_{f,\ benzene(g)} - \Delta H°_{vap} = +30\ kJ\cdot mols^{-1} - (1\ mol)(30.8\ kJ\cdot mol^{-1})$$
$$= -1\ kJ$$

6.43 The combustion reaction of diamond is reversed and added to the combustion reaction of graphite to give the desired reaction:

$$\begin{array}{ll} C(gr) + O_2(g) \longrightarrow CO_2(g) & \Delta H° = -393.51\ kJ \\ CO_2(g) \longrightarrow C(dia) + O_2(g) & \Delta H° = +395.41\ kJ \\ \hline C(gr) \longrightarrow C(dia) & \Delta H° = +1.90\ kJ \end{array}$$

6.45 The first reaction is doubled, reversed, and added to the second to give the desired total reaction:

$$\begin{array}{ll} 2[SO_2(g) \longrightarrow S(s) + O_2(g)] & (2)[+296.83\ kJ] \\ 2\ S(s) + 3\ O_2(g) \longrightarrow 2\ SO_3(g) & -791.44\ kJ \\ \hline 2\ SO_2(g) + O_2(g) \longrightarrow 2\ SO_3(g) & \end{array}$$

$\Delta H° = (2)(+296.83\ kJ \cdot mol^{-1}) - (791.44\ kJ \cdot mol^{-1}) = -197.78\ kJ$

6.47 The equations are arranged to add to the desired reaction. The second equation is multiplied by 4 and added to the first:

$$\begin{array}{ll} P_4(s) + 6\ Cl_2(g) \longrightarrow 4\ PCl_3(l) & -1278.8\ kJ \\ 4[PCl_3(l) + Cl_2(g) \longrightarrow PCl_5(s)] & 4[-124\ kJ] \\ \hline P_4(s) + 10\ Cl_2(g) \longrightarrow 4\ PCl_5(s) & \end{array}$$

$\Delta H° = -1278.8\ kJ + 4\ (-124\ kJ) = -1775\ kJ$ or $-1.775\ MJ$

6.49 First, write the balanced equations for the reaction given:

$$\begin{array}{ll} C_2H_2(g) + \frac{5}{2}\ O_2(g) \longrightarrow 2\ CO_2(g) + H_2O(l) & \Delta H° = -1300\ kJ \\ C_2H_6(g) + \frac{7}{2}\ O_2(g) \longrightarrow 2\ CO_2(g) + 3\ H_2O(l) & \Delta H° = -1560\ kJ \\ H_2(g) + \frac{1}{2}\ O_2(g) \longrightarrow H_2O(l) & \Delta H° = -286\ kJ \end{array}$$

The second equation is reversed and added to the first, plus two times the third:

$$\begin{array}{ll} C_2H_2(g) + \frac{5}{2}\ O_2(g) \longrightarrow 2\ CO_2(g) + H_2O(l) & \Delta H° = -1300\ kJ \\ 2\ CO_2(g) + 3\ H_2O(l) \longrightarrow C_2H_6(g) + \frac{7}{2}\ O_2(g) & \Delta H° = +1560\ kJ \\ 2[H_2(g) + \frac{1}{2}\ O_2(g) \longrightarrow H_2O(l)] & 2[\Delta H° = -286\ kJ] \\ \hline C_2H_2(g) + 2\ H_2(g) \longrightarrow C_2H_6(g) & \end{array}$$

$\Delta H° = -1300\ kJ + 1560\ kJ + 2\ (-286\ kJ) = -312\ kJ$

6.51 The desired reaction may be obtained by reversing the first reaction and multiplying it by 2, reversing the second reaction, and adding these to the third:

$$\begin{array}{ll} 2[NH_4Cl(s) \longrightarrow NH_3(g) + HCl(g)] & 2[\Delta H° = +176.0\ kJ] \\ 2\ NH_3(g) \longrightarrow N_2(g) + 3\ H_2(g) & \Delta H° = +92.22\ kJ \\ N_2(g) + 4\ H_2(g) + Cl_2(g) \longrightarrow 2\ NH_4Cl(s) & \Delta H° = -628.86\ kJ \\ \hline H_2(g) + Cl_2(g) \longrightarrow 2\ HCl(g) & \end{array}$$

$\Delta H° = 2(+176.0\ kJ) + 92.22\ kJ - 628.86\ kJ = -184.6\ kJ$

6.53 (a) $K(s) + \frac{1}{2} Cl_2(g) + \frac{3}{2} O_2(g) \longrightarrow KClO_3(s)$ $\Delta H° = -397.73$ kJ

(b) $2\ C(gr) + \frac{5}{2} H_2(g) + \frac{1}{2} N_2(g) + O_2(g)$

$\longrightarrow H_2NCH_2COOH(s)$ $\Delta H° = -532.9$ kJ

(c) $2\ Al(s) + \frac{3}{2} O_2(g) \longrightarrow Al_2O_3(s)$ $\Delta H° = -1675.7$ kJ

6.55 From Appendix 2A, $\Delta H°_f$ (NO) = +90.25 kJ.

The reaction we want is

$N_2(g) + \frac{5}{2} O_2(g) \longrightarrow N_2O_5(g)$

Adding the first reaction to half of the second gives

$2\ NO(g) + O_2(g) \longrightarrow 2\ NO_2(g)$ $\Delta H° = -114.1$ kJ

$2\ NO_2(g) + \frac{1}{2} O_2(g) \longrightarrow N_2O_5(g)$ $\Delta H° = -55.1$ kJ

$2\ NO(g) + \frac{3}{2} O_2(g) \longrightarrow N_2O_5(g)$ -169.2 kJ

The enthalpy of this reaction equals the enthalpy of formation of N_2O_5 (g) minus twice the enthalpy of formation of NO, so we can write

$-169.2\ kJ = \Delta H°_f (N_2O_5) - 2(+90.25\ kJ)$

$\Delta H°_f (N_2O_5) = +11.3\ kJ$

6.57 The enthalpy of the reaction

$PCl_3(l) + Cl_2(g) \longrightarrow PCl_5(s)$ $\Delta H° = -124$ kJ

is $\Delta H°_r = \Sigma\ \Delta H°_f$ (products) $- \Sigma\ \Delta H°_f$ (reactants)

$-124\ kJ = \Delta H°_f (PCl_5, s) - \Delta H°_f (PCl_3, l)$

Remember that the standard enthalpy of formation of $Cl_2(g)$ will be 0 by definition because this is an element in its reference state. From the Appendix we find that $\Delta H°_f (PCl_3, l) = -319.7\ kJ \cdot mol^{-1}$.

$-124\ kJ = \Delta H°_f (PCl_5, s) - (-319.7\ kJ)$

$\Delta H°_f (PCl_5, s) = -444\ kJ \cdot mol^{-1}$

6.59 (a) The balanced equation is

$SO_2(g) + \frac{1}{2} O_2(g) \longrightarrow SO_3(g)$

From the Appendix we find that the standard enthalpies of formation of $SO_2(g)$ and $SO_3(g)$ are $-296.83\ kJ \cdot mol^{-1}$ and $-395.72\ kJ \cdot mol^{-1}$, respectively. The enthalpy of the reaction is obtained from Hess's law:

$\Delta H°_r = \Sigma\ \Delta H°_f$ (products) $- \Sigma\ \Delta H°_f$ (reactants)

$= \Delta H°_f (SO_3, g) - \Sigma\ \Delta H°_f (SO_2, g)$*

$= -395.72\ kJ \cdot mol^{-1} - (296.83\ kJ \cdot mol^{1})$

$= -98.89\ kJ \cdot mol^{-1}$

---

*remember that the enthalpy of formation of elements in their most stable states is 0 by definition

The enthalpy of reactions for 10.0 g of $SO_2(g)$ will be given by

$$\Delta H = \left(\frac{10.0\ \text{g SO}_2}{64.06\ \text{g}\cdot\text{mol}^{-1}\ \text{SO}_2}\right)(-98.89\ \text{kJ}\cdot\text{mol}^{-1}\ \text{SO}_2) = -15.4\ \text{kJ}$$

(b) The balanced equation is

$$CuO(s) + H_2(g) \longrightarrow Cu(s) + H_2O(l)$$

The enthalpies of formation needed are those of $H_2O(l)$ and $CuO(s)$:

$$\begin{aligned}\Delta H^\circ_r &= \Sigma\, \Delta H^\circ_f\,(\text{products}) - \Sigma\, \Delta H^\circ_f\,(\text{reactants})\\ &= \Delta H^\circ_f\,(H_2O, l) - \Sigma\, \Delta H^\circ_f\,(CuO, S)^*\\ &= -285.83\ \text{kJ}\cdot\text{mol}^{-1} - (-157.3\ \text{kJ}\cdot\text{mol})\\ &= -128.5\ \text{kJ}\cdot\text{mol}^{-1}\end{aligned}$$

Therefore, for 1.00 mol, the enthalpy change will be −128.5 kJ.

6.61 (a) $3\ NO_2(g) + H_2O(l) \longrightarrow 2\ HNO_3(aq) + NO(g)$

$$\begin{aligned}\Delta H^\circ_r &= 2\ \Delta H^\circ_f\,(HNO_3, aq) + \Delta H^\circ_f\,(NO, g - [3\ \Delta H^\circ_f\,(NO_2, g) + \Delta H^\circ_f\,(H_2O, l)]\\ &= 2(-207.36\ \text{kJ}\cdot\text{mol}^{-1}) + 90.25\ \text{kJ}\cdot\text{mol}^{-1}\\ &\qquad - [3(+33.18\ \text{kJ}\cdot\text{mol}^{-1}) + (-285.83\ \text{kJ}\cdot\text{mol}^{-1})]\\ &= -138.18\ \text{kJ}\end{aligned}$$

(b) $B_2O_3(s) + 3\ CaF_2(s) \longrightarrow 2\ BF_3(g) + 3\ CaO(s)$

$$\begin{aligned}\Delta H^\circ_r &= 2\ \Delta H^\circ_f\,(BF_3, g) + 3\ \Delta H^\circ_f\,(CaO, s) - [\Delta H^\circ_f\,(B_2O_3, s) + \Delta H^\circ_f\,(CaF_2, s)]\\ &= 2(-1137.0\ \text{kJ}\cdot\text{mol}^{-1}) + 3(-635.09\ \text{kJ}\cdot\text{mol}^{-1})\\ &\qquad - [-1272.8\ \text{kJ}\cdot\text{mol}^{-1} + 3(-1219.6\ \text{J}\cdot\text{mol}^{-1})]\\ &= +752.3\ \text{kJ}\end{aligned}$$

(c) $H_2S(aq) + 2\ KOH(aq) \longrightarrow K_2S(aq) + 2\ H_2O(l)$

$$\begin{aligned}\Delta H^\circ_r &= \Delta H^\circ_f\,(K_2S, aq) + 2\ \Delta H^\circ_f\,(H_2O, l) - [\Delta H^\circ_f\,(H_2S, aq) + 2\ \Delta H^\circ_f\,(KOH, aq)]\\ &= -417.5\ \text{kJ}\cdot\text{mol}^{-1} + 2(-285.83\ \text{kJ}\cdot\text{mol}^{-1})\\ &\qquad - [-39.7\ \text{kJ}\cdot\text{mol}^{-1} + 2\,(-482.37\ \text{kJ}\cdot\text{mol}^{-1})]\\ &= -38.7\ \text{kJ}\end{aligned}$$

6.63 The lattice energy corresponds to the process

$$CaS(s) \longrightarrow Ca^{2+}(g) + S^{2-}(g)$$

To calculate this, we set up a Born-Haber cycle, which is essentially an application of Hess's law. Perhaps the best way to do this is to write out the reactions that

*remember that the enthalpy of formation of elements in their most stable states is 0 by definition

correspond to each of the pieces of data that we have and then add them up to give the net reaction we want.

$Ca(s) \longrightarrow Ca(g)$	+178 kJ
$S(s) \longrightarrow S(g)$	+279 kJ
$CaS(s) \longrightarrow Ca(s) + S(s)$	+482 kJ
$Ca(g) \longrightarrow Ca^{+}(g) + e^{-}$	+590 kJ
$Ca^{+}(g) \longrightarrow Ca^{2+}(g) + e^{-}$	+1150 kJ
$S(g) + e^{-} \longrightarrow S^{-}(g)$	−200 kJ
$S^{-}(g) + e^{-} \longrightarrow S^{2-}(g)$	+532 kJ
$CaS(s) \longrightarrow Ca^{2+}(g) + S^{2-}(g)$	+3011 kJ

6.65 (a) For the reaction $Na_2O(s) \longrightarrow 2\,Na^{+}(g) + O^{2-}(g)$

$$\Delta H_L = 2\,\Delta H^\circ_f(\text{Na, g}) + \Delta H^\circ_f(\text{O, g}) + 2\,I_1(\text{Na}) - E_{ea1}(\text{O}) - E_{ea2}(\text{O}) - \Delta H_f(\text{Na}_2\text{O(s)l})$$

$$\Delta H_L = 2(107.32\text{ kJ}\cdot\text{mol}^{-1}) + 249\text{ kJ}\cdot\text{mol}^{-1} + 2(494\text{ kJ}\cdot\text{mol}^{-1}) - 141\text{ kJ}\cdot\text{mol}^{-1} + 844\text{ kJ}\cdot\text{mol}^{-1} + 409\text{ kJ}\cdot\text{mol}^{-1}$$

$$\Delta H_L = 2564\text{ kJ}\cdot\text{mol}^{-1}$$

(b) For the reaction $AlCl_3(s) \longrightarrow Al^{3+}(g) + 3\,Cl^{-}(g)$

$$\Delta H_L = \Delta H^\circ_f(\text{Al, g}) + 3\,\Delta H^\circ_f(\text{Cl, g}) + I_1(\text{Al}) + I_2(\text{Al}) + I_3(\text{Al}) - E_{ea}(\text{Cl}) - \Delta H_f(\text{AlCl}_3\text{, s})$$

$$\Delta H_L = 326\text{ kJ}\cdot\text{mol}^{-1} + 3(121\text{ kJ}\cdot\text{mol}^{-1}) + 577\text{ kJ}\cdot\text{mol}^{-1} + 1817\text{ kJ}\cdot\text{mol}^{-1} + 2744\text{ kJ}\cdot\text{mol}^{-1} - 349\text{ kJ}\cdot\text{mol}^{-1} + 704.2\text{ kJ}\cdot\text{mol}^{-1}$$

$$\Delta H_L = 6182\text{ kJ}\cdot\text{mol}^{-1}$$

6.67 The process can be broken down into the following steps:

	ΔH°, kJ
$Mg^{+}(g) + Cl^{-}(g) \longrightarrow MgCl(s)$	−717*
$Mg(s) \longrightarrow Mg(g)$	+148
$Mg(g) \longrightarrow Mg^{+}(g) + e^{-}$	+736
$\frac{1}{2}\,Cl_2(g) \longrightarrow Cl(g)$	+122
$Cl(g) + e^{-} \longrightarrow Cl^{-}(g)$	−349
$Mg(s) + \frac{1}{2}\,Cl_2(g) \longrightarrow MgCl(s)$	$\Delta H_f = -60$

* from the assumption that ΔH_L is the same as that of KCl

	$\Delta H°$, kJ
$Mg^{+}(g) + 2\ Cl^{-}(g) \longrightarrow MgCl_2(s)$	-2524
$Mg(s) \longrightarrow Mg(g)$	$+148$
$Mg(g) \longrightarrow Mg^{2+}(g) + 2e^{-}$	$+2186$
$Cl_2(g) \longrightarrow 2\ Cl(g)$	$+244$
$2(Cl(g) + e^{-} \longrightarrow Cl^{-}, g)$	$2(-349)$
$Mg(s) + Cl_2(g) \longrightarrow MgCl_2(s)$	$\Delta H_f = -644$

Furthermore, for $2\ MgCl(s) \longrightarrow MgCl_2(s) + Mg(s)$:

$\Delta H° = -644\ kJ - 2(-60\ kJ) = -524\ kJ$

Thus, $MgCl_2$ is the far more likely chloride of magnesium.

6.69

(a)	break:	1 mol N—N triple bonds	$(1\ mol)(944\ kJ \cdot mol^{-1})$
		3 mol F—F bonds	$(3\ mol)(158\ kJ \cdot mol^{-1})$
	form:	6 mol N—F bonds	$(6\ mol)(-195\ kJ \cdot mol^{-1})$
		Total	+248 kJ
(b)	break:	1 mol C=C bonds	$(1\ mol)(612\ kJ \cdot mol^{-1})$
		1 mol O—H bonds	$(1\ mol)(463\ kJ \cdot mol^{-1})$
	form:	1 mol C—C bonds	$-(1\ mol)(348\ kJ \cdot mol^{-1})$
		1 mol C—O bonds	$-(1\ mol)(360\ kJ \cdot mol^{-1})$
		1 mol C—H bonds	$-(1\ mol)(412\ kJ \cdot mol^{-1})$
		Total	−45 kJ
(c)	break:	1 mol C—H bonds	$(1\ mol)(412\ kJ \cdot mol^{-1})$
		1 mol Cl—Cl bonds	$(1\ mol)(242\ kJ \cdot mol^{-1})$
	form:	1 mol C—Cl bonds	$-(1\ mol)(338\ kJ \cdot mol^{-1})$
		1 mol H—Cl bonds	$-(1\ mol)(431\ kJ \cdot mol^{-1})$
		Total	−115 kJ

6.71 The value that we want is given simply by the difference between having 3 isolated C=C bonds with 3 isolated C—C single bonds, versus 6 of the resonance-stabilized bonds:

3 C=C bonds + 3 C—C bonds = 3(348 kJ) + 3(612 kJ) = 2880 kJ

6 resonance-stabilized bonds = 6(518 kJ) = 3108 kJ

As can be seen, the 6 resonance-stabilized bonds are more stable by ca. 228 kJ.

6.73 (a) The enthalpy of vaporization is the enthalpy change associated with the conversion $C_6H_6(l) \longrightarrow C_6H_6(g)$ at constant pressure. The value at 298.2 K will be given by

$$\Delta H^\circ_{\text{vaporization at 298 K}} = \Delta H^\circ_f(C_6H_6, g) - \Delta H^\circ_f(C_6H_6, l)$$
$$= 82.93\ kJ\cdot mol^{-1} - (49.0\ kJ\cdot mol^{-1})$$
$$= 33.93\ kJ\cdot mol^{-1}$$

(b) In order to take into account the difference in temperature, we need to use the heat capacities of the reactants and products in order to raise the temperature of the system to 353.2 K. We can rewrite the reactions as follows, to emphasize temperature, and combine them according to Hess's law:

$$C_6H_6(l)_{\text{at 298 K}} \longrightarrow C_6H_6(g)_{\text{at 298 K}} \quad \Delta H^\circ = 33.93\ kJ$$
$$C_6H_6(l)_{\text{at 298 K}} \longrightarrow C_6H_6(l)_{\text{at 353.2 K}} \quad \Delta H^\circ = (1\ mol)(353.2\ K - 298.2\ K)(136.1\ J\cdot mol^{-1}\cdot K^{-1}) = 7.48\ kJ$$
$$C_6H_6(g)_{\text{at 298 K}} \longrightarrow C_6H_6(g)_{\text{at 353.2 K}} \quad \Delta H^\circ = (1\ mol)(353.2\ K - 298.2\ K)(81.67\ J\cdot mol^{-1}\cdot K^{-1}) = 4.49\ kJ$$

To add these together to get the overall equation at 353.2 K, we must reverse the second equation:

$$C_6H_6(l)_{\text{at 298 K}} \longrightarrow C_6H_6(g)_{\text{at 298 K}} \quad \Delta H^\circ = 33.93\ kJ$$
$$C_6H_6(l)_{\text{at 353.2 K}} \longrightarrow C_6H_6(l)_{\text{at 298 K}} \quad \Delta H^\circ = -7.48\ kJ$$
$$C_6H_6(g)_{\text{at 298 K}} \longrightarrow C_6H_6(g)_{\text{at 353.2 K}} \quad \Delta H^\circ = 4.49\ kJ$$
$$\overline{C_6H_6(l)_{\text{at 353.2 K}} \longrightarrow C_6H_6(g)_{\text{at 353.2 K}} \quad \Delta H^\circ = 30.94\ kJ}$$

(c) The value in the table is 30.8 $kJ\cdot mol^{-1}$ for the enthalpy of vaporization of benzene. The value is close to that calculated as corrected by heat capacities. At least part of the error can be attributed to the fact that heat capacities are not strictly constant with temperature.

6.75 (a) and (b) $w = \Delta U - q = 400\ J - 600\ J = -200\ J$. Work was done by the system, as indicated by the negative value.

6.77 This process involves 5 separate steps: (1) raising the temperature of the ice from −30.27°C to 0.00°C, (2) melting the ice at 0.00°C, (3) raising the temperature of the liquid water from 0.00°C to 100.00°C, (4) vaporizing the water at 100.00°C, and (5) raising the temperature of the water vapor from 100.00°C to 150.35°C.

Step 1: $\Delta H = (27.96\ g)(2.03\ J\cdot(°C)^{-1}\cdot g^{-1})(0.00°C - (-30.27°C)) = 1.72\ kJ$

Step 2: $\Delta H = \left(\dfrac{27.96\ g}{18.02\ g\cdot mol^{-1}}\right)(6.01\ kJ\cdot mol^{-1}) = 9.32\ kJ$

Step 3: $\Delta H = (27.96\ g)(4.18\ J\cdot(°C)^{-1}\cdot g^{-1})(100.00°C - 0.00°C) = 11.7\ kJ$

Step 4: $\Delta H = \left(\dfrac{27.96\text{ g}}{18.02\text{ g}\cdot\text{mol}^{-1}}\right)(40.7\text{ kJ}\cdot\text{mol}^{-1}) = 63.1\text{ kJ}$

Step 5: $\Delta H = (27.96\text{ g})(2.01\text{ J}\cdot(°\text{C})^{-1}\cdot\text{g}^{-1})(150.35°\text{C} - 100.00°\text{C}) = 2.83\text{ kJ}$

The total heat required $= 1.72\text{ kJ} + 9.32\text{ kJ} + 11.7\text{ kJ} + 63.1\text{ kJ} + 2.83\text{ kJ}$

$= 88.7\text{ kJ}$

6.79 (a) mass of fat burned in 1 hour $= \dfrac{2000\text{ kJ}}{38\text{ kJ}\cdot\text{g}^{-1}} = 53\text{ g}$

(b) he would need to run $\dfrac{500\text{ g}}{53\text{ g}\cdot\text{hr}^{-1}} = 9.4\text{ hr}$

6.81 First, we need to calculate how much energy from the sunshine will be hitting the surface of the ethanol, so we convert the rate to $\text{kJ}\cdot\text{cm}^{-2}\cdot\text{s}^{-1}$:

$$1\text{ kJ}\cdot\text{m}^{-2}\cdot\text{s}^{-1}\left(\frac{1\text{ m}}{100\text{ cm}}\right)^2 = 1\times 10^{-4}\text{ kJ}\cdot\text{cm}^{-2}\cdot\text{s}^{-1}$$

$$(1\times 10^{-4}\text{ kJ}\cdot\text{cm}^{-2}\cdot\text{s}^{-1})(50.0\text{ cm}^2)\left(10\text{ min}\times\frac{60\text{ s}}{\text{min}}\right) = 3\text{ kJ}$$

The enthalpy of vaporization of ethanol is 43.5 $\text{kJ}\cdot\text{mol}^{-1}$ (see Table 6.2). We will assume that the enthalpy of vaporization is approximately the same at ambient conditions as it would be at the boiling point of ethanol.

$$\left(\frac{3\text{ kJ}}{43.5\text{ kJ}\cdot\text{mol}^{-1}}\right)(46.07\text{ g}\cdot\text{mol}^{-1}) = 3\text{ g}$$

6.83 The heat gained by the water in the ice cube will be equal to the heat lost by the initial sample of hot water. The enthalpy change for the water in the ice cube will be composed of two terms: the heat to melt the ice at 0°C and the heat required to raise the ice from 0°C to the final temperature.

$$\begin{aligned}\text{heat (ice cube)} &= \left(\frac{50.0\text{ g}}{18.02\text{ g}\cdot\text{mol}^{-1}}\right)(6.01\times 10^3\text{ J}\cdot\text{mol}^{-1})\\ &\quad + (50.0\text{ g})(4.184\text{ J}\cdot(°\text{C})^{-1}\cdot\text{g}^{-1})(T_f - 0°)\\ &= 1.67\times 10^4\text{ J} + (209\text{ J}\cdot(°\text{C})^{-1})(T_f - 0°)\\ \text{heat (water)} &= (400\text{ g})(4.184\text{ J}\cdot(°\text{C})^{-1}\cdot\text{g}^{-1})(T_f - 45°)\\ &= (1.67\times 10^3\text{ J}\cdot(°\text{C})^{-1})(T_f - 45°)\end{aligned}$$

Setting these equal:

$$-(1.67\times 10^3\text{ J}\cdot(°\text{C})^{-1})T_f + 7.5\times 10^4\text{ J} = 1.67\times 10^4\text{ J} + (209\text{ J}\cdot(°\text{C})^{-1})T_f$$

Solving for T_f:

$$T_f = \frac{5.8\times 10^4\text{ J}}{1.88\times 10^3\text{ J}\cdot(°\text{C})^{-1}} = 31\text{ °C}$$

6.85 (a) $CH_3OH(l) + \frac{3}{2} O_2(g) \longrightarrow CO_2(g) + 2 H_2O(l)$

(b) Assuming no heat is lost to the surroundings, we may write:

total heat = heat (water) + heat (pyrex) + heat (methanol combustion) = 0

$$(209 \text{ g } H_2O)(4.184 \text{ J}\cdot(°C)^{-1}\cdot g^{-1})(100°C - 20°C)$$
$$+ (50.0 \text{ g Pyrex})(0.78 \text{ J}\cdot(°C)^{-1}\cdot g^{-1})(100°C - 20°C)$$
$$+ (-726 \times 10^3 \text{ J}\cdot mol^{-1})\left(\frac{m_{methanol}}{32.04 \text{ g}\cdot mol^{-1} \text{ methanol}}\right) = 0$$

$m_{methanol} = 3.23$ g

(c) The enthalpy of the combustion reaction will be

$$\Delta H°_{combustion} = \Delta H°_f (CO_2, g) + 2 \Delta H°_f (H_2O, l) - \Delta H°_f (\text{methanol})$$
$$-726 \text{ kJ} = (1)(-393.51 \text{ kJ}\cdot mol^{-1}) + (2)(-285.83 \text{ kJ}\cdot mol^{-1}) - (1)(\Delta H°_f (\text{methanol}))$$
$$\Delta H°_f (\text{methanol}) = -239 \text{ kJ}\cdot mol^{-1}$$

6.87 (a) The reaction enthalpy is obtained by Hess's law:

$$\Delta H°_r = \Delta H°_f (CO, g) - \Delta H°_f (H_2O, g)$$
$$\Delta H°_r = (1)(-110.53 \text{ kJ}\cdot mol^{-1}) - (1)(-241.82 \text{ kJ}\cdot mol^{-1})$$
$$\Delta H°_r = +131.29 \text{ kJ}\cdot mol^{-1}$$

endothermic

(b) The number of moles of H_2 produced is obtained from the ideal gas law:

$$n = \frac{PV}{RT} = \frac{\left(\frac{500 \text{ Torr}}{760 \text{ Torr}\cdot atm^{-1}}\right)(200 \text{ L})}{(0.082\,06 \text{ L}\cdot atm\cdot K^{-1}\cdot mol^{-1})(338 \text{ K})} = 4.74 \text{ mol}$$

The enthalpy change accompanying the production of this amount of hydrogen will be given by

$$\Delta H = (4.74 \text{ mol})(131.29 \text{ kJ}\cdot mol^{-1}) = 623 \text{ kJ}$$

6.89 $$\Delta H° = 12 \Delta H°_f(H_2O, l) - [4(\Delta H°_f(HNO_3, l) + 5(\Delta H°_f(N_2H_4, l)]$$
$$= 12(-285.83 \text{ kJ}\cdot mol^{-1}) - [4(-174.10 \text{ kJ}\cdot mol^{-1}) + 5 (+50.63 \text{ kJ}\cdot mol^{-1})]$$
$$= -2987 \text{ kJ}\cdot mol^{-1}$$

6.91 (a) The enthalpy of combustion of $H_2(g)$ is the same as the enthalpy of formation of liquid water: $\Delta H°_f (H_2O, l) = -285.83$ kJ·mol^{-1}, whereas the enthalpy of combustion of octane is -5471 kJ·mol^{-1}. On a per gram basis, the heat released per gram of H_2 is

$$-285.83 \text{ kJ}\cdot mol^{-1} \div 2.0158 \text{ g}\cdot mol^{-1} = -142 \text{ kJ}\cdot g^{-1}$$

whereas that of octane is

$$-5471 \text{ kJ}\cdot mol^{-1} \div 114.23 \text{ g}\cdot mol^{-1} = -47.89 \text{ kJ}\cdot g^{-1}$$

Hydrogen produces more heat on a per gram basis as does the organic liquid, and it would be significantly cleaner to burn.

(b) The use of hydrogen as a fuel is limited by its storage problems. Normally, hydrogen is a gas, and to use it as a fuel requires carrying a large quantity of hydrogen under pressure. Much research is aimed at creating safe but also economical hydrogen storage tanks (certain metals such as Pd absorb hydrogen). Safety is a major issue because $H_2(g)$ can react explosively with air. This hazard is a serious concern for automobiles involved in accidents.

6.93 (a) First we must balance the chemical reaction:

$C_6H_6(l) + \frac{15}{2}\,O_2(g) \longrightarrow 6\,CO_2(g) + 3\,H_2O(g)$

For 1 mol $C_6H_6(l)$ burned, the change in the number of moles of gas is

$(9.00 - 7.50)\text{ mol} = +2.50\text{ mol} = \Delta n$

$$w = -P\Delta V = -P\left(\frac{\Delta nRT}{P}\right) = -\Delta nRT$$

$w = -(+2.50\text{ mol})(8.314\text{ J}\cdot\text{K}^{-1}\cdot\text{mol}^{-1})(298\text{ K}) = -6.19 \times 10^3\text{ J} = -6.19\text{ kJ}$

(b) $\Delta H_c = 6(-393.51\text{ kJ}\cdot\text{mol}^{-1}) + 3(-241.82\text{ kJ}\cdot\text{mol}^{-1}) - (+49.0\text{ kJ}\cdot\text{mol}^{-1})$

$= -3135.5\text{ kJ}$

(c) $\Delta U° = \Delta H° + w = (-3135.5 - 6.19)\text{ kJ} = -3141.7\text{ kJ}$

6.95 (a) The heat given off by the reaction, which was absorbed by the calorimeter, is given by

$\Delta H = -(525.0\text{ J}\cdot(°\text{C})^{-1})(20.0°\text{C} - 18.6°\text{C}) = -0.74\text{ kJ}$

This, however, is not all the heat produced, as the 100.0 mL of solution resulting from mixing also absorbed some heat. If we assume that the volume of NaOH and HNO_3 are negligible compared to the volume of water present and that the density of the solution is $1.00\text{ g}\cdot\text{mL}^{-1}$, then the change in heat of the solution is given by

$\Delta H = -(20.0°\text{C} - 18.6°\text{C})(4.18\text{ J}\cdot(°\text{C})^{-1}\cdot\text{g}^{-1})(100.0\text{ g}) = -0.59\text{ kJ}$

The total heat given off will be $-0.74\text{ kJ} + (-0.59\text{ kJ}) = -1.33\text{ kJ}$

(b) This heat is for the reaction of $(0.500\text{ M})(0.0500\text{ L}) = 0.0250\text{ mol } HNO_3$, so the amount of heat produced per mole of HNO_3 will be given by

$$\frac{-1.33\text{ kJ}}{0.0250\text{ mol}} = -53.2\text{ kJ}\cdot\text{mol}^{-1}$$

6.97 (a)

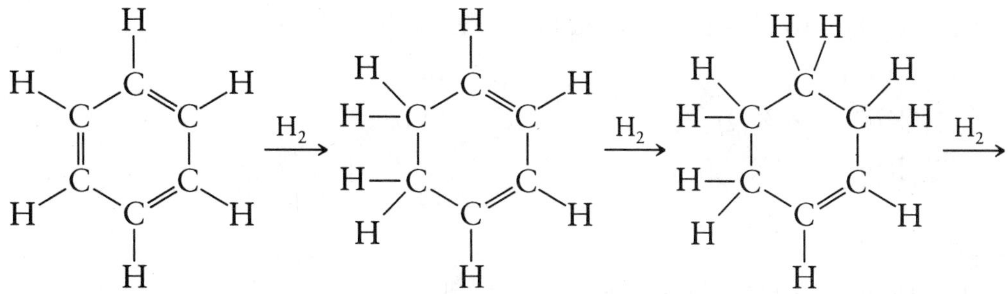

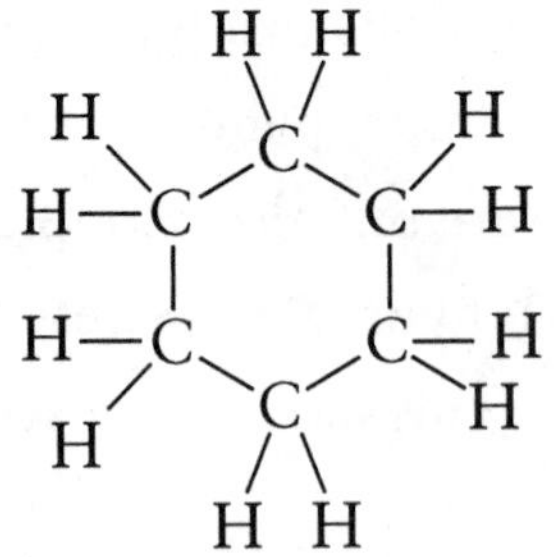

(b) From bond enthalpies, each step is identical, as the number and types of bonds broken and formed are the same:

break:	1 mol C=C bonds	612 kJ
	1 mol H—H bonds	436 kJ
form:	1 mol C—C bonds	−348 kJ
	2 mol C—H bonds	2(−412 kJ)
Total:		−124 kJ

The total energy change should be equal to the sum of the three steps or $3(-124\text{ kJ}) = -372\text{ kJ}$.

(c) The Hess's law calculation using standard enthalpies of formation is easily performed on the composite reaction:

$$C_6H_6(l) + 3\,H_2(g) \longrightarrow C_6H_{12}(l)$$

$$\begin{aligned}\Delta H^\circ_r &= \Sigma\,\Delta H^\circ_f\,(\text{products}) - \Sigma\,\Delta H^\circ_f\,(\text{reactants})\\ &= \Delta H^\circ_f\,(\text{cyclohexane}) - \Delta H^\circ_f\,(\text{benzene})^*\\ &= -156.4\text{ kJ}\cdot\text{mol}^{-1} - (+49.0\text{ kJ}\cdot\text{mol}^{-1})\\ &= -205.4\text{ kJ}\cdot\text{mol}^{-1}\end{aligned}$$

(d) The hydrogenation of benzene is much less exothermic than predicted by bond enthalpy estimations. Part of this difference can be due to the inherent inaccuracy of using average values, but the difference is so large that this cannot be the complete explanation. As may be expected, the resonance energy of benzene makes it more stable than would be expected by treating it as a set of 3 isolated double and 3 isolated single bonds. The difference in these two values ($-205\text{ kJ} - (-372\text{ kJ}) = 167\text{ kJ}$) is a measure of how much more stable benzene is than the Kekulé structure would predict.

6.99 Equations for these various processes can be rearranged and summed according to Hess's law to give the desired lattice enthalpy of KBr(s).

Reaction	ΔH°
(1) $K(s) \longrightarrow K(g)$	+89.2 kJ
(2) $K(g) \longrightarrow K^+(g) + e^-$	+425.0 kJ
(3) $Br_2(l) \longrightarrow Br_2(g)$	+30.9

*remember that $\Delta H^\circ_f\,(H_2, g) = 0$ by definition

(4) $Br_2(g) \longrightarrow 2\ Br(g)$ +192.9

(5) $Br(g) + e^- \longrightarrow Br^-(g)$ −331.0

(6) $K(s) + \frac{1}{2} Br_2(l) \longrightarrow KBr(s)$ −394

Multiply (3) and (4) by $\frac{1}{2}$, reverse (6) and then sum:

$K(s) \longrightarrow K(g)$	+89.2 kJ
$K(g) \longrightarrow K^+(g) + e^-$	+425.0 kJ
$\frac{1}{2} Br_2(l) \longrightarrow \frac{1}{2} Br_2(g)$	$\frac{1}{2}$(+30.9 kJ)
$\frac{1}{2} Br_2(g) \longrightarrow Br(g)$	$\frac{1}{2}$(+192.9 kJ)
$Br(g) + e^- \longrightarrow Br^-(g)$	−331.0 kJ
$KBr(s) \longrightarrow K(s) + \frac{1}{2} Br_2(l)$	+394 kJ
$KBr(s) \longrightarrow K^+(g) + Br^-(g)$	689 $kJ \cdot mol^{-1}$

6.101 (a) The combustion reaction is

$C_{60}(s) + 60\ O_2(g) \rightarrow 60\ CO_2(g)$

The enthalpy of formation of $C_{60}(s)$ will be given by

$\Delta H^\circ_c = 60\ \Delta H^\circ_f(CO_2, g) - \Delta H^\circ_f(C_{60}, s)$

$-25\ 937\ kJ = 60\ mol \times (-393.51\ kJ \cdot mol^{-1}) - \Delta H^\circ_f(C_{60}, s)$

$\Delta H^\circ_f(C_{60}, s) = +2326\ kJ \cdot mol^{-1}$

(b) The bond enthalpy calculation is

$60\ C(gr) \longrightarrow 60\ C(g)$	$(60)(+717\ kJ \cdot mol^{-1})$
Form 60 mol C—C bonds	$-60\ (348\ kJ \cdot mol^{-1})$
Form 30 mol C=C bonds	$-30\ (612\ kJ \cdot mol^{-1})$
$C_{60}(g) \longrightarrow C_{60}(s)$	−233 kJ
$60\ C(gr) \longrightarrow C_{60}(s)$	+3547 kJ

(c) From the experimental data, the enthalpy of formation of C_{60} shows that it is *more* stable by (3547 kJ − 2326 kJ) = 1221 kJ than predicted by the isolated bond model.

(d) 1221 kJ ÷ 60 = 20 kJ per carbon atom

(e) 150 kJ ÷ 6 = 25 kJ per carbon atom

(f) Although the comparison of the stabilization of benzene with that of C_{60} should be treated with caution, it does appear that there is slightly less stabilization per carbon atom in C_{60} than in benzene. This fits with expectations, as the C_{60} molecule is forced by its geometry to be curved. This means that the overlap of the *p*-orbitals, which gives rise to the delocalization that results in resonance, will not be as favorable as in the planar beneze molecule. Another perspective on this is obtained by noting that the C atoms in C_{60} are forced to be partially sp^3 hybridized because they cannot be rigorously planar as required by sp^2 hybridization.

CHAPTER 7
THERMODYNAMICS: THE SECOND AND THIRD LAWS

7.1 (a) HBr because Br is more massive and contains more elementary particles than F in HF; (b) NH_3 because it has greater complexity, being a molecule rather than a single atom; (c) I_2(l) because molecules in liquids are more randomly oriented than molecules in solids; (d) 1.0 mol Ar(g) at 1.00 atm because it will occupy a larger volume than 1.0 mol of Ar(g) at 2.00 atm.

7.3 It is easy to order H_2O in its various phases because entropy will increase when going from a solid to a liquid to a gas. The main question concerns where to place C(s) in this order, and that will essentially become a question of whether C(s) should have more or less entropy than H_2O(s), because we would automatically expect C(s) to have less entropy than any liquid. Because water is a molecular substance held together in the solid phase by weak hydrogen bonds and C(s), which we will take to be the standard state from graphite, the carbon is more rigidly held in place and will have less entropy. C(s) < H_2O(s) < H_2O(l) < H_2O(g)

7.5 (a) The number of moles of gas decreases, so we would expect entropy to decrease.
(b) The conversion of any substance from a solid to a gas should increase entropy due to the increased disorder in the gaseous state.
(c) Entropy should decrease because it decreases as temperature is lowered. At higher temperatures there is more molecular motion, which increases disorder.

7.7 (a) In the standard state, bromine is a liquid and iodine is a solid. If the two substances had the same state (i.e., both were gases or both liquids) we would expect iodine to have the higher entropy due to its larger mass and consequently larger number of fundamental particles. However, because the compounds are in different states, we would expect the liquid to have a higher entropy than the solid.
(b) When we consider the two structures, it is clear that pentene will have more flexibility in its framework than cyclopentane, which will be comparatively rigid. Therefore, we predict pentene to have a higher entropy.
(c) Ethene (or ethylene) is a gas and polyethylene is a solid, so we automatically

expect ethene to have a higher entropy. Also, for the same mass a sample of ethene will be composed of many small molecules, whereas polyethylene will be made up of fewer but larger molecules.

7.9 (a) $\Delta S° = \frac{q}{T} = \frac{\Delta H°}{T} = \frac{1.00 \text{ mol} \times (-6.01 \text{ kJ}\cdot\text{mol}^{-1})}{273.2 \text{ K}} = -22.0 \text{ J}\cdot\text{K}^{-1}$

(b) $\Delta S = \frac{q}{T} = \frac{\Delta H}{T} = \frac{\frac{50.0 \text{ g}}{46.07 \text{ g}\cdot\text{mol}^{-1}} \times 43.5 \text{ kJ}\cdot\text{mol}^{-1}}{351.5 \text{ K}} = +134 \text{ J}\cdot\text{K}^{-1}$

7.11 (a) $H_2(g) + \frac{1}{2} O_2(g) \longrightarrow H_2O(l)$

$\Delta S°_f = \Sigma S°_{\text{products}} - \Sigma S°_{\text{reactants}}$

$= 69.91 \text{ J}\cdot\text{K}^{-1}\cdot\text{mol}^{-1} - [130.68 \text{ J}\cdot\text{K}^{-1}\cdot\text{mol}^{-1} + \frac{1}{2}(205.14 \text{ J}\cdot\text{K}^{-1}\cdot\text{mol}^{-1})]$

$= -163.34 \text{ J}\cdot\text{K}^{-1}\cdot\text{mol}^{-1}$

The entropy change is negative because the number of moles of gas has decreased by 1.5. Note that the absolute entropies of the elements are not 0, and that the entropy change for the reaction in which a compound is formed from the elements is also not 0.

(b) $CO(g) + \frac{1}{2} O_2(g) \longrightarrow CO_2(g)$

$\Delta S°_r = 213.74 \text{ J}\cdot\text{K}^{-1}\cdot\text{mol}^{-1} - [197.67 \text{ J}\cdot\text{K}^{-1}\cdot\text{mol}^{-1} + \frac{1}{2}(205.14 \text{ J}\cdot\text{K}^{-1}\cdot\text{mol}^{-1})]$

$= -86.5 \text{ J}\cdot\text{K}^{-1}\cdot\text{mol}^{-1}$

The entropy change is negative because the number of moles of gas has decreased by 0.5.

(c) $CaCO_3(s) \longrightarrow CaO(s) + CO_2(g)$

$\Delta S°_r = 39.75 \text{ J}\cdot\text{K}^{-1}\cdot\text{mol}^{-1} + 213.74 \text{ J}\cdot\text{K}^{-1}\cdot\text{mol}^{-1} - [+92.9 \text{ J}\cdot\text{K}^{-1}\cdot\text{mol}^{-1}]$

$= +160.6 \text{ J}\cdot\text{K}^{-1}\cdot\text{mol}^{-1}$

The entropy change is positive because the number of moles of gas has increased by 1.

(d) $4 \text{ KClO}_3(s) \longrightarrow 3 \text{ KClO}_4(s) + \text{KCl}(s)$

$\Delta S°_r = 3(151.0 \text{ J}\cdot\text{K}^{-1}\cdot\text{mol}^{-1}) + 82.59 \text{ J}\cdot\text{K}^{-1}\cdot\text{mol}^{-1} - [4(143.1 \text{ J}\cdot\text{K}^{-1}\cdot\text{mol}^{-1})]$

$= -36.8 \text{ J}\cdot\text{K}^{-1}\cdot\text{mol}^{-1}$

It is not immediately obvious, but the four moles of solid products are more ordered than the four moles of solid reactants.

7.13 (a) The relationship to use is $dS = \frac{dq}{T}$. At constant pressure, we can substitute $dq = n\, C_P\, dT$:

$$dS = \frac{n\, C_p\, dT}{T}$$

Upon integration this gives $\Delta S = n\, Cp \ln \frac{T_2}{T_1}$. The answer is calculated by simply plugging in the known quantities. Remember that for an ideal monatomic gas $C_P = \frac{5}{2} R$:

$$\Delta S = (1.00 \text{ mol})(\tfrac{5}{2} \times 8.314 \text{ J}\cdot\text{K}^{-1}\cdot\text{mol}^{-1}) \ln \frac{408.6 \text{ K}}{328.9 \text{ K}} = 4.51 \text{ J}\cdot\text{K}^{-1}$$

(b) A similar analysis using C_V gives $\Delta S = n\, C_V \ln \frac{T_2}{T_1}$, where C_V for a monatomic ideal gas is $\frac{3}{2} R$:

$$\Delta S = (1.00 \text{ mol})(\tfrac{3}{2} \times 8.314 \text{ J}\cdot\text{K}^{-1}\cdot\text{mol}^{-1}) \ln \frac{408.6 \text{ K}}{328.9 \text{ K}} = 2.71 \text{ J}\cdot\text{K}^{-1}$$

7.15 Because the process is isothermal and reversible, the relationship $dS = \frac{dq}{T}$ can be used. Because the process is isothermal, $\Delta U = 0$ and hence $q = -w$, where $w = -P dV$. Making this substitution we obtain

$$dS = \frac{P\, dV}{T} = \frac{nRT}{TV}\, dV = \frac{nR}{V}\, dV$$

$$\therefore \Delta S = nR \ln \frac{V_2}{V_1}$$

Substituting the known quantities, we obtain

$$\Delta S = (9.73 \text{ mol})(8.314 \text{ J}\cdot\text{K}^{-1}\cdot\text{mol}^{-1}) \ln \frac{9.645 \text{ L}}{3.358 \text{ L}}$$

$$= 85.4 \text{ J}\cdot\text{K}^{-1}$$

7.17 CH_3D. Deuterium is similar in size to the proton, so its replacement of hydrogen in methane would not be expected to change the overall nature of the crystal lattice significantly; however, its presence would allow four different orientations in the solid, because the position of the D atom could be at any of the locations where the H atoms were in the original methane molecule. This residual disorder would give CH_3D a higher residual entropy at 0 K.

7.19 (a) The boiling point of a liquid may be obtained from the relationship $\Delta S_{vap} = \frac{\Delta H_{vap}}{T_B}$, or $T_B = \frac{\Delta H_{vap}}{\Delta S_{vap}}$. This relationship should be rigorously true if we have the actual enthalpy and entropy of vaporization. The data in the Appendix, however, are for 298 K. Thus, calculation of $\Delta H°_{vap}$ or $\Delta S°_{vap}$, using the enthalpy and entropy differences between the gas and liquid forms at 298 K, will give a good approximation of these quantities but the values will not be exact. For ethanal(l) ⟶ ethanal(g), the data in the appendix give

$$\Delta H_{vap} \cong -166.19\ kJ\cdot mol^{-1} - (-192.30)\ kJ\cdot mol^{-1} = 26.11\ kJ\cdot mol^{-1}$$

$$\Delta S_{vap} \cong 250.3\ J\cdot K^{-1}\cdot mol^{-1} - 160.2\ J\cdot K^{-1}\cdot mol^{-1} = 90.1\ J\cdot K^{-1}\cdot mol^{-1}$$

$$T_B = \frac{26.11 \times 10^3\ J\cdot mol^{-1}}{90.1\ J\cdot K^{-1}\cdot mol^{-1}} = 290\ K$$

(b) The boiling point of ethanal is 20.8°C or 293.9 K.

(c) These numbers are in very good agreement.

(d) Differences arise partly because the enthalpy and entropy of vaporization are slightly different from the values calculated at 298 K, but the boiling point of ethanal is not 298 K.

7.21 Trouton's rule indicates that the entropy of vaporization for a number of organic liquids is approximately 85 $J\cdot K^{-1}\cdot mol^{-1}$. Using this information and the relationship $T_B = \frac{\Delta H°_{vap}}{\Delta S°_{vap}} = \frac{21.51 \times 10^3\ J\cdot mol^{-1}}{85\ J\cdot K^{-1}\cdot mol^{-1}} = 253\ K$. The experimental boiling point of dimethyl ether is 248 K, which is in reasonably close agreement, given the nature of the approximation.

7.23 (a) The total entropy change is given by $\Delta S_{tot} = \Delta S_{surr} + \Delta S$. ΔS for an isothermal, reversible process is calculated from $\Delta S = \frac{q_{rev}}{T} = \frac{-w_{rev}}{T} = nR \ln \frac{V_2}{V_1}$. To do the calculation we need the value of n, which is obtained by use of the ideal gas law: $(3.67\ atm)(2.59\ L) = n(0.082\ 06\ L\cdot atm\cdot K^{-1}\cdot mol^{-1})(323\ K)$; $n = 0.359$ mol.

$\Delta S = (0.359\ mol)(8.314\ J\cdot K^{-1}\cdot mol^{-1}) \ln \frac{8.89\ L}{2.59\ L} = +3.68\ J\cdot K^{-1}$. Because the process is reversible, $\Delta S_{tot} = 0$, so $\Delta S_{surr} = -\Delta S = -3.68\ J\cdot K^{-1}$.

(b) For the irreversible process, ΔS is the same, $+3.68\ J\cdot K^{-1}$. No work is done in free expansion (see Section 6.6) so $w = 0$. Because $\Delta U = 0$, it follows that $q = 0$. Therefore, no heat is transferred into the surroundings, and their entropy is unchanged: $\Delta S_{surr} = 0$. The total change in entropy is therefore $\Delta S_{tot} = +3.68\ J\cdot K^{-1}$.

7.25 (a) $\Delta S_{surr} = \frac{-\Delta H}{T} = \frac{-(-1 \times 10^{-3}\ J)}{2 \times 10^{-7}\ K} = 5 \times 10^3\ J\cdot K^{-1}$

(b) $\Delta S_{surr} = \frac{-(-1\ J)}{310\ K} = 3 \times 10^{-3}\ J\cdot K^{-1}$

(c) $\Delta S_{\text{surr}} = \dfrac{-(-20\text{ J})}{3.5\text{ K}} = 5.7\text{ J}\cdot\text{K}^{-1}$

7.27 (a) rate of entropy generation $= \dfrac{\Delta S_{\text{surr}}}{time} = -\dfrac{q_{\text{rev}}}{time\cdot T}$

$$= -\frac{\text{rate of heat generation}}{T}$$

$$= \frac{-(100\text{ J}\cdot\text{s}^{-1})}{293\text{ K}} = 0.341\text{ J}\cdot\text{K}^{-1}\cdot\text{s}^{-1}$$

(b) $\Delta S_{\text{day}} = (0.341\text{ J}\cdot\text{K}^{-1}\cdot\text{s}^{-1})(60\text{ sec}\cdot\text{min}^{-1})(60\text{ min}\cdot\text{hr}^{-1})(24\text{ hr}\cdot\text{day}^{-1})$

$= 29.5\text{ kJ}\cdot\text{K}^{-1}\cdot\text{day}^{-1}$

(c) Less, because in the equation $\Delta S = \dfrac{-\Delta H}{T}$, if T is larger, ΔS is smaller.

7.29 (a) $\Delta S = \dfrac{q_{\text{rev}}}{T} = \dfrac{5\text{ J}}{298\text{ K}} = 0.02\text{ J}\cdot\text{K}^{-1}$

(b) $\Delta S = \dfrac{5\text{ J}}{373\text{ K}} = 0.01\text{ J}\cdot\text{K}^{-1}$

(c) The entropy change is smaller at higher temperatures, because the matter is already more chaotic. The same amount of heat has a greater effect on entropy changes when transferred at lower temperatures.

7.31 The change in entropy of the surroundings will be obtained from $\Delta S_{\text{surr}} = \dfrac{-\Delta H_{\text{system}}}{T}$. $\Delta S_{\text{surr}} = \dfrac{-1.96\text{ kJ}}{298\text{ K}} = -6.58\text{ J}\cdot\text{K}^{-1}$. The total change in entropy will be negative, so we predict the process to be nonspontaneous.

7.33 Exothermic reactions tend to be spontaneous because the result is an increase in the entropy of the surroundings. Using the mathematical relationship $\Delta G_r = \Delta H_r - T\Delta S_r$, it is clear that if ΔH_r is large compared to ΔS_r, then the reaction will generally be spontaneous.

7.35 (a) The change in entropy will be given by

$$\Delta S_{\text{surr}} = \frac{-\Delta H_{\text{system}}}{T} = \frac{-1.00\text{ mol} \times 8.2 \times 10^3\text{ J}\cdot\text{mol}^{-1}}{111.7\text{ K}} = -73\text{ J}\cdot\text{K}^{-1}$$

$$\Delta S_{\text{system}} = \frac{\Delta H_{\text{system}}}{T} = \frac{1.00\text{ mol} \times 8.2 \times 10^3\text{ J}\cdot\text{mol}^{-1}}{111.7\text{ K}} = +73\text{ J}\cdot\text{K}^{-1}$$

(b) $\Delta S_{surr} = \dfrac{-\Delta H_{system}}{T} = \dfrac{-1.00 \text{ mol} \times 4.60 \times 10^3 \text{ J}\cdot\text{mol}^{-1}}{158.7 \text{ K}} = -29.0 \text{ J}\cdot\text{K}^{-1}$

$\Delta S_{system} = \dfrac{\Delta H_{system}}{T} = \dfrac{1.00 \text{ mol} \times 4.60 \times 10^3 \text{ J}\cdot\text{mol}^{-1}}{158.7 \text{ K}} = +29.0 \text{ J}\cdot\text{K}^{-1}$

(c) $\Delta S_{surr} = \dfrac{-\Delta H_{system}}{T} = \dfrac{-(1.00 \text{ mol} \times -4.60 \times 10^3 \text{ J}\cdot\text{mol}^{-1})}{158.7 \text{ K}} = +29.0 \text{ J}\cdot\text{K}^{-1}$

$\Delta S_{system} = \dfrac{\Delta H_{system}}{T} = \dfrac{1.00 \text{ mol} \times -4.60 \times 10^3 \text{ J}\cdot\text{mol}^{-1}}{158.7 \text{ K}} = -29.0 \text{ J}\cdot\text{K}^{-1}$

7.37 (a) $\Delta H°_r = 3(-824.2 \text{ kJ}\cdot\text{mol}^{-1}) - [2(-1118.4 \text{ kJ}\cdot\text{mol}^{-1})] = -235.8 \text{ kJ}\cdot\text{mol}^{-1}$

$\Delta S°_r = 3(87.40 \text{ J}\cdot\text{K}^{-1}\cdot\text{mol}^{-1}) - [2(146.4 \text{ J}\cdot\text{K}^{-1}\cdot\text{mol}^{-1}) + \frac{1}{2}(205.14 \text{ J}\cdot\text{K}^{-1}\cdot\text{mol}^{-1})]$

$= -133.17 \text{ J}\cdot\text{K}^{-1}\cdot\text{mol}^{-1}$

$\Delta G°_r = 3(-742.2 \text{ kJ}\cdot\text{mol}^{-1}) - [2(-1015.4 \text{ kJ}\cdot\text{mol}^{-1})] = -195.8 \text{ kJ}\cdot\text{mol}^{-1}$

ΔG_r may also be calculated from $\Delta H°_r$ and $\Delta S°_r$ (the numbers calculated differ slightly from the two methods due to rounding differences):

$\Delta G°_r = \Delta H°_r - T\Delta S°_r$

$= -235.8 \text{ kJ}\cdot\text{mol}^{-1} - (298 \text{ K})(-133.17 \text{ J}\cdot\text{K}^{-1}\cdot\text{mol}^{-1})/(100 \text{ J}\cdot\text{kJ}^{-1})$

$= -196.1 \text{ kJ}\cdot\text{mol}^{-1}$

(b) $\Delta H°_r = -1208.09 \text{ kJ}\cdot\text{mol}^{-1} - [-1219.6 \text{ kJ}\cdot\text{mol}^{-1}] = 11.5 \text{ kJ}\cdot\text{mol}^{-1}$

$\Delta S°_r = -80.8 \text{ J}\cdot\text{K}^{-1}\cdot\text{mol}^{-1} - [68.87 \text{ J}\cdot\text{K}^{-1}\cdot\text{mol}^{-1}] = -149.7 \text{ J}\cdot\text{K}^{-1}\cdot\text{mol}^{-1}$

$\Delta G°_r = -1111.15 \text{ kJ}\cdot\text{mol}^{-1} - [-1167.3 \text{ kJ}\cdot\text{mol}^{-1}] = +56.2 \text{ kJ}\cdot\text{mol}^{-1}$

or

$\Delta G°_r = \Delta H°_r - T\Delta S°_r$

$= 11.5 \text{ kJ}\cdot\text{mol}^{-1} - (298 \text{ K})(-149.7 \text{ J}\cdot\text{K}^{-1}\cdot\text{mol}^{-1})/(1000 \text{ J}\cdot\text{kJ}^{-1})$

$= 56.1 \text{ kJ}\cdot\text{mol}^{-1}$

(c) $\Delta H°_r = 9.16 \text{ kJ}\cdot\text{mol}^{-1} - [2(33.18 \text{ kJ}\cdot\text{mol}^{-1})] = -57.20 \text{ kJ}\cdot\text{mol}^{-1}$

$\Delta S°_r = 304.29 \text{ J}\cdot\text{K}^{-1}\cdot\text{mol}^{-1} - [2(240.06 \text{ J}\cdot\text{K}^{-1}\cdot\text{mol}^{-1})]$

$= -175.83 \text{ J}\cdot\text{K}^{-1}\cdot\text{mol}^{-1}$

$\Delta G°_r = 97.89 \text{ kJ}\cdot\text{mol}^{-1} - [2(51.31 \text{ kJ}\cdot\text{mol}^{-1})] = -4.73 \text{ kJ}\cdot\text{mol}^{-1}$

or

$\Delta G°_r = \Delta H°_r - T\Delta S°_r$

$= -57.2 \text{ kJ}\cdot\text{mol}^{-1} - (298 \text{ K})(-175.83 \text{ J}\cdot\text{K}^{-1}\cdot\text{mol}^{-1})/(1000 \text{ J}\cdot\text{kJ}^{-1})$

$= -4.80 \text{ kJ}\cdot\text{mol}^{-1}$

7.39 (a) $\frac{1}{2} N_2(g) + \frac{3}{2} H_2(g) \longrightarrow NH_3(g)$

$\Delta H°_r = \Delta H°_f(NH_3) = -46.11\ kJ\cdot mol^{-1}$

$\Delta S°_r = S°_m(NH_3, g) - [\frac{1}{2}S°_m(N_2, g) + \frac{3}{2}S°_m(H_2, g)]$

$= 192.45\ J\cdot K^{-1}\cdot mol^{-1} - [\frac{1}{2}(191.61\ J\cdot K^{-1}\cdot mol^{-1}) + \frac{3}{2}(130.68\ J\cdot K^{-1}\cdot mol^{-1})]$

$= -99.38\ J\cdot K^{-1}\cdot mol^{-1}$

$\Delta G°_r = -46.11\ kJ\cdot mol^{-1} - (298\ K)(-99.38\ J\cdot K^{-1}\cdot mol^{-1})/(1000\ J\cdot kJ^{-1})$

$= -16.49\ kJ\cdot mol^{-1}$

(b) $H_2(g) + \frac{1}{2}\ O_2(g) \longrightarrow H_2O(g)$

$\Delta H°_r = \Delta H°_f(H_2O, g) = -241.82\ kJ\cdot mol^{-1}$

$\Delta S°_r = S°_m(H_2O, g) - [S°_m(H_2, g) + \frac{1}{2}S°_m(O_2, g)]$

$= 188.83\ J\cdot K^{-1}\cdot mol^{-1} - [130.68\ J\cdot K^{-1}\cdot mol^{-1} + \frac{1}{2}(205.14\ J\cdot K^{-1}\cdot mol^{-1})]$

$= -44.42\ J\cdot K^{-1}\cdot mol^{-1}$

$\Delta G°_r = -241.82\ kJ\cdot mol^{-1} - (298\ K)(-44.42\ J\cdot K^{-1}\cdot mol^{-1})/(1000\ J\cdot kJ^{-1})$

$= -228.58\ kJ\cdot mol^{-1}$

(c) C(s), graphite $+ \frac{1}{2}\ O_2(g) \longrightarrow CO(g)$

$\Delta H°_r = \Delta H°_f(CO, g) = -110.53\ kJ\cdot mol^{-1}$

$\Delta S°_r = S°_m(CO, g) - [S°_m(C, s) + \frac{1}{2}S°_m(O_2, g)]$

$= 197.67\ J\cdot K^{-1}\cdot mol^{-1} - [5.740\ J\cdot K^{-1}\cdot mol^{-1} + \frac{1}{2}(205.14\ J\cdot K^{-1}\cdot mol^{-1})]$

$= 89.36\ J\cdot K^{-1}\cdot mol^{-1}$

$\Delta G°_r = -110.53\ kJ\cdot mol^{-1} - (298\ K)(89.36\ J\cdot K^{-1}\cdot mol^{-1})/(1000\ J\cdot kJ^{-1})$

$= -137.2\ kJ\cdot mol^{-1}$

(d) $\frac{1}{2}\ N_2(g) + O_2(g) \longrightarrow NO_2(g)$

$\Delta H°_r = \Delta H°_f(NO_2) = 33.18\ kJ\cdot mol^{-1}$

$\Delta S°_r = S°_m(NO_2, g) - [\frac{1}{2}S°_m(N_2, g) + S°_m(O_2, g)]$

$= 240.06\ J\cdot K^{-1}\cdot mol^{-1} - [\frac{1}{2}(191.61\ J\cdot K^{-1}\cdot mol^{-1}) + 205.14\ J\cdot K^{-1}\cdot mol^{-1}]$

$= -60.89\ J\cdot K^{-1}\cdot mol^{-1}$

$\Delta G°_r = 33.18\ kJ\cdot mol^{-1} - (298\ K)(-60.89\ J\cdot K^{-1}\cdot mol^{-1})/(1000\ J\cdot kJ^{-1})$

$= 51.33\ kJ\cdot mol^{-1}$

7.41 Use the relationship $\Delta G°_r = \Sigma\ \Delta G°_f(\text{products}) - \Sigma\ \Delta G°_r(\text{reactants})$:

(a) $\Delta G°_r = 2\Delta G°_f(SO_3, g) - [2\Delta G°_f(SO_2, g)]$

$= 2(-371.06\ kJ\cdot mol^{-1}) - [2(-300.19\ kJ\cdot mol^{-1})]$

$= -141.74\ kJ\cdot mol^{-1}$

The reaction is spontaneous.

(b) $\Delta G°_r = \Delta G°_f(CaO, s) + \Delta G°_f(CO_2, g) - \Delta G°_f(CaCO_3, s)$

$= (-604.03\ kJ\cdot mol^{-1}) + (-394.36\ kJ\cdot mol^{-1}) - (-1128.8\ kJ\cdot mol^{-1})$

$= +130.41\ kJ\cdot mol^{-1}$

The reaction is not spontaneous.

(c) $\Delta G°_r = \Delta G°_f(SbCl_3, g) - \Delta G°_f(SbCl_5, g)$

$= (-301.2\ kJ\cdot mol^{-1}) - (-334.29\ kJ\cdot mol^{-1})$

$= +33.1\ kJ\cdot mol^{-1}$

The reaction is not spontaneous.

(d) $\Delta G°_r = 16\Delta G°_f(CO_2, g) + 18\Delta G°_f(H_2O, l) - [2\Delta G°_f(C_8H_{18}, l)]$

$= 16(-394.36\ kJ\cdot mol^{-1}) + 18(-237.13\ kJ\cdot mol^{-1}) - [2(6.4\ kJ\cdot mol^{-1})]$

$= -10\,590.9\ kJ\cdot mol^{-1}$

The reaction is spontaneous.

7.43 The standard free energies of formation of the compounds are: (a) $PCl_5(g)$, $-305.0\ kJ\cdot mol^{-1}$; (b) HCN(g), $+124.7\ kJ\cdot mol^{-1}$; (c) NO(g), $+86.55\ kJ\cdot mol^{-1}$; (d) $SO_2(g)$, $-300.19\ kJ\cdot mol^{-1}$. Those compounds with a positive free energy of formation are unstable with respect to the elements. Thus (a) and (d) are thermodynamically stable.

7.45 To understand what happens to $\Delta G°_r$ as temperature is raised, we use the relationship $\Delta G°_r = \Delta H°_r - T\Delta S°_r$. From this it is clear that the free energy of the reaction becomes less favorable (more positive) as temperature increases, only if $\Delta S°_r$ is a negative number. Therefore, we have only to find out whether the standard entropy of formation of the compound is a negative number. This is calculated for each compound as follows:

(a) $P(s) + \frac{5}{2} Cl_2(g) \longrightarrow PCl_5(g)$

$\Delta S°_r = S°_m(PCl_5, g) - [S°_m(P, s) + \frac{5}{2}S°_m(Cl_2, g)]$

$= 364.6\ J\cdot K^{-1}\cdot mol^{-1} - [41.09\ J\cdot K^{-1}\cdot mol^{-1} + \frac{5}{2}(223.07\ J\cdot K^{-1}\cdot mol^{-1})]$

$= -234.2\ J\cdot K^{-1}\cdot mol^{-1}$

The compound is less stable at higher temperatures.

(b) C(s), graphite $+ \frac{1}{2} N_2(g) + \frac{1}{2} H_2(g) \longrightarrow HCN(g)$

$\Delta S°_r = S°_m(HCN, g) - [S°_m(C, s) + \frac{1}{2}S°_m(N_2, g) + \frac{1}{2}S°_m(H_2, g)]$

$= 201.78\ J\cdot K^{-1}\cdot mol^{-1} - [5.740\ J\cdot K^{-1}\cdot mol^{-1} + \frac{1}{2}(191.61\ J\cdot K^{-1}\cdot mol^{-1})$
$+ \frac{1}{2}(130.68\ J\cdot K^{-1}\cdot mol^{-1})]$

$= +34.90\ J\cdot K^{-1}\cdot mol^{-1}$

HCN(g) is more stable at higher T.

(c) $\frac{1}{2} N_2(g) + \frac{1}{2} O_2(g) \longrightarrow NO(g)$

$\Delta S°_r = S°_m(NO, g) - [\frac{1}{2}S°_m(N_2, g) + \frac{1}{2}S°_m(O_2, g)]$

$= 210.76\ J \cdot K^{-1} \cdot mol^{-1} - [\frac{1}{2}(191.61\ J \cdot K^{-1} \cdot mol^{-1}) + \frac{1}{2}(205.14\ J \cdot K^{-1} \cdot mol^{-1})]$

$= +12.38\ J \cdot K^{-1} \cdot mol^{-1}$

NO(g) is more stable as T increases.

(d) $S(s) + O_2(g) \longrightarrow SO_2(g)$

$\Delta S°_r = S°(SO_2, g) - [S°(S, s) + S°(O_2, g)]$

$= 248.22\ J \cdot K^{-1} \cdot mol^{-1} - [31.80\ J \cdot K^{-1} \cdot mol^{-1} + 205.14\ J \cdot K^{-1} \cdot mol^{-1}]$

$= +11.28\ J \cdot K^{-1} \cdot mol^{-1}$

$SO_2(g)$ is more stable at T increases.

7.47 To answer this question, we need to calculate $\Delta G°$ and $\Delta S°$ for the reaction $4\ KClO_3(s) \longrightarrow 3\ KClO_4(s) + KCl(s)$

From the data in Appendix 2A, these values can be calculated:

$\Delta H°_r = 3\Delta H°_f(KClO_4, s) + \Delta H°_f(KCl, s) - [4\Delta H°_f(KClO_3, s)]$

$= 3(-432.75\ kJ \cdot mol^{-1}) + (-436.75) - [4(-397.73\ kJ \cdot mol^{-1})]$

$= -144.08\ kJ \cdot mol^{-1}$

$\Delta S°_r = 3S°_m(KClO_4, s) + S°_m(KCl, s) - [4S°_m(KClO_3, s)]$

$= 3(151.0\ J \cdot K^{-1} \cdot mol^{-1}) + 82.59\ J \cdot K^{-1} \cdot mol^{-1} - [4(143.1\ J \cdot K^{-1} \cdot mol^{-1})]$

$= -36.8\ J \cdot K^{-1} \cdot mol^{-1}$

$\Delta G°_r = 3\Delta G°_f(KClO_4, s) + \Delta G°_f(KCl, s) - [4\Delta G°_f(KClO_3, s)]$

$= 3(-303.09\ kJ \cdot mol^{-1}) + (-409.14\ kJ \cdot mol^{-1}) - [4(-296.25\ kJ \cdot mol^{-1})]$

$= -133.41\ kJ \cdot mol^{-1}$

The standard free energy of the reaction is negative, which means there is a thermodynamic tendency for the reaction to occur. To determine the effect of temperature, we need to look at the entropy change in the reaction. This is obtained from the relationship $\Delta G°_r = \Delta H°_r - T\Delta S°_r$. Because $\Delta S°$ is a negative number, the reaction will be less favorable at higher temperatures. The reaction will be favored by lower temperatures.

7.49 In order to find $\Delta G°_r$ at a temperature other than 298 K, we must first calculate $\Delta H°_r$ and $\Delta S°_r$ and then use the relationship $\Delta G°_r = \Delta H°_r + T\Delta S°_r$ to calculate $\Delta G°_r$.

(a) $\Delta H°_r = 2\Delta H°_f(BF_3, g) + 3\Delta H°_f(H_2O, l) - [\Delta H°_f(B_2O_3, s) + 6\Delta H°_f(HF, g)]$

$= 2(-1137.0\ kJ \cdot mol^{-1}) + 3(-285.83\ kJ \cdot mol^{-1})$
$- [(-1272.8\ kJ \cdot mol^{-1}) + 6(-271.1\ kJ \cdot mol^{-1})]$

$= -232.1\ kJ \cdot mol^{-1}$

$$\Delta S^\circ_r = 2S^\circ_m(BF_3, g) + 3S^\circ_m(H_2O, l) - [S^\circ_m(B_2O_3, S) + 6S^\circ_m(HF, g)]$$

$$= 2(254.12\ J\cdot K^{-1}\cdot mol^{-1}) + 3(69.91\ J\cdot K^{-1}\cdot mol^{-1}) - [53.97\ J\cdot K^{-1}\cdot mol^{-1} + 6(173.78\ J\cdot K^{-1}\cdot mol^{-1})]$$

$$= -378.68\ J\cdot K^{-1}\cdot mol^{-1}$$

$$\Delta G^\circ_r = -232.1\ J\cdot K^{-1}\cdot mol^{-1} - (353\ K)(-378.68\ J\cdot K^{-1}\cdot mol^{-1})/(1000\ J\cdot kJ^{-1})$$

$$= -98.42\ kJ\cdot mol^{-1}$$

In order to determine over what range the reaction will be spontaneous, we consider the relative signs of ΔH°_r and ΔS°_r and their effect on ΔG°_r. Because ΔH°_r is negative and ΔS°_r is also negative, we expect the reaction to be spontaneous at low temperatures, where the term $T\Delta S^\circ_r$ will be less than ΔH°_r. To find the temperature of the cutoff, we calculate the temperature at which $\Delta G^\circ_r = 0$. For this reaction, that temperature is

$$\Delta G^\circ_r = 0 = -232.1\ kJ\cdot mol^{-1} - (T)(-378.68\ J\cdot K^{-1}\cdot mol^{-1})/(1000\ J\cdot kJ^{-1})$$

$$T = 612.9\ K$$

The reaction should be spontaneous below 612.9 K.

(b) $$\Delta H^\circ_r = \Delta H^\circ_f(CaCl_2, aq) + \Delta H^\circ_f(C_2H_2, g) - [\Delta H^\circ_f(CaC_2, s) + 2\Delta H^\circ_f(HCl, aq)]$$

$$= (-877.1\ kJ\cdot mol^{-1}) + 226.73\ kJ\cdot mol^{-1} - [(-59.8\ kJ\cdot mol^{-1}) + 2(-167.16\ kJ\cdot mol^{-1})]$$

$$= -256.3\ kJ\cdot mol^{-1}$$

$$\Delta S^\circ_r = S^\circ_m(CaCl_2, aq) + S^\circ_m(C_2H_2, g) - [S^\circ_m(CaC_2, s) + 2S^\circ_m(HCl, aq)]$$

$$= 59.8\ J\cdot K^{-1}\cdot mol^{-1} + 200.94\ J\cdot K^{-1}\cdot mol^{-1} - [69.96\ J\cdot K^{-1}\cdot mol^{-1} + 2(56.5\ J\cdot K^{-1}\cdot mol^{-1})]$$

$$= +77.8\ J\cdot K^{-1}\cdot mol^{-1}$$

$$\Delta G^\circ_r = -256.2\ kJ\cdot mol^{-1} - (353\ K)(+77.8\ J\cdot K^{-1}\cdot mol^{-1})/(1000\ J\cdot kJ^{-1})$$

$$= -283.7\ kJ\cdot mol^{-1}$$

Because ΔH°_r is negative and ΔS°_r is positive, the reaction will be spontaneous at all temperatures.

(c) $$\Delta H^\circ_r = \Delta H^\circ_f(C\ (s), diamond) = +1.895\ kJ\cdot mol^{-1}$$

$$\Delta S^\circ_r = S^\circ_m(C\ (s), diamond) - S^\circ_m(C\ (s), graphite)$$

$$= +2.377\ J\cdot K^{-1}\cdot mol^{-1} - 5.740\ J\cdot K^{-1}\cdot mol^{-1}$$

$$= -3.363\ J\cdot K^{-1}$$

$$\Delta G^\circ_r = +1.895\ kJ\cdot mol^{-1} - (353\ K)(-3.363\ J\cdot K^{-1}\cdot mol^{-1})/(1000\ J\cdot kJ^{-1})$$

$$= +3.082\ kJ\cdot mol^{-1}$$

Because ΔH°_r is positive and ΔS°_r is negative, the reaction will be nonspontaneous at all temperatures. Note: this calculation is for atmospheric pressure. Diamond can be produced from graphite at elevated pressures and high temperatures.

7.51 The Cl_2 molecule has not only translational energy but also rotational and vibrational energy. The disordering due to this energy contributes to the overall energy disorder. The molecule also has twice as many atoms. However, matter is more organized in the molecule, so that the molar entropy of the molecule is not twice that of the atom.

7.53 (a) Entropy should decrease because the number of moles of gas is less on the product side of the reaction.
(b) Entropy should increase because the dissolution of the solid copper phosphate will increase the randomness of the copper and phosphate ions.
(c) Entropy should decrease as the total number of moles decreases.

7.55 (a) $2\ H_2O_2(l) \longrightarrow 2\ H_2O(l) + O_2(g)$

$$\Delta S^\circ_r = 2S^\circ_m(H_2O,\ l) + S^\circ_m(O_2,\ g) - 2S^\circ_m(H_2O_2,\ l)$$
$$= 2(69.91\ J\cdot K^{-1}\cdot mol^{-1}) + 205.14\ J\cdot K^{-1}\cdot mol^{-1} - 2(109.6\ J\cdot K^{-1}\cdot mol^{-1})$$
$$= +125.8\ J\cdot K^{-1}\cdot mol^{-1}$$

$$\Delta H^\circ_r = 2\Delta H^\circ_f(H_2O,\ l) - 2\Delta H^\circ_f(H_2O_2,\ l)$$
$$= 2(-285.83\ kJ\cdot mol^{-1}) - 2(-187.78\ kJ\cdot mol^{-1})$$
$$= -196.10\ kJ\cdot mol^{-1}$$

$$\Delta G^\circ_r = 2\Delta G^\circ_f(H_2O,\ l) - 2\Delta G^\circ_f(H_2O_2,\ l)$$
$$= 2(-237.13\ kJ\cdot mol^{-1}) - 2(-120.35\ kJ\cdot mol^{-1})$$
$$= -233.56\ kJ\cdot mol^{-1}$$

ΔG°_r can also be calculated from ΔS°_r and ΔH°_r using the relationship:

$$\Delta G^\circ_r = \Delta H^\circ_r - T\Delta S^\circ_r$$
$$= -196.1\ kJ\cdot mol^{-1} - (298\ K)(+125.8\ J\cdot K^{-1}\cdot mol^{-1})/1000\ J\cdot kJ^{-1})$$
$$= -233.6\ kJ\cdot mol^{-1}$$

(b) $2\ F_2(g) + 2\ H_2O(l) \longrightarrow 4\ HF\ (aq) + O_2(g)$

$$\Delta S^\circ_r = 4S^\circ_m(HF,\ aq) + S^\circ_m(O_2,\ g) - [2S^\circ_m(F_2,\ g) + 2S^\circ_m(H_2O,\ l)]$$
$$= 4\ (-88.7\ J\cdot K^{-1}\cdot mol^{-1}) + 205.14\ J\cdot K^{-1}\cdot mol^{-1}$$
$$- [2(202.78\ J\cdot K^{-1}\cdot mol^{-1}) + 2(69.91\ J\cdot K^{-1}\cdot mol^{-1})]$$
$$= +14.6\ J\cdot K^{-1}\cdot mol^{-1}$$

$$\Delta H^\circ_r = 4\Delta H^\circ_f(HF,\ aq) - [2\Delta H^\circ_f(H_2O,\ l)]$$
$$= 4(-330.08\ kJ\cdot mol^{-1}) - [2(-285.83\ kJ\cdot mol^{-1})]$$
$$= -748.66\ kJ\cdot mol^{-1}$$

$$\Delta G^\circ_r = 4 \times \Delta G^\circ_f(HF,\ aq) - [2 \times \Delta G^\circ_f(H_2O,\ l)]$$
$$= 4(-296.82\ kJ\cdot mol^{-1}) - [2(-237.13\ kJ\cdot mol^{-1})]$$
$$= -713.02\ kJ\cdot mol^{-1}$$

$\Delta G°_r$ can also be calculated from $\Delta S°_r$ and $\Delta H°_r$ using the relationship:

$$\Delta G°_r = \Delta H°_r - T\Delta S°_r$$

$$= -748.66\ \text{kJ}\cdot\text{mol}^{-1} - (298\ \text{K})(14.6\ \text{J}\cdot\text{K}^{-1}\cdot\text{mol}^{-1})/(1000\ \text{J}\cdot\text{kJ}^{-1})$$

$$= -753.01\ \text{kJ}\cdot\text{mol}^{-1}$$

7.57 (a) The value can be estimated from

$$\Delta H°_{vap} = T\Delta S°_{vap}$$

$$\Delta H°_{vap} = (353\ \text{K})(85\ \text{J}\cdot\text{mol}^{-1}\cdot\text{K}^{-1})$$

$$= +30\ \text{kJ}\cdot\text{mol}^{-1}$$

(b) $\Delta S°_{surr} = -\dfrac{\Delta H°_{system}}{T}$

$$\Delta S_{surr} = -\left(\frac{10\ \text{g}}{78.11\ \text{g}\cdot\text{mol}^{-1}}\right)\left(\frac{30\ \text{kJ}\cdot\text{mol}^{-1}}{353\ \text{K}}\right) = -11\ \text{J}\cdot\text{K}^{-1}$$

7.59 According to Trouton's rule, the entropy of vaporization of an organic liquid is approximately a constant of 85 $\text{J}\cdot\text{mol}^{-1}\cdot\text{K}^{-1}$. The relationship between entropy of fusion, enthalpy of fusion,and melting point is given by $\Delta S°_{fus} = \dfrac{\Delta H°_{fus}}{T_{fus}}$.

For Pb: $\Delta S°_{fus} = \dfrac{5100\ \text{J}}{600\ \text{K}} = 8.50\ \text{J}\cdot\text{K}^{-1}$

For Hg: $\Delta S°_{fus} = \dfrac{2290\ \text{J}}{234\ \text{K}} = 9.79\cdot\text{K}^{-1}$

For Na: $\Delta S°_{fus} = \dfrac{2640\ \text{J}}{371\ \text{K}} = 7.12\ \text{J}\cdot\text{K}^{-1}$

These numbers are reasonably close but clearly much smaller than the value associated with Trouton's rule.

7.61 We first calculate $\Delta G°_r$ using data from Appendix 2A:

$$\Delta G°_r = 6\Delta G°_f(\text{Fe}_2\text{O}_3, \text{s}) - [4\Delta G°_f(\text{Fe}_3\text{O}_4, \text{s})]$$

$$\Delta G°_r = 6(-742.2\ \text{kJ}\cdot\text{mol}^{-1}) - [4(-1015.4\ \text{kJ}\cdot\text{mol}^{-1})]$$

$$= -391.6\ \text{kJ}\cdot\text{mol}^{-1}$$

Because $\Delta G°_r$ is negative, the process is spontaneous at 25°C.

7.63 (a) The simplest way to calculate $\Delta G°_r$ is to use the $\Delta G°_f$ values from Appendix 2A, but because we will want to calculate the temperature when $\Delta G°_r = 0$, we will need to calculate $\Delta H°_r$ and $\Delta S°_r$.

$$\Delta G^\circ_r = \Delta G^\circ_f(CO, g) - [\Delta G^\circ_f(H_2O, g)]$$
$$= (-137.17\ kJ \cdot mol^{-1}) - [-228.57\ kJ \cdot mol^{-1}]$$
$$= +91.40\ kJ \cdot mol^{-1}$$
$$\Delta H^\circ_r = \Delta H^\circ_f(CO, g) - [\Delta H^\circ_f(H_2O, g)]$$
$$= -110.53\ kJ \cdot mol^{-1} - [-241.82\ kJ \cdot mol^{-1}]$$
$$= +131.29\ kJ \cdot mol^{-1}$$
$$\Delta S^\circ_r = S^\circ_m(CO, g) + S^\circ_m(H_2, g) - [S^\circ_m(H_2O, g) + S^\circ_m(C\ (s, graphite)]$$
$$= 197.67\ J \cdot K^{-1} \cdot mol^{-1} + 130.68\ J \cdot K^{-1} \cdot mol^{-1} - [188.83\ J \cdot K^{-1} \cdot mol^{-1} + 5.740\ J \cdot K^{-1} \cdot mol^{-1}]$$
$$= +133.78\ J \cdot K^{-1} \cdot mol^{-1}$$
$$\Delta G^\circ_r = \Delta H^\circ_r - T\Delta S^\circ_r$$
$$= +131.29\ kJ \cdot mol^{-1} - (298.2\ K)(133.78\ J \cdot K^{-1} \cdot mol^{-1})/(1000\ J \cdot kJ^{-1})$$
$$= +91.40\ kJ \cdot mol^{-1}$$

(b) $\Delta G^\circ_r = \Delta H^\circ_r - T\Delta S^\circ_r$
$$= +131.29\ kJ \cdot mol^{-1} - (T)(133.78\ J \cdot K^{-1} \cdot mol^{-1})/(1000\ J \cdot kJ^{-1}) = 0$$
$$T = 981.39\ K$$

Even though the calculation will allow us to give T to five significant figures, the fact that ΔH_r and ΔS_r vary somewhat with temperature, make reporting that number of figures unreasonable. A better answer, given the limitations of the data, is 980 K.

7.65 For the first step, $q = +50$ J, $w = 0$ (since volume is constant). This means that $\Delta U = +50$ J. Because the overall process returns the system to its original conditions, ΔU_{tot} must be equal to 0, so that for the second step ΔU must be equal to -50 J. From the relationships $\Delta U = q + w$ and $w = -P\Delta V$, we can write $\Delta U = -50\ J = -5\ J - (1\ atm)\Delta V$ because for the second step q is 5 J and the system works against a constant opposing pressure of 1 atm. In order to solve the equation we must convert J to L·atm:

$$-45\ J\left(\frac{0.082\ 06\ L \cdot atm \cdot K^{-1} \cdot mol^{-1}}{8.314\ J \cdot K^{-1} \cdot mol^{-1}}\right) = -(1\ atm)(\Delta V)$$
$$\Delta V = +0.44\ L$$

Because ΔV is positive, the process is expansion.

7.67 (a) Because the enthalpy change for dissolution is positive, the entropy change of the surroundings must be a negative number $\left(\Delta S^\circ_{surr} = -\frac{\Delta H^\circ_{system}}{T}\right)$. Because spontaneous processes are accompanied by an increase in entropy, the change in enthalpy does not favor the dissolution process. (b) In order for the process to be spontaneous (because it occurs readily, we know it is spontaneous), the entropy change of the system must be positive. (c) Locational disorder is dominant. (d) Because the surroundings participate in the solution process only as a source of heat, the entropy change of the surroundings is primarily a result of the dispersal of thermal motion. (e) The driving force for the dissolution is the dispersal of matter, resulting in an overall positive ΔS.

7.69 The values are calculated simply from the Hess's law relationship that the sums of the various energy quantities for the products minus the similar sum for the reactants will give the overall change in the state function desired:

$\text{C(s), graphite} \longrightarrow \text{C(s), diamond}$

$\Delta H° = \Delta H°_f(\text{C(s), diamond}) = +1.895 \text{ kJ}\cdot\text{mol}^{-1}$

$\Delta S° = \Delta S°_m(\text{C(s), diamond}) - \Delta S°_m(\text{C(s), graphite})$

$= 2.377 \text{ J}\cdot\text{K}^{-1}\cdot\text{mol}^{-1} - 5.740 \text{ J}\cdot\text{K}^{-1}\cdot\text{mol}^{-1}$

$= -3.363 \text{ J}\cdot\text{K}^{-1}\cdot\text{mol}^{-1}$

$\Delta G° = \Delta G°_f(\text{C(s), diamond}) = +2.900 \text{ kJ}\cdot\text{mol}^{-1}$

Notice that the values used for the entropy calculation are absolute entropies, not $\Delta S°$ values and that the $S°_m$ value for C(s), graphite, is not 0. Graphite has a delocalized structure similar to that of benzene, whereas in diamond, all the carbon atoms are bonded to four other carbon atoms in a very rigid lattice. It is not surprising that the change in entropy upon going from graphite to diamond would decrease, because we would expect graphite to have a higher molar entropy than diamond. Similarly, one can compare the bond formation and breaking that accompanies a change from graphite to diamond. Given the actual numbers, the change is clearly small however. In graphite, the carbon atom is bonded to three other carbon atoms with delocalized bonds (we can approximate this very roughly, using the values of 518 $\text{kJ}\cdot\text{mol}^{-1}$ of delocalized C—C bonds, as given in Table 6.7). In diamond, the carbon atom is bonded to four other carbon atoms by single C—C bonds (approximately by 348 $\text{kJ}\cdot\text{mol}^{-1}$). Even though these approximations overestimate the effect of delocalization on the C—C bond strength in graphite, the trend is expected—the three delocalized bonds in C(s) graphite are actually slightly more exothermic than the four C—C single bonds in diamond, making the standard enthalpy change a positive number. (Notice that if one considered the C—C bonds in graphite to be localized, the opposite prediction would have been made.) The standard free energy change for this reaction is positive as follows, from a positive standard enthalpy change and a negative standard entropy change for the reaction.

7.71 The entries all correspond to aqueous ions. The fact that they are negative is due to the reference point that has been established. Because ions cannot actually be separated and measured independently, a reference point that defines $S°_m(H^+, aq) = 0$ has been established. This definition is then used to calculate the standard entropies for the other ions. The fact that they are negative will arise in part because the solvated ion $M(H_2O)_x^{n+}$ will be more ordered than the isolated ion and solvent molecules ($M^{n+} + x\ H_2O$).

7.73 In this reaction, we expect the change in enthalpy to be close to 0 (it should not be exactly 0) because the number and types of bonds that are broken are equal to the number and types of bonds that are formed. The free energy change should therefore be dominated by the entropy change of the reaction. We would expect the entropy change of the reaction to be positive, because the mixed complex $InCl_2Br$ would be expected to have a larger disorder than a molecule composed of all the same types of halide ions, InX_3. Consequently, we predict a positive value for $\Delta S°_r$ and a negative value of $\Delta G°_r$.

7.75 (a) In order to calculate the free energy at different temperatures, we need to know $\Delta H°$ and $\Delta S°$ for the process: $H_2O(l) \longrightarrow H_2O(g)$

$$\Delta H°_r = \Delta H°_f(H_2O, g) - \Delta H°_f(H_2O, l)$$
$$= (-241.82\ kJ\cdot mol^{-1}) - [-285.83\ kJ\cdot mol^{-1}]$$
$$= 44.01\ kJ\cdot mol^{-1}$$
$$\Delta S°_r = S°_m(H_2O, g) - S°_m(H_2O, l)$$
$$= 188.83\ J\cdot K^{-1}\cdot mol^{-1} - [69.91\ J\cdot K^{-1}\cdot mol^{-1}]$$
$$= 118.92\ J\cdot K^{-1}\cdot mol^{-1}$$
$$\Delta G°_r = \Delta H°_r - T\Delta S°_r$$
$$= 44.01\ kJ\cdot mol^{-1} - T(118.92\ J\cdot K^{-1}\cdot mol^{-1})/(1000\ J\cdot kJ^{-1})$$

T(K)	$\Delta G°_r$(kJ)
298	8.57 kJ
373	−0.35 kJ
423	−6.29 kJ

The reaction goes from being nonspontaneous near room temperature to being spontaneous above 100°C.

(b) The value at 100°C should be exactly 0, because this is the normal boiling point of water.

(c) The discrepancy arises because the enthalpy and entropy values calculated from the tables are not rigorously constant with temperature. Better values would be obtained using the actual enthalpy and entropy of vaporization measured at the boiling point.

7.77 The dehydrogenation of cyclohexane to benzene follows the following equation:

$C_6H_{12}(l) \longrightarrow C_6H_6(l) + 3\ H_2(g)$

We can confirm that this process is nonspontaneous by calculating the ΔG°_r for the process, using data in Appendix 2A:

$$\begin{aligned}\Delta G^\circ_r &= \Delta G^\circ_f(C_6H_6, l) - \Delta G^\circ_f(C_6H_{12}, l)\\ &= 124.3\ kJ\cdot mol^{-1} - 26.7\ kJ\cdot mol^{-1}\\ &= +97.6\ kJ\cdot mol^{-1}\end{aligned}$$

The reaction of ethene with hydrogen can be examined similarly:

$C_2H_2(g) + H_2(g) \longrightarrow C_2H_6(g)$

$$\begin{aligned}\Delta G^\circ_r &= \Delta G^\circ_f(C_2H_6, g) - \Delta G^\circ_f(C_2H_2, g)\\ &= (-32.82\ kJ\cdot mol^{-1}) - 68.15\ kJ\cdot mol^{-1}\\ &= -100.97\ kJ\cdot mol^{-1}\end{aligned}$$

We can now combine these two reactions so that $C_2H_2(g)$ accepts the hydrogen that is formed in the dehydrogenation reaction:

$$\begin{array}{ll} C_6H_{12}(l) \longrightarrow C_6H_6(l) + 3\ H_2(g) & \Delta G^\circ_r = +97.6\ kJ\cdot mol^{-1}\\ +\ 3[C_2H_2(g) + H_2(g) \longrightarrow C_2H_6(g)] & \Delta G^\circ_r = 3(-100.97\ kJ\cdot mol^{-1})\\ \hline C_6H_{12}(l) + 3\ C_2H_2(g) \longrightarrow C_6H_6(l) + 3\ C_2H_6(g) & \Delta G^\circ_r = -205.13\ kJ\cdot mol^{-1}\end{array}$$

We can see that by combining these two reactions, the overall process becomes spontaneous. Essentially, we are using the energy of the favorable reaction to drive the nonfavorable process.

7.79 (a) For the three reactions, the calculation of ΔG°, ΔH°, and ΔS° are as follows:

$1 \rightleftharpoons 2$

$$\begin{aligned}\Delta G^\circ_r &= \Delta G^\circ_f(2) - \Delta G^\circ_f(1)\\ &= 62.97\ kJ\cdot mol^{-1} - 65.86\ kJ\cdot mol^{-1}\\ &= -2.89\ kJ\cdot mol^{-1}\\ \Delta H^\circ_r &= \Delta H^\circ_f(2) - \Delta H^\circ_f(1)\\ &= (-11.17\ kJ\cdot mol^{-1}) - (-6.99\ kJ\cdot mol^{-1})\\ &= -4.18\ kJ\cdot mol^{-1}\\ \Delta G^\circ_r &= \Delta H^\circ_r - T\Delta S^\circ_r\\ -2.89\ kJ\cdot mol^{-1} &= -4.18\ kJ\cdot mol^{-1} - (298\ K)(\Delta S^\circ_r)/(1000\ J\cdot kJ^{-1})\\ \Delta S^\circ_r &= -4.33\ J\cdot K^{-1}\cdot mol^{-1}\end{aligned}$$

$1 \rightleftharpoons 3$

$$\Delta G^\circ_r = \Delta G^\circ_f(3) - \Delta G^\circ_f(1)$$
$$= 58.07\ \text{kJ}\cdot\text{mol}^{-1} - 65.86\ \text{kJ}\cdot\text{mol}^{-1}$$
$$= -7.79\ \text{kJ}\cdot\text{mol}^{-1}$$
$$\Delta H^\circ_f = \Delta H^\circ_f(3) - \Delta H^\circ_f(1)$$
$$= (-16.90\ \text{kJ}\cdot\text{mol}^{-1}) - (-6.99\ \text{kJ}\cdot\text{mol}^{-1})$$
$$= -9.91\ \text{kJ}\cdot\text{mol}^{-1}$$
$$\Delta G^\circ_r = \Delta H^\circ_r - T\Delta S^\circ_r$$
$$-7.79\ \text{kJ}\cdot\text{mol}^{-1} = -9.91\ \text{kJ}\cdot\text{mol}^{-1} - (298\ \text{K})(\Delta S^\circ_r)/(1000\ \text{J}\cdot\text{kJ}^{-1})$$
$$\Delta S^\circ_r = -7.11\ \text{J}\cdot\text{K}^{-1}\cdot\text{mol}^{-1}$$

$2 \rightleftharpoons 3$

$$\Delta G^\circ_r = \Delta G^\circ_f(3) - \Delta G^\circ_f(2)$$
$$= 58.07\ \text{kJ}\cdot\text{mol}^{-1} - 62.97\ \text{kJ}\cdot\text{mol}^{-1}$$
$$= -4.90\ \text{kJ}\cdot\text{mol}^{-1}$$
$$\Delta H^\circ_r = \Delta H^\circ_f(3) - \Delta H^\circ_f(2)$$
$$= (-16.90\ \text{kJ}\cdot\text{mol}^{-1}) - (-11.17\ \text{kJ}\cdot\text{mol}^{-1})$$
$$= -5.73\ \text{kJ}\cdot\text{mol}^{-1}$$
$$\Delta G^\circ_r = \Delta H^\circ_r - T\Delta S^\circ_r$$
$$-4.90\ \text{kJ}\cdot\text{mol}^{-1} = -5.73\ \text{kJ}\cdot\text{mol}^{-1} - (298\ \text{K})(\Delta S^\circ_r)/(1000\ \text{J}\cdot\text{kJ}^{-1})$$
$$\Delta S^\circ_r = -2.78\ \text{J}\cdot\text{K}^{-1}\cdot\text{mol}^{-1}$$

(b) The most stable of the three compounds is 2-methylpropene.

(c) Because ΔS° is also equal to the difference in the S°_m values for the compounds, we can examine those values to place the three compounds in order of their relative absolute entropies. The ordering is $S^\circ_m(1) > S^\circ_m(2) > S^\circ_m(3)$.

7.81 We need to calculate ΔH°_r and ΔS°_r for each process.

(a) $HCOOH(l) \longrightarrow H_2(g) + CO_2(g)$

$$\Delta H^\circ_r = \Delta H^\circ_f(CO_2, g) - \Delta H^\circ_f(HCOOH, l)$$
$$= (-393.51\ \text{kJ}\cdot\text{mol}^{-1}) - (-424.72\ \text{kJ}\cdot\text{mol}^{-1})$$
$$= +31.21\ \text{kJ}\cdot\text{mol}^{-1}$$
$$\Delta S^\circ_r = S^\circ_m(H_2, g) + S^\circ(CO_2, g) - [S^\circ_m(HCOOH, l)]$$
$$= 130.68\ \text{J}\cdot\text{K}^{-1}\cdot\text{mol}^{-1} + 213.74\ \text{J}\cdot\text{K}^{-1}\cdot\text{mol}^{-1} - [128.95\ \text{J}\cdot\text{K}^{-1}\cdot\text{mol}^{-1}]$$
$$= +215.47\ \text{J}\cdot\text{K}^{-1}\cdot\text{mol}^{-1}$$

The reaction will become spontaneous above the temperature at which $\Delta G^\circ_r = 0$:

$$0 = 31.21\ \text{kJ}\cdot\text{mol}^{-1} - T(215.47\ \text{J}\cdot\text{K}^{-1}\cdot\text{mol}^{-1})/(1000\ \text{J}\cdot\text{kJ}^{-1})$$
$$T = 145\ \text{K}$$

(b) $CH_3COOH(l) \longrightarrow CH_4(g) + CO_2(g)$

$$\Delta H°_r = \Delta H°_f(CH_4, g) + \Delta H°_f(CO_2, g) - \Delta H°_f(CH_3COOH, l)$$
$$= (-74.81\ kJ\cdot mol^{-1}) + (-393.51\ kJ\cdot mol^{-1}) - (-484.5\ kJ\cdot mol^{-1})$$
$$= +16.18\ kJ\cdot mol^{-1}$$
$$\Delta S°_r = S°_m(CH_4, g) + S°(CO_2, g) - [S°_m(CH_3COOH, l)]$$
$$= 186.26\ J\cdot K^{-1}\cdot mol^{-1} + 213.74\ J\cdot K^{-1}\cdot mol^{-1} - [159.8\ J\cdot K^{-1}\cdot mol^{-1}]$$
$$= 240.2\ J\cdot K^{-1}\cdot mol^{-1}$$

The reaction will become spontaneous above the temperature at which $\Delta G°_r = 0$:

$$0 = 16.18\ kJ\cdot mol^{-1} - T(240.2\ J\cdot K^{-1}\cdot mol^{-1})/(1000\ J\cdot kJ^{-1})$$
$$T = 67\ K$$

(c) $C_6H_5COOH(s) \longrightarrow C_6H_6(l) + CO_2(g)$

$$\Delta H°_r = \Delta H°_f(C_6H_6, l) + \Delta H°_f(CO_2, g) - [\Delta H°_f(C_6H_5COOH, s)]$$
$$= +49.0\ kJ\cdot mol^{-1} + (-393.51\ kJ\cdot mol^{-1}) - [-385.1\ kJ\cdot mol^{-1}]$$
$$= +40.6\ kJ\cdot mol^{-1}$$
$$\Delta S°_r = S°_m(C_6H_6, l) + S°(CO_2, g) - [S°_m(C_6H_5COOH, s)]$$
$$= +173.3\ J\cdot K^{-1}\cdot mol^{-1} + 213.74\ J\cdot K^{-1}\cdot mol^{-1} - (167.6\ J\cdot K^{-1}\cdot mol^{-1})$$
$$= +219.4\ J\cdot K^{-1}\cdot mol^{-1}$$

The reaction will become spontaneous above the temperature at which $\Delta G°_r = 0$:

$$0 = 40.58\ kJ\cdot mol^{-1} - T(219.4\ J\cdot K^{-1}\cdot mol^{-1})/(1000\ J\cdot kJ^{-1})$$
$$T = 185\ K$$

It is clear from these calculations that all of these carboxylic acids are thermodynamically unstable with respect to decomposition to produce $CO_2(g)$. A consideration of the parameters shows that this is driven by the entropy increase in the production of the gas, because the enthalpy of the reaction is all cases is endothermic.

CHAPTER 8
PHYSICAL EQUILIBRIA

8.1 Dynamic equilibrium is a condition in which the forward process and its reverse are occurring simultaneously at equal rates. Processes occur at the molecular level; the macroscopic system appears unchanged. In a static equilibrium, no process is occurring at any level.

8.3 In a 1.0 L vessel at 20°C, there will be 17.5 Torr of water vapor. The ideal gas law can be used to calculate the mass of water present:

$$PV = nRT$$

let m = mass of water

$$\left(\frac{17.5\ \text{Torr}}{760\ \text{Torr}\cdot\text{atm}^{-1}}\right)(1.0\ \text{L}) = \left(\frac{m}{18.02\ \text{g}\cdot\text{mol}^{-1}}\right)(0.082\ 06\ \text{L}\cdot\text{atm}\cdot\text{K}^{-1}\cdot\text{mol}^{-1})(293\ \text{K})$$

$$m = \frac{(17.5\ \text{Torr})(1.0\ \text{L})(18.02\ \text{g}\cdot\text{mol}^{-1})}{(760\ \text{Torr}\cdot\text{atm}^{-1})(0.082\ 06\ \text{L}\cdot\text{atm}\cdot\text{K}^{-1}\cdot\text{mol}^{-1})(293\ \text{K})}$$

$$m = 0.017\ \text{g}$$

8.5 The volume of the bathroom (neglecting internal fixtures) is $4.0 \times 3.0 \times 3.0\ \text{m}^3 = 36\ \text{m}^3$ or $36\ \text{m}^3 \times 1000\ \text{L}\cdot\text{m}^{-3} = 36\ 000\ \text{L}$. The ideal gas law can be used to calculate the mass of water present:

$$PV = nRT$$

let m = mass of water

$$\left(\frac{7.4\ \text{kPa}}{101.325\ \text{kPa}\cdot\text{atm}^{-1}}\right)(36\ 000\ \text{L})$$

$$= \left(\frac{m}{18.02\ \text{g}\cdot\text{mol}^{-1}}\right)(0.082\ 06\ \text{L}\cdot\text{atm}\cdot\text{K}^{-1}\cdot\text{mol}^{-1})(313\ \text{K})$$

$$m = \frac{(7.4\ \text{kPa})(36\ 000\ \text{L})(18.02\ \text{g}\cdot\text{mol}^{-1})}{(101.325\ \text{kPa}\cdot\text{atm}^{-1})(0.082\ 06\ \text{L}\cdot\text{atm}\cdot\text{K}^{-1}\cdot\text{mol}^{-1})(313\ \text{K})}$$

$$m = 1.8 \times 10^3\ \text{g}$$

8.7 (a) 99.2°C; (b) 99.7°C

8.9 (a) The quantities ΔH°_{vap} and ΔS°_{vap} can be calculated using the relationship

$$\ln P = -\frac{\Delta H^\circ_{vap}}{R}\cdot\frac{1}{T} + \frac{\Delta S^\circ_{vap}}{R}$$

Because we have two temperatures with corresponding vapor pressures, we can set up two equations with two unknowns and solve for ΔH°_{vap} and ΔS°_{vap}. If the equation is used as is, P must be expressed in atm, which is the standard reference state. Remember that the value used for P is really activity, which for pressure is P divided by the reference state of 1 atm, so that the quantity inside the ln term is dimensionless.

$$8.314\ \mathrm{J\cdot K^{-1}\cdot mol^{-1}} \times \ln\frac{58\ \mathrm{Torr}}{760\ \mathrm{Torr}} = -\frac{\Delta H^\circ_{vap}}{250.4\ \mathrm{K}} + \Delta S^\circ_{vap}$$

$$8.314\ \mathrm{J\cdot K^{-1}\cdot mol^{-1}} \times \ln\frac{512\ \mathrm{Torr}}{760\ \mathrm{Torr}} = -\frac{\Delta H^\circ_{vap}}{298.2\ \mathrm{K}} + \Delta S^\circ_{vap}$$

which give, upon combining terms,

$-21.39\ \mathrm{J\cdot K^{-1}\cdot mol^{-1}} = -0.003\ 994\ \mathrm{K^{-1}} \times \Delta H^\circ_{vap} + \Delta S^\circ_{vap}$

$-3.284\ \mathrm{J\cdot K^{-1}\cdot mol^{-1}} = -0.003\ 353\ \mathrm{K^{-1}} \times \Delta H^\circ_{vap} + \Delta S^\circ_{vap}$

Subtracting one equation from the other will eliminate the ΔS°_{vap} term and allow us to solve for ΔH°_{vap}

$-18.11\ \mathrm{J\cdot K^{-1}\cdot mol^{-1}} = -0.000\ 641 \times \Delta H^\circ_{vap}$

$\Delta H^\circ_{vap} = +28.3\ \mathrm{kJ\cdot mol^{-1}}$

(b) We can then use ΔH°_{vap} to calculate ΔS°_{vap} using either of the two equations:

$-21.39\ \mathrm{J\cdot K^{-1}\cdot mol^{-1}} = -0.003\ 994\ \mathrm{K^{-1}} \times (+28\ 200\ \mathrm{J\cdot mol^{-1}}) + \Delta S^\circ_{vap}$

$\Delta S^\circ_{vap} = 91.2\ \mathrm{J\cdot K^{-1}\cdot mol^{-1}}$

$-3.284\ \mathrm{J\cdot K^{-1}\cdot mol^{-1}} = -0.003\ 353\ \mathrm{K^{-1}} \times (+28\ 200\ \mathrm{J\cdot mol^{-1}}) + \Delta S^\circ_{vap}$

$\Delta S^\circ_{vap} = 91.3\ \mathrm{J\cdot K^{-1}\cdot mol^{-1}}$

(c) The ΔG°_{vap} is calculated using $\Delta G^\circ_r = \Delta H^\circ_r - T\Delta S^\circ_r$

$\Delta G^\circ_r = +28.3\ \mathrm{kJ\cdot mol^{-1}} - (298\ \mathrm{K})(91.2\ \mathrm{J\cdot K^{-1}\cdot mol^{-1}})/(1000\ \mathrm{J\cdot kJ^{-1}})$

$\Delta G^\circ_r = +1.1\ \mathrm{kJ\cdot mol^{-1}}$

(d) The boiling point can be calculated using one of several methods. The easiest to use is the one developed in the last chapter:

$\Delta G^\circ_{vap} = \Delta H^\circ_{vap} - T_B\,\Delta S^\circ_{vap} = 0$

$$\Delta H^\circ_{vap} = T_B\,\Delta S^\circ_{vap} \text{ or } T_B = \frac{\Delta H^\circ_{vap}}{\Delta S^\circ_{vap}}$$

$$T_B = \frac{28.2\ \mathrm{kJ\cdot mol^{-1}} \times 1000\ \mathrm{J\cdot kJ^{-1}}}{91.2\ \mathrm{J\cdot K^{-1}\cdot mol^{-1}}} = 309\ \mathrm{K} \text{ or } 36^\circ\mathrm{C}$$

Alternatively, we could use the relationship $\ln\frac{P_2}{P_1} = -\frac{\Delta H^\circ_{vap}}{R}\left[\frac{1}{T_2} - \frac{1}{T_1}\right]$. Here, we would substitute, in one of the known vapor pressure points, the value of the enthalpy of vaporization and the condition that $P = 1$ atm at the normal boiling point.

8.11 (a) The quantities $\Delta H°_{vap}$ and $\Delta S°_{vap}$ can be calculated using the relationship

$$\ln P = -\frac{\Delta H°_{vap}}{R}\cdot\frac{1}{T} + \frac{\Delta S°_{vap}}{R}$$

Because we have two temperatures with corresponding vapor pressures (we know that the vapor pressure = 1 atm at the boiling point), we can set up two equations with two unknowns and solve for $\Delta H°_{vap}$ and $\Delta S°_{vap}$. If the equation is used as is, P must be expressed in atm, which is the standard reference state. Remember that the value used for P is really activity, which for pressure is P divided by the reference state of 1 atm, so that the quantity inside the ln term is dimensionless.

$$8.314\ \text{J}\cdot\text{K}^{-1}\cdot\text{mol}^{-1} \times \ln 1 = -\frac{\Delta H°_{vap}}{292.7\ \text{K}} + \Delta S°_{vap}$$

$$8.314\ \text{J}\cdot\text{K}^{-1}\cdot\text{mol}^{-1} \times \ln\frac{359\ \text{Torr}}{760\ \text{Torr}} = -\frac{\Delta H°_{vap}}{273.2\ \text{K}} + \Delta S°_{vap}$$

which give, upon combining terms,

$$0\ \text{J}\cdot\text{K}^{-1}\cdot\text{mol}^{-1} = -0.003\,416\ \text{K}^{-1} \times \Delta H°_{vap} + \Delta S°_{vap}$$

$$-6.235\ \text{J}\cdot\text{K}^{-1}\cdot\text{mol}^{-1} = -0.003\,660\ \text{K}^{-1} \times \Delta H°_{vap} + \Delta S°_{vap}$$

Subtracting one equation from the other will eliminate the $\Delta S°_{vap}$ term and allow us to solve for $\Delta H°_{vap}$

$$+6.235\ \text{J}\cdot\text{K}^{-1}\cdot\text{mol}^{-1} = +0.000\,244\ \text{K}^{-1} \times \Delta H°_{vap}$$

$$\Delta H°_{vap} = +25.6\ \text{kJ}\cdot\text{mol}^{-1}$$

(b) We can then use $\Delta H°_{vap}$ to calculate $\Delta S°_{vap}$ using either of the two equations:

$$0 = -0.003\,416\ \text{K}^{-1} \times (+25\,600\ \text{J}\cdot\text{mol}^{-1}) + \Delta S°_{vap}$$

$$\Delta S°_{vap} = 87.4\ \text{J}\cdot\text{K}^{-1}\cdot\text{mol}^{-1}$$

$$-6.235\ \text{J}\cdot\text{K}^{-1}\cdot\text{mol}^{-1} = -0.003\,660\ \text{K}^{-1} \times (+25\,600\ \text{J}\cdot\text{mol}^{-1}) + \Delta S°_{vap}$$

$$\Delta S°_{vap} = 87.5\ \text{J}\cdot\text{K}^{-1}\cdot\text{mol}^{-1}$$

(c) The vapor pressure at another temperature is calculated using

$$\ln\frac{P_2}{P_1} = -\frac{\Delta H°_{vap}}{R}\left[\frac{1}{T_2} - \frac{1}{T_1}\right]$$

We need to insert the calculated value of the enthalpy of vaporization and one of the known vapor pressure points:

$$\ln\frac{P_{\text{at 8.5°C}}}{1\ \text{atm}} = -\frac{25\,600\ \text{J}\cdot\text{mol}^{-1}}{8.314\ \text{J}\cdot\text{K}^{-1}\cdot\text{mol}^{-1}}\left[\frac{1}{291.7\ \text{K}} - \frac{1}{292.7\ \text{K}}\right]$$

$$P_{\text{at 8.5°C}} = 1.0\ \text{atm or } 7.6 \times 10^2\ \text{Torr}$$

8.13 The relationship of choice is

$$\ln\frac{P_2}{P_1} = -\frac{\Delta H^\circ_{vap}}{R}\left[\frac{1}{T_2} - \frac{1}{T_1}\right]$$

because the Table contains the enthalpy of vaporization and the normal boiling point (vapor pressure = 1 atm or 760 Torr).

$$\ln\frac{P_2}{1} = -\frac{5.93 \times 10^4\ \text{J}\cdot\text{mol}^{-1}}{8.314\ \text{J}\cdot\text{K}^{-1}\cdot\text{mol}^{-1}}\left[\frac{1}{523\ \text{K}} - \frac{1}{629.7\ \text{K}}\right]$$

$P_2 = 0.099$ atm or 75 Torr

8.15 (a) vapor; (b) liquid; (c) vapor; (d) vapor

8.17 (a) 2.4 K; (b) ~10 atm; (c) 5.5 K; (d) no

8.19 (a) At the lower pressure triple point, liquid helium-I and -II are in equilibrium with helium gas; at the higher pressure triple point, liquid helium-I and -II are in equilibrium with solid helium. (b) helium-I

8.21 Its triple point is higher than room temperature. From this information only, no conclusion can be drawn about its triple point pressure.

8.23 (a) KCl is an ionic solid so water would be the best choice; (b) CCl_4 is nonpolar so the best choice is benzene; (c) CH_3COOH is polar so water is the better choice.

8.25 (a) hydrophilic, because NH_2 is polar and has a lone pair and H atoms that can participate in hydrogen bonding to water molecules; (b) hydrophobic, because the CH_3 group is not very polar; (c) hydrophobic, because the Br group is not very polar; (d) hydrophilic, because the carboxylic acid group has lone pairs on oxygen and an acidic proton that can participate in hydrogen bonding to water molecules.

8.27 (a) The solubility of O_2(g) in water is $1.3 \times 10^{-3}\ \text{mol}\cdot\text{L}^{-1}\cdot\text{atm}^{-1}$.

$$\text{solubility at 50 kPa} = \frac{50\ \text{kPa}}{101.325\ \text{kPa}\cdot\text{atm}^{-1}} \times 1.3 \times 10^{-3}\ \text{mol}\cdot\text{L}^{-1}\cdot\text{atm}^{-1}$$
$$= 6.4 \times 10^{-4}\ \text{mol}\cdot\text{L}^{-1}$$

(b) The solubility of CO_2(g) in water is $2.3 \times 10^{-2}\ \text{mol}\cdot\text{L}^{-1}\cdot\text{atm}^{-1}$.

$$\text{solubility at 500 Torr} = \frac{500\ \text{Torr}}{760\ \text{Torr}\cdot\text{atm}^{-1}} \times 2.3 \times 10^{-2}\ \text{mol}\cdot\text{L}^{-1}\cdot\text{atm}^{-1}$$
$$= 1.5 \times 10^{-2}\ \text{mol}\cdot\text{L}^{-1}$$

(c) solubility = $0.10 \text{ atm} \times 2.3 \times 10^{-2} \text{ mol}\cdot\text{L}^{-1}\cdot\text{atm}^{-1} = 2.3 \times 10^{-3} \text{ mol}\cdot\text{L}^{-1}$.

8.29 The question essentially asks how much CO_2 is dissolved in the solution. This we can calculate using the solubility from Table 8.5. This produces an answer in moles, which can be converted into volume by using the ideal gas equation:

$$PV = nRT \text{ or } V = \frac{nRT}{P}$$

$$V_{CO_2} = (3.00 \text{ atm})(0.355 \text{ L})(2.3 \times 10^{-2} \text{ mol}\cdot\text{L}^{-1}\cdot\text{atm}^{-1}) \left(\frac{0.082\,06 \text{ L}\cdot\text{atm}\cdot\text{K}^{-1}\cdot\text{mol}^{-1}}{1 \text{ atm}}\right)(293 \text{ K})$$

$$= 0.59 \text{ L}$$

8.31 (a) By Henry's law, the concentration of CO_2 in solution will double. (b) No change in the equilibrium will occur; the partial pressure of CO_2 is unchanged and the concentration is unchanged.

8.33 (a) $4 \text{ mg}\cdot\text{L}^{-1} \times 1000 \text{ mL}\cdot\text{L}^{-1} \times 1.00 \text{ g}\cdot\text{mL}^{-1} \times 1 \text{ kg}/1000 \text{ g}$
$= 4 \text{ mg}\cdot\text{L}^{-1}$ or 4 ppm

(b) The solubility of O_2 (g) in water is $1.3 \times 10^{-3} \text{ mol}\cdot\text{L}^{-1}\cdot\text{atm}^{-1}$ which can be converted to parts per million as follows:

In 1.00 L (corresponding to 1 kg) of solution there will be 1.3×10^{-3} mol O_2

$1.3 \times 10^{-3} \text{ mol}\cdot\text{kg}^{-1}\cdot\text{atm}^{-1}\ O_2 \times 32.00 \text{ g}\cdot\text{mol}^{-1}\ O_2 \times 10^3 \text{ mg}\cdot\text{g}^{-1}$
$= 42 \text{ mg}\cdot\text{kg}^{-1}\cdot\text{atm}^{-1}$ or $42 \text{ ppm}\cdot\text{atm}^{-1}$

$4 \text{ mg} \div 41 \text{ mg}\cdot\text{kg}^{-1}\cdot\text{atm}^{-1} = 0.1 \text{ atm}$

(c) $P = \dfrac{0.1 \text{ atm}}{0.21} = 0.5 \text{ atm}$

8.35 Ion hydration enthalpies for ions of the same charge within a group of elements become less negative moving down the group. This trend parallels increasing ion size. Hydration energy is an ion-dipole interaction, so it should be higher for small, high charge density ions.

8.37 (a) Because it is exothermic, the enthalpy change must be negative; (b) $Li_2SO_4(s) \rightleftharpoons 2Li^+(aq) + SO_4^{2-}(aq) + \text{heat}$; (c) Given that $\Delta H_L + \Delta H_{Hydration} = \Delta H$ of solution, the enthalpy of hydration should be larger. If the lattice energy were greater, the overall process would be endothermic.

8.39 To answer this question we must first determine the molar enthalpies of solution and multiply this by the number of moles of solid dissolved to get the actual amount of heat released.

(a) $\Delta H°_{\text{hydration}} = +3.9$

$$\Delta H = \frac{10.0 \text{ g NaCl}}{58.44 \text{ g}\cdot\text{mol}^{-1}} \times (+3.9 \text{ kJ}\cdot\text{mol}^{-1}) = 0.67 \text{ kJ or } 6.7 \times 10^2 \text{ J}$$

(b) $\Delta H°_{\text{hydration}} = -0.6$

$$\Delta H = \frac{10.0 \text{ g NaBr}}{102.90 \text{ g}\cdot\text{mol}^{-1}} \times (-0.6 \text{ kJ}\cdot\text{mol}^{-1}) = 0.06 \text{ kJ or } -6 \times 10 \text{ J}$$

(c) $\Delta H°_{\text{hydration}} = -329 \text{ kJ}\cdot\text{mol}^{-1}$

$$\Delta H = \frac{10.0 \text{ g AlCl}_3}{133.33 \text{ g}\cdot\text{mol}^{-1}} \times (-329 \text{ kJ}\cdot\text{mol}^{-1}) = -24.7 \text{ kJ}$$

(d) $\Delta H°_{\text{hydration}} = 25.7 \text{ k}\cdot\text{mol}^{-1}$

$$\Delta H = \frac{10.0 \text{ g NH}_4\text{NO}_3}{80.04 \text{ g}\cdot\text{mol}^{-1}} \times (+25.7 \text{ kJ}\cdot\text{mol}^{-1}) = +3.21 \text{ kJ}$$

8.41 From the tables we find that the $\Delta H°_L = 2153 \text{ kJ}\cdot\text{mol}^{-1}$, and hydration enthalpy of $Sr^{2+} = -1524 \text{ kJ}\cdot\text{mol}^{-1}$, hydration enthalpy of $Cl^- = -340 \text{ kJ}\cdot\text{mol}^{-1}$.

$$\begin{aligned}\Delta H_{\text{hydration}}(\text{SrCl}_2) &= 1 \text{ mol} \times \Delta H_{\text{hydration}}(\text{Sr}^{2+}) + 2 \text{ mol} \times \Delta H_{\text{hydration}}(\text{Cl}^-)\\ &= 1 \text{ mol} \times (-1524 \text{ kJ}\cdot\text{mol}^{-1}) + 2 \text{ mol} \times (-340 \text{ kJ}\cdot\text{mol}^{-1})\\ &= -2204 \text{ kJ}\end{aligned}$$

This corresponds to 1 mol $SrCl_2$ so it can also be written as $-2204 \text{ kJ}\cdot\text{mol}^{-1}$.

$$\begin{aligned}\Delta H°_{\text{solution}} &= \Delta H°_L + \Delta H_{\text{hydration}}\\ &= 2153 \text{ kJ}\cdot\text{mol}^{-1} + (-2204 \text{ kJ}\cdot\text{mol}^{-1})\\ &= -51 \text{ kJ}\cdot\text{mol}^{-1}\end{aligned}$$

8.43 All the enthalpies of solution are positive. Those of the alkali metal chlorides increase as the cation becomes larger and less strongly hydrated by water. All the alkali metal chlorides are soluble in water, but AgCl is not. AgCl has a relatively large positive enthalpy of solution. When dissolving is highly endothermic, the small increase in disorder due to solution formation may not be enough to compensate for the decrease in disorder for the surroundings, and a solution does not form. This is the case for AgCl.

8.45 (a) $P = x_{\text{solvent}} \times P_{\text{pure solvent}}$

At 100°C, the normal boiling point of water, the vapor pressure of water is 1.00 atm. If the mole fraction of sucrose is 0.100, then the mole fraction of water is 0.900

$P = 0.900 \times 1.000 \text{ atm} = 0.900 \text{ atm or } 684 \text{ Torr}$

(b) First, the molality must be converted to mole fraction. If the molality is 0.100 $mol \cdot kg^{-1}$, then there will be 0.100 mol sucrose per 1000 g of water.

$$x_{H_2O} = \frac{n_{H_2O}}{n_{H_2O} + n_{sucrose}} = \frac{\dfrac{1000\text{ g}}{18.02\text{ g}\cdot\text{mol}^{-1}}}{\dfrac{1000\text{ g}}{18.02\text{ g}\cdot\text{mol}^{-1}} + 0.100\text{ mol}} = 0.998$$

$P = 0.998 \times 1.000\text{ atm} = 0.998\text{ atm}$ or 758 Torr

8.47 (a) The vapor pressure of water at 0°C is 4.58 Torr. The concentration of the solution must be converted to mole fraction in order to perform the calculation. A solution that is 1.00% ethylene glycol by mass will contain 1.00 g of ethylene glycol per 99.00 g of water.

$$x_{H_2O} = \frac{n_{H_2O}}{n_{H_2O} + n_{\text{ethylene glycoal}}} = \frac{\dfrac{99.00\text{ g}}{18.02\text{ g}\cdot\text{mol}^{-1}}}{\dfrac{99.00\text{ g}}{18.02\text{ g}\cdot\text{mol}^{-1}} + \dfrac{1.00\text{ g}}{62.07\text{ g}\cdot\text{mol}^{-1}}} = 0.997$$

$P = x_{\text{solvent}} \times P_{\text{pure solvent}}$

$P = 0.997 \times 4.58\text{ Torr} = 4.57\text{ Torr}$

(b) The vapor pressure of water is 355.26 Torr at 80°C. The concentration given in $mol \cdot kg^{-1}$ must be converted to mole fraction.

$$x_{H_2O} = \frac{n_{H_2O}}{n_{H_2O} + n_{Na^+} + n_{OH^-}} = \frac{\dfrac{1000\text{ g}}{18.02\text{ g}\cdot\text{mol}^{-1}}}{\dfrac{1000\text{ g}}{18.02\text{ g}\cdot\text{mol}^{-1}} + 2 \times 0.100\text{ mol}} = 0.9964$$

$P = 0.9964 \times 355.26\text{ Torr} = 354.0\text{ Torr}$

(c) At 10°C, the vapor pressure of water is 9.21 Torr. The concentration must be expressed in terms of mole fraction:

$$x_{H_2O} = \frac{n_{H_2O}}{n_{H_2O} + n_{\text{urea}}} = \frac{\dfrac{100\text{ g}}{18.02\text{ g}\cdot\text{mol}^{-1}}}{\dfrac{100\text{ g}}{18.02\text{ g}\cdot\text{mol}^{-1}} + \dfrac{6.60\text{ g}}{60.06\text{ g}\cdot\text{mol}^{-1}}} = 0.981$$

$P = 0.981 \times 9.21\text{ Torr} = 9.03\text{ Torr}$

The change in the vapor pressure will therefore be $9.21 - 9.03 = 0.18$ Torr

8.49 (a) From the relationship $P = x_{\text{solvent}} \times P_{\text{pure solvent}}$ we can calculate the mole fraction of the solvent:

$94.8\text{ Torr} = x_{\text{solvent}} \times 100.0\text{ Torr}$

$x_{\text{solvent}} = 0.948$

The mole fraction of the unknown compound will be 1.000 − 0.948 = 0.052.

(b) The molar mass can be calculated by using the definition of mole fraction for either the solvent or solute. In this case, the math is slightly easier if the definition of mole fraction of the solvent is used:

$$x_{\text{solvent}} = \frac{n_{\text{solvent}}}{n_{\text{unknown}} + n_{\text{solvent}}}$$

$$0.948 = \frac{\dfrac{100\ \text{g}}{78.11\ \text{g}\cdot\text{mol}^{-1}}}{\dfrac{100\ \text{g}}{78.11\ \text{g}\cdot\text{mol}^{-1}} + \dfrac{8.05\ \text{g}}{\text{M}_{\text{unknown}}}}$$

$$\text{M}_{\text{unknown}} = \frac{8.05\ \text{g}}{\left[\left(\dfrac{100\ \text{g}}{78.11\ \text{g}\cdot\text{mol}^{-1}}\right)\Big/0.948\right] - \left(\dfrac{100\ \text{g}}{78.11\ \text{g}\cdot\text{mol}^{-1}}\right)} = 115\ \text{g}\cdot\text{mol}^{-1}$$

8.51 (a) $\Delta T_b = ik_b m$

Because sucrose is a nonelectrolyte, $i = 1$.

$\Delta T_b = 0.51\ \text{K}\cdot\text{kg}\cdot\text{mol}^{-1} \times 0.10\ \text{mol}\cdot\text{kg}^{-1} = 0.051\ \text{K}$ or 0.051°C

The boiling point will be 100.000°C + 0.051°C.

(b) $\Delta T_b = ik_b m$

For NaCl, $i = 2$.

$\Delta T_b = 2 \times 0.51\ \text{K}\cdot\text{kg}\cdot\text{mol}^{-1} \times 0.22\ \text{mol}\cdot\text{kg}^{-1} = 0.22\ \text{K}$ or 0.22°C

The boiling point will be 100.00°C + 0.22°C = 100.22°C.

(c) $\Delta T_b = ik_b m$

$$\Delta T_b = 2 \times 0.51\ \text{K}\cdot\text{kg}\cdot\text{mol}^{-1} \times \frac{\left(\dfrac{0.230\ \text{g}}{25.94\ \text{g}\cdot\text{mol}^{-1}}\right)}{0.100\ \text{kg}} = 0.090\ \text{K or } 0.090\text{°C}$$

The boiling point will be 100.000°C + 0.090°C or 100.090°C.

8.53 (a) Pure water has a vapor pressure of 760.00 Torr at 100°C. The mole fraction of the solution can be determined from $P = x_{\text{solvent}} \times P_{\text{pure solvent}}$.

751 Torr = $x_{\text{solvent}} \times 760.00$ Torr

$x_{\text{solvent}} = 0.988$

The mole fraction needs to be converted to molality:

$$x_{\text{solvent}} = 0.988 = \frac{n_{H_2O}}{n_{H_2O} + n_{\text{solute}}}$$

Because the absolute amount of solution is not important, we can assume that the total number of moles = 1.00.

$$0.988 = \frac{n_{H_2O}}{1.00\ \text{mol}}$$

$n_{H_2O} = 0.988$ mol; $n_{solute} = 0.012$ mol

$$\text{molality} = \frac{0.012 \text{ mol}}{\left(\frac{0.988 \text{ mol} \times 18.02 \text{ g}\cdot\text{mol}^{-1}}{1000 \text{ g}\cdot\text{kg}^{-1}}\right)} = 0.67 \text{ mol}\cdot\text{kg}^{-1}$$

Knowing the mole fraction, one can calculate ΔT_b:

$\Delta T_b = k_b m$

$\Delta T_b = 0.51 \text{ K}\cdot\text{mol}\cdot\text{kg}^{-1} \times 0.67 \text{ mol}\cdot\text{kg}^{-1} = 0.34 \text{ K or } 0.34°\text{C}$

Boiling point = 100.00°C + 0.34°C = 100.34°C

(b) The procedure is the same as in (a). Pure benzene has a vapor pressure of 760.00 Torr at 80.1°C. The mole fraction of the solution can be determined from $P = x_{solvent} \times P_{pure\ solvent}$.

740 Torr = $x_{solvent} \times 760.00$ Torr

$x_{solvent} = 0.974$

The mole fraction needs to be converted to molality:

$$x_{solvent} = 0.974 = \frac{n_{benzene}}{n_{benzene} + n_{solute}}$$

Because the absolute amount of solution is not important, we can assume that the total number of moles = 1.00:

$$0.974 = \frac{n_{benzene}}{1.00 \text{ mol}}$$

$n_{benzene} = 0.974$ mol; $n_{solute} = 0.026$ mol

$$\text{molality} = \frac{0.026 \text{ mol}}{\left(\frac{0.974 \text{ mol} \times 78.11 \text{ g}\cdot\text{mol}^{-1}}{1000 \text{ g}\cdot\text{kg}^{-1}}\right)} = 0.34 \text{ mol}\cdot\text{kg}^{-1}$$

Knowing the mole fraction, one can calculate ΔT_b:

$\Delta T_b = k_b m$

$\Delta T_b = 2.53 \text{ K}\cdot\text{kg}\cdot\text{mol}^{-1} \times 0.34 \text{ mol}\cdot\text{kg}^{-1} = 0.86 \text{ K or } 0.86\ °\text{C}$

Boiling point = 80.1°C + 0.86°C = 81.0°C

8.55 $\Delta T_b = 61.51°\text{C} - 61.20°\text{C} = 0.31°\text{C or } 0.31 \text{ K}$

$\Delta T_b = k_b \times \text{molality}$

$0.31 \text{ K} = 4.95 \text{ K}\cdot\text{kg}\cdot\text{mol}^{-1} \times \text{molality}$

$$0.31 \text{ K} = 4.95 \text{ K}\cdot\text{kg}\cdot\text{mol}^{-1} \times \frac{\left(\frac{1.05 \text{ g}}{M_{unknown}}\right)}{0.100 \text{ kg}}$$

$$\frac{0.100 \text{ kg} \times 0.31 \text{ K}}{4.95 \text{ K}\cdot\text{kg}\cdot\text{mol}^{-1}} = \frac{1.05 \text{ g}}{M_{unknown}}$$

$$M_{unknown} = \frac{1.05 \text{ g} \times 4.95 \text{ K}\cdot\text{kg}\cdot\text{mol}^{-1}}{0.100 \text{ kg} \times 0.31 \text{ K}} = 1.7 \times 10^2 \text{ g}\cdot\text{mol}^{-1}$$

8.57 (a) $\Delta T_f = ik_f m$

Because sucrose is a nonelectrolyte, $i = 1$.

$\Delta T_f = 1.86\ \text{K}\cdot\text{kg}\cdot\text{mol}^{-1} \times 0.10\ \text{mol}\cdot\text{kg}^{-1} = 0.19\ \text{K or } 0.19°\text{C}$

The freezing point will be 0.000°C − 0.19°C = −0.19°C.

(b) $\Delta T_f = ik_f m$

For NaCl, $i = 2$.

$\Delta T_f = 2 \times 1.86\ \text{K}\cdot\text{kg}\cdot\text{mol}^{-1} \times 0.22\ \text{mol}\cdot\text{kg}^{-1} = 0.82\ \text{K or } 0.82°\text{C}$

The freezing point will be 0.00°C − 0.82°C = −0.82°C.

(c) $\Delta T_f = ik_f m$

$i = 2$ for LiF.

$$\Delta T_f = 2 \times 1.86\ \text{K}\cdot\text{kg}\cdot\text{mol}^{-1} \times \frac{\left(\dfrac{0.120\ \text{g}}{25.94\ \text{g}\cdot\text{mol}^{-1}}\right)}{0.100\ \text{kg}} = 0.172\ \text{K or } 0.172°\text{C}$$

The freezing point will be 0.000°C − 0.172°C = −0.172°C.

8.59 $\Delta T_f = 179.8°\text{C} - 177.3°\text{C} = 2.5°\text{C or } 2.5\ \text{K}$

$\Delta T_f = k_f m$

$2.5\ \text{K} = (39.7\ \text{K}\cdot\text{kg}\cdot\text{mol}^{-1})m$

$$2.5\ \text{K} = (39.7\ \text{K}\cdot\text{kg}\cdot\text{mol}^{-1})\frac{\left(\dfrac{1.14\ \text{g}}{M_{\text{unknown}}}\right)}{0.100\ \text{kg}}$$

$$\frac{0.100\ \text{kg} \times 2.0\ \text{K}}{39.7\ \text{K}\cdot\text{kg}\cdot\text{mol}^{-1}} = \frac{1.14\ \text{g}}{M_{\text{unknown}}}$$

$$M_{\text{unknown}} = \frac{1.14\ \text{g} \times 39.7\ \text{K}\cdot\text{kg}\cdot\text{mol}^{-1}}{0.100\ \text{kg} \times 2.5\ \text{K}} = 1.8 \times 10^2\ \text{g}\cdot\text{mol}^{-1}$$

8.61 (a) First, calculate the molality using the change in boiling point, and then use that value to calculate the change in freezing point.

$\Delta T_b = k_b m$

$\Delta T_b = 82.0°\text{C} - 80.1°\text{C} = 1.9°\text{C or } 1.9\ \text{K}$

$1.9\ \text{K} = 2.53\ \text{K}\cdot\text{kg}\cdot\text{mol}^{-1} \times \text{molality}$

$\text{molality} = 0.75\ \text{mol}\cdot\text{kg}^{-1}$

$\Delta T_f = k_f m$

$\Delta T_f = 5.12\ \text{K}\cdot\text{kg}\cdot\text{mol}^{-1} \times 0.75\ \text{mol}\cdot\text{kg}^{-1}$

$\Delta T_f = 3.84\ \text{K or } 3.84°\text{C}$

The freezing point will be 5.5°C − 3.84°C = 1.7°C

(b) $\Delta T_f = k_f m$

Because the freezing point of water = 0.00°C, the freezing point of the solution equals the freezing point depression.

$3.04\ \text{K} = 1.86\ \text{K}\cdot\text{kg}\cdot\text{mol}^{-1} \times \text{molality}$

$\text{molality} = 1.63\ \text{mol}\cdot\text{kg}^{-1}$

(c) $\Delta T_f = k_f m$

$1.94\ \text{K} = (1.86\ \text{K}\cdot\text{kg}\cdot\text{mol}^{-1})m$

$$1.94\ \text{K} = 1.86\ \text{K}\cdot\text{kg}\cdot\text{mol}^{-1} \times \frac{n_{\text{solute}}}{\text{kg (solvent)}}$$

$$1.94\ \text{K} = 1.86\ \text{K}\cdot\text{kg}\cdot\text{mol}^{-1} \times \frac{n_{\text{solute}}}{0.200\ \text{kg}}$$

$n_{\text{solute}} = 0.209\ \text{mol}$

8.63 (a) A 1.00% aqueous solution of NaCl will contain 1.00 g of NaCl for 99.0 g of water. To use the freezing point depression equation, we need the molality of the solution:

$$\text{molality} = \frac{\left(\dfrac{1.00\ \text{g}}{58.44\ \text{g}\cdot\text{mol}^{-1}}\right)}{0.0990\ \text{kg}} = 0.173\ \text{mol}\cdot\text{kg}^{-1}$$

$\Delta T_f = i\,k_f m$

$\Delta T_f = i\,(1.86\ \text{K}\cdot\text{kg}\cdot\text{mol}^{-1})(0.173\ \text{mol}\cdot\text{kg}^{-1}) = 0.593\ \text{K}$

$i = 1.84$

(b) molality of all solute species (undissociated NaCl(aq) plus $Na^+(aq) + Cl^-(aq)$)
$= 1.84 \times 0.173\ \text{mol}\cdot\text{kg}^{-1} = 0.318\ \text{mol}\cdot\text{kg}^{-1}$

(c) If all the NaCl had dissociated, the total molality in solution would have been $0.346\ \text{mol}\cdot\text{kg}^{-1}$, giving an i value equal to 2. If no dissociation had taken place, the molality in solution would have equaled $0.173\ \text{mol}\cdot\text{kg}^{-1}$.

$$\text{NaCl(aq)} \rightleftharpoons Na^+(aq) + Cl^-(aq)$$

$0.173\ \text{mol}\cdot\text{kg}^{-1} - x \qquad x \qquad x$

$0.173\ \text{mol}\cdot\text{kg}^{-1} - x + x + x = 0.318\ \text{mol}\cdot\text{kg}^{-1}$

$0.173\ \text{mol}\cdot\text{kg}^{-1} + x = 0.318\ \text{mol}\cdot\text{kg}^{-1}$

$x = 0.145\ \text{mol}\cdot\text{kg}^{-1}$

$$\%\ \text{dissociation} = \frac{0.145\ \text{mol}\cdot\text{kg}^{-1}}{0.173\ \text{mol}\cdot\text{kg}^{-1}} \times 100 = 83.8\%$$

8.65 For an electrolyte that dissociates into two ions, the van't Hoff i factor will be 1 plus the degree of dissociation, in this case 0.075. This can be readily seen for the general case MX. Let A = initial concentration of MX (if none is dissociated) and let Y = the concentration of MX that subsequently dissociates:

$$\begin{array}{lll} MX(aq) \longrightarrow & M^{n+}(aq) + & X^{n-}(aq) \\ A - Y & Y & Y \end{array}$$

The total concentration of solute species is (A − Y) + Y + Y = A + Y

The value of i will then be equal to A + Y or 1.075.

The freezing point change is then easy to calculate:

$$\begin{aligned} \Delta T_f &= ik_f m \\ &= 1.075 \times 1.86\ \text{K·kg·mol}^{-1} \times 0.10\ \text{mol·kg}^{-1} \\ &= 0.20 \end{aligned}$$

Freezing point of the solution will be 0.00°C − 0.20°C = −0.20°C.

8.67 (a) $\Pi = iRT \times \text{molarity}$

$$\begin{aligned} &= 1 \times 0.082\,06\ \text{L·atm·K}^{-1}\text{·mol}^{-1} \times 293\ \text{K} \times 0.010\ \text{mol·L}^{-1} \\ &= 0.24\ \text{atm or } 1.8 \times 10^2\ \text{Torr} \end{aligned}$$

(b) Because HCl is a strong acid, it should dissociate into two ions, H^+ and Cl^-, so $i = 2$.

$$\begin{aligned} \Pi &= 2 \times 0.082\,06\ \text{L·atm·K}^{-1}\text{·mol}^{-1} \times 293\ \text{K} \times 1.0\ \text{mol·L}^{-1} \\ &= 48\ \text{atm} \end{aligned}$$

(c) $CaCl_2$ should dissociate into 3 ions in solution, therefore $i = 3$.

$$\begin{aligned} \Pi &= 3 \times 0.082\,06\ \text{L·atm·K}^{-1}\text{·mol}^{-1} \times 293\ \text{K} \times 0.010\ \text{mol·L}^{-1} \\ &= 0.72\ \text{atm or } 5.5 \times 10^2\ \text{Torr} \end{aligned}$$

8.69 The polypeptide is a nonelectrolyte, so $i = 1$.

$\Pi = iRT \times \text{molarity}$

$$\Pi = \frac{3.74\ \text{Torr}}{760\ \text{Torr·atm}^{-1}} = 1 \times 0.082\,06\ \text{L·atm·K}^{-1}\text{·mol}^{-1} \times 300\ \text{K} \times \frac{\left(\dfrac{0.40\ \text{g}}{M_{\text{unknown}}}\right)}{1.0\ \text{L}}$$

$$\begin{aligned} M_{\text{unknown}} &= \frac{0.082\,06\ \text{L·atm·K}^{-1}\text{·mol}^{-1} \times 300\ \text{K} \times 0.40\ \text{g} \times 760\ \text{Torr·atm}^{-1}}{3.74\ \text{Torr} \times 1.0\ \text{L}} \\ &= 2.0 \times 10^3\ \text{g·mol}^{-1} \end{aligned}$$

8.71 We assume the polymer to be a nonelectrolyte, so $i = 1$.

$\Pi = iRT \times \text{molarity}$

$$\Pi = \frac{5.4\ \text{Torr}}{760\ \text{Torr·atm}^{-1}} = 1 \times 0.082\,06\ \text{L·atm·K}^{-1}\text{·mol}^{-1} \times 293\ \text{K} \times \frac{\left(\dfrac{0.10\ \text{g}}{M_{\text{unknown}}}\right)}{0.100\ \text{L}}$$

$$\begin{aligned} M_{\text{unknown}} &= \frac{0.082\,06\ \text{L·atm·K}^{-1}\text{·mol}^{-1} \times 293\ \text{K} \times 0.10\ \text{g} \times 760\ \text{Torr·atm}^{-1}}{5.4\ \text{Torr} \times 0.100\ \text{L}} \\ &= 3.4 \times 10^3\ \text{g·mol}^{-1} \end{aligned}$$

8.73 (a) $C_{12}H_{22}O_{11}$ should be a nonelectrolyte, so $i = 1$.

$\Pi = iRT \times \text{molarity}$

$= 1 \times 0.082\,06\ \text{L}\cdot\text{atm}\cdot\text{K}^{-1}\cdot\text{mol}^{-1} \times 293\ \text{K} \times 0.050\ \text{mol}\cdot\text{L}^{-1}$

$= 1.2\ \text{atm}$

(b) NaCl dissociates to give 2 ions in solution, so $i = 2$.

$\Pi = iRT \times \text{molarity}$

$= 2 \times 0.082\,06\ \text{L}\cdot\text{atm}\cdot\text{K}^{-1}\cdot\text{mol}^{-1} \times 293\ \text{K} \times 0.0010\ \text{mol}\cdot\text{L}^{-1}$

$= 0.048\ \text{atm}$ or 36 Torr

(c) AgCN dissociates in solution to give two ions (Ag^+ and CN^-), so $i = 2$.

$\Pi = iRT \times \text{molarity}$

We must assume that the AgCN does not significantly affect either the volume or density of the solution, which is reasonable given the very small amount of it that dissolves.

$$\Pi = 2 \times 0.082\,06\ \text{L}\cdot\text{atm}\cdot\text{K}^{-1}\cdot\text{mol}^{-1} \times 293\ \text{K} \times \frac{\left(\dfrac{2.3 \times 10^{-5}\ \text{g}}{133.89\ \text{g}\cdot\text{mol}^{-1}}\right)}{\left(\dfrac{100\ \text{g H}_2\text{O}}{1\ \text{g}\cdot\text{cm}^{-3}\ \text{H}_2\text{O} \times 1000\ \text{cm}^3\cdot\text{L}^{-1}}\right)}$$

$= 8.3 \times 10^{-5}\ \text{atm}$

8.75 (a) To determine the vapor pressure of the solution, we need to know the mole fraction of each component.

$$x_{\text{benzene}} = \frac{1.50\ \text{mol}}{1.50\ \text{mol} + 0.50\ \text{mol}} = 0.75$$

$x_{\text{toluene}} = 1 - x_{\text{benzene}} = 0.25$

$P_{\text{total}} = (0.75 \times 94.6\ \text{Torr}) + (0.25 \times 29.1\ \text{Torr}) = 78.2\ \text{Torr}$

The vapor phase composition will be given by

$$x_{\text{benzene in vapor phase}} = \frac{P_{\text{benzene}}}{P_{\text{total}}} = \frac{0.75 \times 94.6\ \text{Torr}}{78.2\ \text{Torr}} = 0.91$$

$x_{\text{toluene in vapor phase}} = 1 - 0.91 = 0.09$

The vapor is richer in the more volatile benzene, as expected.

(b) The procedure is the same as in (a) but the number of moles of each component must be calculated first:

$$n_{\text{benzene}} = \frac{15.0\ \text{g}}{78.11\ \text{g}\cdot\text{mol}^{-1}} = 0.192$$

$$n_{\text{toluene}} = \frac{65.3\ \text{g}}{92.14\ \text{g}\cdot\text{mol}^{-1}} = 0.709$$

$$x_{\text{benzene}} = \frac{0.192 \text{ mol}}{0.192 \text{ mol} + 0.709 \text{ mol}} = 0.213$$

$x_{\text{toluene}} = 1 - x_{\text{benzene}} = 0.787$

$P_{\text{total}} = (0.213 \times 94.6 \text{ Torr}) + (0.787 \times 29.1 \text{ Torr}) = 43.0 \text{ Torr}$

The vapor phase composition will be given by

$$x_{\text{benzene in vapor phase}} = \frac{P_{\text{benzene}}}{P_{\text{total}}} = \frac{0.213 \times 94.6 \text{ Torr}}{43.0 \text{ Torr}} = 0.469$$

$x_{\text{toluene in vapor phase}} = 1 - 0.469 = 0.531$

8.77 To calculate this quantity, we must first find the mole fraction of each that will be present in the mixture. This value is obtained from the relationship

$$P_{\text{total}} = [x_{\text{1,1-dichloroethane}} \times P_{\text{pure 1,1-dichloroethane}}] + [x_{\text{1,1-dichlorotetrafluoroethane}} \times P_{\text{pure 1,1-dichlorotetrafluoroethane}}]$$

$157 \text{ Torr} = [x_{\text{1,1-dichloroethane}} \times 228 \text{ Torr}] + [x_{\text{1,1-dichlorotetrafluoroethane}} \times 79 \text{ Torr}]$

$157 \text{ Torr} = [x_{\text{1,1-dichloroethane}} \times 228 \text{ Torr}] + [(1 - x_{\text{1,1-dichloroethane}}) \times 79 \text{ Torr}]$

$157 \text{ Torr} = 79 \text{ Torr} + [(x_{\text{1,1-dichloroethane}} \times 228 \text{ Torr}) - (x_{\text{1,1-dichloroethane}} \times 79 \text{ Torr})]$

$78 \text{ Torr} = x_{\text{1,1-dichloroethane}} \times 149 \text{ Torr}$

$x_{\text{1,1-dichloroethane}} = 0.52$

$x_{\text{1,1-dichlorotetrafluoroethane}} = 1 - 0.52 = 0.48$

To calculate the number of grams of 1,1-dichloroethane, we use the definition of mole fraction. Mathematically, it is simpler to use the $x_{\text{1,1-dichlorotetrafluoroethane}}$ definition:

$$n_{\text{1,1-dichlorotetrafluoroethane}} = \frac{100.0 \text{ g}}{170.92 \text{ g} \cdot \text{mol}^{-1}} = 0.5851 \text{ mol}$$

$$x_{\text{1,1-dichlorotetrafluoroethane}} = \frac{0.5851 \text{ mol}}{0.5851 \text{ mol} + \dfrac{m}{98.95 \text{ g} \cdot \text{mol}^{-1}}} = 0.48$$

$$0.5851 \text{ mol} = 0.48 \times \left[0.5851 \text{ mol} + \frac{m}{98.95 \text{ g} \cdot \text{mol}^{-1}}\right]$$

$$0.5851 \text{ mol} - (0.48 \times 0.5851 \text{ mol}) = 0.48 \times \left[\frac{m}{98.95 \text{ g} \cdot \text{mol}^{-1}}\right]$$

$0.3042 \text{ mol} = 0.004\,851 \text{ mol} \cdot \text{g}^{-1} \times m$

$m = 63 \text{ g}$

8.79 Raoult's Law applies to the vapor pressure of the mixture, so positive deviation means that the vapor pressure is higher than expected for an ideal solution. Negative deviation means that the vapor pressure is lower than expected for an ideal solution. Negative deviation will occur when the interactions between the different molecules are somewhat stronger than the interactions between molecules of the same kind. (a) For methanol and ethanol, we expect the types of intermolecular attractions in the mixture to be similar to those in the component liquids, so that an ideal solution is predicted. (b) For HF and H_2O, the possibility of intermolecular hydrogen bonding between water and HF would suggest that negative deviation would be observed, which is the case. HF and H_2O form an azeotrope that boils at 111°C, a temperature higher than the boiling pont of either HF (19.4°C) or water. (c) Because hexane is nonpolar and water is polar with hydrogen bonding, we would expect a mixture of these two to exhibit positive deviation (the interactions between the different molecules would be weaker than the intermolecular forces between like molecules). Hexane and water do form an azeotrope that boils at 61.6°C, a temperature below the boiling point of either hexane or water.

8.81 (a) stronger; (b) low; (c) high; (d) weaker; (e) weak, low; (f) low; (g) strong, high

8.83 (a, b) Viscosity and surface tension decrease with increasing temperature; at high temperatures the molecules readily move away from their neighbors because of increased kinetic energy. (c) Evaporation rate and vapor pressure increase with increasing temperature because the kinetic energy of the molecules increases with temperature, and the molecules are more likely to escape into the gas phase.

8.85 If the external pressure is lowered, water boils at a lower temperature. The boiling temperature is the temperature at which the vapor pressure equals the external pressure. Vapor pressure increases with temperature. If the external pressure is reduced, the vapor pressure of the water will reach that value at a lower temperature; thus, lowering the external pressure lowers the boiling temperature.

8.87 (a) The quantities ΔH°_{vap} and ΔS°_{vap} can be calculated using the relationship

$$\ln P = \frac{\Delta H^\circ_{vap}}{R} \cdot \frac{1}{T} + \frac{\Delta S^\circ_{vap}}{R}$$

Because we have two temperatures with corresponding vapor pressures, we can set up two equations with two unknowns and solve for ΔH°_{vap} and ΔS°_{vap}. If the equation is used as is, P must be expressed in atm, which is the standard reference state. Remember that the value used for P is really activity which, for pressure, is P

divided by the reference state of 1 atm, so that the quantity inside the ln term is dimensionless.

$$8.314\ \text{J}\cdot\text{K}^{-1}\cdot\text{mol}^{-1} \times \ln\frac{35\ \text{Torr}}{760\ \text{Torr}} = -\frac{\Delta H^\circ_{\text{vap}}}{161.2\ \text{K}} + \Delta S^\circ_{\text{vap}}$$

$$8.314\ \text{J}\cdot\text{K}^{-1}\cdot\text{mol}^{-1} \times \ln\frac{253\ \text{Torr}}{760\ \text{Torr}} = -\frac{\Delta H^\circ_{\text{vap}}}{189.6\ \text{K}} + \Delta S^\circ_{\text{vap}}$$

which give, upon combining terms,

$$-25.59\ \text{J}\cdot\text{K}^{-1}\cdot\text{mol}^{-1} = -0.006\,203\ \text{K}^{-1} \times \Delta H^\circ_{\text{vap}} + \Delta S^\circ_{\text{vap}}$$

$$-9.145\ \text{J}\cdot\text{K}^{-1}\cdot\text{mol}^{-1} = -0.005\,274\ \text{K}^{-1} \times \Delta H^\circ_{\text{vap}} + \Delta S^\circ_{\text{vap}}$$

Subtracting one equation from the other will eliminate the $\Delta S^\circ_{\text{vap}}$ term and allow us to solve for $\Delta H^\circ_{\text{vap}}$:

$$-16.45\ \text{J}\cdot\text{K}^{-1}\cdot\text{mol}^{-1} = -0.000\,929 \times \Delta H^\circ_{\text{vap}}$$

$$\Delta H^\circ_{\text{vap}} = +17.7\ \text{kJ}\cdot\text{mol}^{-1}$$

(b) We can then use $\Delta H^\circ_{\text{vap}}$ to calculate $\Delta S^\circ_{\text{vap}}$ using either of the two equations:

$$-25.59\ \text{J}\cdot\text{K}^{-1}\cdot\text{mol}^{-1} = -0.006\,203\ \text{K}^{-1} \times (+17\,700\ \text{J}\cdot\text{mol}^{-1}) + \Delta S^\circ_{\text{vap}}$$

$$\Delta S^\circ_{\text{vap}} = 84\ \text{J}\cdot\text{K}^{-1}\cdot\text{mol}^{-1}$$

$$-9.145\ \text{J}\cdot\text{K}^{-1}\cdot\text{mol}^{-1} = -0.005\,274\ \text{K}^{-1} \times (+17\,700\ \text{J}\cdot\text{mol}^{-1}) + \Delta S^\circ_{\text{vap}}$$

$$\Delta S^\circ_{\text{vap}} = 84\ \text{J}\cdot\text{K}^{-1}\cdot\text{mol}^{-1}$$

(c) The $\Delta G^\circ_{\text{vap}}$ is calculated using $\Delta G^\circ_{\text{r}} = \Delta H^\circ_{\text{r}} - T\Delta S^\circ_{\text{r}}$

$$\Delta G^\circ_{\text{r}} = +17.7\ \text{kJ}\cdot\text{mol}^{-1} - (298\ \text{K})(84\ \text{J}\cdot\text{K}^{-1}\cdot\text{mol}^{-1})/(1000\ \text{J}\cdot\text{kJ}^{-1})$$

$$\Delta G^\circ_{\text{r}} = -7.3\ \text{kJ}\cdot\text{mol}^{-1}$$

Notice that the standard $\Delta G^\circ_{\text{r}}$ is negative, so that the vaporization of arsine is spontaneous, which is as expected; under those conditions arsine is a gas at room temperature.

(d) The boiling point can be calculated from one of several methods. The easiest to use is that developed in the last chapter:

$$\Delta G^\circ_{\text{vap}} = \Delta H^\circ_{\text{vap}} - T_B\Delta S^\circ_{\text{vap}} = 0$$

$$\Delta H^\circ_{\text{vap}} = T_{\text{B}}\,\Delta S^\circ_{\text{vap}} \text{ or } T_{\text{B}} = \frac{\Delta H^\circ_{\text{vap}}}{\Delta S^\circ_{\text{vap}}}$$

$$T_{\text{B}} = \frac{17.7\ \text{kJ}\cdot\text{mol}^{-1} \times 1000\ \text{J}\cdot\text{kJ}^{-1}}{84\ \text{J}\cdot\text{K}^{-1}\cdot\text{mol}^{-1}} = 211\ \text{K}$$

Alternatively, we could use the relationship $\ln\frac{P_2}{P_1} = -\frac{\Delta H^\circ_{\text{vap}}}{R}\left[\frac{1}{T_2} - \frac{1}{T_1}\right]$. Here, we would substitute, in one of the known vapor pressure points, the value of the enthalpy of vaporization and the condition that $P = 1$ atm at the normal boiling point.

8.89 (a) The quantities $\Delta H^\circ_{\text{vap}}$ and $\Delta S^\circ_{\text{vap}}$ can be calculated, using the relationship

$$\ln P = -\frac{\Delta H^\circ_{vap}}{R}\cdot\frac{1}{T} + \frac{\Delta S^\circ_{vap}}{R}$$

Because we have two temperatures with corresponding vapor pressures (we know that the vapor pressure = 1 atm at the boiling point), we can set up two equations with two unknowns and solve for ΔH°_{vap} and ΔS°_{vap}. If the equation is used as is, P must be expressed in atm which is the standard reference state. Remember that the value used for P is really activity which, for pressure, is P divided by the reference state of 1 atm, so that the quantity inside the ln term is dimensionless.

$$8.314\ \text{J}\cdot\text{K}^{-1}\cdot\text{mol}^{-1} \times \ln 1 = -\frac{\Delta H^\circ_{vap}}{315.58\ \text{K}} + \Delta S^\circ_{vap}$$

$$8.314\ \text{J}\cdot\text{K}^{-1}\cdot\text{mol}^{-1} \times \ln \frac{140\ \text{Torr}}{760\ \text{Torr}} = -\frac{\Delta H^\circ_{vap}}{273.2\ \text{K}} + \Delta S^\circ_{vap}$$

which give, upon combining terms,

$0\ \text{J}\cdot\text{K}^{-1}\cdot\text{mol}^{-1} = -0.003\ 169\ \text{K}^{-1} \times \Delta H^\circ_{vap} + \Delta S^\circ_{vap}$

$-14.06\ \text{J}\cdot\text{K}^{-1}\cdot\text{mol}^{-1} = -0.003\ 660\ \text{K}^{-1} \times \Delta H^\circ_{vap} + \Delta S^\circ_{vap}$

Subtracting one equation from the other will eliminate the ΔS°_{vap} term and allow us to solve for ΔH°_{vap}.

$-14.06\ \text{J}\cdot\text{K}^{-1}\cdot\text{mol}^{-1} = -0.000\ 491\ \text{K}^{-1} \times \Delta H^\circ_{vap}$

$\Delta H^\circ_{vap} = +28.6\ \text{kJ}\cdot\text{mol}^{-1}$

(b) We can then use ΔH°_{vap} to calculate ΔS°_{vap} using either of the two equations:

$0 = -0.003\ 169\ \text{K}^{-1} \times (+28\ 600\ \text{J}\cdot\text{mol}^{-1}) + \Delta S^\circ_{vap}$

$\Delta S^\circ_{vap} = 90.6\ \text{J}\cdot\text{K}^{-1}\cdot\text{mol}^{-1}$

$-14.06\ \text{J}\cdot\text{K}^{-1}\cdot\text{mol}^{-1} = -0.003\ 660\ \text{K}^{-1} \times (+28\ 600\ \text{J}\cdot\text{mol}^{-1}) + \Delta S^\circ_{vap}$

$\Delta S^\circ_{vap} = 90.6\ \text{J}\cdot\text{K}^{-1}\cdot\text{mol}^{-1}$

(c) The ΔG°_{vap} is calculated using $\Delta G^\circ_r = \Delta H^\circ_r - T\Delta S^\circ_r$

$\Delta G^\circ_r = +28.6\ \text{kJ}\cdot\text{mol}^{-1} - (298\ \text{K})(90.6\ \text{J}\cdot\text{K}^{-1}\cdot\text{mol}^{-1})/(1000\ \text{J}\cdot\text{kJ}^{-1})$

$\Delta G^\circ_r = +1.60\ \text{kJ}\cdot\text{mol}^{-1}$

(d) The vapor pressure at another temperature is calculated using

$$\ln\frac{P_2}{P_1} = -\frac{\Delta H^\circ_{vap}}{R}\left[\frac{1}{T_2} - \frac{1}{T_1}\right]$$

We need to insert the calculated value of the enthalpy of vaporization and one of the known vapor pressure points:

$$\ln\frac{P_{at\ 25.0^\circ C}}{1\ \text{atm}} = -\frac{28\ 600\ \text{J}\cdot\text{mol}^{-1}}{8.314\ \text{J}\cdot\text{K}^{-1}\cdot\text{mol}^{-1}}\left[\frac{1}{298.2\ \text{K}} - \frac{1}{315.58\ \text{K}}\right]$$

$P_{at\ 25.0^\circ C} = 0.53\ \text{atm or}\ 4.0 \times 10^2\ \text{Torr}$

8.91 Table 6.2 contains the enthalpy of vaporization and the boiling point of methanol (at which the vapor pressure = 1 atm). Using this data and the equation

$$\ln \frac{P_2}{P_1} = -\frac{\Delta H°_{vap}}{R}\left[\frac{1}{T_2} - \frac{1}{T_1}\right]$$

$$\ln \frac{P_{25.0°C}}{1} = -\frac{35\ 300\ J\cdot mol^{-1}}{8.314\ J\cdot K^{-1}\cdot mol^{-1}}\left[\frac{1}{298.2\ K} - \frac{1}{337.2\ K}\right]$$

$P_{25.0°C} = 0.19$ atm or 1.5×10^2 Torr

8.93 The pressure increase would bring CO_2 into the solid region.

8.95 (a) $\frac{25.0\ \text{Torr}}{31.83\ \text{Torr}} \times 100 = 78.5\%$; (b) At 25°C the vapor pressure of water is only 23.76 Torr, so some of the water vapor in the air would condense as dew or fog.

8.97 (a) Molar mass of $CaCl_2\cdot 6\ H_2O = 219.08\ g\cdot mol^{-1}$, which consists of 110.98 g $CaCl_2$ and 108.10 g of water.

$$m_{CaCl_2} = \frac{x\ \text{mol}\ CaCl_2\cdot 6\ H_2O}{0.250\ \text{kg} + x\ (6 \times 0.018\ 02\ \text{kg}\ H_2O)}$$

Note: 18.02 g H_2O = 0.018 02 kg H_2O = 1.000 mol H_2O

x = number of moles of $CaCl_2\cdot 6\ H_2O$ needed to prepare a solution of molality m_{CaCl_2}, in which each mole of $CaCl_2\cdot 6\ H_2O$ produces 6(0.018 02 kg) of water as solvent (assuming we begin with 0.250 kg H_2O).

For a 0.10 *m* solution of $CaCl_2\cdot 6\ H_2O$,

$$0.10\ m = \frac{x}{0.250\ \text{kg} + x(6)(0.018\ 02\ \text{kg}\ H_2O)}$$

$x = 0.0250$ mol + x(0.0108 mol)

x − 0.0108 x = 0.0250 mol

0.989 x = 0.0250 mol

x = 0.0253 mol $CaCl_2\cdot 6\ H_2O$

$$\therefore (0.0253\ \text{mol}\ CaCl_2\cdot 6\ H_2O) \times \left(\frac{219.08\ \text{g}\ CaCl_2\cdot 6\ H_2O}{1\ \text{mol}\ CaCl_2\cdot 6\ H_2O}\right) =$$

5.54 g $CaCl_2\cdot 6\ H_2O$

(b) Molar mass of $NiSO_4\cdot 6\ H_2O = 262.86\ g\cdot mol^{-1}$, which consists of 154.77 g $NiSO_4$ and 108.09 g H_2O.

$$m_{NiSO_4} = \frac{x\ \text{mol}\ NiSO_4\cdot 6\ H_2O}{0.500\ \text{kg} + x\ (6 \times 0.018\ 02\ \text{kg}\ H_2O)}$$

where x = number of moles of $NiSO_4\cdot 6\ H_2O$ needed to prepare a solution of molality m_{NiSO_4}, in which each mole of $NiSO_4\cdot 6\ H_2O$ produces 6(0.018 02 kg) of water as solvent (assuming we begin with 0.500 kg H_2O) for a 0.22 *m* solution of $NiSO_4\cdot 6\ H_2O$,

$$0.22\ m = \frac{x}{0.500\ \text{kg} + x(6)(0.018\ 02\ \text{kg H}_2\text{O})}$$

$$x = 0.11\ \text{mol} + x(0.0238\ \text{mol})$$

$$x - 0.0238\ x = 0.11\ \text{mol}$$

$$0.976\ x = 0.11\ \text{mol}$$

$$x = 0.11\ \text{mol NiSO}_4\cdot 6\ \text{H}_2\text{O}$$

$$\therefore (0.11\ \text{mol NiSO}_4\cdot 6\ \text{H}_2\text{O}) \times \left(\frac{262.86\ \text{g NiSO}_4\cdot 6\ \text{H}_2\text{O}}{1\ \text{mol NiSO}_4\cdot 6\ \text{H}_2\text{O}}\right) = 29\ \text{g NiSO}_4\cdot 6\ \text{H}_2\text{O}$$

8.99 Assume 100 g of solution:

$$16.0\%\ C_{12}H_{22}O_{11}(aq)\ \text{is} \equiv \frac{16.0\ \text{g}\ C_{12}H_{22}O_{11}}{100\ \text{g soln}}$$

(a) $$\text{molality} = \left(\frac{16.0\ \text{g}\ C_{12}H_{22}O_{11}}{100\ \text{g soln}}\right)\left(\frac{1.0635\ \text{g soln}}{1\ \text{mL soln}}\right)\left(\frac{1\ \text{mL}}{10^{-3}\ \text{L}}\right)\left(\frac{1\ \text{mol}\ C_{12}H_{22}O_{11}}{342.30\ \text{g}\ C_{12}H_{22}O_{11}}\right)$$

$$= 0.497\ \text{mol}\cdot\text{L}^{-1}$$

(b) The solution contains 16.0 g $C_{12}H_{22}O_{11}$ in 84.0 g H_2O:

$$n_{C_{12}H_{22}O_{11}} = (16.0\ \text{g}\ C_{12}H_{22}O_{11})\left(\frac{1\ \text{mol}\ C_{12}H_{22}O_{11}}{342.30\ \text{g}\ C_{12}H_{22}O_{11}}\right)$$

$$= 0.0467\ C_{12}H_{22}O_{11}$$

$$n_{H_2O} = (84.0\ \text{g H}_2\text{O})\left(\frac{1\ \text{mol H}_2\text{O}}{18.02\ \text{g H}_2\text{O}}\right)$$

$$= 4.66\ \text{mol H}_2\text{O}$$

$$x_{H_2O} = \frac{4.66\ \text{mol H}_2\text{O}}{4.66\ \text{mol H}_2\text{O} + 0.0467\ \text{mol}\ C_{12}H_{22}O_{11}} = 0.990$$

$$P_{H_2O} = x_{H_2O}P_{\text{pure H}_2\text{O}} = 0.990 \times 17.54\ \text{Torr} = 17.4\ \text{Torr}$$

(c) $$\Delta T_b = ik_b m = (0.51\ \text{K}\cdot\text{kg}\cdot\text{mol}^{-1})\left(\frac{0.0468\ \text{mol}\ C_{12}H_{22}O_{11}}{0.0840\ \text{kg H}_2\text{O}}\right)$$

$$= 0.28\ \text{K or } 0.28°\text{C}$$

$$\text{b.p.} = 100.00°\text{C} + 0.28°\text{C} = 100.28°\text{C}$$

8.101 (a) At 30°C, the vapor pressure of pure water = 31.83 Torr. According to Raoult's law, $P = x_{\text{solvent}}\cdot P_{\text{pure}}$. To calculate the x_{solvent}: 0.50 m NaCl gives 0.50 moles Na^+ and 0.50 moles Cl^- per kg solvent. The number of moles of solvent is 1000 kg ÷ 18.02 $\text{g}\cdot\text{mol}^{-1}$ = 55.49 mol. The total number of moles = 0.50 mol + 0.50 mol + 55.49 mol = 56.49 mol.

$$x_{\text{solvent}} = \frac{55.49\ \text{mol}}{56.49\ \text{mol}} = 0.9823$$

$P = x_{\text{solvent}} \cdot P_{\text{pure}} = 0.9823 \times 31.83 \text{ Torr} = 31.26 \text{ Torr}$

(b) At 100°C, $P_{\text{pure}} = 760$ Torr: $P = x_{\text{solvent}} \cdot P_{\text{pure}} = 0.9823 \times 760 \text{ Torr} = 747$ Torr

At 0°C, $P_{\text{pure}} = 4.58$ Torr; $P = x_{\text{solvent}} \cdot P_{\text{pure}} = 0.9823 \times 4.58 \text{ Torr} = 4.50 \text{ Torr}$

8.103 (a) $\Delta T_f = 1.20°\text{C} - 5.5°\text{C} = -4.3°\text{C}$

$$\Delta T_f = ik_f m = ik_f \frac{\left(\dfrac{m_{\text{solute}}}{M_{\text{solute}}}\right)}{\text{kg solvent}}$$

$$M_{\text{solute}} = \frac{ik_f m_{\text{solute}}}{(\text{kg solvent})(\Delta T_f)} = \frac{(1)(5.12 \text{ K}\cdot\text{kg}\cdot\text{mol}^{-1})(10.0 \text{ g})}{(0.0800 \text{ kg})(4.3°\text{C})}$$

$$= 1.5 \times 10^2 \text{ g}\cdot\text{mol}^{-1}$$

(b) The empirical formula mass $= 73.4 \text{ g}\cdot\text{mol}^{-1}$; the experimental formula mass is roughly double this value, so the molecular formula is $C_6H_4Cl_2$.

(c) molar mass $= 146.99 \text{ g}\cdot\text{mol}^{-1}$

8.105 (a) The elemental analysis yields the following:

element	% by mass	mol in 100 g	ratio to smallest moles (0.542 for N)
C	59.0	4.91	9.04
O	26.2	1.64	3.02
H	7.10	7.04	13.0
N	7.60	0.542	1

The empirical formula is $C_9H_{13}O_3N$ (formula mass ~ 183 $\text{g}\cdot\text{mol}^{-1}$).

(b) and (c)

$$\Delta T_f = ik_f m = ik_f \frac{\left(\dfrac{\text{mass}_{\text{solute}}}{M_{\text{solute}}}\right)}{\text{kg solvent}}$$

$$M_{\text{solute}} = \frac{ik_f \text{mass}_{\text{solute}}}{(\text{kg solvent})(\Delta T_f)} = \frac{(1)(5.12 \text{ K}\cdot\text{kg}\cdot\text{mol}^{-1})(0.64 \text{ g})}{(0.036 \text{ kg})(0.50°\text{C})}$$

$$= 1.8 \times 10^2 \text{ g}\cdot\text{mol}^{-1}$$

The molecular formula is the same as the empirical formula, $C_9H_{13}O_3N$. Molar mass of epinephrine $= 183.20 \text{ g}\cdot\text{mol}^{-1}$.

8.107 $\Pi = n\dfrac{RT}{V} = \dfrac{m}{M}\dfrac{RT}{V}$

where m is the mass of unknown compound.

$$M = \frac{mRT}{V\Pi}$$

$$M = \frac{(0.166\ \text{g})(0.082\ 06\ \text{L}\cdot\text{atm}\cdot\text{K}^{-1}\cdot\text{mol}^{-1})(293\ \text{K})}{(0.010\ \text{L})\left(\dfrac{1.2\ \text{Torr}}{760\ \text{Torr}\cdot\text{atm}^{-1}}\right)} = 2.5 \times 10^5\ \text{g}\cdot\text{mol}^{-1}$$

8.109 Hypotonic solutions will cause a net flow of solvent into the cells to equalize the osmotic pressure. The cells will burst and die (hemolysis). Hypertonic solutions will cause a net flow of solvent out of the cells to equalize the osmotic pressure. The cells will shrink and die.

8.111 Because molar mass = $(iRT\ \text{mass}_{\text{solute}})/(\Pi \times V_{solution})$, the determination of the molar mass of a solute requires a measurement of mass, volume, temperature, and osmotic pressure. Osmotic pressures are generally large and can be determined quite accurately, yielding accurate molar masses. Boiling point elevations and freezing point depressions are usually small and not very accurate, so molar mass determinations based upon these measures often are not accurate.

8.113 (a) To determine the vapor pressure of the solution, we need to know the mole fraction of each component.

$$x_{\text{diethyl ether}} = \frac{0.75\ \text{mol}}{0.75\ \text{mol} + 0.50\ \text{mol}} = 0.60$$

$$x_{\text{ethyl methyl ether}} = 1 - x_{\text{diethyl ether}} = 0.40$$

$$P_{\text{total}} = (0.60 \times 185\ \text{Torr}) + (0.40 \times 554\ \text{Torr}) = 3.3 \times 10^2\ \text{Torr}$$

The vapor phase composition will be given by

$$x_{\text{diethyl ether in vapor phase}} = \frac{P_{\text{diethyl ether}}}{P_{\text{total}}} = \frac{0.60 \times 185\ \text{Torr}}{3.3 \times 10^2\ \text{Torr}} = 0.34$$

$$x_{\text{ethyl methyl ether in vapor phase}} = 1 - 0.34 = 0.66$$

The vapor is richer in the more volatile ethyl methyl ether, as expected.

(b) The procedure is the same as in (a) but the number of moles of each component must be calculated first:

$$n_{\text{diethyl ether}} = \frac{25.0\ \text{g}}{74.12\ \text{g}\cdot\text{mol}^{-1}} = 0.337$$

$$n_{\text{ethyl methyl ether}} = \frac{35.0\ \text{g}}{60.10\ \text{g}\cdot\text{mol}^{-1}} = 0.582$$

$$x_{\text{diethyl ether}} = \frac{0.337\ \text{mol}}{0.337\ \text{mol} + 0.582\ \text{mol}} = 0.367$$

$$x_{\text{ethyl methyl ether}} = 1 - x_{\text{diethyl ether}} = 0.633$$

$$P_{\text{total}} = (0.367 \times 185\ \text{Torr}) + (0.633 \times 554\ \text{Torr}) = 419\ \text{Torr}$$

The vapor phase composition will be given by

$$x_{\text{diethyl ether in vapor phase}} = \frac{P_{\text{diethyl ether}}}{P_{\text{total}}} = \frac{0.367 \times 185 \text{ Torr}}{419 \text{ Torr}} = 0.162$$

$$x_{\text{ethyl methyl ether in vapor phase}} = 1 - 0.162 = 0.838$$

8.115 (a) The partial pressure remains the same. Some of the ethanol will condense to return the ethanol(l) $\rightleftharpoons$ ethanol(g) reaction to the equilibrium point. (b) The total pressure will be the sum of the pressures due to the ethanol vapor and to the air. The air will not condense, so its pressure should follow the ideal gas law. If the total pressure initially is 750 Torr and 58.9 Torr is due to ethanol vapor, the 750 Torr − 58.9 Torr = 691 Torr will be due to the air. If the volume is halved at constant temperature, then the pressure due to the air will be doubled, or 2 × 691 Torr = 1.38×10^3 Torr. The total pressure will then be 1.38×10^3 Torr + 58.9 Torr = 1.44×10^3 Torr.

8.117 When a drop of aqueous solution containing $Ca(HCO_3)_2$ seeps through a cave ceiling, it encounters a situation where the partial pressure of CO_2 is reduced and the reaction $Ca(HCO_3)_2(aq) \rightleftharpoons CaCO_3(s) + CO_2(g) + H_2O(l)$ occurs. The concentration of CO_2 decreases as the CO_2 escapes as a gas, with $CaCO_3$ precipitating and forming a column that extends downward from the ceiling to form a stalactite. Stalagmite formation is similar, except the drop falls to the floor and the precipitate grows upward.

8.119 (a) Given $d = 1.00 \text{ g·cm}^{-3} = 1.00 \text{ g·mL}^{-1}$,

$$\therefore \frac{1.54 \text{ g}}{100 \text{ mL solution}} = 15.4 \text{ g·L}^{-1}$$

$$15.4 \text{ g·L}^{-1} \times \frac{1 \text{ mol } Li_2CO_3}{73.89 \text{ g } Li_2CO_3} = 0.208 \text{ mol·L}^{-1}$$

$$\Pi = iRT \times \text{molarity}$$

$$= 3 \times 0.082\,06 \text{ L·atm·K}^{-1}\text{·mol}^{-1} \times 273 \text{ K} \times 0.208 \text{ mol·L}^{-1} = 14.0 \text{ atm}$$

(b) $P_{\text{solution}} = x_{\text{solvent}} \times P_{\text{pure solvent}}$

$$751 \text{ Torr} = x_{\text{solvent}} \times 760 \text{ Torr}$$

$$x_{\text{solvent}} = 0.988$$

$$x_{\text{solute}} = 1 - 0.988 = 0.012$$

Consider a solution of 0.012 mol solute in 0.988 mol H_2O. Assume that the volume of solution equals the volume of H_2O. Then

$$V = 0.988 \text{ mol } H_2O \times \frac{18.02 \text{ g } H_2O}{1.00 \text{ mol } H_2O} \times \frac{1.00 \text{ mL } H_2O}{1.00 \text{ g } H_2O} = 17.8 \text{ mL} = 0.0178 \text{ L}$$

$$\text{molarity of solution} = \frac{0.012\ \text{mol}}{0.0178\ \text{L}} = 0.67\ \text{mol·L}^{-1}$$

Assuming that $i = 1$,

$\Pi = 1 \times 0.082\,06\ \text{L·atm·K}^{-1}\text{·mol}^{-1} \times 373\ \text{K} \times 0.67\ \text{mol·L}^{-1} = 21\ \text{atm}$

(c) $\Delta T = i\,k_b \times$ molality, assume $i = 1$, then

$$\text{molality} = \frac{\Delta T_b}{k_b} = \frac{1\ \text{K}}{0.51\ \text{K·kg·mol}^{-1}} = 2.0\ \text{mol·kg}^{-1}$$

Given that $d_{\text{solution}} = d_{\text{water}} = 1.00\ \text{g·cm}^{-3} = 1.00\ \text{g·mL}^{-1}$, we have

$$2.0\ \text{mol·kg}^{-1} \times \frac{1\ \text{kg}}{1000\ \text{g}} \times \frac{1.00\ \text{g}}{1.00\ \text{mL}} \times \frac{1000\ \text{mL}}{1.00\ \text{L}} = 2.0\ \text{mol·L}^{-1}$$

Then $\Pi = iRT \times$ molarity

$= 1 \times 0.08206\ \text{L·atm·K}^{-1}\text{·mol}^{-1} \times 374\ \text{K} \times 2.0\ \text{mol·L}^{-1}$

$= 61\ \text{atm}$

8.121 (a) Because the osmotic pressure is calculated from $\Pi = iRT \times$ molarity (both compounds are nonelectrolytes, so $i = 1$), the solution with the higher concentration will have the higher osmotic pressure, which in this case is the solution of urea, $CO(NH_2)_2$.
(b) The more concentrated solution, $CO(NH_2)_2$, will become more dilute with the passage of H_2O molecules through the membrane. (c) Pressure must be applied to the more concentrated solution, $CO(NH_2)_2$, to equilibrate the water flow. (d) $\Pi = iRT \times \Delta M$ (where ΔM is the difference in molar concentration between the two solutions):

$\Pi = 1 \times 0.082\,06\ \text{L·atm·K}^{-1}\text{·mol}^{-1} \times 298\ \text{K} \times 0.030\ \text{mol·L}^{-1} = 0.73\ \text{atm}$

8.123 (a) The data in Table 6.2 can be used to obtain the heat of vaporization and boiling point of benzene.

$C_6H_6(l) \rightleftharpoons C_6H_6(g)$

$\Delta H°_{\text{vap}} = 30.8\ \text{kJ·mol}^{-1}$

The boiling point is 353.2 K or 80.0°C, at which the vapor pressure = 1 atm. To derive the general equation, we start with the expression that $\Delta G°_{\text{vap}} = -RT \ln P$, where P is the vapor pressure of the solvent. Because $\Delta G°_{\text{vap}} = \Delta H°_{\text{vap}} - T\Delta S°_{\text{vap}}$, this is the relationship to use to determine the temperature dependence of ln P:

$\Delta H°_{\text{vap}} - T\Delta S°_{\text{vap}} = -RT \ln P$

This equation can be rearranged to give:

$$\ln P = -\frac{\Delta H°_{\text{vap}}}{R}\cdot\frac{1}{T} + \frac{\Delta S°_{\text{vap}}}{R}$$

Because we do not have enough data to calculate $\Delta S°_{\text{vap}}$ from the Table or Appendix, we can use the alternate form of the equation, which relates the vapor pressure at two points to the corresponding temperatures. This equation is obtained by subtracting one specific point from another as shown:

$$\ln P_2 = -\frac{\Delta H^\circ_{vap}}{R} \cdot \frac{1}{T_2} + \frac{\Delta S^\circ_{vap}}{R}$$

$$-\left[\ln P_1 = -\frac{\Delta H^\circ_{vap}}{R} \cdot \frac{1}{T_1} + \frac{\Delta S^\circ_{vap}}{R}\right]$$

$$\ln \frac{P_2}{P_1} = -\frac{\Delta H^\circ_{vap}}{R}\left[\frac{1}{T_2} - \frac{1}{T_1}\right]$$

Because we know ΔH°_{vap} and one point, we can introduce those values:

$$\ln \frac{P}{1} = -\frac{30\ 800\ \mathrm{J \cdot mol^{-1}}}{8.314\ \mathrm{J \cdot K^{-1} \cdot mol^{-1}}}\left[\frac{1}{T} - \frac{1}{353.2\ \mathrm{K}}\right]$$

$$\ln P = -\frac{3.70 \times 10^3\ \mathrm{K}}{T} + 10.5$$

(b) The appropriate quantities to plot to get a straight line are $\ln P$ versus $\frac{1}{T}$.

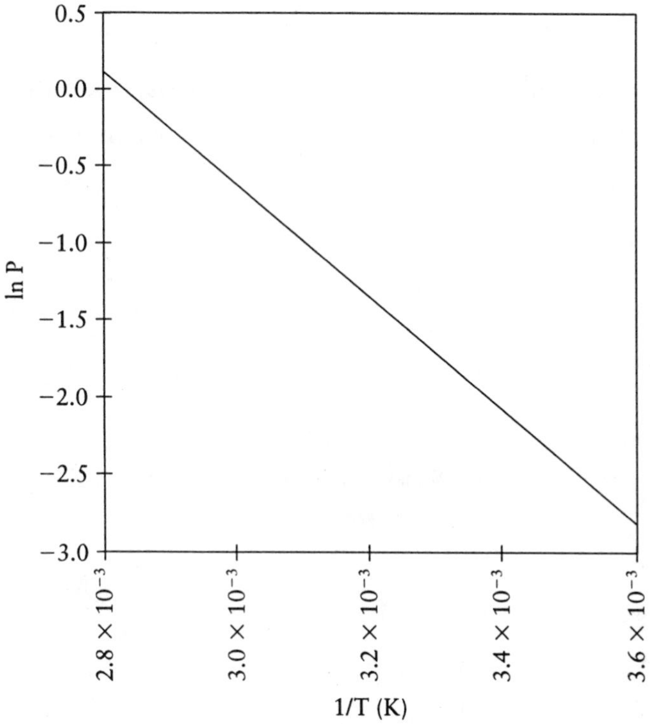

(c) The solution will boil when the vapor pressure equals the external pressure. Simply substitute the numbers, once the equation has been derived:

$$\ln P = -\frac{3.70 \times 10^3\ \mathrm{K}}{T} + 10.5$$

$$\ln 0.655 = -\frac{3.70 \times 10^3\ \mathrm{K}}{T} + 10.5$$

$T = 339$ K or 66°C

(d) The equation of the line is written in such a form that the constant 10.5 is equal to $\frac{\Delta S^\circ_{vap}}{R}$.

$$\frac{\Delta S^\circ_{vap}}{R} = 10.5$$

$\Delta S^\circ_{vap} = 10.5 \times 8.314\ \text{J}\cdot\text{K}^{-1}\cdot\text{mol}^{-1} = 87.3\ \text{J}\cdot\text{K}^{-1}\cdot\text{mol}^{-1}$

Because $\Delta S^\circ_{vap} = S^\circ_m(\text{benzene, g}) - S^\circ_m(\text{benzene, l})$

we can now calculate the standard molar entropy of benzene(g)

$87.3\ \text{J}\cdot\text{K}^{-1}\cdot\text{mol}^{-1} = S^\circ_m(\text{benzene, g}) - 173.3\ \text{J}\cdot\text{K}^{-1}\cdot\text{mol}^{-1}$

$S^\circ_m(\text{benzene, g}) = 260.6\ \text{J}\cdot\text{K}^{-1}\cdot\text{mol}^{-1}$

8.125 The equation to give vapor pressure as a function of temperature is

$$\ln P = -\frac{\Delta H^\circ_{vap}}{R}\cdot\frac{1}{T} + \frac{\Delta S^\circ_{vap}}{R}$$

$$\ln P = -\frac{57\,814\ \text{J}\cdot\text{mol}^{-1}}{8.314\ \text{J}\cdot\text{K}^{-1}\cdot\text{mol}^{-1}}\cdot\frac{1}{T} + \frac{124\ \text{J}\cdot\text{mol}^{-1}}{8.314\ \text{J}\cdot\text{K}^{-1}\cdot\text{mol}^{-1}}$$

$$\ln P = -\frac{6954\ \text{K}}{T} + 14.9$$

In order fot the substance to boil, we need to reduce the external pressure on the sample to its vapor pressure at that temperature. Thus, the problem becomes simply a matter of calculating the vapor pressure at 80°C:

$$\ln P = -\frac{6954\ \text{K}}{353\ \text{K}} + 14.9 = -4.80$$

$P = 0.008\,23$ atm or 6.26 Torr

The pressure needs to be reduced to 6.26 Torr.

8.127 The critical pressures and temperatures of the compounds are as follows:

Compound	T_C (°C)	P_C (atm)
Methane, CH_4	−82.1	45.8
Methyl amine, CH_3NH_2	156.9	40.2
Ammonia, NH_3	132.5	112.5
Tetrafluoromethane, CF_4	−45.7	41.4

In order to access the supercritical fluid state, we must have conditions in excess of the critical temperature and pressure. Given the rating of the autoclave, ammonia would not be suitable because one could not access the supercritical state due to the pressure limitation. Methylamine would not be suitable for a room temperature extraction because its T_C is too high. Either methane or tetrafluoromethane would be suitable for this application.

8.129 Compound (a). The stationary phase is more polar than the liquid phase, so the more polar compound of (a) and (b) should be attracted more strongly to the stationary phase and should remain on the column longer. Because compound (a) has two carboxylic acid units (COOH), it will be more polar and will have the greater value of k.

CHAPTER 9
CHEMICAL EQUILIBRIA

9.1 (a) At the molecular level, chemical equilibrium is a dynamic process, and interconversion of reactants and products continues to occur. At the macroscopic level, no concentration change in products or reactants is observable at equilibrium, because rates of product and reactant formation are equal. (b) At a fixed temperature, the equilibrium constant is truly a constant, independent of concentrations. However, if the concentration of a reactant is increased, the concentrations of other chemical species involved in the reaction will change and more product will form. Concentrations adjust because the equilibrium constant remains the same.

9.3 The concentrations of reactants and products stop changing when equilibrium has been attained. Equilibrium occurs when the forward and reverse rates of reaction are equal, a condition that will always be reached eventually, given enough time.

9.5 The calculations for any dilution or concentration process is calculated from the relationship: $G_m = G°_m + RT \ln Q$, where Q in this case represents the concentration or pressure that is different from the standard state value of 1 bar or $mol \cdot L^{-1}$. We never know the absolute values of G, but we can calculate the change in free energy from one state to another. Thus, we can write this relationship for conditions 1 and 2, and then find the difference between them:

$$G_{m,1} = G°_m + RT \ln P_1$$

$$G_{m,2} = G°_m + RT \ln P_2$$

$$G_{m,2} - G_{m,1} = G°_m + RT \ln P_2 - (G°_m + RT \ln P_1)$$

$$\Delta G_m = RT \ln \frac{P_2}{P_1}$$

A similar expression can be written for concentrations in place of pressures. Note that the nature of the compound is irrelevant—an identical free energy change is observed regardless of what type of gas is compressed or expanded, or what type of solution (solute or solvent) is diluted or concentrated.

(a) $$\Delta G_m = (8.314 \text{ J}\cdot\text{K}^{-1}\cdot\text{mol}^{-1})(298 \text{ K})\left(\ln \frac{25.00}{0.500}\right)/(1000 \text{ J}\cdot\text{kJ}^{-1})$$

$$= 9.69 \text{ kJ}\cdot\text{mol}^{-1}$$

(b) $\Delta G_m = (8.314\ \text{J}\cdot\text{K}^{-1}\cdot\text{mol}^{-1})(298\ \text{K})\left(\ln\dfrac{5.00}{10.00}\right)/(1000\ \text{J}\cdot\text{kJ}^{-1})$

$= -1.72\ \text{kJ}\cdot\text{mol}^{-1}$

(c) $\Delta G_m = (8.314\ \text{J}\cdot\text{K}^{-1}\cdot\text{mol}^{-1})(298\ \text{K})\left(\ln\dfrac{10.00}{20.00}\right)/(1000\ \text{J}\cdot\text{kJ}^{-1})$

$= -1.72\ \text{kJ}\cdot\text{mol}^{-1}$

(d) $\Delta G_m = (8.314\ \text{J}\cdot\text{K}^{-1}\cdot\text{mol}^{-1})(298\ \text{K})\left(\ln\dfrac{10.00}{20.00}\right)/(1000\ \text{J}\cdot\text{kJ}^{-1})$

$= -1.72\ \text{kJ}\cdot\text{mol}^{-1}$

9.7 See explanation for Exercise 9.5.

(a) $\Delta G_m = (8.314\ \text{J}\cdot\text{K}^{-1}\cdot\text{mol}^{-1})(298\ \text{K})\left(\ln\dfrac{0.350}{0.450}\right)/(1000\ \text{J}\cdot\text{kJ}^{-1})$

$= -0.623\ \text{kJ}\cdot\text{mol}^{-1}$

(b) $\Delta G_m = (8.314\ \text{J}\cdot\text{K}^{-1}\cdot\text{mol}^{-1})(298\ \text{K})\left(\ln\dfrac{0.700}{0.900}\right)/1000\ \text{J}\cdot\text{kJ}^{-1})$

$= -0.623\ \text{kJ}\cdot\text{mol}^{-1}$

(c) $\Delta G_m = (8.314\ \text{J}\cdot\text{K}^{-1}\cdot\text{mol}^{-1})(298\ \text{K})\left(\ln\dfrac{0.350}{0.175}\right)/(1000\ \text{J}\cdot\text{kJ}^{-1})$

$= 1.72\ \text{kJ}\cdot\text{mol}^{-1}$

(d) $\Delta G_m = (8.314\ \text{J}\cdot\text{K}^{-1}\cdot\text{mol}^{-1})(298\ \text{K})\left(\ln\dfrac{0.350}{0.175}\right)/(1000\ \text{J}\cdot\text{kJ}^{-1})$

$= 1.72\ \text{kJ}\cdot\text{mol}^{-1}$

9.9 To answer these questions, we will first calculate $\Delta G°$ for each reaction and then use that value in the expression $\Delta G° = -RT \ln K$.

(a) $2\ H_2(g) + O_2(g) \longrightarrow 2\ H_2O(g)$

$\Delta G°_r = 2(\Delta G°_f(H_2O,\ g) = 2(-228.57\ \text{kJ}\cdot\text{mol}^{-1}) = -457.14\ \text{kJ}$

$°G° = -RT \ln K$ or

$$\ln K = -\frac{\Delta G°}{RT}$$

$$\ln K = -\frac{-457\ 140\ \text{J}}{(8.314\ \text{J}\cdot\text{K}^{-1}\cdot\text{mol}^{-1})(298\ \text{K})} = +184.51$$

$K = 1 \times 10^{80}$

(b) $2\ CO(g) + O_2(g) \longrightarrow 2\ CO_2(g)$

$\Delta G°_r = 2 \times \Delta G°_f(CO_2,\ g) - [2 \times \Delta G°_f(CO,\ g)]$

$= 2(-394.36\ \text{kJ}\cdot\text{mol}^{-1}) - [2(-137.17\ \text{kJ}\cdot\text{mol}^{-1})]$

$= -514.38\ \text{kJ}$

$$\ln K = -\frac{-514\ 380\ \text{J}}{(8.314\ \text{J}\cdot\text{K}^{-1}\cdot\text{mol}^{-1})(298\ \text{K})} = +208$$

$K = 1 \times 10^{90}$

(c) $CaCO_3(s) \longrightarrow CaO(s) + CO_2(g)$

$$\Delta G°_r = \Delta G°_f\,(\text{CaO, s}) + \Delta G°_f\,(\text{CO}_2,\text{ g}) - [\Delta G°_f\,(\text{CaCO}_3\text{ (s)})]$$

$$= (-604.03\ \text{kJ}\cdot\text{mol}^{-1}) + (-394.36\ \text{kJ}\cdot\text{mol}^{-1}) - [-1128.8\ \text{kJ}\cdot\text{mol}^{-1}]$$

$$= +130.4\ \text{kJ}$$

$$\ln K = -\frac{+130\ 400\ \text{J}}{(8.314\ \text{J}\cdot\text{K}^{-1}\cdot\text{mol}^{-1})(298\ \text{K})} = -52.6$$

$K = 1 \times 10^{-23}$

9.11 (a) $\Delta G°_r = -RT\ln K = -(8.314\ \text{J}\cdot\text{K}^{-1}\cdot\text{mol}^{-1})(400\ \text{K})\ln 41 = -12\ \text{kJ}\cdot\text{mol}^{-1}$

(b) $\Delta G°_r = -RT\ln K = -(8.314\ \text{J}\cdot\text{K}^{-1}\cdot\text{mol}^{-1})(700\ \text{K})\ln 3.0 \times 10^4$

$$= -60\ \text{kJ}\cdot\text{mol}^{-1}$$

9.13 First we must calculate K for the reaction, which can be done using data from Appendix 2A:

$\Delta G°_r = 2 \times \Delta G°_f\,(\text{NO, g}) = 2 \times 86.55\ \text{kJ}\cdot\text{mol}^{-1} = 173.1\ \text{kJ}\cdot\text{mol}^{-1}$

$\Delta G° = -RT\ln K$

$$\ln K = -\frac{\Delta G^0}{RT}$$

$$\ln K = -\frac{+173\ 100\ \text{J}\cdot\text{mol}^{-1}}{(8.314\ \text{J}\cdot\text{K}^{-1}\cdot\text{mol}^{-1})(298\ \text{K})} = -69.9$$

$K = 4 \times 10^{-31}$

Because $Q > K$, the reaction will tend to proceed to produce reactants.

9.15 The free energy at a specific set of conditions is given by

$\Delta G_r = \Delta G°_r + RT\ln Q$

$\Delta G_r = -RT\ln K + RT\ln Q$

$$\Delta G_r = -RT\ln K + RT\ln\frac{[\text{I}]^2}{[\text{I}_2]}$$

$$= -(8.314\ \text{J}\cdot\text{K}^{-1}\cdot\text{mol}^{-1})(1200\ \text{K})\ln 0.068$$

$$+ (8.314\ \text{J}\cdot\text{K}^{-1}\cdot\text{mol}^{-1})(1200\ \text{K})\ln\frac{(0.0084)^2}{(0.026)}$$

$$= -32\ \text{kJ}\cdot\text{mol}^{-1}$$

Because ΔG_r is negative, the reaction will be spontaneous to produce I.

9.17 The free energy at a specific set of conditions is given by

$$\Delta G_r = \Delta G^\circ_r + RT \ln Q$$

$$\Delta G_r = -RT \ln K + RT \ln Q$$

$$\Delta G_r = -RT \ln K + RT \ln \frac{[NH_3]^2}{[N_2][H_2]^3}$$

$$= -(8.314\ J\cdot K^{-1}\cdot mol^{-1})(400\ K) \ln 41$$

$$+ (8.314\ J\cdot K^{-1}\cdot mol^{-1})(400\ K) \ln \frac{(63)^2}{(1.0)(4.2)^3}$$

$$= +0.89\ kJ\cdot mol^{-1}$$

Because ΔG_r is positive, the reaction will proceed to form reactants.

9.19 (a) $K_C = \dfrac{[COCl][Cl]}{[CO][Cl_2]}$; (b) $K_C = \dfrac{[HBr]^2}{[H_2][Br_2]}$; (c) $K_C = \dfrac{[SO_2]^2[H_2O]^2}{[H_2S]^2[O_2]^3}$

9.21 (a) $K = \dfrac{P_{NO}{}^2 P_{Cl_2}}{P_{NOCl}{}^2} = 1.8 \times 10^{-2}$

$$K = \left(\frac{T}{12.027\ K}\right)^{\Delta n} K_C$$

$$K_C = \left(\frac{12.027\ K}{T}\right)^{\Delta n} K = \left(\frac{12.027\ K}{500\ K}\right)^{(3-2)} 1.8 \times 10^{-2} = 4.3 \times 10^{-4}$$

(b) $K = P_{CO_2} = 167$

$$K_C = \left(\frac{12.027\ K}{T}\right)^{\Delta n} K = \left(\frac{12.027\ K}{1073\ K}\right)^{(1)} 167 = 1.87$$

9.23 (a) Because the volume is the same, the number of moles of O_2 is larger in the second experiment. (b) Because K_C is a constant and the denominator is larger in the second case, the numerator must also be larger; so the concentration of O_2 is larger in the second case. (c) Although $[O_2]^3/[O_3]^2$ is the same, $[O_2]/[O_3]$ will be different, a result seen by solving for K_C in each case. (d) Because K_C is a constant, $[O_2]^3/[O_3]^2$ is the same. (e) Because $[O_2]^3/[O_3]^2$ is the same, its reciprocal must be the same.

9.25 For the reaction written as $N_2(g) + 3\ H_2(g) \rightleftharpoons 2\ NH_3(g)$ Eq. 1

$$K = \frac{P_{NH_3}{}^2}{P_{N_2} P_{H_2}{}^3} = 41$$

(a) For the reaction written as $2\ NH_3(g) \rightleftharpoons N_2(g) + 3\ H_2(g)$ Eq. 2

$$K = \frac{P_{N_2} P_{H_2}^{\ 3}}{P_{NH_3}^{\ 2}}$$

This is $\frac{1}{K_{Eq.\ 1}} = \frac{1}{41} = 0.024$

(b) For the reaction written as $\frac{1}{2}\ N_2(g) + \frac{3}{2}\ H_2(g) \rightleftharpoons NH_3(g)$ Eq. 3

$$K_{Eq.\ 3} = \frac{P_{NH_3}}{P_{N_2}^{\ 3/2} P_{H_2}^{\ 3/2}} = \sqrt{K_{Eq.\ 1}} = \sqrt{41} = 6.4$$

Note that Eq. 3 = $\frac{1}{2}$ Eq. 1 and thus $K_{Eq.\ 3} = K_{Eq.\ 1}^{\ 1/2}$.

(c) For the reaction written as $2\ N_2(g) + 6\ H_2(g) \rightleftharpoons 4\ NH_3(g)$ Eq. 4

$$K_{Eq.\ 4} = \frac{P_{NH_3}^{\ 4}}{P_{N_2}^{\ 2} P_{H_2}^{\ 6}} = K_{Eq.\ 1}^{\ 2} = 41^2 = 1.7 \times 10^3$$

Note that Eq. 4 = 2 Eq. 1 and thus $K_{Eq.\ 3} = K_{Eq.\ 1}^{\ 2}$.

9.27 $K_C = \frac{[HI]^2}{[H_2][I_2]}$

for condition 1, $K_C = \frac{(0.0137)^2}{(6.47 \times 10^{-3})(0.594 \times 10^{-3})} = 48.8$

for condition 2, $K_C = \frac{(0.0169)^2}{(3.84 \times 10^{-3})(1.52 \times 10^{-3})} = 48.9$

for condition 3, $K_C = \frac{(0.0100)^2}{(1.43 \times 10^{-3})(1.43 \times 10^{-3})} = 48.9$

9.29 All the equilibria are heterogeneous, because in each case more than one phase is involved in the equilibrium.

(a) $Q = (P_{Cl_2})^{-3}$; (b) $Q = P_{NH_3}^{\ 2} P_{CO_2}$; (c) $Q = P_{O_2}$

9.31 (a) $Q_C = \frac{1}{[Cl_2]}$; (b) $Q_C = [N_2O][H_2O]^2$; (c) $Q_C = [CO_2]$

9.33 $H_2(g) + I_2(g) \rightleftharpoons 2\ HI(g) \qquad K_C = 160$

$$K_C = \frac{[HI]^2}{[H_2][I_2]}$$

$$160 = \frac{(2.21 \times 10^{-3})^2}{[H_2](1.46 \times 10^{-3})}$$

$$[H_2] = \frac{(2.21 \times 10^{-3})^2}{(160)(1.46 \times 10^{-3})}$$

$[H_2] = 2.1 \times 10^{-5}$

9.35 $K = \dfrac{P_{PCl_3}P_{Cl_2}}{P_{PCl_5}}$

$$25 = \frac{P_{PCl_3}(0.20)}{0.15}$$

$$P_{PCl_3} = \frac{(25)(0.15)}{0.20} = 19$$

$$P_{PCl_3} = 19 \text{ bar}$$

9.37 (a) $K_C = \dfrac{[HI]^2}{[H_2][I_2]} = 160$

$$Q_C = \frac{[2.4 \times 10^{-3}]^2}{[4.8 \times 10^{-3}][2.4 \times 10^{-3}]} = 0.50$$

(b) $Q_C \neq K_C$, ∴ the system is not at equilibrium

(c) Because $Q_C < K_C$, more products will need to be formed.

9.39 (a) $K_C = \dfrac{[SO_3]^2}{[SO_2]^2[O_2]} = 1.7 \times 10^6$

$$Q_C = \frac{\left[\dfrac{1.0 \times 10^{-4}}{0.500}\right]^2}{\left[\dfrac{1.20 \times 10^{-3}}{0.500}\right]^2\left[\dfrac{5.0 \times 10^{-4}}{0.500}\right]} = 6.9$$

(b) Because $Q_C < K_C$, more products will tend to form, which will result in the formation of more SO_3.

9.41 $\dfrac{1.90 \text{ g HI}}{127.91 \text{ g}\cdot\text{mol}^{-1}} = 0.0148 \text{ mol HI}$

$$2\,HI(g) \rightleftharpoons H_2(g) + I_2$$

$$0.0172 \text{ mol} - 2x \qquad x \qquad x$$

$$0.0148 \text{ mol} = 0.0172 \text{ mol} - 2x$$

$$x = 0.0012 \text{ mol}$$

$$K_C = \frac{\left[\dfrac{0.0012}{2.00}\right]\left[\dfrac{0.0012}{2.00}\right]}{\left[\dfrac{0.0148}{2.00}\right]^2} = \frac{\left[\dfrac{0.0012}{2.00}\right]^2}{\left[\dfrac{0.0148}{2.00}\right]^2} = 0.0066 \text{ or } 6.6 \times 10^{-3}$$

9.43 $\dfrac{25.0 \text{ g } NH_4(NH_2CO_2)}{78.07 \text{ g}\cdot\text{mol}^{-1}\ NH_4(NH_2CO_2)} = 0.320 \text{ mol } NH_4(NH_2CO_2)$

$$\frac{0.0174 \text{ g } CO_2}{44.01 \text{ g}\cdot\text{mol}^{-1} CO_2} = 3.95 \times 10^{-4} \text{ mol } CO_2$$

2 mol NH_3 are formed per mol of CO_2, so mol $NH_3 = 2 \times 3.95 \times 10^{-4}$
$= 7.90 \times 10^{-4}$

$$K_C = [NH_3]^2[CO_2] = \left(\frac{7.90 \times 10^{-4}}{0.250}\right)^2\left(\frac{3.95 \times 10^{-4}}{0.250}\right) = 1.58 \times 10^{-8}$$

9.45 For each set of conditions, we need to calculate the final concentrations of reactants present. Because the stoichiometry is one mole $CH_3COOC_2H_5$ produced per mole of CH_3COOH and per mole of C_2H_5OH consumed and the amount of water produced equals the amount of $CH_3COOC_2H_5$ produced, the values are easy to obtain:

	$[CH_3COOH]$	$[C_2H_5OH]$	$[CH_3COOC_2H_5]$	$[H_2O]$
final	$1.00 - 0.171 = 0.829$	$0.180 - 0.171 = 0.009$	0.171	0.171
final	$1.00 - 0.667 = 0.333$	$1.00 - 0.667 = 0.333$	0.667	0.667
final	$1.00 - 0.966 = 0.034$	$8.00 - 0.966 = 7.03$	0.966	0.966

$$K_C = \frac{[CH_3COOC_2H_5][H_2O]}{[CH_3COOH][C_2H_5OH]}$$

$$\#1\ K_C = \frac{[0.171][0.171]}{[0.829][0.009]} = 4$$

$$\#2\ K_C = \frac{[0.667][0.667]}{[0.333][0.333]} = 4.01$$

$$\#3\ K_C = \frac{[0.966][0.966]}{[0.034][7.03]} = 3.9$$

These numbers are identical within the experimental error.

9.47 (a) The balanced equation is: $Cl_2(g) \rightleftharpoons 2\ Cl(g)$

The initial concentration of $Cl_2(g)$ is $\dfrac{0.0020 \text{ mol } Cl_2}{2.0 \text{ L}} = 0.0010 \text{ mol}\cdot\text{L}^{-1}$

Concentration ($\text{mol}\cdot\text{L}^{-1}$)	$Cl_2(g)$	$\rightleftharpoons$	$2\ Cl(g)$
initial	0.0010		0
change	$-x$		$+2x$
equilibrium	$0.0010 - x$		$+2x$

$$K_C = \frac{[Cl]^2}{[Cl_2]} = \frac{(2x)^2}{(0.0010 - x)} = 1.2 \times 10^{-7}$$

$$4x^2 = (1.2 \times 10^{-7})(0.0010 - x)$$

$$4x^2 + (1.2 \times 10^{-7})x - (1.2 \times 10^{-10}) = 0$$

$$x = \frac{-(1.2 \times 10^{-7}) \pm \sqrt{(1.2 \times 10^{-7})^2 - 4(4)(-1.2 \times 10^{-10})}}{2 \cdot 4}$$

$$x = \frac{-(1.2 \times 10^{-7}) \pm 4.4 \times 10^{-5}}{8}$$

$x = -5.5 \times 10^{-6}$ or $+5.5 \times 10^{-6}$

The negative answer is not meaningful, so we choose $x = +5.5 \times 10^{-6}\ \text{mol} \cdot \text{L}^{-1}$. The concentration of Cl_2 is essentially unchanged because $0.0010 - 5.5 \times 10^{-6} \cong 0.0010$. The concentration of Cl atoms is $2 \times (5.5 \times 10^{-6}) = 1.1 \times 10^{-5}\ \text{mol} \cdot \text{L}^{-1}$. The percentage decomposition of Cl_2 is given by

$$\frac{5.5 \times 10^{-6}}{0.0010} \times 100 = 0.55\%$$

(b) The balanced equation is: $F_2(g) \rightleftharpoons 2\ F(g)$

The problem is worked in an identical fashion to (a) but the equilibrium constant is now 1.2×10^{-4}.

The initial concentration of $F_2(g)$ is $\dfrac{0.0020 \text{ mol } F_2}{2.0 \text{ L}} = 0.0010\ \text{mol} \cdot \text{L}^{-1}$

Concentration ($\text{mol} \cdot \text{L}^{-1}$)	$F_2(g)$	$\rightleftharpoons$	$2\ F(g)$
initial	0.0010		0
change	$-x$		$+2x$
equilibrium	$0.0010 - x$		$+2x$

$$K_C = \frac{[F]^2}{[F_2]} = \frac{(2x)^2}{(0.0010 - x)} = 1.2 \times 10^{-4}$$

$$4x^2 = (1.2 \times 10^{-4})(0.0010 - x)$$

$$4x^2 + (1.2 \times 10^{-4})x - (1.2 \times 10^{-7}) = 0$$

$$x = \frac{-(1.2 \times 10^{-4}) \pm \sqrt{(1.2 \times 10^{-4})^2 - 4(4)(-1.2 \times 10^{-7})}}{2 \cdot 4}$$

$$x = \frac{-(1.2 \times 10^{-4}) \pm 1.4 \times 10^{-3}}{8}$$

$x = -1.9 \times 10^{-4}$ or $+1.6 \times 10^{-4}$

The negative answer is not meaningful, so we choose $x = +1.6 \times 10^{-4}\ \text{mol} \cdot \text{L}^{-1}$. The concentration of F_2 is $0.0010 - 1.6 \times 10^{-4} = 8 \times 10^{-4}\ \text{mol} \cdot \text{L}^{-1}$. The concentration of F atoms is $2 \times (1.6 \times 10^{-4}) = 3.2 \times 10^{-4}\ \text{mol} \cdot \text{L}^{-1}$. The percentage decomposition of F_2 is given by

$$\frac{1.6 \times 10^{-4}}{0.0010} \times 100 = 16\%$$

(c) Cl_2 is more stable. This can be seen even without the aid of the calculation from the larger equilibrium constant for the dissociation for F_2 compared to Cl_2.

9.49

Concentration ($mol\cdot L^{-1}$)	$2\,HBr(g)$ $\rightleftharpoons$	$H_2(g)$ +	$Br_2(g)$
initial	1.2×10^{-3}	0	0
change	$-2x$	$+x$	$+x$
final	$1.2 \times 10^{-3} - 2x$	$+x$	$+x$

$$K_C = \frac{[H_2][Br_2]}{[HBr]^2}$$

$$7.7 \times 10^{-11} = \frac{(x)(x)}{(1.2 \times 10^{-3} - 2x)^2} = \frac{x^2}{(1.2 \times 10^{-3} - 2x)^2}$$

$$\sqrt{7.7 \times 10^{-11}} = \sqrt{\frac{x^2}{(1.2 \times 10^{-3} - 2x)^2}}$$

$$\frac{x}{(1.2 \times 10^{-3} - 2x)} = 8.8 \times 10^{-6}$$

$$x = (8.8 \times 10^{-6})(1.2 \times 10^{-3} - 2x)$$

$$x + 2(8.8 \times 10^{-6})x = (8.8 \times 10^{-6})(1.2 \times 10^{-3})$$

$$x \cong 1.1 \times 10^{-8}$$

$[Br_2] = [H_2] = 1.1 \times 10^{-8}\ mol\cdot L^{-1}$; the concentration of HBr is essentially unaffected by the formation of Br_2 and H_2.

The percentage decomposition is given by

$$\frac{2\,(1.1 \times 10^{-8}\ mol\cdot L^{-1})}{1.2 \times 10^{-3}\ mol\cdot L^{-1}} \times 100 = 1.8 \times 10^{-3}\%$$

9.51 (a) Concentration of PCl_5 initially $= \dfrac{\left(\dfrac{1.0\ g\ PCl_5}{208.22\ g\cdot mol^{-1}\ PCl_5}\right)}{0.250\ L} = 0.019\ mol\cdot L^{-1}$

Concentration ($mol\cdot L^{-1}$)	$PCl_5(g)$ $\rightleftharpoons$	$PCl_3(g)$ +	Cl_2
initial	0.019	0	0
change	$-x$	$+x$	$+x$
final	$0.019 - x$	$+x$	$+x$

$$K_C = \frac{[PCl_3][Cl_2]}{[PCl_5]} = \frac{(x)(x)}{(0.019 - x)} = \frac{x^2}{(0.019 - x)}$$

$$\frac{x^2}{(0.019 - x)} = 1.1 \times 10^{-2}$$

$$x^2 = (1.1 \times 10^{-2})(0.019 - x)$$

$$x^2 + (1.1 \times 10^{-2})x - 2.1 \times 10^{-4} = 0$$

$$x = \frac{-(1.1 \times 10^{-2}) \pm \sqrt{(1.1 \times 10^{-2})^2 - (4)(1)(-2.1 \times 10^{-4})}}{2\cdot 1}$$

$$x = \frac{-(1.1 \times 10^{-2}) \pm 0.031}{2\cdot 1} = +0.010 \text{ or } -0.021$$

The negative root is not meaningful, so we choose $x = 0.010\ \text{mol·L}^{-1}$.
$[PCl_3] = [Cl_2] = 0.010\ \text{mol·L}^{-1}$; $[PCl_5] = 0.009\ \text{mol·L}^{-1}$.

(b) The percentage decomposition is given by

$$\frac{0.010}{0.019} \times 100 = 53\%$$

9.53 Starting concentration of $NH_3 = \dfrac{0.400\ \text{mol}}{2.00\ \text{L}} = 0.200\ \text{mol·L}^{-1}$

Concentration (mol·L^{-1})	$NH_4HS(s)$ $\rightleftharpoons$	$NH_3(g)$ +	$H_2S(g)$
initial	—	0.200	0
change	—	$+x$	$+x$
final	—	$0.200 + x$	$+x$

$K_C = [NH_3][H_2S] = (0.200 + x)(x)$
$1.6 \times 10^{-4} = (0.200 + x)(x)$
$x^2 + 0.200x - 1.6 \times 10^{-4} = 0$

$$x = \frac{-(+0.200) \pm \sqrt{(+0.200)^2 - (4)(1)(-1.6 \times 10^{-4})}}{2 \cdot 1}$$

$$x = \frac{-0.200 \pm 0.2016}{2 \cdot 1} = +0.0008 \text{ or } -0.2008$$

The negative root is not meaningful, so we choose $x = 8 \times 10^{-4}\ \text{mol·L}^{-1}$ (note that in order to get this number we have had to ignore our normal significant figure conventions).

$[NH_3] = +0.200\ \text{mol·L}^{-1} + 8 \times 10^{-4}\ \text{mol·L}^{-1} = 0.200\ \text{mol·L}^{-1}$
$[H_2S] = 8 \times 10^{-4}\ \text{mol·L}^{-1}$

Alternatively, we could have assumed that $x \ll 0.2$, the $0.200x = 1.6 \times 10^{-4}$,
$x = 8.0 \times 10^{-4}$

9.55 The initial concentrations of PCl_5 and PCl_3 are calculated as follows:

$[PCl_5] = \dfrac{0.200\ \text{mol}}{4.00\ \text{L}} = 0.0500\ \text{mol·L}^{-1}$; $[PCl_3] = \dfrac{0.600\ \text{mol}}{4.00\ \text{L}} = 0.150\ \text{mol·L}^{-1}$

Concentrations (mol·L^{-1})	PCl_5 $\rightleftharpoons$	$PCl_3(g)$ +	$Cl_2(g)$
initial	0.0500	0.150	0
change	$-x$	$+x$	$+x$
final	$0.0500 - x$	$0.150 + x$	$+x$

$$K_C = \frac{[PCl_3][Cl_2]}{[PCl_5]} = \frac{(0.150 + x)(x)}{(0.0500 - x)} = 33.3$$

$x^2 + 0.150x = (33.3)(0.0500 - x)$

$x^2 + 33.45x - 1.665 = 0$

$$x = \frac{-33.45 \pm \sqrt{(33.45)^2 - (4)(1)(-1.665)}}{(2)(1)} = \frac{-33.4 \pm 33.6}{2} = 0.0497$$

The negative root has no physical meaning and so it can be discarded.

$[PCl_5] = 0.0500\ mol \cdot L^{-1} - 0.0497\ mol \cdot L^{-1} = 3 \times 10^{-4}\ mol \cdot L^{-1}$

$[PCl_3] = 0.150\ mol \cdot L^{-1} + 0.0497\ mol \cdot L^{-1} = 0.200\ mol \cdot L^{-1}$

$[Cl_2] = 0.0497\ mol \cdot L^{-1}$

9.57 The initial concentrations of N_2 and O_2 are equal at $0.114\ mol \cdot L^{-1}$ because the vessel has a volume of 1.00 L.

Concentrations ($mol \cdot L^{-1}$)	$N_2(g)$	+	$O_2(g)$	$\rightleftharpoons$	$2\ NO(g)$
initial	0.114		0.114		0
change	$-x$		$-x$		$+2x$
final	$0.114 - x$		$0.114 - x$		$+2x$

$$K_C = \frac{[NO]^2}{[N_2][O_2]} = \frac{(2x)^2}{(0.114 - x)(0.114 - x)} = \frac{(2x)^2}{(0.114 - x)^2}$$

$$1.00 \times 10^{-5} = \frac{(2x)^2}{(0.114 - x)^2}$$

$$\sqrt{1.00 \times 10^{-5}} = \sqrt{\frac{(2x)^2}{(0.114 - x)^2}}$$

$$3.16 \times 10^{-3} = \frac{(2x)}{(0.114 - x)}$$

$2x = (3.16 \times 10^{-3})(0.114 - x)$

$2.003\ 16x = 3.60 \times 10^{-4}$

$x = 1.8 \times 10^{-4}$

$[NO] = 2x = 2 \times 1.8 \times 10^{-4} = 3.6 \times 10^{-4}$; the concentrations of N_2 and O_2 remain essentially unchanged at $0.114\ mol \cdot L^{-1}$.

9.59 The initial concentrations of H_2 and I_2 are:

$$[H_2] = \frac{0.400\ mol}{3.00\ L} = 0.133\ mol \cdot L^{-1};\ [I_2] = \frac{1.60\ mol}{3.00\ L} = 0.533\ mol \cdot L^{-1}$$

Concentrations ($mol \cdot L^{-1}$)	$H_2(g)$	+	$I_2(g)$	$\rightleftharpoons$	$2\ HI(g)$
initial	0.133		0.533		0
change	$-x$		$-x$		$+2x$
final	$0.133 - x$		$0.533 - x$		$+2x$

At equilibrium 60.0% of the H_2 had reacted, so 40.0% of the H_2 remains:

$(0.400)(0.133\ \text{mol}\cdot\text{L}^{-1}) = 0.133\ \text{mol}\cdot\text{L}^{-1} - x$

$x = 0.133\ \text{mol}\cdot\text{L}^{-1} - (0.400)(0.133\ \text{mol}\cdot\text{L}^{-1})$

$x = 0.080\ \text{mol}\cdot\text{L}^{-1}$

at equilibrium: $[H_2] = 0.133\ \text{mol}\cdot\text{L}^{-1} - 0.080\ \text{mol}\cdot\text{L}^{-1} = 0.053\ \text{mol}\cdot\text{L}^{-1}$

$[I_2] = 0.533\ \text{mol}\cdot\text{L}^{-1} - 0.080\ \text{mol}\cdot\text{L}^{-1} = 0.453\ \text{mol}\cdot\text{L}^{-1}$

$[HI] = 2 \times 0.080\ \text{mol}\cdot\text{L}^{-1} = 0.16\ \text{mol}\cdot\text{L}^{-1}$

$$K_C = \frac{[HI]^2}{[H_2][I_2]} = \frac{0.16^2}{(0.053)(0.453)} = 1.1$$

9.61 Initial concentrations of CO and O_2 are given by

$$[CO] = \frac{\left(\dfrac{0.28\ \text{g CO}}{28.01\ \text{g}\cdot\text{mol}^{-1}\ \text{CO}}\right)}{2.0\ \text{L}} = 5.0 \times 10^{-3}\ \text{mol}\cdot\text{L}^{-1}$$

$$[O_2] = \frac{\left(\dfrac{0.032\ \text{g } O_2}{32.00\ \text{g}\cdot\text{mol}^{-1}\ O_2}\right)}{2.0\ \text{L}} = 5.0 \times 10^{-4}\ \text{mol}\cdot\text{L}^{-1}$$

Concentration ($\text{mol}\cdot\text{L}^{-1}$)	$2\ CO(g)$	+	$O_2(g)$	$\rightleftharpoons$	$2\ CO_2(g)$
initial	5.0×10^{-3}		5.0×10^{-4}		0
change	$-2x$		$-x$		$+2x$
final	$5.0 \times 10^{-3} - 2x$		$5.0 \times 10^{-4} - x$		$+2x$

$$K_C = \frac{[CO_2]^2}{[CO]^2[O_2]}$$

$$= \frac{(2x)^2}{(5.0 \times 10^{-3} - 2x)^2(5.0 \times 10^{-4} - x)}$$

$$= \frac{4x^2}{(4x^2 - 0.020x + 2.5 \times 10^{-5})(5.0 \times 10^{-4} - x)}$$

$$0.66 = \frac{4x^2}{-4x^3 + 0.022x^2 - 3.5 \times 10^{-5}x + 1.25 \times 10^{-8}}$$

$4x^2 = (0.66)(-4x^3 + 0.022x^2 - 3.5 \times 10^{-5}x + 1.25 \times 10^{-8})$

$6.06x^2 = -4x^3 + 0.022x^2 - 3.5 \times 10^{-5}x + 1.25 \times 10^{-8}$

$0 = -4x^3 - 6.04x^2 - 3.5 \times 10^{-5}x + 1.25 \times 10^{-8}$

$x = 4.3 \times 10^{-5}$

$[CO_2] = 8.6 \times 10^{-5}\ \text{mol}\cdot\text{L}^{-1}$; $[CO] = 4.9 \times 10^{-3}\ \text{mol}\cdot\text{L}^{-1}$

$[O_2] = 4.6 \times 10^{-4}\ \text{mol}\cdot\text{L}^{-1}$

9.63 Concentrations

(mol·L^{-1})	CH_3COOH	+ C_2H_5OH $\rightleftharpoons$	$CH_3COOC_2H_5$	+ H_2O
initial	0.32	6.30	0	0
change	$-x$	$-x$	$+x$	$+x$
final	$0.32 - x$	$6.30 - x$	$+x$	$+x$

$$K_C = \frac{[CH_3COOC_2H_5][H_2O]}{[CH_3COOH][C_2H_5OH]} = \frac{(x)(x)}{(0.32 - x)(6.30 - x)} = \frac{x^2}{x^2 - 6.62x + 2.02}$$

$$4.0 = \frac{x^2}{x^2 - 6.62x + 2.02}$$

$$4.0x^2 - 26.48x + 8.08 = x^2$$

$$3.0x^2 - 26.48x + 8.08 = 0$$

$$x = \frac{-(-26.48) \pm \sqrt{(-26.48)^2 - (4)(3.0)(8.08)}}{(2)(3.0)} = \frac{+26.48 \pm 24.58}{6.0}$$

$x = 8.51$ or 0.317

The root 8.51 is meaningless because it is larger than the concentration of acetic acid and ethanol, so the value 0.317 is chosen. The equilibrium concentration of the product ester is, therefore, 0.317 mol·L^{-1}. The numbers can be confirmed by placing them into the equilibrium expression:

$$K_C = \frac{[CH_3COOC_2H_5][H_2O]}{[CH_3COOH][C_2H_5OH]} = \frac{(0.317)(0.317)}{(0.320 - 0.317)(6.300 - 0.317)} = 5.6$$

Note: This number does not appear to agree well with the given value of $K_C = 4.0$. If 0.316 is used, the agreement is better, giving a quotient of 4.1. If 0.315 is used, the quotient is 3.3. The better answer is thus 0.316 mol·L^{-1}. The discrepancy is caused by rounding errors in places that are really beyond the accuracy of the measurement. Given that K_C is only given to two significant figures, the best report of the concentration of ester would be 0.32 mol·L^{-1}, even though this value will not satisfy the equilibrium expression as well as 0.316 mol·L^{-1}.

9.65 (a) According to Le Chatelier's principle, an increase in the partial pressure of CO_2 will result in creation of reactants, which will decrease the H_2 partial pressure.

(b) According to Le Chatelier's principle, if the CO pressure is reduced, the reaction will shift to form more CO, which will decrease the pressure of CO_2.

(c) According to Le Chatelier's principle, if the concentration of CO is increased then the reaction will proceed to form more products, which will result in a higher pressure of H_2.

(d) The equilibrium constant for the reaction is unchanged, because it is unaffected by any change in concentration.

9.67 (a) According to Le Chatelier's principle, increasing the concentration of NO will cause the reaction to form reactants in order to reduce the concentration of NO; the amount of water will decrease.

(b) For the same reason as in (a), the amount of O_2 will increase.

(c) According to Le Chatelier's principle, removing water will cause the reaction to shift towards products, resulting in the formation of more NO.

(d) According to Le Chatelier's principle, removing a reactant will cause the reaction to shift in the direction to replace the removed substance; the amount of NH_3 should increase.

(e) According to Le Chatelier's principle, adding ammonia will shift the reaction to the right, but the equilibrium constant, which is a constant, will not be affected.

(f) According to Le Chatelier's principle, removing NO will cause the formation of more products; the amount of NH_3 will decrease.

(g) According to Le Chatelier's principle, adding reactants will promote the formation of products; the amount of oxygen will decrease.

9.69 As per Le Chatelier's principle, whether increasing the pressure on a reaction will affect the distribution of species within an equilibrium mixture of gases depends largely upon the difference in the number of moles of gases between the reactant and product sides of the equation. If there is a net increase in the amount of gas, then applying pressure will shift the reaction towards reactants in order to remove the stress applied by increasing the pressure. Similarly, if there is a net decrease in the amount of gas, applying pressure will cause the formation of products. If the number of moles of gas is the same on the product and reactant side, then changing the pressure will have little or no effect on the equilibrium distribution of species present. Using this information, we can apply it to the specific reactions given. The answers are: (a) reactants; (b) reactants; (c) reactants; (d) no change (there are the same number of moles of gas on both sides of the equation); (e) reactants.

9.71 (a) If the pressure of NO (a product) is increased, the reaction will shift to form more reactants; the pressure of NH_3 should increase.

(b) If the pressure of NH_3 (a reactant) is decreased, then the reaction will shift to form more reactants; the pressure of O_2 should increase.

9.73 If a reaction is exothermic, raising the temperature will tend to shift the reaction towards reactants, whereas if the reaction is endothermic, a shift toward products will be observed. For the specific examples given, (a) and (b) are endothermic (the values for (b) can be calculated, but we know that it requires energy to break an X—X bond so those processes will all be endothermic) and raising the temperature should favor the formation of products; (c) and (d) are exothermic and raising the temperature should favor the formation of reactants.

9.75 Even though numbers are given, we do not need to do a calculation to answer this qualitative question. Because the equilibrium constant for the formation of ammonia is smaller at the higher temperature, raising the temperature will favor the formation of reactants. Less ammonia will be present at higher temperature, assuming no other changes occur to the system (i.e., the volume does not change, no reactants or products are added or removed from the container, etc.).

9.77 (a) $K = \frac{P_{SO_2}}{P_{SO_3}{}^2 P_{O_2}} = 3.0 \times 10^4$

$$K = \left(\frac{T}{12.027\ \text{K}}\right)^{\Delta n} K_C$$

$$K_C = \left(\frac{12.027\ \text{K}}{T}\right)^{\Delta n} K$$

$$\Delta n = 2 - [2 + 1] = -1$$

$$K_C = \left(\frac{12.027\ \text{K}}{700\ \text{K}}\right)^{-1} (3.0 \times 10^4) = 1.7 \times 10^6$$

(b) The reaction written in (b) is 1/2 the reverse of the reaction written in (a), so $K_C(\text{b}) = K_C(\text{a})^{-1/2}$:

$$K_C(\text{b}) = \frac{1}{\sqrt{K_C(\text{a})}} = \frac{1}{\sqrt{1.7 \times 10^6}} = 7.7 \times 10^{-4}$$

9.79 To answer this question we must calculate Q:

$$Q = \frac{[NH_3]^2}{[N_2][H_2]^3} = \frac{(0.500)^2}{(3.00)(2.00)^3} = 0.0104$$

Because $Q \neq K$, the system is not at equilibrium and because $Q < K$, the reaction will proceed to produce more products.

9.81 $K_C = \frac{[BrCl]^2}{[Cl_2][Br_2]}$

$$0.031 = \frac{0.097^2}{(0.22)[Br_2]}$$

$$[Br_2] = \frac{0.097^2}{(0.22)(0.031)} = 1.4\ \text{mol} \cdot \text{L}^{-1}$$

9.83 We can calculate changes according to the reaction stoichiometry:

Amount (mol)	$CO(g)$	+	$3\ H_2(g)$	$\rightleftharpoons$	$CH_4(g)$	+	$H_2O(g)$
initial	2.00		3.00		0		0
change	$-x$		$-3x$		$+x$		$+x$
final	$2.00 - x$		$3.00 - 3x$		0.478		$+x$

According the stoichiometry, 0.478 mol = x; therefore, at equilibrium, there are 2.00 mol − 0.478 mol = 1.52 mol CO, 3.00 − 3(0.478 mol) = 1.57 mol H_2, and 0.478 mol H_2O. To employ the equilibrium expression, we need either concentrations or pressures; because K_C is given, we will choose to express these as concentrations. This calculation is easy because $V = 10.0$ L:

$[CO] = 0.152\ mol\cdot L^{-1}$; $[H_2] = 0.157\ mol\cdot L^{-1}$; $[CH_4] = 0.0478\ mol\cdot L^{-1}$; $[H_2O] = 0.0478\ mol\cdot L^{-1}$

$$K_C = \frac{[CH_4][H_2O]}{[CO][H_2]^3} = \frac{(0.0478)(0.0478)}{(0.152)(0.157)^3} = 3.88$$

9.85 First, we calculate the initial concentrations of each species:

$$[SO_2] = [NO] = \frac{0.100\ mol}{5.00\ L} = 0.0200\ mol\cdot L^{-1};$$

$$[NO_2] = \frac{0.200\ mol}{5.00\ L} = 0.0400\ mol\cdot L^{-1};\ [SO_3] = \frac{0.150\ mol}{5.00\ L} = 0.0300\ mol\cdot L^{-1}$$

We can use these values to calculate Q in order to see which direction the reactions will go:

$Q = \dfrac{(0.0200)(0.0300)}{(0.0200)(0.0400)} = 0.75$. Because $Q < K_C$, the reaction will proceed to produce more products.

Concentration ($mol\cdot L^{-1}$)	$SO_2(g)$	+	$NO_2(g)$	$\rightleftharpoons$	$NO(g)$	+	$SO_3(g)$
initial	0.0200		0.0400		0.0200		0.0300
change	$-x$		$-x$		$+x$		$+x$
final	$0.0200 - x$		$0.0400 - x$		$0.0200 + x$		$0.0300 + x$

$$K_C = \frac{[NO][SO_3]}{[SO_2][NO_2]}$$

$$85.0 = \frac{(0.0200 + x)(0.0300 + x)}{(0.0200 - x)(0.0400 - x)}$$

$$= \frac{x^2 + 0.0500x + 0.000\ 600}{x^2 - 0.0600x + 0.000\ 800}$$

$$85.0(x^2 - 0.0600x + 0.000\ 800) = x^2 + 0.0500x + 0.000\ 600$$

$85.0x^2 - 5.10x + 0.0680 = x^2 + 0.0500x + 0.000\,600$

$84.0x^2 - 5.15x + 0.0674 = 0$

$$x = \frac{-(-5.15) \pm \sqrt{(-5.15)^2 - (4)(84.0)(0.0674)}}{(2)(84.0)} = \frac{+5.15 \pm 1.97}{168}$$

$x = +0.0424$ or $+0.0189$

The root 0.0424 is not meaningful because it is larger than the concentration of NO_2. The root of choice is therefore 0.0189.

At equilibrium:

$[SO_2] = 0.0200\ mol \cdot L^{-1} - 0.0189\ mol \cdot L^{-1} = 0.0011\ mol \cdot L^{-1}$

$[NO_2] = 0.0400\ mol \cdot L^{-1} - 0.0189\ mol \cdot L^{-1} = 0.0211\ mol \cdot L^{-1}$

$[NO] = 0.0200\ mol \cdot L^{-1} + 0.0189\ mol \cdot L^{-1} = 0.0389\ mol \cdot L^{-1}$

$[SO_3] = 0.0300\ mol \cdot L^{-1} + 0.0189\ mol \cdot L^{-1} = 0.0489\ mol \cdot L^{-1}$

To check, we can put these numbers back into the equilibrium constant expression:

$$K_C = \frac{[NO][SO_3]}{[SO_2][NO_2]}$$

$$\frac{(0.0389)(0.0489)}{(0.0011)(0.0211)} = 82.0$$

Compared to $K_C = 85.0$, this is reasonably good agreement given the nature of the calculation. We can check to see, by trial and error, if a better answer could be obtained. Because the K_C value is low for the concentrations we calculated, we can choose to alter x slightly so that this ratio becomes larger. If we let x = 0.0190, the concentrations of NO and SO_3 are increased to 0.0390 and 0.0490, and the concentrations of SO_2 and NO_2 are decreased to 0.0010 and 0.0200 (the stoichiometry of the reaction is maintained by calculating the concentrations in this fashion). Then the quotient calculates to be 91.0, which is further from the value for K_C than the original answer. So, although the agreement is not the best with the numbers we obtained, it is the best possible, given the limitation on the number of significant figures we are allowed to use in the calculation.

9.87 (a) The initial concentrations of O_2 and N_2O are $0.0560\ mol \cdot L^{-1}$ and $0.0200\ mol \cdot L^{-1}$; the final concentration of $NO_2 = 0.0200\ mol \cdot L^{-1}$ because $V = 1.00$ L.

Concentrations ($mol \cdot L^{-1}$)	$2\ N_2O(g)$	+	$3\ O_2(g)$	$\rightleftharpoons$	$4\ NO_2(g)$
initial	0.0200		0.0560		0
change	$-2x$		$-3x$		$+4x$
final	$0.0200 - 2x$		$0.0560 - 3x$		0.0200

Thus $4x = 0.0200\ mol \cdot L^{-1}$, or $x = 0.005\,00\ mol \cdot L^{-1}$.

At equilibrium:

$[N_2O] = 0.0200 \text{ mol·L}^{-1} - 2\,(0.005\,00 \text{ mol·L}^{-1}) = 0.0100 \text{ mol·L}^{-1}$

$[O_2] = 0.0560 \text{ mol·L}^{-1} - 3\,(0.005\,00 \text{ mol·L}^{-1}) = 0.0410 \text{ mol·L}^{-1}$

$[NO_2] = 0.0200 \text{ mol·L}^{-1}$

(b) $K_C = \dfrac{(0.0200)^4}{(0.0100)^2(0.0410)^3} = 23.2$

9.89 (a) The initial concentrations are:

$[PCl_5] = \dfrac{1.50 \text{ mol}}{0.500 \text{ L}} = 3.00 \text{ mol·L}^{-1}$; $[PCl_3] = \dfrac{3.00 \text{ mol}}{0.500 \text{ L}} = 6.00 \text{ mol·L}^{-1}$;

$[Cl_2] = \dfrac{0.500 \text{ mol}}{0.500 \text{ L}} = 1.00 \text{ mol·L}^{-1}$

First calculate Q:

$$Q = \frac{[PCl_5]}{[PCl_3][Cl_2]} = \frac{3.00}{(6.00)(1.00)} = 0.500$$

Because $Q \neq K$, the reaction is not at equilibrium.

(b) Because $Q < K_C$, the reaction will proceed to produce products.

(c)

Concentrations (mol·L^{-1})	$PCl_3(g)$	+	$Cl_2(g)$	$\rightleftharpoons$	$PCl_5(g)$
initial	6.00		1.00		3.00
change	$-x$		$-x$		$+x$
final	$6.00 - x$		$1.00 - x$		$3.00 + x$

$$K_C = \frac{[PCl_5]}{[PCl_3][Cl_2]}$$

$$0.56 = \frac{3.00 + x}{(6.00 - x)(1.00 - x)} = \frac{3.00 + x}{x^2 - 7x + 6.00}$$

$(0.56)(x^2 - 7x + 6.00) = 3.00 + x$

$0.56x^2 - 3.92x + 3.36 = 3.00 + x$

$0.56x^2 - 4.92x + 0.36 = 0$

$$x = \frac{-(-4.92) \pm \sqrt{(-4.92)^2 - (4)(0.56)(0.36)}}{(2)(0.56)} = \frac{+4.92 \pm 4.84}{1.12}$$

$x = 9.2$ or 0.07

Because the root 9.2 is larger than the amount of PCl_3 or Cl_2 available, it is physically meaningless and can be discarded. Thus, $x = 0.071 \text{ mol·L}^{-1}$, giving

$[PCl_5] = 3.00 \text{ mol·L}^{-1} + 0.07 \text{ mol·L}^{-1} = 3.07 \text{ mol·L}^{-1}$

$[PCl_3] = 6.00 \text{ mol·L}^{-1} - 0.07 \text{ mol·L}^{-1} = 5.93 \text{ mol·L}^{-1}$

$[CL_2] = 1.00\ mol\cdot L^{-1} - 0.07\ mol\cdot L^{-1} = 0.93\ mol\cdot L^{-1}$

The number can be checked by substituting them back into the equilibrium constant expression:

$$K_C = \frac{[PCl_5]}{[PCl_3][Cl_2]}$$

$$\frac{(3.07)}{(5.93)(0.93)} \stackrel{?}{=} 0.56$$

$$0.56 \stackrel{\checkmark}{=} 0.56$$

9.91

Pressures (bar)	$2\ HCl(g)$	$\rightleftharpoons$ $H_2(g)$	+ $Cl_2(g)$
initial	0.22	0	0
change	$-2x$	$+x$	$+x$
final	$0.22 - 2x$	$+x$	$+x$

$$K = \frac{P_{H_2}P_{Cl_2}}{P_{HCl}^{\ 2}}$$

$$3.2 \times 10^{-34} = \frac{(x)(x)}{(0.22 - 2x)^2}$$

Because the equilibrium constant is small, assume that $x \ll 0.22$

$$3.2 \times 10^{-34} = \frac{x^2}{(0.22)^2}$$

$$x^2 = (3.2 \times 10^{-34})(0.22)^2$$

$$x = \sqrt{(3.2 \times 10^{-34})(0.22)^2}$$

$$x = \pm 3.9 \times 10^{-18}$$

The negative root is not physically meaningful and can be discarded. x is small compared to 0.22, so the initial assumption was valid. The pressures at equilibrium are

$P_{HCl} = 0.22$ bar; $P_{H_2} = P_{Cl_2} = 3.9 \times 10^{-18}$ bar

The values can be checked by substituting them into the equilibrium expression:

$$\frac{(3.9 \times 10^{-18})(3.9 \times 10^{-18})}{(0.22)^2} \stackrel{?}{=} 3.2 \times 10^{-34}$$

$$3.1 \times 10^{-34} \stackrel{\checkmark}{=} 3.2 \times 10^{-34}$$

$P_{HCl} = 0.22$ bar; $P_{H_2} = P_{Cl_2} = 3.9 \times 10^{-18}$ bar

The numbers agree very well for a calculation of this type.

9.93 (a) The initial pressure of NOCl is calculated as follows:

$$P = \frac{nRT}{V} = \frac{\left(\frac{m}{M}\right)RT}{V}$$

$$= \frac{\left(\frac{30.1\text{ g}}{65.46\text{ g}\cdot\text{mol}^{-1}}\right)(0.082\,06\text{ L}\cdot\text{atm}\cdot\text{K}^{-1}\cdot\text{mol}^{-1})(500\text{ K})\left(\frac{1.013\,25\text{ bar}}{1\text{ atm}}\right)}{0.200\text{ L}}$$

$$= 95.6\text{ bar}$$

Pressures (bar)	$2\,NOCl(g)$ $\rightleftharpoons$	$2\,NO(g)$ +	$Cl_2(g)$
initial	95.6	0	0
change	$-2x$	$+2x$	$+x$
final	$95.6 - 2x$	$+2x$	$+x$

$$K = \frac{P_{NO}{}^2 P_{Cl_2}}{P_{NOCl}{}^2}$$

$$1.13 \times 10^{-3} = \frac{(2x)^2(x)}{(95.6 - 2x)^2}$$

Assume that $x \ll 95.6$:

$$1.13 \times 10^{-3} = \frac{4x^3}{(95.6)^2}$$

$$x^3 = (1.13 \times 10^{-3})(95.6)^2$$

$$x = \sqrt[3]{(1.13 \times 10^{-3})(95.6)^2} = 2.18$$

Because x is not very much less than 95.6, the assumption was not good and the expression should be solved more rigorously:

$$1.13 \times 10^{-3} = \frac{4x^3}{(95.6 - 2x)^2} = \frac{4x^3}{4x^2 - 382.4x + 9139}$$

$$(1.13 \times 10^{-3})(4x^2 - 382.4x + 9139) = 4x^3$$

$$4x^3 - 4.52 \times 10^{-3}x^2 + 0.432x - 10.3 = 0$$

x can be obtained from a graphing calculator or by the trial and error method to give 1.345 bar.

$P_{NOCl} = 95.6\text{ bar} - 2(1.345\text{ bar}) = 92.9\text{ bar}$

$P_{NO} = 2 \times 1.345\text{ bar} = 2.69\text{ bar}$

$P_{Cl_2} = 1.34\text{ bar}$

These numbers can be checked by substituting them into the equilibrium expression:

$$\frac{(2.69)^2(1.34)}{(92.9)^2} \stackrel{?}{=} 1.13 \times 10^{-3}$$

$1.12 \times 10^{-3} \stackrel{\checkmark}{=} 1.13 \times 10^{-3}$

This is good agreement.

(b) $\dfrac{95.6 \text{ bar} - 92.9 \text{ bar}}{95.6 \text{ bar}} \times 100 = 2.8\%$

9.95 (a) According to Le Chatelier's principle, adding a product should cause a shift in the equilibrium towards the reactants side of the equation.

(b) Because there are equal numbers of moles of gas on both sides of the equation, there will be little or no effect upon compressing the system.

(c) If the amount of CO_2 is increased, this will cause the reaction to shift toward the formation of products.

(d) Because the reaction is endothermic, raising the temperature will favor the formation of products.

(e) If the amount of $C_6H_{12}O_6$ is removed, this will cause the reaction to shift toward the formation of products.

(f) Because water is a liquid, it is by definition present at unit concentration, so changing the amount of water will not affect the reaction. As long as the glucose solution is dilute, its concentration can be considered unchanged.

(g) Decreasing the concentration of a reactant will favor the production of more reactants.

9.97 (a) To determine on which side of the equilibrium position the conditions lie, we will calculate Q:

$$[CO] = \frac{0.150 \text{ mol}}{3.00 \text{ L}} = 0.0500 \text{ mol}\cdot\text{L}^{-1};\ [H_2] = \frac{0.0900 \text{ mol}}{3.00 \text{ L}} = 0.0300 \text{ mol}\cdot\text{L}^{-1};$$

$$[CH_3OH] = \frac{0.180 \text{ mol}}{3.00 \text{ L}} = 0.0600 \text{ mol}\cdot\text{L}^{-1}$$

$$Q = \frac{[CH_3OH]}{[CO][H_2]^2} = \frac{0.0600}{(0.0500)(0.0300)^2} = 1.33 \times 10^3$$

Because $Q > K_C$, the reaction will proceed to produce more of the reactants, which means that the concentration of methanol will decrease.

(b)

Concentrations ($\text{mol}\cdot\text{L}^{-1}$)	$CO(g)$	+	$2\ H_2(g)$	$\rightleftharpoons$	$CH_3OH(g)$
initial	0.0500		0.0300		0.0600
change	$+x$		$+2x$		$-x$
final	$0.0500 + x$		$0.0300 + 2x$		$0.0600 - x$

$$K_C = \frac{0.0600 - x}{(0.0500 + x)(0.0300 + 2x)^2} = 1.1 \times 10^{-2}$$

$$0.0600 - x = (1.1 \times 10^{-2})(0.0500 + x)(0.0300 + 2x)^2$$

$$= (5.5 \times 10^{-4} + 1.1 \times 10^{-2}x)(4x^2 + 0.0600x + 9.00 \times 10^{-4})$$

$$= 4.4 \times 10^{-2}x^3 + 2.9 \times 10^{-3}x^2 + 4.3 \times 10^{-5}x + 4.95 \times 10^{-7}$$

$$0 = 4.4 \times 10^{-2}x^3 + 2.9 \times 10^{-3}x^2 + 1.00x - 0.0600$$

This equation can be solved approximately, simply by inspection: it is clear that the x term will be very much larger than the x^3 and the x^2 terms because their coefficients are very small compared to 1.00. This leads to a prediction that $x = 0.0600\ \text{mol}\cdot\text{L}^{-1}$ to within the accuracy of the data. Essentially all of the CH_3OH will react, so that $[CO] = 0.0500\ \text{mol}\cdot\text{L}^{-1} + 0.0600\ \text{mol}\cdot\text{L}^{-1} = 0.110\ \text{mol}\cdot\text{L}^{-1}$; $[H_2] = 0.0300\ \text{mol}\cdot\text{L}^{-1} + 2\,(0.0600\ \text{mol}\cdot\text{L}^{-1}) = 0.150\ \text{mol}\cdot\text{L}^{-1}$. The mathematical situation is odd in that clearly a $[CH_3OH] = 0$ will not satisfy the equilibrium constant. Knowing that the methanol concentration is very small compared to the CO and H_2 concentrations, we can now back-calculate to get a concentration value that will satisfy the equilibrium expression:

$$K_C = \frac{y}{(0.110)(0.150)^2} = 1.1 \times 10^{-2}$$

$$y = (1.1 \times 10^{-2})(0.110)(0.150)^2 = 3 \times 10^{-5}$$

Alternatively, the cubic equation can be solved with the aid of a graphing calculator like the one supplied on the CD accompanying this book. The equilibrium calculation by this method gives a value of 0.5997 for x; the final answer for y is the same, using that method.

9.99 Because we want the equilibrium constant at two temperatures, we will need to calculate ΔH°_r and ΔS°_r for each reaction:

(a) $NH_4Cl \rightleftharpoons NH_3(g) + HCl(g)$

$$\Delta H^\circ_r = \Delta H^\circ_f(NH_3, g) + \Delta H^\circ_f(HCl, g) - \Delta H^\circ_f(NH_4Cl, s)$$

$$\Delta H^\circ_r = (-46.11\ \text{kJ}\cdot\text{mol}^{-1}) + (-92.31\ \text{kJ}\cdot\text{mol}^{-1}) - (-314.43\ \text{kJ}\cdot\text{mol}^{-1})$$

$$\Delta H^\circ_r = 176.01\ \text{kJ}\cdot\text{mol}^{-1}$$

$$\Delta S^\circ_r = S^\circ\,(NH_3, g) + S^\circ\,(HCl, g) - S^\circ\,(NH_4Cl, s)$$

$$\Delta S^\circ_r = 192.45\ \text{J}\cdot\text{K}^{-1}\cdot\text{mol}^{-1} + 186.91\ \text{J}\cdot\text{K}^{-1}\cdot\text{mol}^{-1} - 94.6\ \text{J}\cdot\text{K}^{-1}\cdot\text{mol}^{-1}$$

$$\Delta S^\circ_r = 284.8\ \text{J}\cdot\text{K}^{-1}\cdot\text{mol}^{-1}$$

$$\Delta G^\circ_r = \Delta H^\circ_r - T\Delta S^\circ_r$$

At 298 K:

$$\Delta G^\circ_{r(298\ K)} = 176.01\ kJ - (298\ K)(284.8\ J\cdot K^{-1})/(1000\ J\cdot kJ^{-1})$$
$$= 91.14\ kJ\cdot mol^{-1}$$
$$\Delta G^\circ_{r(298\ K)} = -RT \ln K$$
$$\ln K = -\frac{\Delta G^\circ_{r(298\ K)}}{RT}$$
$$= -\frac{91\,140\ J}{(8.314\ J\cdot K^{-1})(298\ K)} = -36.8$$
$K = 1 \times 10^{-16}$

At 423 K:

$$\Delta G^\circ_{r(423\ K)} = 176.01\ kJ - (423\ K)(284.8\ J\cdot K^{-1})/(1000\ J\cdot kJ^{-1})$$
$$= 55.54\ kJ\cdot mol^{-1}$$
$$\Delta G^\circ_{r(423\ K)} = -RT \ln K$$
$$\ln K = -\frac{\Delta G^\circ_{r(423\ K)}}{RT}$$
$$= -\frac{55\,540\ J}{(8.314\ J\cdot K^{-1})(423\ K)} = -15.8$$
$K = 1 \times 10^{-7}$

(b) $H_2(g) + D_2O(l) \rightleftharpoons D_2(g) + H_2O(l)$

$$\Delta H^\circ_r = \Delta H^\circ_f(H_2O, l) - [\Delta H^\circ_f(D_2O, l)]$$
$$\Delta H^\circ_r = (-285.83\ kJ\cdot mol^{-1}) - [-294.60\ kJ\cdot mol^{-1}]$$
$$\Delta H^\circ_r = 8.77\ kJ\cdot mol^{-1}$$
$$\Delta S^\circ_r = S^\circ(D_2, g) + S^\circ(H_2O, l) - [S^\circ(H_2, g) + S^\circ(D_2O, l)]$$
$$\Delta S^\circ_r = 144.96\ J\cdot K^{-1}\cdot mol^{-1} + 69.91\ J\cdot K^{-1}\cdot mol^{-1} - [130.68\ J\cdot K^{-1}\cdot mol^{-1} + 75.94\ J\cdot K^{-1}\cdot mol^{-1}]$$
$$\Delta S^\circ_r = 8.25\ J\cdot K^{-1}\cdot mol^{-1}$$

At 298 K:

$$\Delta G^\circ_{r(298\ K)} = 8.77\ kJ\cdot mol^{-1} - (298\ K)(8.25\ J\cdot K^{-1}\cdot mol^{-1})/(1000\ J\cdot kJ^{-1})$$
$$= 6.31\ kJ\cdot mol^{-1}$$
$$\Delta G^\circ_{r(298\ K)} = -RT \ln K$$
$$\ln K = -\frac{\Delta G^\circ_{r(298\ K)}}{RT}$$
$$= -\frac{6310\ J}{(8.314\ J\cdot K^{-1})(298\ K)} = -2.55$$
$K = 7.8 \times 10^{-2}$

At 423 K:

$$\Delta G^\circ_{r(423\ K)} = 8.77\ kJ\cdot mol^{-1} - (423\ K)(8.25\ J\cdot K^{-1}\cdot mol^{-1})/(1000\ J\cdot kJ^{-1})$$
$$= 5.28\ kJ\cdot mol^{-1}$$

$$\ln K = -\frac{5280\text{ J}}{(8.314\text{ J}\cdot\text{K}^{-1})(423\text{ K})} = -1.50$$

$K = 0.22$

9.101 (a) $\Delta H°_r = -[2\Delta H°_f(O_3, g)]$

$= -[2(142.7\text{ kJ}\cdot\text{mol}^{-1})]$

$= -285.4\text{ kJ}\cdot\text{mol}^{-1}$

$\Delta S°_r = 3S°(O_2, g) - [2S°(O_3, g)]$

$= 3(205.14\text{ J}\cdot\text{K}^{-1}\cdot\text{mol}^{-1}) - [2(238.93\text{ J}\cdot\text{K}^{-1}\cdot\text{mol}^{-1})]$

$= 137.56\text{ J}\cdot\text{K}^{-1}\cdot\text{mol}^{-1}$

At 298 K:

$\Delta G°_{r(298\text{ K})} = -285.4\text{ kJ}\cdot\text{mol}^{-1} - (298\text{ K})(137.56\text{ J}\cdot\text{K}^{-1}\cdot\text{mol}^{-1})/(1000\text{ J}\cdot\text{kJ}^{-1})$

$= -326.4\text{ kJ}\cdot\text{mol}^{-1}$

or directly from $\Delta G°_f$ values

$\Delta G°_r = -2\Delta G°_f(O_3, g)$

$= -2(163.2\text{ kJ}\cdot\text{mol}^{-1})$

$= -326.4\text{ kJ}\cdot\text{mol}^{-1}$

(b) $\Delta G°_{r(298\text{ K})} = -RT\ln K$

$$\ln K = -\frac{\Delta G°_{r(298\text{ K})}}{RT}$$

$$= -\frac{-326\,390\text{ J}}{(8.314\text{ J}\cdot\text{K}^{-1})(298\text{ K})} = +132$$

$K = 10^{57}$

The equilibrium constant for the decomposition of ozone to oxygen is extremely large, making this process extremely favorable. Note: The presence of ozone in the upper atmosphere is largely due to kinetic factors that inhibit the decomposition; however, in the presence of a suitable catalyst, this process should occur extremely readily.

9.103 In order to solve this problem, we will manipulate the equations with the known K's so that we can combine them to give the desired overall reaction:

First, reverse equation (1) and multiply it by $\frac{1}{2}$:

$$H_2(g) + \tfrac{1}{2}O_2(g) \rightleftharpoons H_2O(g) \qquad K_4 = \frac{1}{(1.6 \times 10^{-11})^{1/2}} \qquad (4)$$

Multiply equation (2) by $\frac{1}{2}$ also:

$$CO_2(g) \rightleftharpoons CO(g) + \tfrac{1}{2}O_2(g) \qquad K_5 = (1.3 \times 10^{-10})^{1/2} \qquad (5)$$

Adding equations (4) and (5) gives the desired reaction. The resultant equilibrium constant will be the product of the K's for (4) and (5):

$$CO_2(g) + H_2(g) \rightleftharpoons H_2O(g) + CO(g) \qquad (3)$$

$$K_5 = \left(\frac{1.3 \times 10^{-10}}{1.6 \times 10^{-11}}\right)^{1/2} = 2.9$$

(b) To obtain the K value for a net equation from two (or more) others, the K's are multiplied, but $\Delta G°_r$'s are added:

$$2\ H_2O(g) \rightleftharpoons 2\ H_2(g) + O_2(g) \qquad (1)$$

$$\begin{aligned}\Delta G°_r &= -RT \ln K \\ &= -(8.314\ \text{J}\cdot\text{K}^{-1}\cdot\text{mol}^{-1})(1565\ \text{K}) \ln 1.6 \times 10^{-11} \\ &= +3.2 \times 10^2\ \text{kJ}\cdot\text{mol}^{-1}\end{aligned}$$

$$2\ CO_2(g) \rightleftharpoons 2\ CO(g) + O_2(g) \qquad (2)$$

$$\begin{aligned}\Delta G°_r &= -RT \ln K \\ &= -(8.314\ \text{J}\cdot\text{K}^{-1}\cdot\text{mol}^{-1})(1565\ \text{K}) \ln 1.3 \times 10^{-10} \\ &= +3.0 \times 10^2\ \text{kJ}\cdot\text{mol}^{-1}\end{aligned}$$

The corresponding values for (4) and (5) are $\Delta G°_{r(4)} = -\frac{1}{2}(3.2 \times 10^2\ \text{kJ}) = -1.6 \times 10^2\ \text{kJ}$; $\Delta G°_{r(5)} = \frac{1}{2}(3.0 \times 10^2\ \text{kJ}) = 1.5 \times 10^2\ \text{kJ}$. Summing these two values will give $\Delta G°_{r(3)}$: $-1.6 \times 10^2\ \text{kJ} + 1.5 \times 10^2\ \text{kJ} = -10\ \text{kJ}\cdot\text{mol}^{-1}$.

$$\ln K = -\frac{\Delta G°_r}{RT} = -\frac{-10\ 000\ \text{J}}{(8.314\ \text{J}\cdot\text{K}^{-1}\cdot\text{mol}^{-1})(1565\ \text{K})} = +0.8$$

$K = 2$

This value is in reasonable agreement with the one obtained in (a), given the problems in significant figures and due to rounding errors.

9.105

Pressure	$PCl_5(g) \rightleftharpoons$	$PCl_3(g)$ +	$Cl_2(g)$
initial	n		
change	$-n\alpha$	$+n\alpha$	$+n\alpha$
final	$n(1-\alpha)$	$+n\alpha$	$+n\alpha$

$$K = \frac{(n\alpha)(n\alpha)}{n(1-\alpha)} = \frac{n^2\alpha^2}{n(1-\alpha)} = \frac{n\alpha^2}{(1-\alpha)}$$

$$P = n(1-\alpha) + n\alpha + n\alpha = n(1+\alpha)$$

$$n = \frac{P}{1+\alpha}$$

$$K = \frac{n\alpha^2}{(1-\alpha)} = \left(\frac{P}{1+\alpha}\right)\left(\frac{\alpha^2}{1-\alpha}\right) = \frac{P\alpha^2}{1-\alpha^2}$$

$$(1-\alpha^2)K = P\alpha^2$$

$$K - K\alpha^2 = P\alpha^2$$

$$K = P\alpha^2 + K\alpha^2$$

$$\alpha^2 = \frac{K}{P+K}$$

$$\alpha = \sqrt{\frac{K}{P + K}}$$

(a) For the specific conditions $K = 4.96$ and $P = 0.50$ bar,

$$\alpha = \sqrt{\frac{4.96}{0.50 + 4.96}} = 0.953$$

(b) For the specific conditions $K = 4.96$ and $P = 1.00$ bar,

$$\alpha = \sqrt{\frac{4.96}{1.00 + 4.96}} = 0.912$$

9.107 (a) $K_C = \frac{[NO_2]^2}{[N_2O_4]} = \frac{(2.13)^2}{0.405} = 11.2$

(b) If NO_2 is added, the equilibrium will shift to produce more N_2O_4. The amount of NO_2 will be greater than initially present, but less than the $3.13\ mol \cdot L^{-1}$ present immediately upon making the addition. K_C will not be affected.

(c)

Concentrations ($mol \cdot L^{-1}$)	N_2O_4	$\rightleftharpoons$	$2\ NO_2$
initial	0.405		3.13
change	$+x$		$-2x$
final	$0.405 + x$		$3.13 - 2x$

$$11.2 = \frac{(3.13 - 2x)^2}{0.405 + x}$$

$(11.2)(0.405 + x) = (3.13 - 2x)^2$

$11.2x + 4.536 = 4x^2 - 12.52x + 9.797$

$4x^2 - 23.7x + 5.26 = 0$

$$x = \frac{-(-23.7) \pm \sqrt{(-23.7)^2 - (4)(4)(5.26)}}{(2)(4)} = \frac{23.7 \pm 21.9}{8}$$

$x = 5.70$ or 0.23

At equilibrium $[N_2O_4] = 0.405\ mol \cdot L^{-1} + 0.23\ mol \cdot L^{-1} = 0.64\ mol \cdot L^{-1}$

$[NO_2] = 3.13\ mol \cdot L^{-1} - 2(0.23\ mol \cdot L^{-1}) = 2.67\ mol \cdot L^{-1}$

These concentrations are consistent with the predictions in (b).

9.109 To find the vapor pressure, we first calculate the $\Delta G°$ for the conversion of the liquid to the gas at 298 K, using the free energies of formation found in the appendix:

$$\Delta G°_{H_2O(l) \rightarrow H_2O(g)} = \Delta G°_{f(H_2O(g))} - \Delta G°_{f(H_2O(l))}$$
$$= (-228.57\ kJ \cdot mol^{-1}) - [-237.13\ kJ \cdot mol^{-1}]$$
$$= 8.56\ kJ \cdot mol^{-1}$$

$$\Delta G^\circ_{D_2O(l) \rightarrow D_2O(g)} = \Delta G^\circ_{f(D_2O(g))} - \Delta G^\circ_{f(D_2O(l))}$$
$$= (-234.54\ \text{kJ}\cdot\text{mol}^{-1}) - [-243.44\ \text{kJ}\cdot\text{mol}^{-1}]$$
$$= 8.90\ \text{kJ}\cdot\text{mol}^{-1}$$

The equilibrium constant for these processes is the vapor pressure of the liquid: $K = P_{H_2O}$ or $K = P_{D_2O}$

Using $\Delta G^\circ = -RT \ln K$, we can calculate the desired values.

For H_2O:

$$\ln K = -\frac{\Delta G^\circ}{RT} = -\frac{8560\ \text{J}\cdot\text{mol}^{-1}}{(8.314\ \text{J}\cdot\text{K}^{-1}\cdot\text{mol}^{-1})(298\ \text{K})} = -3.45$$

$K = 0.032$ bar

$$0.032\ \text{bar} \times \frac{1\ \text{atm}}{1.013\ 25\ \text{bar}} \times \frac{760\ \text{Torr}}{1\ \text{atm}} = 24\ \text{Torr}$$

For D_2O:

$$\ln K = -\frac{\Delta G^\circ}{RT} = -\frac{8900\ \text{J}\cdot\text{mol}^{-1}}{(8.314\ \text{J}\cdot\text{K}^{-1}\cdot\text{mol}^{-1})(298\ \text{K})} = -3.59$$

$K = 0.028$ bar

$$0.028\ \text{bar} \times \frac{1\ \text{atm}}{1.013\ 25\ \text{bar}} \times \frac{760\ \text{Torr}}{1\ \text{atm}} = 21\ \text{Torr}$$

The answer is that D has a lower zero point energy than H. This makes the $D_2O\cdots D_2O$ "hydrogen bond" stronger than the $H_2O\cdots H_2O$ hydrogen bond. Because the hydrogen bond is stronger, the intermolecular forces are stronger, the vapor pressure is lower, and the boiling point is higher.

Potential energy curves for the O···H and O···D bonds as a function of distance:

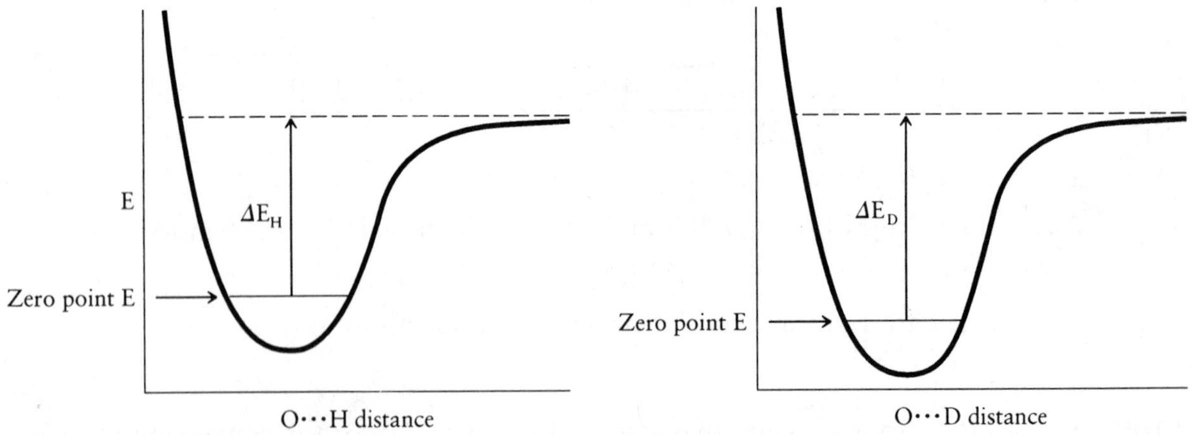

ΔE = energy required to break O···H or O···D bond.

9.111 First, we derive a general relationship that relates ΔG°_f values to nonstandard state conditions. We do this by returning to the fundamental definition of ΔG and ΔG°.

$\Delta G^\circ_f = \Sigma G^\circ_{m(products)} - \Sigma G^\circ_{m(reactants)}$

Similarly, for conditions other than standard state, we can write

$\Delta G_f = \Sigma G_{m(products)} - \Sigma G_{m(reactants)}$

Because $G_m = G^\circ_m + RT \ln Q$,

$\Delta G_f = \Sigma(G^\circ_m + RT \ln P_i)_{(products)} - \Sigma(G^\circ_m + RT \ln P_i)_{(reactants)}$

But because all reactants or products in the same system refer to the same standard state,

$\Delta G_f = \Sigma G^\circ_{m(products)} - \Sigma G^\circ_{m(reactants)} + \Delta n(RT \ln Q_{(reactants)})$

$\Delta G_f = \Delta G^\circ_f + \Delta n(RT \ln Q_{(reactants)})$

The value ΔG_f, which is the nonstandard value for the conditions of 1 atm, 1 $mol \cdot L^{-1}$, etc. becomes the new standard value.

(a) $\frac{1}{2}\, H_2(g) + \frac{1}{2}\, I_2(g) \longrightarrow HI(g)$, $\Delta n = 1\ mol - (\frac{1}{2}\ mol + \frac{1}{2}\ mol) = 0$

1 atm = 1.013 25 bar

$$\Delta G_f = 1.70\ kJ \cdot mol + (0)(8.314\ J \cdot K^{-1} \cdot mol^{-1})(298\ K)(\ln 1.013\ 25)/(1000\ J \cdot kJ^{-1})$$
$$= 1.70\ kJ \cdot mol$$

(b) $C(s) + \frac{1}{2}\, O_2(g) \longrightarrow CO(g)$, $\Delta n = 1\ mol - \frac{1}{2}\ mol = \frac{1}{2}\ mol$

$$\Delta G_f = -137.17\ kJ \cdot mol + (\tfrac{1}{2})(8.314\ J \cdot K^{-1} \cdot mol^{-1})(298\ K)(\ln 1.013\ 25)/(1000\ J \cdot kJ^{-1})$$
$$= -137.15\ kJ \cdot mol^{-1}$$

(c) $C(s) + \frac{1}{2}\, N_2(g) + \frac{1}{2}\, H_2(g) \longrightarrow HCN(g)$, $\Delta n = 1\ mol - (\frac{1}{2}\ mol + \frac{1}{2}\ mol) = 0$

$$1\ Torr \times \frac{1\ atm}{760\ Torr} \times \frac{1.013\ 25\ bar}{1\ atm} = 1.333 \times 10^{-3}\ bar$$

$$\Delta G_f = 124.7\ kJ \cdot mol + (0)(8.314\ J \cdot K^{-1} \cdot mol^{-1})(298\ K)(\ln 1.333 \times 10^{-3})/(1000\ J \cdot kJ^{-1})$$
$$= 124.7\ kJ \cdot mol^{-1}$$

(d) $C(s) + 2\, H_2(g) \longrightarrow CH_4(g)$, $\Delta n = 1\ mol - 2\ mol = -1\ mol$

1 Pa = 10^{-5} bar

$$\Delta G_f = -50.72\ kJ \cdot mol + (-1)(8.314\ J \cdot K^{-1} \cdot mol^{-1})(298\ K)(\ln 10^{-5})/(1000\ J \cdot kJ^{-1})$$
$$= -22.2\ kJ \cdot mol^{-1}$$

9.113 (a)

Compound	$\Delta G^\circ_f (kJ \cdot mol^{-1})$
$CH_3Cl(g)$	−63.01
$CH_3Br(g)$	−28.16

(b) $\Delta G^\circ_r = (-28.16\ \text{kJ}\cdot\text{mol}^{-1}) + (-287.556\ \text{kJ}\cdot\text{mol}^{-1}) - [(-63.01\ \text{kJ}\cdot\text{mol}^{-1}) + (-236.914\ \text{kJ}\cdot\text{mol}^{-1})]$

$= -15.79\ \text{kJ}\cdot\text{mol}^{-1}$

$\Delta G^\circ_r = -RT \ln K,\ K = 584$

(c) BCl_2Br, BCl_3

(d) We can use the equilibrium expression in (b) and the equations already set up in exercise 9.112 (except that the total pressure equation 2 bar = A + B + C + D is no longer valid) to solve this problem. Because there are now six unknowns, we need six equations; we already have three from 9.112. Consistent with that exercise, we will define

$P_{BCl_3} = A$

$P_{BBr_3} = B$

$P_{BCl_2Br} = C$

$P_{BClBr_2} = D$

$P_{CH_3Cl} = E$

$P_{CH_3Br} = F$

(1) $K_1 = \dfrac{C\cdot D}{A\cdot B} = 1.5522$

(2) $K_2 = \dfrac{A\cdot D}{C^2} = 0.5897$

(3) $K_3 = \dfrac{B\cdot C}{D^2} = 1.092$

(4) $CH_3Cl(g) + BBr_3(g) \rightleftharpoons CH_3Br(g) + BClBr_2(g)$

$K_3 = \dfrac{D\cdot F}{E\cdot B} = 584$

(5) From the mass balance we know that

E + F = 1 bar

(6) Also from the mass balance we know

A + B + C + D = 1 bar

Solving these equations simultaneously using a computer program, we obtain

A = 0.191; B = 0.257; C = 0.291; D = 0.262; E = 0.002; F = 0.998

$P_{BCl_3} = 0.191$ bar

$P_{BBr_3} = 0.257$ bar

$P_{BCl_2Br} = 0.291$ bar

$P_{BClBr_2} = 0.262$ bar

$P_{CH_3Cl} = 0.002$ bar

$P_{CH_3Br} = 0.998$ bar

CHAPTER 10
ACIDS AND BASES

10.1 (a) $CH_3NH_3^+$ (b) $NH_2NH_3^+$ (c) H_2CO_3 (d) CO_3^{2-} (e) $C_6H_5O^-$ (f) $CH_3CO_2^-$

10.3 (a) $H_2SO_4(aq) + H_2O(l) \longrightarrow H_3O^+(aq) + HSO_4^-(aq)$

acid$_1$ base$_2$ acid$_2$ base$_1$

(conjugate: acid$_1$ ↔ base$_1$; conjugate: base$_2$ ↔ acid$_2$)

(b) $C_6H_5NH_3^+(aq) + H_2O(l) \rightleftharpoons H_3O^+(aq) + C_6H_5NH_2(aq)$

acid$_1$ base$_2$ acid$_2$ base$_1$

(conjugate: acid$_1$ ↔ base$_1$; conjugate: base$_2$ ↔ acid$_2$)

(c) $H_2PO_4^-(aq) + H_2O(l) \rightleftharpoons H_3O^+(aq) + HPO_4^{2-}(aq)$

acid$_1$ base$_2$ acid$_2$ base$_1$

(conjugate: acid$_1$ ↔ base$_1$; conjugate: base$_2$ ↔ acid$_2$)

(d) $HCOOH(aq) + H_2O(l) \rightleftharpoons H_3O^+(aq) + HCO_2^-(aq)$

acid$_1$ base$_2$ acid$_2$ base$_1$

(conjugate: acid$_1$ ↔ base$_1$; conjugate: base$_2$ ↔ acid$_2$)

(e) $NH_2NH_3^+(aq) + H_2O(l) \rightleftharpoons H_3O^+(aq) + NH_2NH_2(aq)$

acid$_1$ base$_2$ acid$_2$ base$_1$

(conjugate: acid$_1$ ↔ base$_1$; conjugate: base$_2$ ↔ acid$_2$)

10.5 (a) Brønsted acid: HNO_3
Brønsted base: HPO_4^{2-}

(b) conjugate base to HNO_3: NO_3^-
conjugate acid to HPO_4^{2-}: $H_2PO_4^-$

10.7 (a) HCO_3^-, as an acid:

conjugate (HCO_3^- ↔ CO_3^{2-})

$$\underset{\text{acid}_1}{HCO_3^-(aq)} + \underset{\text{base}_2}{H_2O(l)} \rightleftharpoons \underset{\text{acid}_2}{H_3O^+(aq)} + \underset{\text{base}_1}{CO_3^{2-}(aq)}$$

conjugate (H_2O ↔ H_3O^+)

HCO_3^-, as a base:

conjugate (H_2O ↔ OH^-)

$$\underset{\text{acid}_1}{H_2O(l)} + \underset{\text{base}_2}{HCO_3^-(aq)} \rightleftharpoons \underset{\text{acid}_2}{H_2CO_3(aq)} + \underset{\text{base}_1}{OH^-(aq)}$$

conjugate (HCO_3^- ↔ H_2CO_3)

(b) HPO_4^{2-}, as an acid:

conjugate (HPO_4^{2-} ↔ PO_4^{3-})

$$\underset{\text{acid}_1}{HPO_4^{2-}(aq)} + \underset{\text{base}_2}{H_2O(l)} \rightleftharpoons \underset{\text{acid}_2}{H_3O^+(aq)} + \underset{\text{base}_1}{PO_4^{3-}(aq)}$$

conjugate (H_2O ↔ H_3O^+)

HPO_4^{2-}, as a base:

conjugate (H_2O ↔ OH^-)

$$\underset{\text{acid}_1}{H_2O(l)} + \underset{\text{base}_2}{HPO_4^{2-}(aq)} \rightleftharpoons \underset{\text{acid}_2}{H_2PO_4^-(aq)} + \underset{\text{base}_1}{OH^-(aq)}$$

conjugate (HPO_4^{2-} ↔ $H_2PO_4^-$)

10.9 In each case, use $K_w = [H_3O^+][OH^-] = 1.0 \times 10^{-14}$, then

$$[OH^-] = \frac{K_w}{[H_3O^+]} = \frac{1.0 \times 10^{-14}}{[H_3O^+]}$$

(a) $[OH^-] = \dfrac{1.0 \times 10^{-14}}{0.20} = 5.0 \times 10^{-14}\ \text{mol}\cdot\text{L}^{-1}$

(b) $[OH^-] = \dfrac{1.0 \times 10^{-14}}{1.0 \times 10^{-4}} = 1.0 \times 10^{-10}\ \text{mol}\cdot\text{L}^{-1}$

(c) $[OH^-] = \dfrac{1.0 \times 10^{-14}}{3.1 \times 10^{-2}} = 3.2 \times 10^{-3}\ \text{mol}\cdot\text{L}^{-1}$

10.11 (a) $K_w = 2.5 \times 10^{-14} = [H_3O^+][OH^-] = x^2$, where $x = [H_3O^+] = [OH^-]$

$x = \sqrt{2.5 \times 10^{-14}} = 1.6 \times 10^{-7}\ \text{mol}\cdot\text{L}^{-1}$

$\text{pH} = -\log[H_3O^+] = 6.80$

(b) $[OH^-] = [H_3O^+] = 1.6 \times 10^{-7}\ mol \cdot L^{-1}$
$pOH = -\log[OH^-] = 6.80$

10.13 Because $Ba(OH)_2$ is a strong base, $Ba(OH)_2\ (aq) \rightleftharpoons Ba^{2+}(aq) + 2\ OH^-(aq)$, 100%.
Then, $[Ba(OH)_2]_0 = [Ba^{2+}]$, $[OH^-] = 2 \times [Ba(OH)_2]$, where $[Ba(OH)_2]_0$ = nominal concentration of $Ba(OH)_2$.

$$\text{moles of } Ba(OH)_2 = \frac{0.50\ g}{171.36\ g \cdot mol^{-1}} = 2.9 \times 10^{-3}\ mol$$

$$[Ba(OH)_2]_0 = \frac{2.9 \times 10^{-3}\ mol}{0.100\ L} = 2.9 \times 10^{-2}\ mol \cdot L^{-1} = [Ba^{2+}]$$

$$[OH^-] = 2 \times [Ba(OH)_2]_0 = 2 \times 2.9 \times 10^{-2}\ mol \cdot L^{-1} = 5.8 \times 10^{-2}\ mol \cdot L^{-1}$$

$$[H_3O^+] = \frac{K_w}{[OH^-]} = \frac{1.0 \times 10^{-14}}{5.8 \times 10^{-2}} = 1.7 \times 10^{-13}\ mol \cdot L^{-1}$$

10.15 Because $pH = -\log[H_3O^+]$, $\log[H_3O^+] = -pH$. Taking the antilogs of both sides gives $[H_3O^+] = 10^{-pH}\ mol \cdot L^{-1}$
(a) $[H_3O^+] = 10^{-3.3} = 5 \times 10^{-4}\ mol \cdot L^{-1}$
(b) $[H_3O^+] = 10^{-6.7}\ mol \cdot L^{-1} = 2 \times 10^{-7}\ mol \cdot L^{-1}$
(c) $[H_3O^+] = 10^{-4.4}\ mol \cdot L^{-1} = 4 \times 10^{-5}\ mol \cdot L^{-1}$
(d) $[H_3O^+] = 10^{-5.3}\ mol \cdot L^{-1} = 5 \times 10^{-6}\ mol \cdot L^{-1}$
(e) (b) $<$ (d) $<$ (c) $<$ (a)

10.17 $[acid]_0$ = nominal concentration of acid
$[base]_0$ = nominal concentration of base
(a) $[HNO_3]_0 = [H_3O^+] = 0.010\ mol \cdot L^{-1}$; HNO_3 is a strong acid.
$pH = -\log(0.010) = 2.00$, $pOH = 14.00 - 2.00 = 12.00$
(b) $[HCl]_0 = [H_3O^+] = 0.022\ mol \cdot L^{-1}$; HCl is a strong acid.
$pH = -\log(0.022) = 1.66$, $pOH = 14.00 - 0.66 = 12.34$
(c) $[OH^-] = 2 \times [Ba(OH)_2] = 2 \times 1.0 \times 10^{-3}\ M = 2.0 \times 10^{-3}\ mol \cdot L^{-1}$
$pOH = -\log(2.0 \times 10^{-3}) = 2.70$, $pH = 14.00 - 2.70 = 11.30$
(d) $[KOH]_0 = [OH^-]$; KOH is a strong base.

$$[OH^-] = \left(\frac{10.0\ mL}{250\ mL}\right) \times (0.22\ mol \cdot L^{-1}) = 8.8 \times 10^{-3}\ mol \cdot L^{-1}$$

$pOH = -\log(8.8 \times 10^{-3}) = 2.06$, $pH = 14.00 - 2.06 = 11.94$
(e) $[NaOH]_0 = [OH^-]$; NaOH is a strong base.

$$\text{number of moles of NaOH} = \frac{0.0140\ g}{40.00\ g \cdot mol^{-1}} = 3.50 \times 10^{-4}\ mol$$

$$[\text{NaOH}]_0 = \frac{3.50 \times 10^{-4}\ \text{mol}}{0.250\ \text{L}} = 1.40 \times 10^{-3}\ \text{mol·L}^{-1} = [\text{OH}^-]$$

$$\text{pOH} = -\log(1.40 \times 10^{-3}) = 2.85, \quad \text{pH} = 14.00 - 2.85 = 11.15$$

(f) $[\text{HBr}]_0 = [H_3O^+]$; HBr is a strong acid.

$$[H_3O^+] = \left(\frac{50\ \text{mL}}{250\ \text{mL}}\right) \times (0.000\,43\ \text{mol·L}^{-1}) = 8.6 \times 10^{-5}\ \text{mol·L}^{-1}$$

$$\text{pH} = -\log(8.6 \times 10^{-5}) = 4.07, \quad \text{pOH} = 14.00 - 4.07 = 9.93$$

10.19 $pK_{a1} = -\log K_{a1}$; therefore, after taking antilogs, $K_{a1} = 10^{-pK_{a1}}$

Acid	pK_{a1}	K_{a1}
(a) H_3PO_4	2.12	7.6×10^{-3}
(b) H_3PO_3	2.00	0.010
(c) H_2SeO_3	2.46	3.5×10^{-3}
(d) H_2SeO_4	1.92	0.012

(e) The larger K_{a1}, the stronger the acid; therefore $H_2SeO_3 < H_3PO_4 < H_3PO_3 < H_2SeO_4$

10.21 The weakest acid has the strongest conjugate base and vice versa.

(a) $HSeO_3^-$, hydrogen selenite ion

(b) $HSeO_4^-$, hydrogen selenate ion

(c) $HSeO_3^-$, $K_b = \dfrac{K_w}{K_a} = \dfrac{1.0 \times 10^{-14}}{3.5 \times 10^{-3}} = 2.9 \times 10^{-12}$

$HSeO_4^-$, $K_b = \dfrac{1.0 \times 10^{-14}}{0.012} = 8.3 \times 10^{-13}$

(d) $HSeO_3^-$, strongest base corresponds to highest pH.

10.23

(a) $HClO_2(aq) + H_2O(l) \rightleftharpoons H_3O^+(aq) + ClO_2^-(aq)$ $\qquad K_a = \dfrac{[H_3O^+][ClO_2^-]}{[HClO_2]}$

$ClO_2^-(aq) + H_2O(l) \rightleftharpoons HClO_2(aq) + OH^-(aq)$ $\qquad K_b = \dfrac{[HClO_2][OH^-]}{[ClO_2^-]}$

(b) $HCN(aq) + H_2O(l) \rightleftharpoons H_3O^+(aq) + CN^-(aq)$ $\qquad K_a = \dfrac{[H_3O^+][CN^-]}{[HCN]}$

$CN^-(aq) + H_2O(l) \rightleftharpoons HCN(aq) + OH^-(aq)$ $\qquad K_b = \dfrac{[HCN][OH^-]}{[CN^-]}$

(c) $C_6H_5OH(aq) + H_2O(l) \rightleftharpoons H_3O^+(aq) + C_6H_5O^-(aq)$ $\qquad K_a = \dfrac{[H_3O^+][C_6H_5O^-]}{[C_6H_5OH]}$

$C_6H_5O^-(aq) + H_2O(l) \rightleftharpoons C_6H_5OH(aq) + OH^-(aq)$ $\qquad K_b = \dfrac{[C_6H_5OH][OH^-]}{[C_6H_5O^-]}$

10.25 Decreasing pK_a will correspond to increasing acid strength because pK_a = $-\log K_a$. The pK_a values (given in parentheses) determine the following ordering:

$(CH_3)_2NH_2^+$ (14.00 − 3.27 = 10.73) < $^+NH_3OH$ (14.00 − 7.97 = 6.03)
< HNO_2 (3.37) < $HClO_2$ (2.00)

Remember that the pK_a for the conjugate acid of a weak base will be given by pK_a + pK_b = 14.

10.27 Decreasing pK_b will correspond to increasing base strength because pK_b = $-\log K_b$. The pK_b values (given in parentheses) determine the following ordering:

F^- (14.00 − 3.45 = 10.55) < CH_3COO^- (14.00 − 4.75 = 0.25)
< C_5H_5N (8.75) ≪ NH_3 (4.75)

Remember that the pK_b for the conjugate base of a weak acid will be given by pK_a + pK_b = 14.

10.29 Any acid whose conjugate base lies above water in Table 10.3 will be a strong acid, that is, the conjugate base of the acid will be a weaker base than water, and so water will accept the H^+ preferentially. Based upon this information, we obtain the following analysis: (a) $HClO_3$, strong; (b) H_2S, weak; (c) HSO_4^-, weak (Note: Even though H_2SO_4 is a strong acid, HSO_4^- is a weak acid. Its conjugate base is SO_4^{2-}); (d) $CH_3NH_3^+$, weak acid.

10.31 Any base whose conjugate acid lies below water in Table 10.3 will be a strong base, that is, the conjugate acid of the base will be a weaker acid than water, and so water will preferentially protonate the base. Based upon this information, we obtain the following analysis: (a) O^{2-}, strong; (b) Br^-, weak; (c) HSO_4^-, weak; (d) HCO_3^-, weak.

10.33 For oxoacids, the greater the number of highly electronegative O atoms attached to the central atom, the stronger the acid. This effect is related to the increased oxidation number of the central atom as the number of O atoms increases. Therefore, HIO_3 is the stronger acid, with the lower pK_a.

10.35 (a) HCl is the stronger acid, because its bond strength is much weaker than the bond in HF, and bond strength is the dominant factor in determining the strength of binary acids.

(b) $HClO_2$ is stronger; there is one more O atom attached to the Cl atom in $HClO_2$ than in HClO. The additional O in $HClO_2$ helps to pull the electron of the H atom out of the H—O bond. The oxidation state of Cl is higher in $HClO_2$ than in HClO.

(c) $HClO_2$ is stronger; Cl has a greater electronegativity than Br, making the H—O bond $HClO_2$ more polar than in $HBrO_2$.

(d) $HClO_4$ is stronger; Cl has a greater electronegativity than P (see Example 10.4 in the text).

(e) HNO_3 is stronger. The explanation is the same as that for part (b). HNO_3 has one more O atom.

(f) H_2CO_3 is stronger; C has greater electronegativity than Ge. See part (c).

10.37 (a) The $—CCl_3$ group that is bonded to the carboxyl group, —COOH, in trichloroacetic acid, is more electron withdrawing than the $—CH_3$ group in acetic acid. Thus, trichloroacetic acid is the stronger acid.

(b) The $—CH_3$ group in acetic acid has electron-donating properties, which means that it is less electron withdrawing than the —H attached to the carboxyl group in formic acid, HCOOH. Thus, formic acid is a slightly stronger acid than acetic acid. However, it is not nearly as strong as trichloroacetic acid. The order is

$CCl_3COOH \gg HCOOH > CH_3COOH$

10.39 The larger the K_a, the stronger the corresponding acid. 2,4,6-Trichlorophenol is the stronger acid because the chlorine atoms have a greater electron-withdrawing power than the hydrogen atoms present in the unsubstituted phenol.

10.41 The larger the pK_a of an acid, the stronger the corresponding conjugate base; hence, the order is aniline < ammonia < methylamine < ethylamine. Although we should not draw conclusions from such a small data set, we might suggest the possibility that

(1) arylamines < ammonia < alkylamines

(2) methyl < ethyl < etc.

(Arylamines are amines in which the nitrogen of the amine is attached to a benzene ring.)

10.43 (a)

Concentration ($mol\cdot L^{-1}$)	CH_3COOH	+ H_2O $\rightleftharpoons$	H_3O^+	+ $CH_3CO_2^-$
initial	0.15	—	0	0
change	$-x$	—	$+x$	$+x$
equilibrium	$0.15 - x$	—	x	x

$$K_a = 1.8 \times 10^{-5} = \frac{[H_3O^+][CH_3CO_2^-]}{[CH_3COOH]} = \frac{x^2}{0.15 - x} \approx \frac{x^2}{0.15}$$

$x = [H_3O^+] = 1.6 \times 10^{-3}\ mol\cdot L^{-1}$

$pH = -\log(1.6 \times 10^{-3}) = 2.80,\quad pOH = 14.00 - 2.80 = 11.20$

(b) The equilibrium table for (b) is similar to that for (a).

$$K_a = 3.0 \times 10^{-1} = \frac{[H_3O^+][CCl_3CO_2^-]}{[CCl_3COOH]} = \frac{x^2}{0.15 - x}$$

or $x^2 + 3.0 \times 10^{-1}x - 0.045 = 0$

$$x = \frac{-3.0 \times 10^{-1} \pm \sqrt{(3.0 \times 10^{-1})^2 - (4)(-0.045)}}{2} = 0.11, -0.41$$

The negative root is not possible and can be eliminated.

$x = [H_3O^+] = 0.11\ mol\cdot L^{-1}$

$pH = -\log(0.11) = 0.96,\quad pOH = 14.00 - 0.96 = 13.04$

(c)

Concentration ($mol\cdot L^{-1}$)	$HCOOH$	+ H_2O $\rightleftharpoons$	H_3O^+	+ HCO_2^-
initial	0.15	—	0	0
change	$-x$	—	$+x$	$+x$
equilibrium	$0.15 - x$	—	x	x

$$K_a = \frac{[H_3O^+][HCO_2^-]}{[HCOOH]} = \frac{x\cdot x}{0.15 - x} \approx \frac{x^2}{0.15} = 1.8 \times 10^{-4}$$

$x = [H_3O^+] = \sqrt{0.15 \times 1.8 \times 10^{-4}} = 5.2 \times 10^{-3}\ mol\cdot L^{-1}$

$pH = -\log(5.2 \times 10^{-3}) = 2.28,\quad pOH = 14.00 - 2.28 = 11.72$

10.45 (a)

Concentration ($mol\cdot L^{-1}$)	H_2O	+ NH_3 $\rightleftharpoons$	NH_4^+	+ OH^-
initial	—	0.10	0	0
change	—	$-x$	$+x$	$+x$
equilibrium	—	$0.10 - x$	x	x

$$K_b = \frac{[NH_4^+][OH^-]}{[NH_3]} = \frac{x\cdot x}{0.10 - x} \approx \frac{x^2}{0.10} = 1.8 \times 10^{-5}$$

$x = [OH^-] = \sqrt{0.10 \times 1.8 \times 10^{-5}} = 1.3 \times 10^{-3}\ mol\cdot L^{-1}$

$pOH = -\log(1.3 \times 10^{-3}) = 2.89,\quad pH = 14.00 - 2.89 = 11.11$

$$\text{percentage protonation} = \frac{1.3 \times 10^{-3}}{0.10} \times 100\% = 1.3\%$$

(b)

Concentration ($mol \cdot L^{-1}$)	NH_2OH	+ H_2O ⇌	$^+NH_3OH$	+ OH^-
initial	0.0178	—	0	0
change	$-x$	—	$+x$	$+x$
equilibrium	$0.0178 - x$	—	x	x

$$K_b = 1.1 \times 10^{-8} = \frac{x^2}{0.0178 - x} \approx \frac{x^2}{0.0178}$$

$x = [OH^-] = 1.4 \times 10^{-5}\ mol \cdot L^{-1}$

$pOH = -\log(1.4 \times 10^{-5}) = 4.85, \quad pH = 14.00 - 4.85 = 9.15$

$$\text{percentage protonation} = \frac{1.4 \times 10^{-5}}{0.017} \times 100\% = 0.082\%$$

(c)

Concentration ($mol \cdot L^{-1}$)	$(CH_3)_3N$	+ H_2O ⇌	$(CH_3)_3NH^+$	+ OH^-
initial	0.20	—	0	0
change	$-x$	—	$+x$	$+x$
equilibrium	$0.20 - x$	—	$+x$	$+x$

$$6.5 \times 10^{-5} = \frac{x^2}{0.20 - x}$$

assume $x \ll 0.20$

then $x = 3.6 \times 10^{-3}\ mol \cdot L^{-1}$

$[OH^-] = 3.6 \times 10^{-3}\ mol \cdot L^{-1}$

$pOH = -\log(3.6 \times 10^{-3}) = 2.44, \quad pH = 14.00 - 2.44 = 11.56$

$$\text{percentage protonation} = \frac{3.6 \times 10^{-3}}{0.20} \times 100\% = 1.8\%$$

(d) $pK_b = 14.00 - pK_a = 14.00 - 8.21 = 5.79, \quad K_b = 1.6 \times 10^{-6}$

$\text{codeine} + H_2O \rightleftharpoons \text{codeineH}^+ + OH^-$

$$K_b = 1.6 \times 10^{-6} = \frac{x^2}{0.020 - x} \approx \frac{x^2}{0.020}$$

$x = [OH^-] = 1.8 \times 10^{-4}\ mol \cdot L^{-1}$

$pOH = -\log(1.8 \times 10^{-4}) = 3.74, \quad pH = 14.00 - 3.74 = 10.26$

$$\text{percentage protonation} = \frac{1.8 \times 10^{-4}}{0.020} \times 100\% = 0.90\%$$

10.47 (a) $HClO_2 + H_2O \rightleftharpoons H_3O^+ + ClO_2^-$

$[H_3O^+] = [ClO_2^-] = 10^{-pH} = 10^{-1.2} = 0.06\ mol \cdot L^{-1}$

$$K_a = \frac{[H_3O^+][ClO_2^-]}{[HClO_2]} = \frac{(0.06)^2}{0.10 - 0.06} = 0.09 \text{ (1 sf)}$$

$pK_a = -\log(0.09) = 1.0$

(b) $C_3H_7NH_2 + H_2O \rightleftharpoons C_3H_7NH_3^+ + OH^-$

$pOH = 14.00 - 11.86 = 2.14$

$[C_3H_7NH_3^+] = [OH^-] = 10^{-2.14} = 7.2 \times 10^{-3}\ mol \cdot L^{-1}$

$$K_b = \frac{[C_3H_7NH_3^+][OH^-]}{[C_3H_7NH_2]} = \frac{(7.2 \times 10^{-3})^2}{0.10 - 7.2 \times 10^{-3}} = 5.6 \times 10^{-4}$$

$pK_b = -\log(5.6 \times 10^{-4}) = 3.25$

10.49 (a) $pH = 4.60$, $[H_3O^+] = 10^{-pH} = 10^{-4.60} = 2.5 \times 10^{-5}\ mol \cdot L^{-1}$

Let x = nominal concentration of HClO, then

Concentration	$HClO$	+	H_2O	$\rightleftharpoons$	H_3O^+	+	ClO^-
nominal	x		—		0		0
equilibrium	$x - 2.5 \times 10^{-5}$		—		2.5×10^{-5}		2.5×10^{-5}

$$K_a = 3.0 \times 10^{-8} = \frac{(2.5 \times 10^{-5})^2}{x - 2.5 \times 10^{-5}}$$

$$\text{Solve for } x;\ x = \frac{(2.5 \times 10^{-5})^2 + (2.5 \times 10^{-5})(3.0 \times 10^{-8})}{3.0 \times 10^{-8}}$$

$$= 2.1 \times 10^{-2}\ mol \cdot L^{-1} = 0.021\ mol \cdot L^{-1}$$

(b) $pOH = 14.00 - pH = 14.00 - 10.20 = 3.80$

$[OH^-] = 10^{-pOH} = 10^{-3.80} = 1.6 \times 10^{-4}$

Let x = nominal concentration of NH_2NH_2, then

Concentration	NH_2NH_2	+	H_2O	$\rightleftharpoons$	$NH_2NH_3^+$	+	OH^-
nominal	x		—		0		0
equilibrium	$x - 1.6 \times 10^{-4}$		—		1.6×10^{-4}		1.6×10^{-4}

$$K_b = 1.7 \times 10^{-6} = \frac{(1.6 \times 10^{-4})^2}{x - 1.6 \times 10^{-4}}$$

Solve for x; $x = 1.5 \times 10^{-2}\ mol \cdot L^{-1}$

10.51

Concentration ($mol \cdot L^{-1}$)	C_6H_5COOH	+	H_2O	$\rightleftharpoons$	H_3O^+	+	$C_6H_5CO_2^-$
initial	0.110		—		0		0
change	$-x$		—		$+x$		$+x$
equilibrium	$0.110 - x$		—		x		x

$$x = 0.024 \times 0.110\ \text{mol·L}^{-1} = [H_3O^+] = [C_6H_5CO_2^-]$$

$$K_a = \frac{[H_3O^+][C_6H_5COO^-]}{[C_6H_5COOH]} = \frac{(0.024 \times 0.110)^2}{(1 - 0.024) \times 0.110} = 6.3 \times 10^{-5}$$

$$pH = -\log(2.6 \times 10^{-3}) = 2.58$$

10.53 H_2O + octylamine $\rightleftharpoons$ octylamineH^+ + OH^-

The change in the concentration of octylamine is $x = 0.067 \times 0.10 = 0.0067\ \text{mol·L}^{-1}$. Thus the equilibrium table is

Concentration (mol·L^{-1})	H_2O	+	octylamine	$\rightleftharpoons$	octylamineH^+	+	OH^-
initial	—		0.100		0		0
change	—		−0.0067		+0.0067		+0.0067
equilibrium	—		0.100 − 0.0067		0.0067		0.0067

The equilibrium concentrations are

[octylamine] = $0.100 - 0.067 \times 0.10 = 0.093\ \text{mol·L}^{-1}$

[OH^-] = [octylamineH^+] = $0.0067\ \text{mol·L}^{-1}$

$pOH = -\log(0.0067) = 2.17$, $pH = 14.00 - 2.17 = 11.83$

$$K_b = \frac{[\text{octylamineH}^+][OH^-]}{[\text{octylamine}]} = \frac{(6.7 \times 10^{-3})^2}{0.093} = 4.8 \times 10^{-4}$$

10.55 (a) pH < 7, acidic; $NH_4^+(aq) + H_2O(l) \rightleftharpoons H_3O^+(aq) + NH_3(aq)$

(b) pH > 7, basic; $H_2O(l) + CO_3^{2-}(aq) \rightleftharpoons HCO_3^-(aq) + OH^-(aq)$

(c) pH > 7, basic; $H_2O(l) + F^-(aq) \rightleftharpoons HF(aq) + OH^-(aq)$

(d) pH = 7, neutral; K^+ is not an acid, Br^- is not a base

(e) pH < 7, acidic; $Al(H_2O)_6^{3+}(aq) + H_2O(l) \rightleftharpoons H_3O^+(aq) + Al(H_2O)_5OH^{2+}(aq)$

(f) pH < 7, acidic; $Cu(H_2O)_6^{2+}(aq) + H_2O(l) \rightleftharpoons H_3O^+(aq) + Cu(H_2O)_5OH^+(aq)$

10.57 (a) $K_b = \frac{K_w}{K_a} = \frac{1.00 \times 10^{-14}}{1.8 \times 10^{-5}} = 5.6 \times 10^{-10}$

Concentration (mol·L^{-1})	$CH_3CO_2^-(aq)$	+	$H_2O(l)$	$\rightleftharpoons$	$HCH_3CO_2(aq)$	+	$OH^-(aq)$
initial	0.20		—		0		0
change	$-x$		—		$+x$		$+x$
equilibrium	$0.20 - x$		—		x		x

$$K_b = \frac{[HCH_3CO_2][OH^-]}{[CH_3CO_2^-]} = 5.6 \times 10^{-10} = \frac{x^2}{0.20 - x} \approx \frac{x^2}{0.20}$$

$x = 1.1 \times 10^{-5} = [OH^-]$, $pOH = -\log(1.1 \times 10^{-5}) = 4.98$

$pH = 14.00 - pOH = 14.00 - 4.98 = 9.02$

(b) $K_a = \dfrac{K_w}{K_b} = \dfrac{1.00 \times 10^{-14}}{1.8 \times 10^{-5}} = 5.6 \times 10^{-10}$

Concentration ($mol\cdot L^{-1}$)	$NH_4^+(aq)$ +	$H_2O(l)$ $\rightleftharpoons$	$H_3O^+(aq)$ +	$NH_3(aq)$
initial	0.10	—	0	0
change	$-x$	—	$+x$	$+x$
equilibrium	$0.10 - x$	—	x	x

$$K_a = \frac{[H_3O^+][NH_3]}{[NH_4Cl]} = 5.6 \times 10^{-10} = \frac{x^2}{0.10 - x} \approx \frac{x^2}{0.10}$$

$x^2 = 5.6 \times 10^{-11}$

$x = 7.5 \times 10^{-6}\ mol\cdot L^{-1} = [H_3O^+]$

$pH = -\log(7.5 \times 10^{-6}) = 5.12$

(c)

Concentration ($mol\cdot L^{-1}$)	$Al(H_2O)_6^{3+}(aq)$ +	$H_2O(l)$ $\rightleftharpoons$	$H_3O^+(aq)$ +	$Al(H_2O)_5OH^{2+}(aq)$
initial	0.10	—	0	0
change	$-x$	—	$+x$	$+x$
equilibrium	$0.10 - x$	—	x	x

$$K_a = \frac{[H_3O^+][Al(H_2O)_5OH^{2+}]}{[Al(H_2O)_6^{3+}]} = 1.4 \times 10^{-5} = \frac{x^2}{0.10 - x} \approx \frac{x^2}{0.10}$$

$x^2 = 1.42 \times 10^{-6}$

$x = \sqrt{1.4 \times 10^{-6}} = 1.2 \times 10^{-3}\ mol\cdot L^{-1} = [H_3O^+]$

$pH = -\log(1.2 \times 10^{-3}) = 2.92$

(d)

Concentration ($mol\cdot L^{-1}$)	$H_2O(l)$ +	$CN^-(aq)$ $\rightleftharpoons$	$HCN(aq)$ +	$OH^-(aq)$
initial	—	0.15	0	0
change	—	$-x$	$+x$	$+x$
equilibrium	—	$0.15 - x$	x	x

$$K_b = \frac{K_w}{K_a} = \frac{1.00 \times 10^{-14}}{4.9 \times 10^{-10}} = 2.0 \times 10^{-5} = \frac{[HCN][OH^-]}{[CN^-]} = \frac{x^2}{0.15 - x} \approx \frac{x^2}{0.15}$$

$x^2 = (0.15)(2.0 \times 10^{-5}) = 3.0 \times 10^{-6}$

$x = [OH^-] = 1.7 \times 10^{-3}\ mol\cdot L^{-1}$

$pOH = -\log(1.7 \times 10^{-3}) = 2.77$, $pH = 11.23$

10.59 (a) 250 mL of solution contains 10.0 g $KC_2H_3O_2$, molar mass = 98.14 $g\cdot mol^{-1}$

$$(10.0\text{ g }KC_2H_3O_2) \times \left(\frac{1\text{ mol }KC_2H_3O_2}{98.14\text{ g }KC_2H_3O_2}\right)\left(\frac{1}{0.250\text{ L}}\right) = 0.408\text{ M }KC_2H_3O_2$$

Concentration $(mol\cdot L^{-1})$	$H_2O(l)$ +	$C_2H_3O_2^-(aq)$ $\rightleftharpoons$	$HC_2H_3O_2(aq)$ +	$OH^-(aq)$
initial	—	0.408	0	0
change	—	$-x$	$+x$	$+x$
equilibrium	—	$0.408 - x$	x	x

$$\frac{1.0 \times 10^{-14}}{1.8 \times 10^{-5}} = \frac{x^2}{0.408 - x} \approx \frac{x^2}{0.408}$$

$[OH^-] = 1.5 \times 10^{-5}\ mol\cdot L^{-1}$

$[H_3O^+] = 6.7 \times 10^{-10}\ mol\cdot L^{-1}$

$pH = -\log(6.7 \times 10^{-10}) = 9.17$

(b) 100 mL of solution contains 5.75 g NH_4Br, molar mass = 97.95 $g\cdot mol^{-1}$

$$(5.75\text{ g }NH_4Br)\left(\frac{1\text{ mol }NH_4Br}{97.95\text{ g }NH_4Br}\right)\left(\frac{1}{0.100\text{ L}}\right) = 0.587\text{ M }NH_4Br$$

Concentration $(mol\cdot L^{-1})$	$NH_4^+(aq)$ +	$H_2O(l)$ $\rightleftharpoons$	$NH_3(aq)$ +	$H_3O^+(aq)$
initial	0.587	—	0	0
change	$-x$	—	$+x$	$+x$
equilibrium	$0.587 - x$	—	x	x

$$\frac{1.0 \times 10^{-14}}{1.8 \times 10^{-5}} = \frac{x^2}{0.587 - x} \approx \frac{x^2}{0.587}$$

$[H_3O^+] = 1.8 \times 10^{-5}\ mol\cdot L^{-1}$

$pH = -\log(1.8 \times 10^{-5}) = 4.74$

10.61 (a) $$\frac{0.200\ mol\cdot L^{-1}\ NaCH_3CO_2 \times 2.00\ L}{0.500\ L} = 0.800\ mol\cdot L^{-1}$$

Concentration $(mol\cdot L^{-1})$	$H_2O(l)$ +	$CH_3CO_2^-(aq)$ $\rightleftharpoons$	$CH_3COOH(aq)$ +	$OH^-(aq)$
initial	—	0.800	0	0
change	—	$-x$	$+x$	$+x$
equilibrium	—	$0.800 - x$	x	x

$$K_b = \frac{K_w}{K_a} = \frac{1.00 \times 10^{-14}}{1.8 \times 10^{-5}} = 5.6 \times 10^{-10} = \frac{[CH_3COOH][OH^-]}{[CH_3CO_2^-]}$$

$$5.6 \times 10^{-10} = \frac{x^2}{0.800 - x} \approx \frac{x^2}{0.800}$$

$x^2 = 4.5 \times 10^{-10}$

$x = 2.1 \times 10^{-5}\ \text{mol} \cdot \text{L}^{-1} = [\text{CH}_3\text{COOH}]$

(b) $\left(\frac{5.75\ \text{g NH}_4\text{Br}}{400\ \text{mL}}\right)\left(\frac{1\ \text{mL}}{10^{-3}\ \text{L}}\right)\left(\frac{1\ \text{mol NH}_4\text{Br}}{97.95\ \text{g NH}_4\text{Br}}\right)$

$= 0.147\ (\text{mol NH}_4\text{Br}) \cdot \text{L}^{-1}$

Concentration ($\text{mol} \cdot \text{L}^{-1}$)	$NH_4^+(aq)$	+ $H_2O(l)$ $\rightleftharpoons$	$H_3O^+(aq)$	+ $NH_3(aq)$
initial	0.147	—	0	0
change	$-x$	—	$+x$	$+x$
equilibrium	$0.147 - x$	—	x	x

$$K_a = \frac{K_w}{K_b} = \frac{1.00 \times 10^{-14}}{1.8 \times 10^{-5}} = 5.6 \times 10^{-10} = \frac{[\text{NH}_3][\text{H}_3\text{O}^+]}{[\text{NH}_4^+]}$$

$$5.6 \times 10^{-10} = \frac{x^2}{0.147 - x} \approx \frac{x^2}{0.147}$$

$x^2 = 8.2 \times 10^{-11}$

$x = 9.1 \times 10^{-6}\ \text{mol} \cdot \text{L}^{-1} = [\text{H}_3\text{O}^+]$ and $\text{pH} = -\log(9.1 \times 10^{-6}) = 5.04$

10.63 We can use the relation derived in the text: $[\text{H}_3\text{O}^+]^2 - [\text{HA}]_{\text{initial}}[\text{H}_3\text{O}^+] - K_w = 0$, in which HA is any strong acid.

$[\text{H}_3\text{O}^+]^2 - (6.55 \times 10^{-7})[\text{H}_3\text{O}^+] - (1.00 \times 10^{-14}) = 0$

Solving using the quadratic equation gives $[\text{H}_3\text{O}^+] = 6.70 \times 10^{-7}$. pH = 6.174. This value is slightly lower than the value calculated, based on the acid concentration alone ($\text{pH} = -\log(6.55 \times 10^{-7}) = 6.184$).

10.65 We can use the relation derived in the text: $[\text{H}_3\text{O}^+]^2 + [\text{B}]_{\text{initial}}[\text{H}_3\text{O}^+] - K_w = 0$, in which B is any strong base.

$[\text{H}_3\text{O}^+]^2 + (9.78 \times 10^{-8})[\text{H}_3\text{O}^+] - (1.00 \times 10^{-14}) = 0$

Solving using the quadratic equation gives $[\text{H}_3\text{O}^+] = 6.24 \times 10^{-8}$. pH = 7.205. This value is higher than the value calculated, based on the base concentration alone ($\text{pOH} = -\log(9.78 \times 10^{-8}) = 7.009$).

10.67 (a) In the absence of a significant effect due to the autoprotolysis of water, the pH values of the 1.00×10^{-4} M and 1.00×10^{-6} M HBrO solutions can be calculated as described earlier.

For 1.00×10^{-4} mol·L^{-1}:

Concentration (mol·L^{-1})	HBrO(aq) +	$H_2O(l) \rightleftharpoons$	$H_3O^+(aq)$ +	$BrO^-(aq)$
initial	1.00×10^{-4}	—	0	0
change	$-x$	—	$+x$	$+x$
final	$1.00 \times 10^{-4} - x$	—	$+x$	$+x$

$$K_a = \frac{[H_3O^+][BrO^-]}{[HBrO]}$$

$$2.0 \times 10^{-9} = \frac{(x)(x)}{1.00 \times 10^{-4} - x} = \frac{x^2}{[1.00 \times 10^{-4} - x]}$$

Assume $x \ll 1.00 \times 10^{-4}$

$x^2 = (2.0 \times 10^{-9})(1.00 \times 10^{-4})$

$x = 4.5 \times 10^{-7}$

Because $x < 1\%$ of 1.00×10^{-4}, the assumption was valid. Given this value, the pH is then calculated to be $-\log(4.5 \times 10^{-7}) = 6.35$.

For 1.00×10^{-6} mol·L^{-1}:

Concentration (mol·L^{-1})	HBrO(aq) +	$H_2O(l) \rightleftharpoons$	$H_3O^+(aq)$ +	$BrO^-(aq)$
initial	1.00×10^{-6}	—	0	0
change	$-x$	—	$+x$	$+x$
final	$1.00 \times 10^{-6} - x$	—	$+x$	$+x$

$$K_a = \frac{[H_3O^+][BrO^-]}{[HBrO]}$$

$$2.0 \times 10^{-9} = \frac{(x)(x)}{1.00 \times 10^{-6} - x} = \frac{x^2}{[1.00 \times 10^{-6} - x]}$$

Assume $x \ll 1.00 \times 10^{-6}$

$x^2 = (2.0 \times 10^{-9})(1.00 \times 10^{-6})$

$x = 4.5 \times 10^{-8}$

x is 4.5% of 1.00×10^{-6}, so the assumption is less acceptable. The pH is calculated to be $-\log(4.8 \times 10^{-8}) = 7.32$. Since this predicts a basic solution, it is not reasonable.

(b) To calculate the value taking into account the autoprotolysis of water, we can use equation (22):

$x^3 + K_a x^2 - (K_w + K_a \cdot [HA]_{initial})x - K_w \cdot K_a = 0$, where $x = [H_3O^+]$

To solve the expression, you substitute the values of $K_w = 1.00 \times 10^{-14}$, the initial concentration of acid, and $K_a = 2.0 \times 10^{-9}$ into this equation and then solve the expression either by trial and error or, preferably, using a graphing calculator such as the one found on the CD accompanying this text.

Alternatively, you can use a computer program designed to solve simultaneous equations. Because the unknowns include $[H_3O^+]$, $[OH^-]$, [HBrO], and $[BrO^-]$, you will need four equations. As seen in the text, the pertinent equations are

$$K_a = \frac{[H_3O^+][BrO^-]}{[HBrO]}$$

$$K_w = [H_3O^+][OH^-]$$

$$[H_3O^+] = [OH^-] + [BrO^-]$$

$$[HBrO]_{initial} = [HBrO] + [BrO^-]$$

Using either method should produce the same result.

The values obtained are

$[H_3O^+] = 4.6 \times 10^{-7}\ mol \cdot L^{-1}$, pH = 6.34 (compare to 6.35 obtained in (a))
$[BrO^-] = 4.4 \times 10^{-7}\ mol \cdot L^{-1}$
$[HBrO] \cong 1.0 \times 10^{-5}\ mol \cdot L^{-1}$
$[OH^-] = 2.2 \times 10^{-8}\ mol \cdot L^{-1}$

Similarly, for $[HBrO]_{initial} = 1.00 \times 10^{-6}$:

$[H_3O^+] = 1.1 \times 10^{-7}\ mol \cdot L^{-1}$, pH = 6.96 (compare to 7.32 obtained in (a))
$[BrO^-] = 1.8 \times 10^{-8}\ mol \cdot L^{-1}$
$[HBrO] \cong 9.8 \times 10^{-7}\ mol \cdot L^{-1}$
$[OH^-] = 9.1 \times 10^{-8}\ mol \cdot L^{-1}$

Note that for the more concentrated solution, the effect of the autoprotolysis of water is very small. Notice also that the less concentrated solution is more acidic, due to the autoprotolysis of water, than would be predicted if this effect were not operating.

10.69 (a) In the absence of a significant effect due to the autoprotolysis of water, the pH values of the 8.50×10^{-5} M and 7.37×10^{-6} M HCN solutions can be calculated as described earlier.

For $8.50 \times 10^{-5}\ mol \cdot L^{-1}$:

Concentration ($mol \cdot L^{-1}$)	HCN(aq)	+ $H_2O(l)$ ⇌	$H_3O^+(aq)$	+ $CN^-(aq)$
initial	8.50×10^{-5}	—	0	0
change	$-x$	—	$+x$	$+x$
final	$8.50 \times 10^{-5} - x$	—	$+x$	$+x$

$$K_a = \frac{[H_3O^+][CN^-]}{[HCN]}$$

$$4.9 \times 10^{-10} = \frac{(x)(x)}{8.5 \times 10^{-5} - x} = \frac{x^2}{[8.5 \times 10^{-5} - x]}$$

Assume $x \ll 8.5 \times 10^{-5}$

$x^2 = (4.9 \times 10^{-10})(8.5 \times 10^{-5})$

$x = 2.0 \times 10^{-7}$

Because $x < 1\%$ of 8.50×10^{-5}, the assumption was valid. Given this value, the pH is then calculated to be $-\log(2.0 \times 10^{-7}) = 6.69$.

For 7.37×10^{-6} mol·L^{-1}:

Concentration (mol·L^{-1})	HCN(aq) +	$H_2O(l)$ $\rightleftharpoons$	$H_3O^+(aq)$ +	$CN^-(aq)$
initial	7.37×10^{-6}	—	0	0
change	$-x$	—	$+x$	$+x$
final	$7.37 \times 10^{-6} - x$	—	$+x$	$+x$

$$K_a = \frac{[H_3O^+][CN^-]}{[HCN]}$$

$$4.9 \times 10^{-10} = \frac{(x)(x)}{7.37 \times 10^{-6} - x} = \frac{x^2}{[7.37 \times 10^{-6} - x]}$$

Assume $x \ll 7.37 \times 10^{-6}$

$x^2 = (4.9 \times 10^{-10})(7.37 \times 10^{-6})$

$x = 6.0 \times 10^{-8}$

x is $<1\%$ of 7.37×10^{-6}, so the assumption is still reasonable. The pH is then calculated to be $-\log(6.0 \times 10^{-8}) = 7.22$. This answer is not reasonable because we know HCN is an acid.

(b) To calculate the value, taking into account the autoprotolysis of water, we can use equation (21):

$x^3 + K_a x^2 - (K_w + K_a \cdot [HA]_{initial})x - K_w \cdot K_a = 0$, where $x = [H_3O^+]$

To solve the expression, you substitute the values of $K_w = 1.00 \times 10^{-14}$, the initial concentration of acid, and $K_a = 4.9 \times 10^{-10}$ into this equation and then solve the expression either by trial and error or, preferably, using a graphing calculator such as the one found on the CD accompanying this text.

Alternatively, you can use a computer program designed to solve simultaneous equations. Because the unknowns include $[H_3O^+]$, $[OH^-]$, [HBrO], and $[BrO^-]$, you will need four equations. As seen in the text, the pertinent equations are

$$K_a = \frac{[H_3O^+][CN^-]}{[HCN]}$$

$K_w = [H_3O^+][OH^-]$

$[H_3O^+] = [OH^-] + [CN^-]$

$[HCN]_{initial} = [HCN] + [CN^-]$

Using either method should produce the same result.

For $[HCN] = 8.5 \times 10^{-5}$ mol·L^{-1}, the values obtained are

$[H_3O^+] = 2.3 \times 10^{-7}\ mol\cdot L^{-1}$, pH = 6.64 (compare to 6.69 obtained in (a))
$[CN^-] = 1.8 \times 10^{-7}\ mol\cdot L^{-1}$
$[HCN] \cong 8.5 \times 10^{-5}\ mol\cdot L^{-1}$
$[OH^-] = 4.4 \times 10^{-8}\ mol\cdot L^{-1}$
Similarly, for $[HCN]_{initial} = 7.37 \times 10^{-6}$:
$[H_3O^+] = 1.2 \times 10^{-7}\ mol\cdot L^{-1}$, pH = 6.92 (compare to 7.22 obtained in (a))
$[CN^-] = 3.1 \times 10^{-8}\ mol\cdot L^{-1}$
$[HCN] \cong 7.3 \times 10^{-6}\ mol\cdot L^{-1}$
$[OH^-] = 8.6 \times 10^{-8}\ mol\cdot L^{-1}$
Note that for the more concentrated solution, the effect of the autoprotolysis of water is smaller. Notice also that the less concentrated solution is more acidic, due to the autoprotolysis of water, than would be predicted if this effect were not operating.

10.71 (a) $H_2SO_4(aq) + H_2O(l) \rightleftharpoons H_3O^+(aq) + HSO_4^-(aq)$
$HSO_4^-(aq) + H_2O(l) \rightleftharpoons H_3O^+(aq) + SO_4^{2-}(aq)$
(b) $H_3AsO_4(aq) + H_2O(l) \rightleftharpoons H_3O^+(aq) + H_2AsO_4^-(aq)$
$H_2AsO_4^-(aq) + H_2O(l) \rightleftharpoons H_3O^+(aq) + HAsO_4^{2-}(aq)$
$HAsO_4^{2-}(aq) + H_2O(l) \rightleftharpoons H_3O^+(aq) + AsO_4^{3-}(aq)$
(c) $C_6H_4(COOH)_2(aq) + H_2O(l) \rightleftharpoons H_3O^+(aq) + C_6H_4(COOH)CO_2^-(aq)$
$C_6H_4(COOH)CO_2^-(aq) + H_2O(l) \rightleftharpoons H_3O^+(aq) + C_6H_4(CO_2)_2^{2-}(aq)$

10.73 The initial concentrations of HSO_4^- and H_3O^+ are both $0.15\ mol\cdot L^{-1}$ as a result of the complete ionization of H_2SO_4 in the first step. The second ionization is incomplete.

Concentration ($mol\cdot L^{-1}$)	HSO_4^-	+ H_2O $\rightleftharpoons$	H_3O^+	+ SO_4^{2-}
initial	0.15	—	0.15	0
change	$-x$	—	$+x$	$+x$
equilibrium	$0.15 - x$	—	$0.15 + x$	x

$$K_{a2} = 1.2 \times 10^{-2} = \frac{[H_3O^+][SO_4^{2-}]}{[HSO_4^-]} = \frac{(0.15 + x)(x)}{0.15 - x}$$

$$x^2 + 0.162x - 1.8 \times 10^{-3} = 0$$

$$x = \frac{-0.162 + \sqrt{(0.162)^2 + (4)(1.8 \times 10^{-3})}}{2} = 0.0104\ mol\cdot L^{-1}$$

$[H_3O^+] = 0.15 + x = (0.15 + 0.0104) mol\cdot L^{-1} = 0.16\ mol\cdot L^{-1}$
$pH = -\log(0.16) = 0.80$

10.75 (a) Because $K_{a2} \ll K_{a1}$, the second ionization can be ignored.

Concentration ($mol \cdot L^{-1}$)	H_2CO_3 +	$H_2O \rightleftharpoons$	H_3O^+ +	HCO_3^-
initial	0.010	—	0	0
change	$-x$	—	$+x$	$+x$
equilibrium	$0.010 - x$	—	x	x

$$K_{a1} = \frac{[H_3O^+][HCO_3^-]}{[H_2CO_3]} = \frac{x^2}{0.010 - x} \approx \frac{x^2}{0.010} = 4.3 \times 10^{-7}$$

$x = [H_3O^+] = 6.6 \times 10^{-5}\ mol \cdot L^{-1}$

$pH = -\log(6.6 \times 10^{-5}) = 4.18$

(b) Because $K_{a2} \ll K_{a1}$, the second ionization can be ignored.

Concentration ($mol \cdot L^{-1}$)	$(COOH)_2$ +	$H_2O \rightleftharpoons$	H_3O^+ +	$(COOH)CO_2^-$
initial	0.10	—	0	0
change	$-x$	—	$+x$	$+x$
equilibrium	$0.10 - x$	—	x	x

$$K_{a1} = 5.9 \times 10^{-2} = \frac{[H_3O^+][(COOH)CO_2^-]}{[(COOH)_2]} = \frac{x^2}{0.10 - x}$$

$x^2 + 5.9 \times 10^{-2}x - 5.9 \times 10^{-3} = 0$

$$x = \frac{-5.9 \times 10^{-2} + \sqrt{(5.9 \times 10^{-2})^2 + (4)(5.9 \times 10^{-3})}}{2} = 0.053\ mol \cdot L^{-1}$$

$pH = -\log(0.053) = 1.28$

(c) Because $K_{a2} \ll K_{a1}$, the second ionization can be ignored.

Concentration ($mol \cdot L^{-1}$)	H_2S +	$H_2O \rightleftharpoons$	H_3O^+ +	HS^-
equilibrium	$0.20 - x$	—	x	x

$$K_{a1} = 1.3 \times 10^{-7} = \frac{[H_3O^+][HS^-]}{[H_2S]} = \frac{x^2}{0.20 - x} \approx \frac{x^2}{0.20}$$

$x = [H_3O^+] = 1.6 \times 10^{-4}\ mol \cdot L^{-1}$

$pH = -\log(1.6 \times 10^{-4}) = 3.80$

10.77 (a) The pH is given by $pH = \frac{1}{2}(pK_{a1} + pK_{a2})$. From Table 10.9, we find

$K_{a1} = 1.5 \times 10^{-2} \qquad pK_{a1} = 1.82$

$K_{a2} = 1.2 \times 10^{-7} \qquad pK_{a2} = 6.92$

$pH = \frac{1}{2}(1.82 + 6.92) = 4.37$

(b) The pH of a salt solution of a polyprotic acid is independent of the concentration of the salt, therefore pH = 4.37.

10.79 (a) The pH is given by $pH = \frac{1}{2}(pK_{a1} + pK_{a2})$. For the monosodium salt, the pertinent values are pK_{a1} and pK_{a2}:

$pH = \frac{1}{2}(3.14 + 5.95) = 4.55$

(b) For the disodium salt, the pertinent values are pK_{a2} and pK_{a3}:

$pH = \frac{1}{2}(5.95 + 6.39) = 6.17$

10.81 The equilibrium reactions of interest are

$H_2CO_3(aq) + H_2O(l) \rightleftharpoons H_3O^+(aq) + HCO_3^-(aq) \quad K_{a1} = 4.3 \times 10^{-7}$

$HCO_3^-(aq) + H_2O(l) \rightleftharpoons H_3O^+(aq) + CO_3^{2-}(aq) \quad K_{a2} = 5.6 \times 10^{-11}$

Because the second ionization constant is much smaller than the first, we can assume that the first step dominates:

Concentration ($mol \cdot L^{-1}$)	$H_2CO_3(aq)$ + $H_2O(l)$ $\rightleftharpoons$	$H_3O^+(aq)$ +	$HCO_3^-(aq)$
initial	0.0456	0	0
change	$-x$	$+x$	$+x$
final	$0.0456 - x$	$+x$	$+x$

$$K_{a1} = \frac{[H_3O^+][HCO_3^-]}{[H_2CO_3]}$$

$$4.3 \times 10^{-7} = \frac{(x)(x)}{0.0456 - x} = \frac{x^2}{0.0456 - x}$$

Assume that $x \ll 0.0456$, then

$x^2 = (4.3 \times 10^{-7})(0.0456)$

$x = 1.4 \times 10^{-4}$

Because $x < 1\%$ of 0.0456, the assumption was valid.

$x = [H_3O^+] = [HCO_3^-] = 1.4 \times 10^{-4}\ mol \cdot L^{-1}$

This means that the concentration of H_2CO_3 is $0.0456\ mol \cdot L^{-1} - 0.00014\ mol \cdot L^{-1} = 0.0455\ mol \cdot L^{-1}$. We can then use the other equilibria to determine the remaining concentrations:

$$K_{a2} = \frac{[H_3O^+][CO_3^{2-}]}{[HCO_3^-]}$$

$$5.6 \times 10^{-11} = \frac{(1.4 \times 10^{-4})[CO_3^{2-}]}{(1.4 \times 10^{-4})}$$

$[CO_3^{2-}] = 5.6 \times 10^{-11}\ mol \cdot L^{-1}$

Because $5.6 \times 10^{-11} \ll 1.4 \times 10^{-4}$, the initial assumption that the first ionization would dominate is valid.

To calculate $[OH^-]$, we use the K_w relation:

$K_w = [H_3O^+][OH^-]$

$$[OH^-] = \frac{K_w}{[H_3O^+]} = \frac{1.00 \times 10^{-14}}{1.4 \times 10^{-4}} = 7.1 \times 10^{-11}\ mol \cdot L^{-1}$$

In summary, $[H_2CO_3] = 0.0455\ mol \cdot L^{-1}$, $[H_3O^+] = [HCO_3^-] = 1.4 \times 10^{-4}\ mol \cdot L^{-1}$, $[CO_3^{2-}] = 5.6 \times 10^{-11}\ mol \cdot L^{-1}$, $[OH^-] = 7.1 \times 10^{-11}\ mol \cdot L^{-1}$.

10.83 The equilibrium reactions of interest are now the base forms of the carbonic acid equilibria, so K_b values should be calculated for the following changes:

$$CO_3^{2-}(aq) + H_2O(l) \rightleftharpoons HCO_3^-(aq) + OH^-(aq)$$

$$K_{b1} = \frac{K_w}{K_{a2}} = \frac{1.00 \times 10^{-14}}{5.6 \times 10^{-11}} = 1.8 \times 10^{-4}$$

$$HCO_3^-(aq) + H_2O(l) \rightleftharpoons H_2CO_3(aq) + OH^-(aq)$$

$$K_{b2} = \frac{K_w}{K_{a1}} = \frac{1.00 \times 10^{-14}}{4.3 \times 10^{-7}} = 2.3 \times 10^{-8}$$

Because the second hydrolysis constant is much smaller than the first, we can assume that the first step dominates:

Concentration ($mol \cdot L^{-1}$)	$CO_3^{2-}(aq)$ +	$H_2O(l)$ $\rightleftharpoons$	$HCO_3^-(aq)$ +	$OH^-(aq)$
initial	0.0456	—	0	0
change	$-x$	—	$+x$	$+x$
final	$0.0456 - x$	—	$+x$	$+x$

$$K_{b1} = \frac{[HCO_3^-][OH^-]}{[CO_3^{2-}]}$$

$$1.8 \times 10^{-4} = \frac{(x)(x)}{0.0456 - x} = \frac{x^2}{[0.0456 - x]}$$

Assume that $x \ll 0.0456$, then

$x^2 = (1.8 \times 10^{-4})(0.0456)$

$x = 2.9 \times 10^{-3}$

Because $x > 5\%$ of 0.0456, the assumption was not valid and the full expression should be solved using the quadratic equation:

$x^2 + 1.8 \times 10^{-4}x - (1.8 \times 10^{-4})(0.0456) = 0$

Solving using the quadratic equation gives $x = 0.0028\ mol \cdot L^{-1}$

$x = [HCO_3^-] = [OH^-] = 0.0028\ mol \cdot L^{-1}$

Therefore, $[CO_3^{2-}] = 0.0456\ mol \cdot L^{-1} - 0.0028\ mol \cdot L^{-1} = 0.0428\ mol \cdot L^{-1}$

We can then use the other equilibria to determine the remaining concentrations:

$$K_{b2} = \frac{[H_2CO_3][OH^-]}{[HCO_3^-]}$$

$$2.3 \times 10^{-8} = \frac{[H_2CO_3](0.0028)}{(0.0028)}$$

$[H_2CO_3] = 2.3 \times 10^{-8}\ mol \cdot L^{-1}$

Because $2.3 \times 10^{-8} \ll 0.0028$, the initial assumption that the first hydrolysis would dominate is valid. To calculate $[H_3O^+]$, we use the K_w relation:

$K_w = [H_3O^+][OH^-]$

$$[H_3O^+] = \frac{K_w}{[OH^-]} = \frac{1.00 \times 10^{-14}}{0.0028} = 3.6 \times 10^{-12}\ mol \cdot L^{-1}$$

In summary, $[H_2CO_3] = 2.3 \times 10^{-8}\ mol \cdot L^{-1}$, $[OH^-] = [HCO_3^-] = 0.0028\ mol \cdot L^{-1}$, $[CO_3^{2-}] = 0.0428\ mol \cdot L^{-1}$, $[H_3O^+] = 3.6 \times 10^{-12}\ mol \cdot L^{-1}$

10.85 (a) phosphorus acid: The two pK_a values are 2.00 and 6.59. Because pH = 6.30 lies between pK_{a1} and pK_{a2}, the dominant form will be the singly deprotonated HA^- ion.

(b) oxalic acid: The two pK_a values are 1.23 and 4.19. Because pH = 6.30 lies above the pK_{a2}, the species present in largest concentration will be the doubly deprotonated A^{2-} ion.

(c) hydrosulfuric acid: The two pK_a values are 6.89 and 14.15. Because pH = 6.30 lies below both pK_a values, the species present in highest concentrations will be the fully protonated H_2A form.

10.87 The equilibria present in solution are

$H_2SO_3(aq) + H_2O(l) \rightleftharpoons H_3O^+(aq) + HSO_3^-(aq) \quad K_{a1} = 1.5 \times 10^{-2}$

$HSO_3^-(aq) + H_2O(l) \rightleftharpoons H_3O^+(aq) + SO_3^{2-}(aq) \quad K_{a2} = 1.2 \times 10^{-7}$

The calculation of the desired concentrations follows exactly after the method derived in Eq. 25, substituting H_2SO_3 for H_2CO_3, HSO_3^- for HCO_3^-, and SO_3^{2-} for CO_3^{2-}. First, calculate the quantity f (at pH = 5.50 $[H_3O^+] = 10^{-5.5} = 3.2 \times 10^{-6}\ mol \cdot L^{-1}$):

$$f = [H_3O^+]^2 + [H_3O^+]K_{a1} + K_{a1}K_{a2}$$
$$= (3.2 \times 10^{-6})^2 + (3.2 \times 10^{-6})(1.5 \times 10^{-2}) + (1.5 \times 10^{-2})(1.2 \times 10^{-7})$$
$$= 5.0 \times 10^{-8}$$

The fractions of the species present are then given by

$$\alpha(H_2SO_3) = \frac{[H_3O^+]}{f} = \frac{(3.2 \times 10^{-6})^2}{5.0 \times 10^{-8}} = 2.1 \times 10^{-4}$$

$$\alpha(HSO_3^-) = \frac{[H_3O^+]K_{a1}}{f} = \frac{(3.2 \times 10^{-6})(1.5 \times 10^{-2})}{5.0 \times 10^{-8}} = 0.96$$

$$\alpha(SO_3^{2-}) = \frac{K_{a1}K_{a2}}{f} = \frac{(1.5 \times 10^{-2})(1.2 \times 10^{-7})}{5.0 \times 10^{-8}} = 0.036$$

Thus, in a $0.150\ mol \cdot L^{-1}$ solution at pH 5.50, the dominant species will be HSO_3^- with a concentration of $(0.150\ mol \cdot L^{-1})(0.96) = 0.14\ mol \cdot L^{-1}$. The concentration of H_2SO_3 will be $(2.1 \times 10^{-4})(0.150\ mol\cdot L^{-1}) = 3.2 \times 10^{-5}$ $mol\cdot L^{-1}$ and the concentration of SO_3^{2-} will be $(0.036)(0.150\ mol\cdot L^{-1}) = 0.0054\ mol\cdot L^{-1}$.

10.89 $[H_3O^+] = [OH^-]$

10.91 (a) If a solution is a concentrated solution of an acid with $[H_3O^+] > 1$ M, the pH will be negative.

(b) If a solution is very basic, with $[OH^-] > 1$ M (which means $[H_3O^+] < 10^{-14}$ M), the pH will be greater than 14.

10.93 Both CH_3COOH and HNO_2 are monoprotic acids; therefore, for solutions with the same concentration of acid, the one with the larger K_a will have the greater percentage ionization.

$K_a(HNO_2) = 4.3 \times 10^{-4}$ (greater)

$K_a(CH_3COOH) = 1.8 \times 10^{-5}$

10.95 (a) $C_6H_5OH + H_2O \rightleftharpoons H_3O^+ + C_6H_5O^-$

$NH_4^+ + H_2O \rightleftharpoons H_3O^+ + NH_3$

(b) $pK_a(C_6H_5OH) = -\log(1.3 \times 10^{-10}) = 9.89$

$pK_a(NH_4^+) = -\log(5.6 \times 10^{-10}) = 9.25$

(c) NH_4^+ is the stronger acid; it has the smaller pK_a and the larger K_a.

10.97 Let B = base; $B + H_2O \rightleftharpoons BH^+ + OH^-$

$pOH = 14.00 - 10.05 = 3.95$

The equilibrium concentrations are

$[OH^-] = [BH^+] = 10^{-pOH}\ mol\cdot L^{-1} = 10^{-3.95}\ mol\cdot L^{-1} = 1.1 \times 10^{-4}\ mol\cdot L^{-1}$

initial concentration of B: $[B]_0 = \left(\dfrac{0.150\ g}{0.0500\ L}\right)\left(\dfrac{1\ mol}{31.06\ g}\right) = 0.0966\ mol\cdot L^{-1}$

equilibrium concentration of B: $[B] = (0.0966 - 1.1 \times 10^{-4})\ mol\cdot L^{-1}$

$= 0.0965\ mol\cdot L^{-1}$

Concentration ($mol\cdot L^{-1}$)	H_2O	+	B	$\rightleftharpoons$	BH^+	+	OH^-
initial	—		0.0966		0		0
change	—		-1.1×10^{-4}		$+1.1 \times 10^{-4}$		$+1.1 \times 10^{-4}$
equilibrium	—		$0.0966 - 1.1 \times 10^{-4}$		1.1×10^{-4}		1.1×10^{-4}

$$K_b = \frac{[BH^+][OH^-]}{[B]} = \frac{(1.1 \times 10^{-4})^2}{0.0965} = 1.3 \times 10^{-7}$$

$pK_b = -\log(1.3 \times 10^{-7}) = 6.89, \quad pK_a(BH^+) = 14.00 - 6.89 = 7.11$

$$\text{percentage protonation} = \frac{1.1 \times 10^{-4}}{0.0965} \times 100\% = 0.11\%$$

10.99 (a) The pH is given by $pH = \frac{1}{2}(pK_{a1} + pK_{a2})$. From Table 10.9, we find for oxalic acid:

$K_{a1} = 5.9 \times 10^{-2}$ $\qquad pK_{a1} = 1.23$

$K_{a2} = 6.5 \times 10^{-5}$ $\qquad pK_{a2} = 4.19$

$pH = \frac{1}{2}(1.23 + 4.19) = 2.71$

(b) From Table 10.9, we find for tartaric acid:

$K_{a1} = 6.0 \times 10^{-4}$ $\qquad pK_{a1} = 3.22$

$K_{a2} = 1.5 \times 10^{-5}$ $\qquad pK_{a2} = 4.82$

$pH = \frac{1}{2}(3.22 + 4.82) = 4.02$

10.101 (a) $D_2O + D_2O \rightleftharpoons D_3O^+ + OD^-$

(b) $K_w = [D_3O^+][OD^-] = 1.35 \times 10^{-15}, \quad pK_w = -\log K_w = 14.870$

(c) $[D_3O^+] = [OD^-] = \sqrt{1.35 \times 10^{-15}} = 3.67 \times 10^{-8}\ mol \cdot L^{-1}$

(d) $pD = -\log(3.67 \times 10^{-8}) = 7.435 = pOD$

(e) $pD + pOD = pK_w(D_2O) = 14.870$

10.103 (a)

Concentration ($mol \cdot L^{-1}$)	H_2O +	B $\rightleftharpoons$	BH^+ +	OH^-
initial	—	0.025	0	0
change	—	$-x$	$+x$	$+x$
equilibrium	—	$0.025 - x$	x	x

$pOH = 14.0 - 11.6 = 2.4,$

$x = [OH^-] = [BH^+] = 10^{-2.4}\ mol \cdot L^{-1} =$
$4 \times 10^{-3}\ mol \cdot L^{-1}$

$$K_b = \frac{[BH^+][OH^-]}{[BH]} = \frac{(4 \times 10^{-3})^2}{(0.025 - 4 \times 10^{-3})} = 8 \times 10^{-4}$$

$pK_b = -\log(8 \times 10^{-4}) = 3.1, \quad pK_a = 14.0 - 3.1 = 10.9$

(b) Let T = thiazole

Concentration ($mol \cdot L^{-1}$)	H_2O +	T $\rightleftharpoons$	TH^+ +	OH^-
initial	—	0.0010	0	0
change	—	$-x$	$+x$	$+x$
equilibrium	—	$0.0010 - x$	x	x

$x = 5.2 \times 10^{-5} \times 0.0010\ mol \cdot L^{-1} = [TH^+] \neq [OH^-]$

This is a case where the autoprotolysis of water cannot be ignored. $[OH^-]$ is not the same as [thiazolate ion].

$$K_b = \frac{[TH^+][OH^-]}{[T]}, \quad K_w = [H_3O^+][OH^-] = 1.0 \times 10^{-14}$$

Electroneutrality requires $[TH^+] + [H_3O^+] - [OH^-] = 0$,

giving $[OH^-] = 5.2 \times 10^{-8} + \dfrac{(1.0 \times 10^{-14})}{[OH^-]}$

Let $y = [OH^-]$, then $y^2 - 5.2 \times 10^{-8}y - 1.0 \times 10^{-14} = 0$

$$y = \frac{-5.2 \times 10^{-8} \pm \sqrt{(5.2 \times 10^{-8})^2 + 4.0 \times 10^{-14}}}{2}$$

$$= 1.3 \times 10^{-7}\ \text{mol}\cdot\text{L}^{-1} = [OH^-]$$

(Keeping only the positive root)

$$K_b = \frac{5.2 \times 10^{-8} \times 1.3 \times 10^{-7}}{0.0010} = 6.8 \times 10^{-12}$$

$$[H_3O^+] = \frac{1.0 \times 10^{-14}}{1.3 \times 10^{-7}} = 7.7 \times 10^{-8}\ \text{mol}\cdot\text{L}^{-1}$$

$$\text{pH} = -\log(7.7 \times 10^{-8}) = 7.11$$

10.105 (a)

Concentration (mol·L^{-1})	$B(OH)_3$	+ $2\,H_2O \rightleftharpoons$	H_3O^+	+ $B(OH)_4^-$
initial	1.0×10^{-4}	—	0	0
change	$-x$	—	$+x$	$+x$
equilibrium	$1.0 \times 10^{-4} - x$	—	x	x

$$K_a = 7.2 \times 10^{-10} = \frac{[H_3O^+][B(OH)_4^-]}{[B(OH)_3]} = \frac{x^2}{1.0 \times 10^{-4} - x} \approx \frac{x^2}{1.0 \times 10^{-4}}$$

$x = [H_3O^+] = 2.7 \times 10^{-7}\ \text{mol}\cdot\text{L}^{-1}$

$\text{pH} = -\log(2.7 \times 10^{-7}) = 6.57$

Note: This value of $[H_3O^+]$ is not much different from the value for pure water, $1.0 \times 10^{-7}\ \text{mol}\cdot\text{L}^{-1}$; therefore, it is at the lower limit of safely ignoring the contribution to $[H_3O^+]$ from the autoprotolysis of water. The exercise should be solved by simultaneously considering both equilibria.

Concentration (mol·L^{-1})	$B(OH)_3$	+ $2\,H_2O \rightleftharpoons$	H_3O^+	+ $B(OH)_4^-$
equilibrium	$1.0 \times 10^{-14} - x$	—	x	y

Concentration (mol·L^{-1}) $2\,H_2O \rightleftharpoons$	H_3O^+	+ OH^-
equilibrium —	x	z

Because there are now two contributions to $[H_3O^+]$, $[H_3O^+]$ is no longer equal to $[B(OH)_4^-]$, nor is it equal to $[OH^-]$, as in pure water. To avoid a cubic equation, x will again be ignored relative to 1.0×10^{-4} mol·L^{-1}. This approximation is justified by the approximate calculation above, and because K_a is very small relative to 1.0×10^{-4}. Let a = initial concentration of $B(OH)_3$, then

$$K_a = 7.2 \times 10^{-10} = \frac{xy}{a - x} \approx \frac{xy}{a} \text{ or } y = \frac{aK_a}{x}$$

$$K_w = 1.0 \times 10^{-14} = xz$$

Electroneutrality requires $x = y + z$ or $z = x - y$; hence, $K_w = xz = x(x - y)$.

Substituting for y from above:

$$x \times \left(x - \frac{aK_a}{x}\right) = K_w$$

$$x^2 - aK_a = K_w$$

$$x^2 = K_w + aK_a$$

$$x = \sqrt{K_w + aK_a} = \sqrt{1.0 \times 10^{-14} + 1.0 \times 10^{-4} \times 7.2 \times 10^{-10}}$$

$$x = 2.9 \times 10^{-7} \text{ mol·L}^{-1} = [H_3O^+]$$

$$pH = -\log(2.9 \times 10^{-7}) = 6.54$$

This value is slightly, but measurably, different from the value 6.57 obtained by ignoring the contribution to $[H_3O^+]$ from water.

(b) In this case, the second ionization can safely be ignored; $K_{a2} \ll K_{a1}$.

Concentration (mol·L^{-1})	H_3PO_4	+ $H_2O \rightleftharpoons$	H_3O^+	+ $H_2PO_4^-$
initial	0.015	—	0	0
change	$-x$	—	$+x$	$+x$
equilibrium	$0.015 - x$	—	x	x

$$K_{a1} = 7.6 \times 10^{-3} = \frac{x^2}{0.015 - x}$$

$$x^2 + 7.6 \times 10^{-3}x - 1.14 \times 10^{-4} = 0$$

$$x = [H_3O^+] = \frac{-7.6 \times 10^{-3} + \sqrt{(7.6 \times 10^{-3})^2 + 4.56 \times 10^{-4}}}{2}$$

$$= 7.5 \times 10^{-3} \text{ mol·L}^{-1}$$

$$pH = -\log(7.5 \times 10^{-3}) = 2.12$$

(c) In this case, the second ionization can safely be ignored; $K_{a2} \ll K_{a1}$.

Concentration (mol·L^{-1})	H_2SO_3	+ $H_2O \rightleftharpoons$	H_3O^+	+ HSO_3^-
initial	0.1	—	0	0
change	$-x$	—	$+x$	$+x$
equilibrium	$0.1 - x$	—	x	x

$$K_{a1} = 1.5 \times 10^{-2} = \frac{x^2}{0.10 - x}$$

$$x^2 + 1.5 \times 10^{-2}x - 1.5 \times 10^{-3} = 0$$

$$x = [H_3O^+] = \frac{-1.5 \times 10^{-2} + \sqrt{(1.5 \times 10^{-2})^2 + 6.0 \times 10^{-3}}}{2} = 0.032 \text{ mol}\cdot\text{L}^{-1}$$

$$\text{pH} = -\log(0.032) = 1.49$$

10.107 $T' = 30°\text{C} = 303 \text{ K}, T = 20°\text{C} = 293 \text{ K}$

$$\ln\left(\frac{K'_a}{K_a}\right) = \frac{\Delta H°}{R}\left(\frac{1}{T} - \frac{1}{T'}\right) = \frac{\Delta H°}{R}\left(\frac{T' - T}{TT'}\right)$$

This is the van't Hoff equation.

$$\ln\left(\frac{1.768 \times 10^{-4}}{1.765 \times 10^{-4}}\right) = \frac{\Delta H°}{8.314 \text{ J}\cdot\text{K}^{-1}\cdot\text{mol}^{-1}}\left(\frac{303 \text{ K} - 293 \text{ K}}{293 \text{ K} \times 303 \text{ K}}\right) = 0.0017$$

$$\Delta H° = \frac{0.0017 \times 8.314 \text{ J}\cdot\text{K}^{-1}\cdot\text{mol}^{-1} \times 293 \text{ K} \times 303 \text{ K}}{10 \text{ K}} = 1.3 \times 10^2 \text{ J}\cdot\text{mol}^{-1}$$

10.109 $T' = 10°\text{C} = 283 \text{ K}, T = 50°\text{C} = 323 \text{ K}$

$$\ln\left(\frac{K'_w}{K_w}\right) = \frac{\Delta H°}{R}\left(\frac{1}{T} - \frac{1}{T'}\right) = \frac{\Delta H°}{R}\left(\frac{T' - T}{TT'}\right)$$

$$\ln\left(\frac{K'_w}{K_w}\right) = \ln\left(\frac{10^{-14.10}}{10^{-15.44}}\right) = \ln\left(\frac{\text{antilog}(-14.10)}{\text{antilog}(-15.44)}\right) = 3.09$$

$$= \frac{\Delta H°}{8.314 \text{ J}\cdot\text{K}^{-1}\cdot\text{mol}^{-1}}\left(\frac{40 \text{ K}}{283 \text{ K} \times 323 \text{ K}}\right)$$

$\Delta H° = 5.9 \times 10^4 \text{ J}\cdot\text{mol}^{-1}$, where 1 mol refers to the reaction

$$2\ D_2O \rightleftharpoons D_3O^+ + OD^-$$

$[D_3O^+][OD^-] = x^2 = 10^{-14.39} = \text{antilog}(-14.39) = 4.1 \times 10^{-15}$ at 40°C

$$x = 6.4 \times 10^{-8} \text{ mol}\cdot\text{L}^{-1} = [D_3O^+] = [OD^-]$$

$$\text{pD} = -\log(6.4 \times 10^{-8}) = 7.20 = \text{pOD}$$

$$\text{p}K_{D_2O} = 14.39$$

$$K_{D_2O} = 10^{-14.39} = 4.1 \times 10^{-15}$$

CHAPTER 11
AQUEOUS EQUILIBRIA

11.1 (a) When solid sodium acetate is added to an acetic acid solution, the concentration of H_3O^+ decreases because the equilibrium

$HC_2H_3O_2(aq) + H_2O(l) \rightleftharpoons H_3O^+(aq) + C_2H_3O_2^-(aq)$

shifts to the left to relieve the stress imposed by the increase of $[C_2H_3O_2^-]$ (Le Chatelier's principle).

(b) When HCl is added to a benzoic acid solution, the percentage of benzoic acid that is deprotonated decreases because the equilibrium

$C_6H_5COOH(aq) + H_2O(l) \rightleftharpoons H_3O^+(aq) + C_6H_5CO_2^-(aq)$

shifts to the left to relieve the stress imposed by the increased $[H_3O^+]$ (Le Chatelier's principle).

(c) When solid NH_4Cl is added to an ammonia solution, the concentration of OH^- decreases because the equilibrium

$NH_3(aq) + H_2O(l) \rightleftharpoons NH_4^+(aq) + OH^-(aq)$

shifts to the left to relieve the stress imposed by the increased $[NH_4^+]$ (Le Chatelier's principle). Because $[OH^-]$ decreases, $[H_3O^+]$ increases and pH decreases.

11.3 $K_a = \dfrac{[H_3O^+][A^-]}{[HA]}$; $pK_a = pH - \log\dfrac{[A^-]}{[HA]}$. If $[A^-] = [HA]$, then $pK_a = pH$.

$pH = pK_a = 3.08, \quad K_a = 8.3 \times 10^{-4}$

(b) Let $x = [\text{lactate ion}] = [L^{-1}]$ and $y = [H_3O^+]$

Concentration ($mol\cdot L^{-1}$)	$HL(aq)$	$+ H_2O(l) \rightleftharpoons$	$H_3O^+(aq)$	$+ L^-(aq)$
initial	$2x$	—	—	x
change	$-y$	—	$+y$	$+y$
equilibrium	$2x - y$	—	y	$y + x$

$$K_a = \frac{[H_3O^+][L^-]}{[HL]} = \frac{(y)(y + x)}{(2x - y)} \cong \frac{(y)(x)}{(2x)} = 8.4 \times 10^{-4}$$

$y = 2(8.3 \times 10^{-4}) \cong 1.7 \times 10^{-3}\ mol\cdot L^{-1} \cong [H_3O^+]$

$pH \approx 2.77$

11.5 In each case, the equilibrium involved is

$HSO_4^-(aq) + H_2O(l) \rightleftharpoons H_3O^+(aq) + SO_4^{2-}(aq)$

$HSO_4^-(aq)$ and $SO_4^{2-}(aq)$ are conjugate acid and base; therefore, the pH calculation is most easily performed with the Henderson-Hasselbalch equation:

$$pH = pK_a + \log\left(\frac{[\text{base}]}{[\text{acid}]}\right) = pK_a + \log\left(\frac{[SO_4^{2-}]}{[HSO_4^-]}\right)$$

(a) $pH = 1.92 + \log\left(\frac{0.080 \text{ mol·L}^{-1}}{0.40 \text{ mol·L}^{-1}}\right) = 1.22, \quad pOH = 14.00 - 1.22 = 12.78$

(b) $pH = 1.92 + \log\left(\frac{0.20 \text{ mol·L}^{-1}}{0.40 \text{ mol·L}^{-1}}\right) = 1.62, \quad pOH = 12.38$

(c) $pH = pK_a = 1.92, \quad pOH = 12.08$

See solution to Exercise 11.3.

11.7 $\left(\frac{1.0 \text{ g NaF}}{0.025 \text{ L}}\right)\left(\frac{1 \text{ mol NaF}}{41.99 \text{ g NaF}}\right) = 0.95 \text{ mol·L}^{-1}$

Concentration (mol·L^{-1})	$HF(aq)$ +	$H_2O(l) \rightleftharpoons$	$H_3O^+(aq)$ +	$F^-(aq)$
initial	0.40	—	0	0.95
change	$-x$	—	$+x$	$+x$
equilibrium	$0.40 - x$	—	x	$0.95 + x$

$$K_a = \frac{[H_3O^+][F^-]}{[HF]} = \frac{(x)(0.95 + x)}{(0.40 - x)} \approx \frac{(x)(0.95)}{(0.40)} = 3.5 \times 10^{-4}$$

$x \cong 1.5 \times 10^{-4} \text{ mol·L}^{-1} \cong [H_3O^+]$

$pH = -\log[H_3O^+] = -\log(1.5 \times 10^{-4}) = 3.82$

change in $pH = 3.82 - 1.93 = 1.89$

11.9 (a) $HCN(aq) + H_2O(l) \rightleftharpoons H_3O^+(aq) + CN^-(aq)$

total volume = 100 mL = 0.100 L

moles of HCN = $0.0200 \text{ L} \times 0.050 \text{ mol·L}^{-1} = 1.0 \times 10^{-3}$ mol HCN

moles of NaCN = $0.0800 \text{ L} \times 0.030 \text{ mol·L}^{-1} = 2.4 \times 10^{-3}$ mol NaCN

$$\text{initial } [HCN]_0 = \frac{1.0 \times 10^{-3} \text{ mol}}{0.100 \text{ L}} = 1.0 \times 10^{-2} \text{ mol·L}^{-1}$$

$$\text{initial } [CN^-]_0 = \frac{2.4 \times 10^{-3} \text{ mol}}{0.100 \text{ L}} = 2.4 \times 10^{-2} \text{ mol·L}^{-1}$$

Concentration (mol·L^{-1})	$HCN(aq)$ +	$H_2O(l) \rightleftharpoons$	$H_3O^+(aq)$ +	$CN^-(aq)$
initial	1.0×10^{-2}	—	0	2.4×10^{-2}
change	$-x$	—	$+x$	$+x$
equilibrium	$1.0 \times 10^{-2} - x$	—	x	$2.4 \times 10^{-2} + x$

$$K_a = \frac{[H_3O^+][CN^-]}{[HCN]} = \frac{(x)(2.4 \times 10^{-2} + x)}{(1.0 \times 10^{-2} - x)} \approx \frac{(x)(2.4 \times 10^{-2})}{(1.0 \times 10^{-2})} = 4.9 \times 10^{-10}$$

$x \approx 2.0 \times 10^{-10}\ mol \cdot L^{-1} \approx [H_3O^+]$

$pH = -\log[H_3O^+] = -\log(2.0 \times 10^{-10}) = 9.70$

(b) The solution here is the same as for part (a), except for the initial concentrations:

$$[HCN]_0 = \frac{0.0800\ L \times 0.030\ mol \cdot L^{-1}}{0.100\ L} = 2.4 \times 10^{-2}\ mol \cdot L^{-1}$$

$$[CN^-]_0 = \frac{0.0200\ L \times 0.050\ mol \cdot L^{-1}}{0.100\ L} = 1.0 \times 10^{-2}\ mol \cdot L^{-1}$$

$$K_a = 4.9 \times 10^{-10} = \frac{(x)(1.0 \times 10^{-2})}{(2.4 \times 10^{-2})}$$

$x = [H_3O^+] = 1.2 \times 10^{-9}\ mol \cdot L^{-1}$

$pH = -\log(1.2 \times 10^{-9}) = 8.92$

(c) $[HCN]_0 = [NaCN]_0$ after mixing; therefore,

$$K_a = 4.9 \times 10^{-10} = \frac{(x)[NaCN]_0}{[HCN]_0} = x = [H_3O^+]$$

$pH = pK_a = -\log(4.9 \times 10^{-10}) = 9.31$

11.11 In a solution containing HClO(aq) and ClO^-(aq), the following equilibrium occurs:

$HClO(aq) + H_2O(l) \rightleftharpoons H_3O^+(aq) + ClO^-(aq)$

The ratio $[ClO^-]/[HClO]$ is related to pH, as given by the Henderson-Hasselbalch equation:

$$pH = pK_a + \log\left(\frac{[ClO^-]}{[HClO]}\right), \text{ or}$$

$$\log\left(\frac{[ClO^-]}{[HClO]}\right) = pH - pK_a = 6.50 - 7.53 = -1.03$$

$$\frac{[ClO^-]}{[HClO]} = 9.3 \times 10^{-2}$$

11.13 The rule of thumb we use is that the effective range of a buffer is roughly within plus or minus one pH unit of the pK_a of the acid. Therefore,

(a) $pK_a = 3.08$; pH range, 2–4

(b) $pK_a = 4.19$; pH range, 3–5

(c) $pK_{a3} = 12.68$; pH range, 11.5–13.5

(d) $pK_{a2} = 7.21$; pH range, 6–8

(e) $pK_b = 7.97$, $pK_a = 6.03$; pH range, 5–7

11.15 Choose a buffer system in which the conjugate acid has a pK_a close to the desired pH. Therefore,

(a) $HClO_2$ and $NaClO_2$, pK_a = 2.00

(b) NaH_2PO_4 and Na_2HPO_4, pK_{a2} = 7.21

(c) $CH_2ClCOOH$ and $NaCH_2ClCO_2$, pK_a = 2.85

(d) Na_2HPO_4 and Na_3PO_4, pK_a = 12.68

11.17 (a) $HCO_3^-(aq) + H_2O(l) \rightleftharpoons CO_3^{2-}(aq) + H_3O^+(aq)$

$$K_{a2} = \frac{[H_3O^+][CO_3^{2-}]}{[HCO_3^-]}, \quad pK_{a2} = 10.25$$

$$pH = pK_{a2} + \log\left(\frac{[CO_3^{2-}]}{[HCO_3^-]}\right)$$

$$\log\left(\frac{[CO_3^{2-}]}{[HCO_3^-]}\right) = pH - pK_{a2} = 11.0 - 10.25 = 0.75$$

$$\frac{[CO_3^{2-}]}{[HCO_3^-]} = 5.6$$

(b) $[CO_3^{2-}] = 5.6 \times [HCO_3^-] = 5.6 \times 0.100\ mol \cdot L^{-1} = 0.56\ mol \cdot L^{-1}$

moles of CO_3^{2-} = moles of K_2CO_3 = $0.56\ mol \cdot L^{-1} \times 1\ L = 0.56\ mol$

$$\text{mass of } K_2CO_3 = 0.56\ mol \times \left(\frac{138.21\ g\ K_2CO_3}{1\ mol\ K_2CO_3}\right) = 77\ g\ K_2CO_3$$

(c) $$[HCO_3^-] = \frac{[CO_3^{2-}]}{5.6} = \frac{0.100\ mol \cdot L^{-1}}{5.6} = 1.8 \times 10^{-2}\ mol \cdot L^{-1}$$

moles of HCO_3^- = moles of $KHCO_3$ = $1.8 \times 10^{-2}\ mol \cdot L^{-1} \times 1\ L$

$= 1.8 \times 10^{-2}\ mol$

mass $KHCO_3 = 1.8 \times 10^{-2}\ mol \times 100.12\ g \cdot mol^{-1} = 1.8\ g\ KHCO_3$

(d) $[CO_3^{2-}] = 5.6 \times [HCO_3^-]$

moles HCO_3^- = moles $KHCO_3$ = $0.100\ mol \cdot L^{-1} \times 0.100\ L = 1.00 \times 10^{-2}\ mol$

Because the final total volume is the same for both $KHCO_3$ and K_2CO_3, the number of moles of K_2CO_3 required is $5.6 \times 1.00 \times 10^{-2}\ mol = 5.6 \times 10^{-2}\ mol$. Thus,

$$\text{volume of } K_2CO_3 \text{ solution} = \frac{5.6 \times 10^{-2}\ mol}{0.200\ mol \cdot L^{-1}} = 0.28\ L = 2.8 \times 10^2\ mL$$

11.19 $CH_3COOH(aq) + H_2O(l) \rightleftharpoons H_3O^+(aq) + CH_3CO_2^-(aq)$

$$K_a = \frac{[H_3O^+][CH_3CO_2^-]}{[CH_3COOH]}$$

$$pH = pK_a + \log\left(\frac{[CH_3CO_2^-]}{[CH_3COOH]}\right)$$

(a) $\text{pH} = \text{p}K_a + \log\left(\frac{0.10}{0.10}\right) = \text{p}K_a = 4.75$

(b) 3.0 mmol NaOH = 3.0×10^{-3} mol NaOH (strong base) produces 3.0×10^{-3} mol $CH_3CO_2^-$ from CH_3COOH (assume no volume change).

$0.10\ \text{mol}\cdot\text{L}^{-1} \times 0.100\ \text{L} = 1.0 \times 10^{-2}$ mol CH_3COOH initially

$0.10\ \text{mol}\cdot\text{L}^{-1} \times 0.100\ \text{L} = 1.0 \times 10^{-2}$ mol $CH_3CO_2^-$ initially

After adding NaOH:

$$[CH_3COOH] = \frac{(1.0 \times 10^{-2} - 3.0 \times 10^{-3})\ \text{mol}}{0.100\ \text{L}} = \frac{7 \times 10^{-3}\ \text{mol}}{0.100\ \text{L}}$$

$$= 7 \times 10^{-2}\ \text{mol}\cdot\text{L}^{-1}$$

$$[CH_3CO_2^-] = \frac{(1.0 \times 10^{-2} + 3.0 \times 10^{-3})\ \text{mol}}{0.100\ \text{L}} = \frac{1.3 \times 10^{-2}\ \text{mol}}{0.100\ \text{L}}$$

$$= 1.3 \times 10^{-1}\ \text{mol}\cdot\text{L}^{-1}$$

$$\text{pH} = 4.75 + \log\left(\frac{1.3 \times 10^{-1}}{7 \times 10^{-2}}\right)$$

$\text{pH} = 4.75 + 0.3 = 5.1$ ($\Delta\text{pH} = 0.3$)

(c) 6.0 mmol HNO_3 = 6.0×10^{-3} mol HNO_3 (strong acid) produces 6.0×10^{-3} mol CH_3COOH from $CH_3CO_2^-$ (assume no volume change).

After adding HNO_3:

$$[CH_3COOH] = \frac{(1.0 \times 10^{-2} + 6.0 \times 10^{-3})\ \text{mol}}{0.100\ \text{L}} = \frac{1.6 \times 10^{-2}\ \text{mol}}{0.100\ \text{L}}$$

$$= 1.6 \times 10^{-1}\ \text{mol}\cdot\text{L}^{-1}$$

$$[CH_3CO_2^-] = \frac{(1.0 \times 10^{-2} - 6.0 \times 10^{-3})\ \text{mol}}{0.100\ \text{L}} = \frac{4 \times 10^{-3}\ \text{mol}}{0.100\ \text{L}}$$

$$= 4 \times 10^{-2}\ \text{mol}\cdot\text{L}^{-1}$$

$$\text{pH} = 4.75 + \log\left(\frac{4 \times 10^{-2}}{1.6 \times 10^{-1}}\right)$$

$\text{pH} = 4.75 - 0.6 = 4.2$ ($\Delta\text{pH} = -0.6$)

11.21 (a) $\text{pH} = \text{p}K_a + \log\left(\frac{[CH_3CO_2^-]}{[CH_3COOH]}\right)$ (see Exercise 11.23)

$\text{pH} = \text{p}K_a + \log\left(\frac{0.100}{0.100}\right) = 4.75$ (initial pH)

final pH: $(0.0100\ \text{L})(0.950\ \text{mol}\cdot\text{L}) = 9.50 \times 10^{-3}$ mol NaOH (strong base) produces 9.50×10^{-3} mol $CH_3CO_2^-$ from CH_3COOH

$0.100\ \text{mol}\cdot\text{L}^{-1} \times 0.100\ \text{L} = 1.00 \times 10^{-2}$ mol CH_3COOH initially

$0.100\ \text{mol}\cdot\text{L}^{-1} \times 0.100\ \text{L} = 1.00 \times 10^{-2}$ mol $CH_3CO_2^-$ initially

After adding NaOH:

$$[CH_3COOH] = \frac{(1.00 \times 10^{-2} - 9.5 \times 10^{-3})\ mol}{0.110\ L} = 5 \times 10^{-3}\ mol \cdot L^{-1}$$

$$[CH_3CO_2^-] = \frac{(1.00 \times 10^{-2} + 9.5 \times 10^{-3})\ mol}{0.110\ L} = 1.77 \times 10^{-1}\ mol \cdot L^{-1}$$

$$pH = 4.75 + \log\left(\frac{1.77 \times 10^{-1}\ mol \cdot L^{-1}}{5 \times 10^{-3}\ mol \cdot L^{-1}}\right) = 4.75 + 1.5 = 6.3$$

(b) $(0.0200\ L)(0.100\ mol \cdot L^{-1}) = 2.00 \times 10^{-3}$ mol HNO_3 (strong acid) produces 2.00×10^{-3} mol CH_3COOH from $CH_3CO_2^-$

After adding HNO_3 [see part (a) of this exercise]:

$$[CH_3COOH] = \frac{(1.00 \times 10^{-2} + 2.00 \times 10^{-3})\ mol}{0.120\ L} = 1.00 \times 10^{-1}\ mol \cdot L^{-1}$$

$$[CH_3CO_2^-] = \frac{(1.00 \times 10^{-2} - 2.00 \times 10^{-3})\ mol}{0.120\ L} = 6.7 \times 10^{-2}\ mol \cdot L^{-1}$$

$$pH = 4.75 + \log\left(\frac{6.7 \times 10^{-2}\ mol \cdot L^{-1}}{1.00 \times 10^{-1}\ mol \cdot L^{-1}}\right) = 4.75 - 0.17 = 4.58$$

$$\Delta pH = -0.17$$

11.23 initial $pH = -\log(0.10) = 1.00$

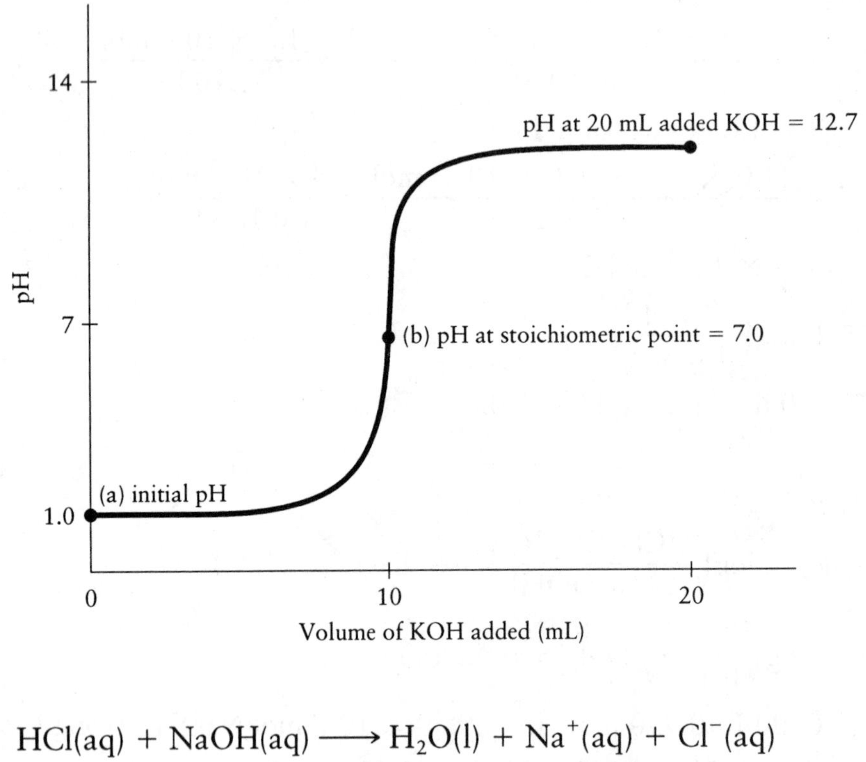

11.25 $HCl(aq) + NaOH(aq) \longrightarrow H_2O(l) + Na^+(aq) + Cl^-(aq)$

(a) $$V_{HCl} = (\tfrac{1}{2})(25.0\ mL)\left(\frac{10^{-3}\ L}{1\ mL}\right)\left(\frac{0.110\ mol\ NaOH}{1\ L}\right)\left(\frac{1\ mol\ HCl}{1\ mol\ NaOH}\right)\left(\frac{1\ L\ HCl}{0.150\ mol\ HCl}\right)$$

$$= 9.17 \times 10^{-3}\ L\ HCl$$

(b) $2 \times 9.17 \times 10^{-3}\ \text{L} = 0.0183\ \text{L}$

(c) volume $= (0.0250 + 0.0183)\ \text{L} = 0.0433\ \text{L}$

$$[\text{Na}^+] = (0.0250\ \text{L})\left(\frac{0.110\ \text{mol NaOH}}{1\ \text{L}}\right)\left(\frac{1\ \text{mol Na}^+}{1\ \text{mol NaOH}}\right)\left(\frac{1}{0.0433\ \text{L}}\right)$$

$$= 0.0635\ \text{mol}\cdot\text{L}^{-1}$$

(d) number of moles of H_3O^+ (from acid) $= (0.0200\ \text{L})\left(\frac{0.150\ \text{mol}}{1\ \text{L}}\right)$

$$= 3.00 \times 10^{-3}\ \text{mol}\ H_3O^+$$

number of moles of OH^- (from base) $= (0.0250\ \text{L})\left(\frac{0.110\ \text{mol Na}^+}{1\ \text{L}}\right)$

$$= 2.75 \times 10^{-3}\ \text{mol}\ OH^-$$

excess $H_3O^+ = (3.00 - 2.75) \times 10^{-3}\ \text{mol} = 2.5 \times 10^{-4}\ \text{mol}\ H_3O^+$

$$[H_3O^+] = \frac{2.5 \times 10^{-4}\ \text{mol}}{0.0450\ \text{L}} = 5.6 \times 10^{-3}\ \text{mol}\cdot\text{L}^{-1}$$

$\text{pH} = -\log(5.6 \times 10^{-3}) = 2.25$

11.27 $\text{HCl(aq)} + \text{NaOH(aq)} \longrightarrow H_2O(l) + Na^+(aq) + Cl^-(aq)$

(a) moles of HCl $= (2.54\ \text{g NaOH})\left(\frac{1\ \text{mol NaOH}}{40.00\ \text{g NaOH}}\right)\left(\frac{1\ \text{mol HCl}}{1\ \text{mol NaOH}}\right)$

$$= 6.35 \times 10^{-2}\ \text{mol HCl}$$

$$\text{volume of HCl} = \frac{6.35 \times 10^{-2}\ \text{mol}}{0.150\ \text{mol}\cdot\text{L}^{-1}} = 0.423\ \text{L} = 423\ \text{mL}$$

(b) moles of $Cl^- = 0.423\ \text{L}\left(\frac{0.150\ \text{mol HCl}}{1\ \text{L}}\right)\left(\frac{1\ \text{mol Cl}^-}{1\ \text{mol HCl}}\right)$

$$= 6.35 \times 10^{-2}\ \text{mol Cl}^-$$

$$[Cl^-] = \frac{6.35 \times 10^{-2}\ \text{mol Cl}^-}{(0.423 + 0.025)\ \text{L}} = 0.142\ \text{mol}\cdot\text{L}^{-1}$$

11.29 mass of pure NaOH $= (0.0158\ \text{L HCl})\left(\frac{0.107\ \text{mol HCl}}{1\ \text{L HCl}}\right)$

$$\left(\frac{1\ \text{mol NaOH}}{1\ \text{mol HCl}}\right)\left(\frac{40.00\ \text{g NaOH}}{1\ \text{mol NaOH}}\right)\left(\frac{200\ \text{mL}}{20.0\ \text{mL}}\right)$$

$= 0.676\ \text{g}$

$$\text{percent purity} = \frac{0.676\ \text{g}}{0.968\ \text{g}} \times 100\% = 69.8\%$$

11.31 (a) $\text{pOH} = -\log(0.110) = 0.959, \quad \text{pH} = 14.00 - 0.959 = 13.04$

(b) initial moles of OH^- (from base) $= (0.0250\ \text{L})\left(\frac{0.110\ \text{mol}}{1\ \text{L}}\right)$

$$= 2.75 \times 10^{-3}\ \text{mol}\ OH^-$$

$$\text{moles of } H_3O^+ \text{ added} = (0.0050\ L)\left(\frac{0.150\ mol}{1\ L}\right) = 7.5 \times 10^{-4}\ mol\ H_3O^+$$

$$\text{excess } OH^- = (2.75 - 0.75) \times 10^{-3}\ mol = 2.00 \times 10^{-3}\ mol\ OH^-$$

$$[OH^-] = \frac{2.00 \times 10^{-3}\ mol}{0.030\ L} = 0.067\ mol\cdot L^{-1}$$

$$pOH = -\log(0.067) = 1.17, \quad pH = 14.00 - 1.17 = 12.83$$

(c) $\text{moles of } H_3O^+ \text{ added} = 2 \times 7.5 \times 10^{-4}\ mol = 1.50 \times 10^{-3}\ mol\ H_3O^+$

$$\text{excess } OH^- = (2.75 - 1.50) \times 10^{-3}\ mol = 1.25 \times 10^{-3}\ mol\ OH^-$$

$$[OH^-] = \frac{1.25 \times 10^{-3}\ mol}{0.035\ L} = 0.036\ mol\cdot L^{-1}$$

$$pOH = -\log(0.036) = 1.44, \quad pH = 14.00 - 1.44 = 12.56$$

(d) $pH = 7.00$

$$V_{HCl} = (2.75 \times 10^{-3}\ mol\ NaOH)\left(\frac{1\ mol\ HCl}{1\ mol\ NaOH}\right)\left(\frac{1\ L\ HCl}{0.150\ mol\ HCl}\right)$$

$$= 0.0183\ L$$

(e) $$[H_3O^+] = (0.0050\ L)\left(\frac{0.150\ mol}{1\ L}\right)\left(\frac{1}{(0.0250 + 0.0183 + 0.0050)\ L}\right)$$

$$= 0.016\ mol\cdot L^{-1}$$

$$pH = -\log(0.016) = 1.80$$

(f) $$[H_3O^+] = \left(\frac{0.010\ L}{0.0533\ L}\right)\left(\frac{0.150\ mol}{1\ L}\right) = 0.028\ mol\cdot L^{-1}$$

$$pH = -\log(0.028) = 1.55$$

11.33 (a)

Concentration ($mol\cdot L^{-1}$)	$CH_3COOH(aq)$ +	$H_2O(l)$ $\rightleftharpoons$	$H_3O^+(aq)$ +	$CH_3CO_2^-(aq)$
initial	0.10	—	0	0
change	$-x$	—	$+x$	$+x$
equilibrium	$0.10 - x$	—	x	x

$$K_a = \frac{[H_3O^+][CH_3CO_2^-]}{[CH_3COOH]} = 1.8 \times 10^{-5} = \frac{x^2}{0.10 - x} \approx \frac{x^2}{0.10}$$

$$x^2 = 1.8 \times 10^{-6}$$

$$x = 1.3 \times 10^{-3}\ mol\cdot L^{-1} = [H_3O^+]$$

$$\text{initial pH} = -\log(1.3 \times 10^{-3}) = 2.89$$

(b) $\text{moles of } CH_3COOH = (0.0250\ L)(0.10\ M)$

$$= 2.50 \times 10^{-3}\ mol\ CH_3COOH$$

$$\text{moles of NaOH} = (0.0100\ L)(0.10\ M) = 1.0 \times 10^{-3}\ mol\ OH^-$$

After neutralization:

$$\frac{1.50 \times 10^{-3} \text{ mol } CH_3COOH}{0.0350 \text{ L}} = 4.29 \times 10^{-2} \text{ mol}\cdot\text{L}^{-1}\ CH_3COOH$$

$$\frac{1.0 \times 10^{-3} \text{ mol } CH_3CO_2^-}{0.0350 \text{ L}} = 2.86 \times 10^{-2} \text{ mol}\cdot\text{L}^{-1}\ CH_3CO_2^-$$

Then, consider equilibrium, $K_a = \frac{[H_3O^+][CH_3CO_2^-]}{[CH_3COOH]}$

Concentration ($\text{mol}\cdot\text{L}^{-1}$)	$CH_3COOH(aq)$ +	$H_2O(l) \rightleftharpoons$	$H_3O^+(aq)$ +	$CH_3CO_2^-(aq)$
initial	4.29×10^{-2}	—	0	2.86×10^{-2}
change	$-x$	—	$+x$	$+x$
equilibrium	$4.29 \times 10^{-2} - x$	—	x	$2.86 \times 10^{-2} + x$

$$1.8 \times 10^{-5} = \frac{(x)(x + 2.86 \times 10^{-2})}{(4.29 \times 10^{-2} - x)}; \text{ assume } +x \text{ and } -x \text{ negligible}$$

$[H_3O^+] = x = 2.7 \times 10^{-5}\ \text{mol}\cdot\text{L}^{-1}$ and $\text{pH} = -\log(2.7 \times 10^{-5}) = 4.56$

(c) Because acid and base concentrations are equal, their volumes are equal at the stoichiometric point. Therefore, 25.0 mL NaOH are required to reach the stoichiometric point and 12.5 mL NaOH are required to reach the half stoichiometric point.

(d) At the half stoichiometric point, $\text{pH} = \text{p}K_a = 4.75$

(e) 25.0 mL; see part (c)

(f) The pH is that of 0.050 M $NaCH_3CO_2$.

Concentration ($\text{mol}\cdot\text{L}^{-1}$)	$H_2O(l)$ +	$CH_3CO_2^-(aq) \rightleftharpoons$	$CH_3COOH(aq)$ +	$OH^-(aq)$
initial	—	0.050	0	0
change	—	$-x$	$+x$	$+x$
equilibrium	—	$0.050 - x$	x	x

$$K_b = \frac{K_w}{K_a} = \frac{1.00 \times 10^{-14}}{1.8 \times 10^{-5}} = 5.6 \times 10^{-10} = \frac{x^2}{0.050 - x} \approx \frac{x^2}{0.050}$$

$x^2 = 2.8 \times 10^{-11}$

$x = 5.3 \times 10^{-6}\ \text{mol}\cdot\text{L}^{-1} = [OH^-]$

$\text{pOH} = 5.28,\quad \text{pH} = 14.00 - 5.28 = 8.72$

11.35 (a) $K_b = \frac{[NH_4^+][OH^-]}{[NH_3]} = 1.8 \times 10^{-5}$

Concentration ($mol\cdot L^{-1}$)	$H_2O(l)$ +	$NH_3(aq)$ $\rightleftharpoons$	$NH_4^+(aq)$ +	$OH^-(aq)$
initial	—	0.15	0	0
change	—	$-x$	$+x$	$+x$
equilibrium	—	$0.15 - x$	x	x

$$1.8 \times 10^{-5} = \frac{x^2}{0.15 - x} \approx \frac{x^2}{0.15}$$

$[OH^-] = x = 1.6 \times 10^{-3}\ mol\cdot L^{-1}$

$pOH = 2.80$, initial $pH = 14.00 - 2.80 = 11.20$

(b) initial moles of $NH_3 = (0.0150\ L)(0.15\ mol\cdot L^{-1}) = 2.3 \times 10^{-3}\ mol\ NH_3$

moles of $HCl = (0.0150\ L)(0.10\ mol\cdot L^{-1}) = 1.5 \times 10^{-3}\ mol\ HCl$

$$\frac{(2.3 \times 10^{-3} - 1.5 \times 10^{-3})\ mol\ NH_3}{0.0300\ L} = 2.7 \times 10^{-2}\ mol\cdot L^{-1}\ NH_3$$

$$\frac{1.5 \times 10^{-3}\ mol\ HCl}{0.0300\ L} = 5.0 \times 10^{-2}\ mol\cdot L^{-1}\ HCl \approx 5.0 \times 10^{-2}\ mol\cdot L^{-1}\ NH_4^+$$

Then, consider the equilibrium.

Concentration ($mol\cdot L^{-1}$)	$H_2O(l)$ +	$NH_3(aq)$ $\rightleftharpoons$	$NH_4^+(aq)$ +	$OH^-(aq)$
initial	—	2.7×10^{-2}	5.0×10^{-2}	0
change	—	$-x$	$+x$	$+x$
equilibrium	—	$2.7 \times 10^{-2} - x$	$5.0 \times 10^{-2} + x$	x

$$K_b = \frac{[NH_4^+][OH^-]}{[NH_3]} = 1.8 \times 10^{-5}$$

$$= \frac{(x)(5.0 \times 10^{-2} + x)}{(2.7 \times 10^{-2} - x)}; \text{ assume } +x \text{ and } -x \text{ are negligible}$$

$[OH^-] = x = 9.7 \times 10^{-6}\ mol\cdot L^{-1}$ and $pOH = 5.01$

Therefore, $pH = 14.00 - 5.01 = 8.99$

(c) At the stoichiometric point, moles of NH_3 = moles of HCl.

$$\text{volume HCl added} = \frac{(0.15\ mol\cdot L^{-1}\ NH_3)(0.0150\ L)}{0.10\ mol\cdot L^{-1}\ HCl} = 0.0225\ L\ HCl$$

Therefore, halfway to the stoichiometric point, volume HCl added = 22.5/2 = 11.25 mL

(d) At half stoichiometric point, $pOH = pK_b$ and $pOH = 4.75$.

Therefore, $pH = 14.00 - 4.75 = 9.25$.

(e) 22.5 mL; see part (c)

(f) $NH_4^+(aq) + H_2O(l) \rightleftharpoons H_3O^+(aq) + NH_3(aq)$

The initial moles of NH_3 have now been converted to moles of NH_4^+ in a $(15 + 22.5 = 37.5)$ mL volume:

$$[NH_4^+] = \frac{2.25 \times 10^{-3}\ \text{mol}}{0.0375\ \text{L}} = 0.060\ \text{mol}\cdot\text{L}^{-1}$$

$$K_a = \frac{K_w}{K_b} = \frac{1.00 \times 10^{-14}}{1.8 \times 10^{-5}} = 5.6 \times 10^{-10}$$

Concentration ($\text{mol}\cdot\text{L}^{-1}$)	$NH_4^+(aq)$	+ $H_2O(l)$ $\rightleftharpoons$	$H_3O^+(aq)$	+ $NH_3(aq)$
initial	0.060	—	0	0
change	$-x$	—	$+x$	$+x$
equilibrium	$0.060 - x$	—	x	x

$$K_a = 5.6 \times 10^{-10} = \frac{x^2}{0.060 - x} \approx \frac{x^2}{0.060}$$

$x = [H_3O^+] = 5.8 \times 10^{-6}\ \text{mol}\cdot\text{L}^{-1}$

$\text{pH} = -\log(5.8 \times 10^{-6}) = 5.24$

11.37 (a) Let HL = lactic acid; then, $HL(aq) + H_2O(l) \rightleftharpoons H_3O^+(aq) + L^-(aq)$

$$K_a = \frac{[H_3O^+][L^-]}{[HL]} = 8.4 \times 10^{-4}$$

Concentration ($\text{mol}\cdot\text{L}^{-1}$)	$HL(aq)$	+ $H_2O(l)$ $\rightleftharpoons$	$H_3O^+(aq)$	+ $L^-(aq)$
initial	0.110	—	0	0
change	$-x$	—	$+x$	$+x$
equilibrium	$0.110 - x$	—	x	x

$$8.4 \times 10^{-4} = \frac{x^2}{0.110 - x} \approx \frac{x^2}{0.110}$$

$[H_3O^+] = x = 9.6 \times 10^{-3}\ \text{mol}\cdot\text{L}^{-1}$

initial $\text{pH} = -\log(9.6 \times 10^{-3}) = 2.02$

(b) initial moles of HL = $(0.0250\ \text{L})(0.110\ \text{M}) = 2.75 \times 10^{-3}$ mol HL

moles of NaOH = $(0.0050\ \text{L})(0.150\ \text{M}) = 7.5 \times 10^{-4}$ mol NaOH

After neutralization:

moles of HL = $2.75 \times 10^{-3}\ \text{mol} - 7.5 \times 10^{-4}\ \text{mol} = 2.00 \times 10^{-3}$ mol

$$[HL] = \frac{2.00 \times 10^{-3}\ \text{mol HL}}{0.0300\ \text{L}} = 6.67 \times 10^{-2}\ \text{mol}\cdot\text{L}^{-1}$$

$$[L^-] = \frac{7.5 \times 10^{-4}\ \text{mol L}^-}{0.0300\ \text{L}} = 2.5 \times 10^{-2}\ \text{mol}\cdot\text{L}^{-1}$$

Then, consider the equilibrium:

Concentration $(mol \cdot L^{-1})$	$HL(aq)$	+ $H_2O(l) \rightleftharpoons$	$H_3O^+(aq)$	+ $L^-(aq)$
initial	6.67×10^{-2}	—	0	2.5×10^{-2}
change	$-x$	—	$+x$	$+x$
equilibrium	$6.67 \times 10^{-2} - x$	—	x	$2.5 \times 10^{-2} + x$

$$K_a = \frac{[H_3O^+][L^-]}{[HL]} = 8.4 \times 10^{-4} = \frac{(x)(x + 2.5 \times 10^{-2})}{(6.67 \times 10^{-2} - x)}; \ +x \text{ and } -x \text{ negligible}$$

$[H_3O^+] = x = 2.2 \times 10^{-3}\ mol \cdot L^{-1}$

$pH = -\log(2.2 \times 10^{-3}) = 2.66$

(c) moles of HL $= (0.0250\ L)(0.110\ mol \cdot L^{-1}) = 2.75 \times 10^{-3}$

moles of NaOH $= (0.0100\ L)(0.150\ mol \cdot L^{-1}) = 1.50 \times 10^{-3}$

After neutralization:

$$[HL] = \frac{1.25 \times 10^{-3}\ mol\ HL}{0.0350\ L} = 3.57 \times 10^{-2}\ mol \cdot L^{-1}$$

$$[L^-] = \frac{1.50 \times 10^{-3}\ mol\ L^-}{0.0350\ L} = 4.29 \times 10^{-2}\ mol \cdot L^{-1}$$

Then, consider the equilibrium:

Concentration $(mol \cdot L^{-1})$	$HL(aq)$	+ $H_2O(l) \rightleftharpoons$	$H_3O^+(aq)$	+ $L^-(aq)$
initial	3.57×10^{-2}	—	0	4.29×10^{-2}
change	$-x$	—	$+x$	$+x$
equilibrium	$3.57 \times 10^{-2} - x$	—	x	$4.29 \times 10^{-2} + x$

$$K_a = \frac{[H_3O^+][L^-]}{[HL]} = 8.4 \times 10^{-4} = \frac{(x)(x + 4.29 \times 10^{-2})}{(3.57 \times 10^{-2} - x)}; \ +x \text{ and } -x \text{ negligible}$$

$[H_3O^+] = x = 7.0 \times 10^{-4}\ mol \cdot L^{-1}$

$pH = -\log(7.0 \times 10^{-4}) = 3.15$

(d) At the stoichiometric point, moles acid = moles base = moles L^-

$$\text{volume base} = \frac{(0.110\ mol \cdot L^{-1}\ acid)(0.0250\ L\ acid)}{0.150\ mol \cdot L^{-1}\ base} = 0.0183\ L\ base$$

$$K_b = \frac{K_w}{K_a} = \frac{1.00 \times 10^{-14}}{8.4 \times 10^{-4}} = 1.2 \times 10^{-11}$$

$$[L^-] = \frac{2.75 \times 10^{-3}\ mol \cdot L^-}{(0.025 + 0.0183)\ L} = 6.35 \times 10^{-2}\ mol \cdot L^{-1}$$

Concentration ($mol \cdot L^{-1}$)	$H_2O(l)$ +	$L^-(aq)$ $\rightleftharpoons$	$HL(aq)$ +	$OH^-(aq)$
initial	—	6.35×10^{-2}	0	0
change	—	$-x$	$+x$	$+x$
equilibrium	—	$6.35 \times 10^{-2} - x$	x	x

$$1.2 \times 10^{-11} = \frac{[HL][OH^-]}{[L^-]} = \frac{x^2}{6.35 \times 10^{-2} - x} \cong \frac{x^2}{6.35 \times 10^{-2}}$$

$[OH^-] = x = 8.73 \times 10^{-7}\ mol \cdot L^{-1}$

$pOH = -\log(8.73 \times 10^{-7}) = 6.06, \quad pH = 14.00 - 6.06 = 7.94$

(e) At 5.0 mL base beyond the stoichiometric point, consider only strong base.

$$\text{molar concentration of } OH^- = \frac{(0.0050\ L)(0.150\ mol \cdot L^{-1})}{0.0250\ L + 0.0234\ L} = 1.6 \times 10^{-2}\ mol \cdot L^{-1}$$

$pOH = -\log(1.6 \times 10^{-2}) = 1.80, \quad pH = 14.00 - 1.80 = 12.20$

(f) At 10.0 mL base beyond stoichiometric point, consider only strong base.

$$\text{molar concentration of } OH^- = \frac{(0.0100\ L)(0.150\ mol \cdot L^{-1})}{0.0250\ L + 0.0284\ L} = 2.81 \times 10^{-2}\ mol \cdot L^{-1}$$

$pOH = -\log(2.81 \times 10^{-2}) = 1.55, \quad pH = 14.00 - 1.55 = 12.45$

11.39 At the stoichiometric point, the volume of solution will have doubled; therefore, the concentration of $CH_3CO_2^-$ will be 0.10 M. The equilibrium is

Concentration ($mol \cdot L^{-1}$)	$CH_3CO_2^-(aq)$ +	$H_2O(l)$ $\rightleftharpoons$	$HCH_3CO_2(aq)$ +	$OH^-(aq)$
initial	0.10	—	0	0
change	$-x$	—	$+x$	$+x$
equilibrium	$0.10 - x$	—	x	x

$$K_b = \frac{K_w}{K_a} = \frac{1.00 \times 10^{-14}}{1.8 \times 10^{-5}} = 5.6 \times 10^{-10}$$

$$K_b = \frac{[HCH_3CO_2][OH^-]}{[CH_3CO_2^-]} = \frac{x^2}{0.10 - x} \approx \frac{x^2}{0.10} = 5.6 \times 10^{-10}$$

$x = 7.5 \times 10^{-6}\ mol \cdot L^{-1} = [OH^-]$

$pOH = -\log(7.5 \times 10^{-6}) = 5.12, \quad pH = 14.00 - 5.12 = 8.88$

From Table 11.2, we see that this pH value lies within the range for phenolphthalein, so that indicator would be suitable; the others would not.

11.41 Exercise 11.33; thymol blue or phenolphthalein; Exercise 11.35: methyl red or bromcresol green; Exercise 11.37: phenol red.

11.43 (a) To reach the first stoichiometric point, we must add enough solution to neutralize one H^+on the H_3AsO_4. To do this, we will require $0.0750\ L \times 0.137\ mol \cdot L^{-1} = 0.0103\ mol\ OH^-$. The volume of base required will be given by the number of moles of base required, divided by the concentration of base solution: $\frac{0.0750\ L \times 0.137\ mol \cdot L^{-1}}{0.275\ mol \cdot L^{-1}} = 0.0374\ L$ or 37.4 mL. (b) and (c) To reach the second stoichiometric point will require double the amount calculated in (a), or 74.8 mL, and the third stoichiometric point will be reached with three times the amount added in (a), or 112 mL.

11.45 (a) The base HPO_3^{2-} is the fully deprotonated form of phosphorous acid H_3PO_3 (the remaining H attached to P is not acidic). It will require an equal number of moles of HNO_3 to react with HPO_3^{2-}, in order to reach the first stoichiometric point (formation of $H_2PO_3^-$). The value will be given by $\frac{0.0550\ L \times 0.195\ mol \cdot L^{-1}}{0.395\ mol \cdot L^{-1}} = 0.0272\ L$ or 27.2 mL. (b) To reach the second stoichiometric point would require double the amount of solution calculated in (a), or 54.4 mL.

11.47 (a) This value is calculated as described in Example 10.12. First we calculate the molarity of the starting phosphorous acid solution: $\frac{0.164\ g}{81.99\ g \cdot mol^{-1}} \Big/ 0.0500\ L = 0.0400\ mol \cdot L^{-1}$. We then use the first acid dissociation of phosphorous acid as the dominant equilibrium. The K_{a1} is 1.0×10^{-2}. Let H_2P represent the fully-protonated phosphorus acid.

Concentration ($mol \cdot L^{-1}$)	$H_2P(aq)$	+ $H_2O(l)$ $\rightleftharpoons$	$HP^-(aq)$	+ $H_3O^+(aq)$
initial	0.0400	—	0	0
change	$-x$	—	$+x$	$+x$
final	$0.0400 - x$	—	$+x$	$+x$

$$K_a = \frac{[H_3O^+][HP^-]}{[H_2P]} = 1.0 \times 10^{-2}$$

$$1.0 \times 10^{-2} = \frac{x \cdot x}{0.0400 - x} = \frac{x^2}{0.0400 - x}$$

If we assume $x \ll 0.0400$, then the equation becomes

$x^2 = (1.0 \times 10^{-2})(0.0400) = 4.00 \times 10^{-4}$

$x = 2.0 \times 10^{-2}$

Because this value is more than 10% of 0.0400, the full quadratic solution should be undertaken. The equation is

$x^2 = (1.0 \times 10^{-2})(0.0400 - x)$ or

$x^2 + (1.0 \times 10^{-2}x) - (4.0 \times 10^{-4}) = 0$

Using the quadratic formula, we obtain $x = 0.016$.

pH = 1.80

(b) First, carry out the reaction between phosphorous acid and the strong base to completion:

$H_2P(aq) + OH^-(aq) \longrightarrow HP^-(aq) + H_2O(l)$

moles of $H_2P = (0.0400\ mol \cdot L^{-1})(0.0500\ L) = 2.00 \times 10^{-3}$ mol

moles of $OH^- = (0.006\ 50\ L)(0.175\ mol \cdot L^{-1}) = 1.14 \times 10^{-3}$ mol

1.14×10^{-3} mol OH^- will react completely with 2.00×10^{-3} mol H_2P to give 1.14×10^{-3} mol HP^- with 8.6×10^{-4} moles of H_2P remaining.

$$[H_2P] = \frac{8.6 \times 10^{-4}\ mol}{0.0565\ L} = 0.0152\ mol \cdot L^{-1}$$

$$[HP^-] = \frac{1.14 \times 10^{-3}\ mol}{0.0565\ L} = 0.0202\ mol \cdot L^{-1}$$

Concentration ($mol \cdot L^{-1}$)	$H_2P(aq)$	+ $H_2O(l)$ $\rightleftharpoons$	$HP^-(aq)$	+ $H_3O^+(aq)$
initial	0.0152	—	0.0202	0
change	$-x$	—	$+x$	$+x$
final	$0.0152 - x$	—	$0.0202 + x$	$+x$

The calculation is performed as in part (a):

$$1.0 \times 10^{-2} = \frac{(0.0202 + x)x}{0.0152 - x}$$

$x^2 + (3.0 \times 10^{-2}x) - (1.5 \times 10^{-4}) = 0$

$x = 4.4 \times 10^{-3}$

pH = 2.36

(c) The total amount of OH^- added will be 6.50 mL + 4.93 mL = 11.43 mL. The moles of $OH^- = (0.011\ 43\ L)(0.175\ mol \cdot L^{-1}) = 0.002\ 00$ mol. This amount of base will exactly neutralize one H^+ present in H_2P, so the problem becomes one of calculating the pH of a solution of NaHP. As presented in section 10.16, the pH of a solution of a salt of a polyprotic acid is given by

$pH = \frac{1}{2}(pK_{a1} + pK_{a2})$

$pH = \frac{1}{2}(2.00 + 6.59) = 4.30$

11.49 (a) The reaction of the base Na_2HPO_4 with the strong acid will be taken to completion first:

$HPO_4^{2-}(aq) + H_3O^+(aq) \longrightarrow H_2PO_4^- + H_2O(l)$

Initially, moles of HPO_4^{2-} = moles of H_3O^+ = $0.0500\ L \times 0.350\ mol\cdot L^{-1}$ = 0.0175 mol

Because this reaction proceeds with no excess base or acid, we are dealing with a solution that can be viewed as being composed of $H_2PO_4^-$. The problem then becomes one of estimating the pH of this solution, which can be done from the relationship:

$pH = \frac{1}{2}(pK_{a1} + pK_{a2})$

$pH = \frac{1}{2}(2.12 + 7.21) = 4.66$

(b) This reaction proceeds as in (a), but there is more strong acid available, so the excess acid will react with $H_2PO_4^-$ to produce H_3PO_4. Addition of the first 50.0 mL of acid solution will convert all the HPO_4^{2-} into $H_2PO_4^-$. The additional 25.0 mL of the strong acid will react with $H_2PO_4^-$:

$H_2PO_4^-(aq) + H_3O^+(aq) \longrightarrow H_3PO_4(aq) + H_2O(l)$

0.0175 mol $H_2PO_4^-$ will react with 0.008 75 mol H_3O^+ to give 0.0088 mol H_3PO_4 with 0.088 mol $H_2PO_4^-$ in excess. The concentrations will be

$[H_3PO_4] = [H_2PO_4^-] = \dfrac{0.0088\ mol}{0.125\ L} = 0.070$. The appropriate relationship to use is then:

Concentration ($mol\cdot L^{-1}$)	$H_3PO_4(aq)$	+ $H_2O(l)$ ⇌	$H_2PO_4^-(aq)$	+ $H_3O^+(aq)$
initial	0.070	—	0.070	0
change	$-x$	—	$+x$	$+x$
final	$0.070 - x$	—	$0.070 + x$	$+x$

$$K_{a2} = \frac{[H_2PO_4^-][H_3O^+]}{[H_3PO_4]}$$

$$\frac{(0.070 + x)x}{(0.070 - x)} = 7.6 \times 10^{-3}$$

Because the equilibrium constant is not small compared to 0.070, the full quadratic solution must be calculated:

$x^2 + 0.070x = 7.6 \times 10^{-3}\ (0.070 - x)$

$x^2 + 0.078x - 5.3 \times 10^{-4} = 0$

$x = 6.3 \times 10^{-3}$

$pH = -\log(6.3 \times 10^{-3}) = 2.20$

11.51 (a) The solubility equilibrium is $AgBr(s) \rightleftharpoons Ag^+(aq) + Br^-(aq)$.

$[Ag^+] = [Br^-] = 8.8 \times 10^{-7}\ mol \cdot L^{-1} = S =$ solubility

$K_{sp} = [Ag^+][Br^-] = (8.8 \times 10^{-7})(8.8 \times 10^{-7}) = 7.7 \times 10^{-13}$

(b) The solubility equilibrium is $PbCrO_4(s) \rightleftharpoons Pb^{2+}(aq) + CrO_4^{\ 2-}(aq)$

$[Pb^{2+}] = 1.3 \times 10^{-7}\ mol \cdot L^{-1} = S, \quad [CrO_4^{\ 2-}] = 1.3 \times 10^{-7}\ mol \cdot L^{-1} = S$

$K_{sp} = [Pb^{2+}][CrO_4^{\ 2-}] = (1.3 \times 10^{-7})(1.3 \times 10^{-7}) = 1.7 \times 10^{-14}$

(c) The solubility equilibrium is $Ba(OH)_2(s) \rightleftharpoons Ba^{2+}(aq) + 2\ OH^-(aq)$

$[Ba^{2+}] = 0.11\ mol \cdot L^{-1} = S, \quad [OH^-] = 0.22\ mol \cdot L^{-1} = 2S$

$K_{sp} = [Ba^{2+}][OH^-]^2 = (0.11)(0.22)^2 = 5.3 \times 10^{-3}$

(d) The solubility equilibrium is $MgF_2(s) \rightleftharpoons Mg^{2+}(aq) + 2F^-(aq)$

$[Mg^{2+}] = 1.2 \times 10^{-3}\ mol \cdot L^{-1} = S, \quad [F^-] = 2.4 \times 10^{-3}\ mol \cdot L^{-1} = 2S$

$K_{sp} = [Mg^{2+}][F^-]^2 = (1.2 \times 10^{-3})(2.4 \times 10^{-3})^2 = 6.9 \times 10^{-9}$

11.53 (a) Equilibrium equation: $Ag_2S(s) \rightleftharpoons 2\ Ag^+(aq) + S^{2-}(aq)$

$K_{sp} = [Ag^+]^2[S^{2-}] = (2S)^2(S) = 4S^3 = 6.3 \times 10^{-51}$

$S = 1.2 \times 10^{-17}\ mol \cdot L^{-1}$

(b) Equilibrium equation: $CuS(s) \rightleftharpoons Cu^{2+}(aq) + S^{2-}(aq)$

$K_{sp} = [Cu^{2+}][S^{2-}] = S \times S = S^2 = 1.3 \times 10^{-36}$

$S = 1.1 \times 10^{-18} mol \cdot L^{-1}$

(c) Equilibrium equation: $CaCO_3(s) \rightleftharpoons Ca^{2+}(aq) + CO_3^{\ 2-}(aq)$

$K_{sp} = [Ca^{2+}][CO_3^{\ 2-}] = S \times S = S^2 = 8.7 \times 10^{-9}$

$S = 9.3 \times 10^{-5}\ mol \cdot L^{-1}$

11.55 $Tl_2CrO_4(s) \rightleftharpoons 2\ Tl^+(aq) + CrO_4^{\ 2-}(aq)$

$[CrO_4^{\ 2-}] = S = 6.3 \times 10^{-5}\ mol \cdot L^{-1}$

$[Tl^+] = 2S = 2(6.3 \times 10^{-5})\ mol \cdot L^{-1}$

$K_{sp} = [Tl^+]^2[CrO_4^{\ 2-}] = (2S)^2 \times (S)$

$K_{sp} = [2(6.3 \times 10^{-5})]^2 \times (6.3 \times 10^{-5}) = 1.0 \times 10^{-12}$

11.57 (a)

Concentration ($mol \cdot L^{-1}$)	$AgCl(s) \rightleftharpoons$	$Ag^+(aq)$ +	$Cl^-(aq)$
initial	—	0	0.20
change	—	$+S$	$+S$
equilibrium	—	S	$S + 0.20$

$K_{sp} = [Ag^+][Cl^-] = (S) \times (S + 0.20) = 1.6 \times 10^{-10}$

Assume S in $(S + 0.20)$ is negligible, so $0.20\ S = 1.6 \times 10^{-10}$

$S = 8.0 \times 10^{-10}\ mol \cdot L^{-1} = [Ag^+] =$ molar solubility of AgCl in 0.20 M NaCl

(b) Concentration ($mol \cdot L^{-1}$) $Hg_2Cl_2(s) \rightleftharpoons Hg_2^{2+}(aq) + 2\ Cl^-(aq)$

	$Hg_2Cl_2(s)$	$Hg_2^{2+}(aq)$	$2\ Cl^-(aq)$
initial	—	0	0.10
change	—	$+S$	$+2S$
equilibrium	—	S	$0.10 + 2S$

$K_{sp} = [Hg_2^{2+}][Cl^-]^2 = (S) \times (2S + 0.10)^2 = 1.3 \times 10^{-18}$

Assume $2S$ in $(2S + 0.10)$ is negligible, so $0.010S = 1.3 \times 10^{-18}$

$S = 1.3 \times 10^{-16}\ mol \cdot L^{-1} = [Hg_2^{2+}]$

$=$ molar solubility of Hg_2Cl_2 in 0.10 M NaCl

(c) Concentration ($mol \cdot L^{-1}$) $PbCl_2(s) \rightleftharpoons Pb^{2+}(aq) + 2\ Cl^-(aq)$

	$PbCl_2(s)$	$Pb^{2+}(aq)$	$2\ Cl^-(aq)$
initial	—	0	$2 \times 0.10 = 0.20$
change	—	$+S$	$+S$
equilibrium	—	S	$S + 0.20$

$K_{sp} = [Pb^{2+}][Cl^-]^2 = S \times (S + 0.20)^2 = 1.6 \times 10^{-5}$

S may not be negligible relative to 0.20, so the full cubic form may be required. We do it both ways:

For $S^3 + 0.40\ S^2 + (4 \times 10^{-2}S) - (1.6 \times 10^{-5}) = 0$, the solution by standard methods is $S = 4.0 \times 10^{-4}\ mol \cdot L^{-1}$.

If S had been neglected, the answer would have been the same, 4.0×10^{-4}, to within two significant figures.

(d) Concentration ($mol \cdot L^{-1}$) $Fe(OH)_2(s) \rightleftharpoons Fe^{2+}(aq) + 2\ OH^-(aq)$

	$Fe(OH)_2(s)$	$Fe^{2+}(aq)$	$2\ OH^-(aq)$
initial	—	1.0×10^{-4}	0
change	—	$+S$	$+2S$
equilibrium	—	$1.0 \times 10^{-4} + S$	$2S$

$K_{sp} = [Fe^{2+}][OH^-]^2 = (S + 1.0 \times 10^{-4}) \times (2S)^2 = 1.6 \times 10^{-14}$

Assume S in $(S + 1.0 \times 10^{-4})$ is negligible, so $4S^2 \times 1.0 \times 10^{-4} = 1.6 \times 10^{-14}$

$S^2 = 4.0 \times 10^{-11}$

$S = 6.3 \times 10^{-6}\ mol \cdot L^{-1} =$ molar solubility of $Fe(OH)_2$ in 1.0×10^{-4} M $FeCl_2$

11.59 (a) $Ag^+(aq) + Cl^-(aq) \rightleftharpoons AgCl(s)$

Concentration ($mol \cdot L^{-1}$)	Ag^+	Cl^-
initial	0	1.0×10^{-5}
change	$+x$	0
equilibrium	x	1.0×10^{-5}

$K_{sp} = [Ag^+][Cl^-] = 1.6 \times 10^{-10} = (x)(1.0 \times 10^{-5})$

$x = [Ag^+] = 1.6 \times 10^{-5}\ mol \cdot L^{-1}$

(b) mass $AgNO_3$

$$= \left(\frac{1.6 \times 10^{-5}\ mol\ AgNO_3}{1\ L}\right)(0.100\ L)\left(\frac{169.88\ g\ AgNO_3}{1\ mol\ AgNO_3}\right)\left(\frac{1\ \mu g}{10^{-6}\ g}\right)$$

$= 2.7 \times 10^2\ \mu g\ AgNO_3$

11.61 $Ni^{2+}(aq) + 2\ OH^-(aq) \rightleftharpoons Ni(OH)_2(s)$

Concentration ($mol \cdot L^{-1}$)	Ni^{2+}	OH^-
initial	0.010	0
change	0	$+x$
equilibrium	0.010	x

$K_{sp} = [Ni^{2+}][OH^-]^2 = 6.5 \times 10^{-18} = (0.010)(x)^2$

$$[OH^-] = x = \sqrt{\frac{6.5 \times 10^{-18}}{0.010}} = 2.5 \times 10^{-8}\ mol \cdot L^{-1}$$

$pOH = -\log(2.5 \times 10^{-8}) = 7.60, \quad pH = 14.00 - 7.60 = 6.40$

11.63 (a) $Ag^+(aq) + Cl^-(aq) \rightleftharpoons AgCl(s), \quad [Ag^+][Cl^-] = K_{sp}$

$$Q_{sp} = \left[\frac{(0.073\ L)(0.0040\ mol \cdot L^{-1})}{0.100\ L}\right]\left[\frac{(0.027\ L)(0.0010\ mol \cdot L^{-1})}{0.100\ L}\right]$$

$= (2.9 \times 10^{-3})(2.7 \times 10^{-4}) = 7.9 \times 10^{-7}$

Will precipitate, because $Q_{sp}\ (7.9 \times 10^{-7}) > K_{sp}\ (1.6 \times 10^{-10})$

(b) $Ca^{2+}(aq) + SO_4^{\ 2-}(aq) \rightleftharpoons CaSO_4(s), \quad [Ca^{2+}][SO_4^{\ 2-}] = K_{sp}$

$$Q_{sp} = \left[\frac{(0.0100\ L)(0.0030\ mol \cdot L^{-1})}{0.111\ L}\right]\left[\frac{(0.0010\ L)(1.0\ mol \cdot L^{-1})}{0.111\ L}\right]$$

$= (2.7 \times 10^{-4})(9.0 \times 10^{-3}) = 2.4 \times 10^{-6}$

Will not precipitate, because $Q_{sp}\ (2.4 \times 10^{-6}) < K_{sp}(2.4 \times 10^{-5})$

11.65 $\left(\frac{1\ mL}{20\ drops}\right) \times 1\ drop = 0.05\ mL = 0.05 \times 10^{-3}\ L = 5 \times 10^{-5}\ L$

and $(5 \times 10^{-5}\ L)(0.010\ mol \cdot L^{-1}) = 5 \times 10^{-7}\ mol\ NaCl = 5 \times 10^{-7}\ mol\ Cl^-$

(a) $Ag^+(aq) + Cl^-(aq) \rightleftharpoons AgCl(s), \quad [Ag^+][Cl^-] = K_{sp}$

$$Q_{sp} = \left[\frac{(0.010\ L)(0.0040\ mol \cdot L^{-1})}{0.010\ L}\right]\left[\frac{5 \times 10^{-7}\ mol}{0.010\ L}\right] = 2 \times 10^{-7}$$

Will precipitate, because $Q_{sp}\ (2 \times 10^{-7}) > K_{sp}\ (1.6 \times 10^{-10})$

(b) $Pb^{2+}(aq) + 2\,Cl^-(aq) \rightleftharpoons PbCl_2(s)$, $[Pb^{2+}][Cl^-]^2 = K_{sp}$

$$Q_{sp} = \left[\frac{(0.0100\ \text{L})(0.0040\ \text{mol}\cdot\text{L}^{-1})}{0.010\ \text{L}}\right]\left[\frac{5 \times 10^{-7}\ \text{mol}}{0.010\ \text{L}}\right]^2 = 1 \times 10^{-11}$$

Will not precipitate, because Q_{sp} $(1 \times 10^{-11}) < K_{sp}$ (1.6×10^{-5})

11.67 (a) $K_{sp}[Ni(OH)_2] < K_{sp}[Mg(OH)_2] < K_{sp}[Ca(OH)_2]$

This is the order for the solubility products of these hydroxides. Thus, order of precipitation is (first to last): $Ni(OH)_2$, $Mg(OH)_2$, $Ca(OH)_2$

(b) $K_{sp}[Ni(OH)_2] = 6.5 \times 10^{-18} = [Ni^{2+}][OH^-]^2$

$$[OH^-]^2 = \frac{6.5 \times 10^{-18}}{0.0010} = 6.5 \times 10^{-15}$$

$[OH^-] = 8.1 \times 10^{-8}$

$pOH = -\log[OH^-] = 7.09 \quad pH \approx 7$

$K_{sp}[Mg(OH)_2] = 1.1 \times 10^{-11} = [Mg^{2+}][OH^-]^2$

$$[OH^-] = \sqrt{\frac{1.1 \times 10^{-11}}{0.0010}} = 1.0 \times 10^{-4}$$

$pOH = -\log(1.0 \times 10^{-4}) = 4.00 \quad pH = 14.00 - 4.00 = 10.00,\ pH \approx 10$

$K_{sp}[Ca(OH)^2] = 5.5 \times 10^{-6} = [Ca^{2+}][OH^-]^2$

$$[OH^-] = \sqrt{\frac{5.5 \times 10^{-6}}{0.0010}} = 7.4 \times 10^{-2}$$

$pOH = -\log(7.4 \times 10^{-2}) = 1.13 \quad pH = 14.00 - 1.13 = 12.87,\ pH \approx 13$

11.69 We wish to separate magnesium ions and barium ions by selective precipitation. Which anion, fluoride or carbonate, would be the better choice for achieving this precipitation? Why?

The K_{sp} values are:

The K_{sp} values are:	MgF_2	6.4×10^{-9}
	BaF_2	1.7×10^{-6}
	$MgCO_3$	1.0×10^{-5}
	$BaCO_3$	8.1×10^{-9}

The difference in these numbers suggests that there is a greater solubility difference between the carbonates, and thus this anion should give a better separation. Because different numbers of ions are involved, it is instructive to convert the K_{sp} values into molar solubility. For the fluorides the reaction is

$$MF_2(s) \rightleftharpoons M^{2+}(aq) + 2\,F^-(aq)$$

Change $\qquad +x \qquad +2x$

$K_{sp} = x(2x)^2$

Solving this for MgF_2 gives 0.0012 M and for BaF_2 gives 0.0075 M.

For the carbonates:

$$MCO_3(s) \rightleftharpoons M^{2+}(aq) + CO_3^{2-}(aq)$$
$$+x \qquad\qquad +x$$

$K_{sp} = x^2$

Solving this for $MgCO_3$ gives 0.0032 M and for $BaCO_3$ gives 9.0×10^{-5} M.

Clearly, the solubility difference is greatest between the two carbonates, and CO_3^{2-} is the better choice of anion.

11.71 $Cu(IO_3)_2$ ($K_{sp} = 1.4 \times 10^{-7}$) is more soluble than $Pb(IO_3)_2$ ($K_{sp} = 2.6 \times 10^{-13}$) so Cu^{2+} will remain in solution until essentially all the $Pb(IO_3)_2$ has precipitated. Thus, we expect very little Pb^{2+} to be left in solution by the time we reach the point at which $Cu(IO_3)_2$ begins to precipitate.

The concentration of IO_3^- at which Cu^{2+} begins to precipitate will be given by

$K_{sp} = [Cu^{2+}][IO_3^-]^2 = 1.4 \times 10^{-7} = [0.0010][IO_3^-]^2$

$[IO_3^-] = 0.012\ \text{mol}\cdot\text{L}^{-1}$

The concentration of Pb in solution when the $[IO_3^-] = 0.012\ \text{mol}\cdot\text{L}^{-1}$ is given by

$K_{sp} = [Pb^{2+}][IO_3^-]^2 = 2.6 \times 10^{-13} = [Pb^{2+}][0.012]^2$

$[Pb^{2+}] = 1.8 \times 10^{-9}\ \text{mol}\cdot\text{L}^{-1}$

11.73 (a) pH = 7.0; $[OH^-] = 1.0 \times 10^{-7}\ \text{mol}\cdot\text{L}^{-1}$

$$Al^{3+}(aq) + 3\ OH^-(aq) \rightleftharpoons Al(OH)_3(s)$$

$[Al^{3+}][OH^-]^3 = K_{sp} = 1.0 \times 10^{-33}$

$S \times (10^{-7})^3 = 1.0 \times 10^{-33}$

$$S = \frac{1.0 \times 10^{-33}}{1 \times 10^{-21}} = 1.0 \times 10^{-12}\ \text{mol}\cdot\text{L}^{-1} = [Al^{3+}]$$

= molar solubility of $Al(OH)_3$ at pH = 7.0

(b) pH = 4.5; pOH = 9.5; $[OH^-] = 3.2 \times 10^{-10}\ \text{mol}\cdot\text{L}^{-1}$

$[Al^{3+}][OH^-]^3 = K_{sp} = 1.0 \times 10^{-33}$

$S \times (3.2 \times 10^{-10})^3 = 1.0 \times 10^{-33}$

$$S = \frac{1.0 \times 10^{-33}}{3.3 \times 10^{-29}} = 3.1 \times 10^{-5}\ \text{mol}\cdot\text{L}^{-1} = [Al^{3+}]$$

= molar solubility of $Al(OH)_3$ at pH = 4.5

(c) pH = 7.0; $[OH^-] = 1.0 \times 10^{-7}\ \text{mol}\cdot\text{L}^{-1}$

$$Zn^{2+}(aq) + 2\ OH^-(aq) \rightleftharpoons Zn(OH)_2(s)$$

$[Zn^{2+}][OH^-]^2 = K_{sp} = 2.0 \times 10^{-17}$

$S \times (1.0 \times 10^{-7})^2 = 2.0 \times 10^{-17}$

$$S = \frac{2.0 \times 10^{-17}}{1.0 \times 10^{-14}} = 2.0 \times 10^{-3}\ \text{mol} \cdot \text{L}^{-1} = [\text{Zn}^{2+}]$$

$= \text{molar solubility of } Zn(OH)_2 \text{ at pH} = 7.0$

(d) $\text{pH} = 6.0$; $\text{pOH} = 8.0$; $[\text{OH}^-] = 1.0 \times 10^{-8}\ \text{mol} \cdot \text{L}^{-1}$

$[\text{Zn}^{2+}][\text{OH}^-]^2 = 2.0 \times 10^{-17} = K_{sp}$

$S \times (1.0 \times 10^{-8})^2 = 2.0 \times 10^{-17}$

$$S = \frac{2.0 \times 10^{-17}}{1.0 \times 10^{-16}} = 2.0 \times 10^{-1} = 0.20\ \text{mol} \cdot \text{L}^{-1} = [\text{Zn}^{2+}]$$

$= \text{molar solubility of } Zn(OH)_2 \text{ at pH} = 6.0$

11.75 $CaF_2(s) \rightleftharpoons Ca^{2+}(aq) + 2F^-(aq) \qquad K_{sp} = 4.0 \times 10^{-11}$

$F^-(aq) + H_2O(l) \rightleftharpoons HF(aq) + OH^-(aq) \qquad K_b(F^-) = 2.9 \times 10^{-11}$

(a) Multiply the second equilibrium equation by 2 and add to the first equilibrium:

$CaF_2(s) + 2\,H_2O(l) \rightleftharpoons Ca^{2+}(aq) + 2\,HF(aq) + 2\,OH^-(aq)$

$K = K_w \cdot K_b{}^2 = (4.0 \times 10^{-11})(2.9 \times 10^{-11})^2 = 3.4 \times 10^{-32}$

(b) The calculation of K_{sp} is complicated by the fact that the anion of the salt is part of a weak base-acid pair. If we wish to solve the equation algebraically, then we need to consider which equilibrium is the dominant one at pH = 7.0, for which $[H_3O^+] = 1 \times 10^{-7}$.

To determine whether F^- or HF is the dominant species at this pH (if either), consider the base hydrolysis reaction:

$F^-(aq) + H_2O(l) \rightleftharpoons HF(aq) + OH^-(aq) \qquad K_b = 2.9 \times 10^{-11}$

$$K_b = \frac{[\text{HF}][\text{OH}^-]}{[\text{F}^-]}$$

$$2.9 \times 10^{-11} = \frac{[\text{HF}][1 \times 10^{-7}]}{[\text{F}^-]}$$

$$\frac{[\text{HF}]}{[\text{F}^-]} = \frac{2.9 \times 10^{-11}}{1 \times 10^{-7}} = 3 \times 10^{-4}$$

Given that the ratio of HF to F^- is on the order of 10^{-4} to 1, the F^- species is still dominant. The appropriate equation to use is thus the original one for the K_{sp} of $CaF_2(s)$:

$CaF_2(s) \rightleftharpoons Ca^{2+}(aq) + 2\,F^-(aq) \qquad K_{sp} = 4.0 \times 10^{-11}$

$K_{sp} = [Ca^{2+}][F^-]^2$

$4.0 \times 10^{-11} = x(2x)^2 = 4x^3$

$x = 2.2 \times 10^{-4}$

molar solubility $= 2.2 \times 10^{-4}\ \text{mol} \cdot \text{L}^{-1}$

(c) At pH = 3.0, $[H_3O^+] = 1 \times 10^{-3}\ mol \cdot L^{-1}$ and $[OH^-] = 1 \times 10^{-11}\ mol \cdot L^{-1}$. Under these conditions

$$K_b = \frac{[HF][OH^-]}{[F^-]}$$

$$2.9 \times 10^{-11} = \frac{[HF][1 \times 10^{-11}]}{[F^-]}$$

$$\frac{[HF]}{[F^-]} = \frac{2.9 \times 10^{-11}}{1 \times 10^{-11}} = 3$$

$$[HF] = 3\ [F^-]$$

As can be seen, at pH 3.0 the amounts of F^- and HF are comparable, so the protonation of F^- to form HF cannot be ignored. The relation $2\ [Ca^{2+}] = [F^-] + [HF]$ is required by the mass balance as imposed by the stoichiometry of the dissolution equilibrium.

$$2\ [Ca^{2+}] = [F^-] + 3\ [F^-]$$

$$2\ [Ca^{2+}] = 4\ [F^-]$$

$$[Ca^{2+}] = 2\ [F^-]$$

Using this with K_{sp} relation:

$$(2\ [F^-])[F^-]^2 = 4.0 \times 10^{-11}$$

$$2\ [F^-]^3 = 4.0 \times 10^{-11}$$

$$[F^-] = 2 \times 10^{-4}$$

$$[Ca^{2+}] = (2)(2 \times 10^{-4}) = 4 \times 10^{-4}\ mol \cdot L^{-1}$$

The solubility is about double that at pH = 7.0.

11.77

$AgBr(s) \rightleftharpoons Ag^+(aq) + Br^-(aq)$	$K_{sp} = 7.7 \times 10^{-13}$
$Ag^+(aq) + 2\ CN^-(aq) \rightleftharpoons Ag(CN)_2^-(aq)$	$K_f = 5.6 \times 10^8$
$AgBr(s) + 2\ CN^-(aq) \rightleftharpoons Ag(CN)_2^-(aq) + Br^-(aq)$	$K = 4.3 \times 10^{-4}$

Hence, $K = \dfrac{[Ag(CN)_2^-][Br^-]}{[CN^-]^2} = 4.3 \times 10^{-4}$

Concentration $(mol \cdot L^{-1})$	$AgBr(s)$	$+\ 2\ CN^-(aq)$	$\rightleftharpoons Ag(CN)_2^-(aq)$	$+\ Br^-(aq)$
initial	—	0.10	0	0
change	—	$-2S$	$+S$	$+S$
equilibrium	—	$0.10 - 2S$	S	S

$$\frac{[Ag(CN)_2^-][Br^-]}{[CN^-]^2} = \frac{S^2}{(0.10 - 2S)^2} = 4.3 \times 10^{-4}$$

$$\frac{S}{0.10 - 2S} = \sqrt{4.3 \times 10^{-4}} = 2.1 \times 10^{-2}$$

$S = (2.1 \times 10^{-3}) - (4.2 \times 10^{-2}S)$

$1.042S = 2.1 \times 10^{-3}$

$S = 2.0 \times 10^{-3}\ \text{mol}\cdot\text{L}^{-1}$ = molar solubility of AgBr

11.79 The two salts can be distinguished by their solubility in NH_3. The equilibria that are pertinent are

$AgCl(s) + 2\ NH_3(aq) \rightleftharpoons Ag(NH_3)_2^+(aq) + Cl^-(aq) \quad K = K_{sp}\cdot K_f = 2.6 \times 10^{-3}$

$AgI(s) + 2\ NH_3(aq) \rightleftharpoons Ag(NH_3)_2^+(aq) + I^-(aq) \quad K = K_{sp}\cdot K_f = 2.4 \times 10^{-9}$

For example, let's consider the solubility of these two salts in 1.00 M NH_3 solution:

For AgCl $K = \dfrac{[Ag(NH_3)_2^+][Cl^-]}{[NH_3]^2} = 2.56 \times 10^{-3}$

Concentration ($\text{mol}\cdot\text{L}^{-1}$)	AgCl(s) +	2 HN_3(aq) $\rightleftharpoons$	$Ag(NH_3)_2^+$(aq) +	Cl^-(aq)
initial	—	1.00	0	0
change	—	$-2x$	$+x$	$+x$
final	—	$1.00 - 2x$	$+x$	$+x$

$$K = \frac{[Ag(NH_3)_2^+][Cl^-]}{[NH_3]^2} = 2.6 \times 10^{-3}$$

$$2.6 \times 10^{-3} = \frac{[x][x]}{[1.00 - 2x]^2} = \frac{x^2}{[1.00 - 2x]^2}$$

$$0.051 = \frac{x}{1.00 - 2x}$$

$x = 0.046$

0.046 mol AgCl will dissolve in 1.00 L of aqueous solution. The molar mass of AgCl is 143.32 $\text{g}\cdot\text{mol}^{-1}$; this corresponds to 0.046 $\text{mol}\cdot\text{L}^{-1} \times 142.32\ \text{g}\cdot\text{mol}^{-1}$ = 6.5 $\text{g}\cdot\text{L}^{-1}$.

For AgI, the same calculation gives $x = 4.9 \times 10^{-5}\ \text{mol}\cdot\text{L}^{-1}$. The molar mass of AgI is 234.77 $\text{g}\cdot\text{mol}^{-1}$, giving a solubility of $4.9 \times 10^{-5}\ \text{mol}\cdot\text{L}^{-1} \times 234.77\ \text{g}\cdot\text{mol}^{-1} = 0.023\ \text{g}\cdot\text{L}^{-1}$.

Thus, we could treat a 0.10 g sample of the compound with 20.0 mL of 1.00 M NH_3. The AgCl would all dissolve, whereas practically none of the AgI would. Note: AgI is also slightly yellow in color, whereas AgCl is white, so an initial distinction could be made based upon the color of the sample.

11.81 In order to use qualitative analyses, the sample must first be dissolved. This can be accomplished by digesting the sample with concentrated HNO_3 and then diluting the resulting solution. HCl or H_2SO_4 could not be used, because some of the metal compounds formed would be insoluble, whereas all of the nitrates would dissolve. Once the sample is dissolved and diluted, an aqueous solution containing chloride ions can be introduced. This should precipitate the Ag^+ as AgCl but would leave the bismuth and nickel in solution, as long as the solution was acidic. The remaining solution can then be treated with H_2S. In acidic solution, Bi_2S_3 will precipitate but NiS will not. Once the Bi_2S_3 has been precipitated, the pH of the solution can be raised by addition of base. Once this is done, NiS should precipitate.

11.83 Let NaB = sodium barbiturate, HB = barbituric acid

$$[B^-] = [NaB] = \left(\frac{10.0\ mg}{250\ mL}\right)\left(\frac{1\ mL}{10^{-3}\ L}\right)\left(\frac{10^{-3}\ g}{1\ mg}\right)\left(\frac{1\ mol}{150\ g}\right)$$

$$= 2.67 \times 10^{-4}\ mol\cdot L^{-1}$$

$pH = 7.71$, $pOH = 6.29$, $[OH^-] = 5.0 \times 10^{-7}\ mol\cdot L^{-1}$

Concentration ($mol\cdot L^{-1}$)	$B^-(aq)$	+	$H_2O(l)$	$\rightleftharpoons$	$OH^-(aq)$	+	$HB(aq)$
initial	2.67×10^{-4}		—		0		0
change	-5.0×10^{-7}		—		$+5.0 \times 10^{-7}$		$+5.0 \times 10^{-7}$
equilibrium	$2.67 \times 10^{-4} - 5.0 \times 10^{-7}$		—		5.0×10^{-7}		5.0×10^{-7}

Assume $[OH^-] = [HB]$

(a) $\%\ \text{protonation} = \dfrac{[HB]}{[B^-]_0} \times 100\% = \dfrac{5.0 \times 10^{-7}}{2.67 \times 10^{-4}} \times 100\% = 0.19\%$

(b) $K_b = \dfrac{K_w}{K_a} = \dfrac{[OH^-][HB]}{[B^-]}$

$$K_a = \frac{K_w[B^-]}{[OH^-][HB]} = \frac{(1.0 \times 10^{-14})(2.67 \times 10^{-4} - 5.0 \times 10^{-7})}{(5.0 \times 10^{-7})^2} = 1.1 \times 10^{-5}$$

11.85 The presence of K^+ and Cl^- ions does not affect the pH. Total volume of solution = 200 mL + 150 mL = 350 mL.

moles of $Na_3PO_4 = 0.20\ L \times 0.27\ mol\cdot L^{-1} = 5.4 \times 10^{-2}$ mol

$$\text{molar concentration of } Na_3PO_4 = \frac{5.4 \times 10^{-2}\ mol}{0.350\ L} = 0.15\ mol\cdot L^{-1}$$

Concentration ($mol\cdot L^{-1}$)	$H_2O(l)$	+	$PO_4^{3-}(aq)$	$\rightleftharpoons$	$HPO_4^{2-}(aq)$	+	$OH^-(aq)$
initial	—		0.15		0		0
change	—		$-x$		$+x$		$+x$
equilibrium	—		$0.15 - x$		x		x

$$K_b = \frac{K_w}{K_a} = \frac{[HPO_4^{2-}][OH^-]}{[PO_4^{3-}]} = \frac{x^2}{0.15 - x}$$

$$\frac{1.0 \times 10^{-14}}{4.2 \times 10^{-13}} = \frac{x^2}{0.15 - x} = 2.4 \times 10^{-2} \quad (x \text{ cannot be neglected})$$

$$x^2 + (2.4 \times 10^{-2}x) - (3.6 \times 10^{-3}) = 0$$

Solving this quadratic yields $[OH^-] = x = 4.9 \times 10^{-2}\ mol \cdot L^{-1}$; pOH $= -\log(4.9 \times 10^{-2}) = 1.31$; pH $= 14.00 - 1.31 = 12.69$.

11.87 The strong acid, HCl, will protonate the HCO_2^- ion.

moles of HCl $= 0.0040\ L \times 0.070\ mol \cdot L^{-1} = 2.8 \times 10^{-4}$ mol HCl (H^+)

moles of $HCO_2^- = 0.0600\ L \times 0.10\ mol \cdot L^{-1} = 6.0 \times 10^{-3}$ mol HCO_2^-

After protonation, there are $(6.0 - 0.28) \times 10^{-3}$ mol HCO_2^- = 5.7×10^{-3} mol HCO_2^- and 2.8×10^{-4} mol HCOOH.

$$[HCO_2^-] = \frac{5.7 \times 10^{-3}\ mol}{0.0640\ L} = 8.9 \times 10^{-2}\ mol \cdot L^{-1}$$

$$[HCOOH] = \frac{2.8 \times 10^{-4}\ mol}{0.0640\ L} = 4.4 \times 10^{-3}\ mol \cdot L^{-1}$$

Concentration ($mol \cdot L^{-1}$)	$HCOOH(aq)$ +	$H_2O(l) \rightleftharpoons$	$H_3O^+(aq)$ +	$HCO_2^-(aq)$
initial	4.4×10^{-3}	—	0	8.9×10^{-2}
change	$-x$	—	$+x$	$+x$
equilibrium	$4.4 \times 10^{-3} - x$	—	x	$8.9 \times 10^{-2} + x$

$$K_a = 1.8 \times 10^{-4} = \frac{[H_3O^+][HCO_2^-]}{[HCOOH]} = \frac{(x)(8.9 \times 10^{-2} + x)}{(4.4 \times 10^{-3} - x)}$$

Assume that $+x$ and $-x$ are negligible; so $1.8 \times 10^{-4} = \frac{(8.9 \times 10^{-2}) \times (x)}{4.4 \times 10^{-3}}$

$= 20.2x$

$[H_3O^+] = x = 8.9 \times 10^{-6}\ mol \cdot L^{-1}$

$[HCOOH] = (4.4 \times 10^{-3}) - (8.9 \times 10^{-6}) \approx 4.4 \times 10^{-3}\ mol \cdot L^{-1}$

pH $= -\log(8.9 \times 10^{-6}) = 5.05$

11.89 (a) pH $= -\log(0.020) = 1.70$

(b) initial moles of $H_3O^+ = 0.020\ L \times 0.020\ mol \cdot L^{-1} = 4.0 \times 10^{-4}$ mol

moles of OH^- added $= 0.005\,00\ L \times 0.035\ mol \cdot L^{-1} = 1.8 \times 10^{-4}$ mol

excess $H_3O^+ = (4.0 \times 10^{-4} - 1.8 \times 10^{-4})$ mol $= 2.2 \times 10^{-4}$ mol

$$[H_3O^+] = \frac{2.2 \times 10^{-4}\ mol}{(0.020 + 0.0050)\ L} = 8.8 \times 10^{-3}\ mol \cdot L^{-1}$$

pH $= -\log(8.8 \times 10^{-3}) = 2.06$

(c) moles of OH^- added $= 2 \times 1.8 \times 10^{-4}$ mol $= 3.6 \times 10^{-4}$ mol

excess $H_3O^+ = (4.0 \times 10^{-4} - 3.6 \times 10^{-4})$ mol $= 4 \times 10^{-5}$ mol

$$[H_3O^+] = \frac{4 \times 10^{-5}\ \text{mol}}{0.030\ \text{L}} = 1 \times 10^{-3}\ \text{mol} \cdot \text{L}^{-1}$$

$pH = -\log(1 \times 10^{-3}) = 3.0$

(d) moles of OH^- added $= 3 \times 1.8 \times 10^{-4} = 5.4 \times 10^{-4}$ mol

excess $OH^- = (5.4 \times 10^{-4} - 4.0 \times 10^{-4})$ mol $= 1.4 \times 10^{-4}$ mol

$$[OH^-] = \frac{1.4 \times 10^{-4}\ \text{mol}}{0.035\ \text{L}} = 4.0 \times 10^{-3}\ \text{mol} \cdot \text{L}^{-1}$$

$pOH = -\log(4.0 \times 10^{-3}) = 2.40,\ \ pH = 14.00 - 2.40 = 11.60$

(e) moles of OH^- added $= 4 \times 1.8 \times 10^{-4}$ mol $= 7.2 \times 10^{-4}$ mol

excess $OH^- = (7.2 \times 10^{-4} - 4.0 \times 10^{-4})$ mol $= 3.2 \times 10^{-4}$ mol

$$[OH^-] = \frac{3.2 \times 10^{-4}\ \text{mol}}{0.040\ \text{L}} = 8.0 \times 10^{-3}\ \text{mol} \cdot \text{L}^{-1}$$

$pOH = -\log(8.0 \times 10^{-3}) = 2.10,\ \ pH = 14.00 - 2.10 = 11.90$

(f) $$\text{volume of KOH} = (4.0 \times 10^{-4}\ \text{mol HCl})\left(\frac{1\ \text{mol KOH}}{1\ \text{mol HCl}}\right)\left(\frac{1\ \text{L KOH}}{0.035\ \text{mol KOH}}\right)$$

$$= 0.011\ \text{L KOH} = 11\ \text{mL KOH}$$

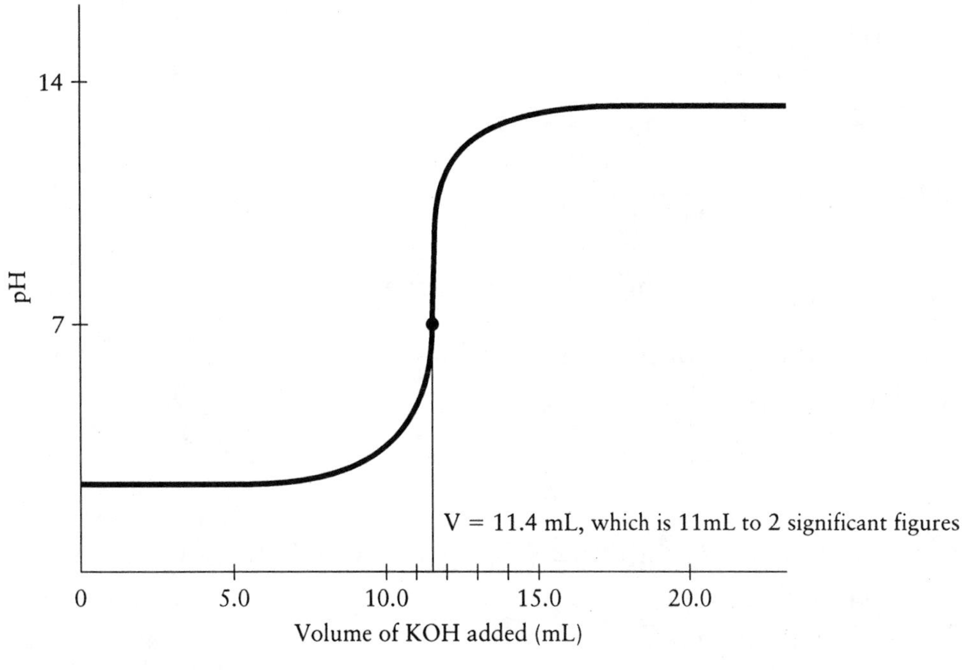

11.91 $HBrO(aq) + OH^-(aq) \rightleftharpoons H_2O(l) + BrO^-(aq)$

(a) at the stoichiometric point, moles of OH^- = moles of HBrO

$$\text{volume NaOH} = \frac{(0.010\ 00\ \text{L})(0.0633\ \text{mol} \cdot \text{L}^{-1})}{0.0400\ \text{mol} \cdot \text{L}^{-1}}$$

1.58×10^{-2} L NaOH

(b) at the stoichiometric point,

$$\frac{\text{mol BrO}^-}{\text{volume}} = \frac{(0.010\,00\text{ L})(0.0633\text{ mol}\cdot\text{L}^{-1})}{0.0258\text{ L}} = 2.5 \times 10^{-2}\text{ mol}\cdot\text{L}^{-1}$$

Concentration $(\text{mol}\cdot\text{L}^{-1})$	$BrO^-(aq)$ +	$H_2O(l)$ ⇌	$HBrO(aq)$ +	$OH^-(aq)$
initial	2.5×10^{-2}	—	0	0
change	$-x$	—	$+x$	$+x$
equilibrium	$2.5 \times 10^{-2} - x$	—	x	x

$$K_b = \frac{K_w}{K_a} = \frac{1.0 \times 10^{-14}}{2.0 \times 10^{-9}} = 5.0 \times 10^{-6} = \frac{x^2}{2.5 \times 10^{-2} - x} \approx \frac{x^2}{2.5 \times 10^{-2}}$$

$[OH^-] = x = 3.5 \times 10^{-4}\text{ mol}\cdot\text{L}^{-1}$

$\text{pOH} = -\log(3.5 \times 10^{-4}) = 3.46, \quad \text{pH} = 14.00 - 3.46 = 10.54$

(c) alizarin yellow R, $pK_{in} = 11.2$

11.93 Let novocaine = N; $N(aq) + H_2O(l) \rightleftharpoons HN^+(aq) + OH^-(aq)$

$$K_b = \frac{[HN^+][OH^-]}{[N]}$$

$pK_a = pK_w - pK_b = 14.00 - 5.05 = 8.95$

$$\text{pH} = pK_a + \log\left(\frac{[N]}{[HN^+]}\right)$$

$$\log\left(\frac{[N]}{[HN^+]}\right) = \text{pH} - pK_a = 7.4 - 8.95 = -1.55$$

Therefore, the ratio of the concentrations of novocaine and its conjugate acid is $[N]/[HN^+] = 10^{-1.55} = 2.8 \times 10^{-2}$.

11.95 $CH_3COOH(aq) + H_2O(l) \rightleftharpoons H_3O^+(aq) + CH_3CO_2^-(aq)$

$$K_a = \frac{[H_3O^+][CH_3CO_2^-]}{[CH_3COOH]}$$

$$\text{pH} = pK_a + \log\frac{[CH_3CO_2^-]}{[CH_3COOH]} = pK_a + \log\left(\frac{0.50}{0.150}\right)$$

$= 4.75 + 0.52 = 5.27$ (initial pH)

(a) $(0.0100\text{ L})(1.2\text{ mol}\cdot\text{L}^{-1}) = 1.2 \times 10^{-2}$ mol HCl added (a strong acid)

Produces 1.2×10^{-2} mol CH_3COOH from $CH_3CO_2^-$ after adding HCl:

$$[CH_3COOH] = \frac{(0.100\ L)(0.150\ mol\cdot L^{-1})}{0.110\ L} + \frac{1.2 \times 10^{-2}\ mol}{0.110\ L}$$
$$= 0.245\ mol\cdot L^{-1}$$
$$[CH_3CO_2^-] = \frac{(0.100\ L)(0.50\ mol\cdot L^{-1})}{0.110\ L} - \frac{1.2 \times 10^{-2}\ mol}{0.110\ L}$$
$$= 0.345\ mol\cdot L^{-1}$$
$$pH = 4.75 + \log\left(\frac{0.345}{0.245}\right) = 4.75 + 0.15 = 4.90 \text{ (after adding HCl)}$$

(b) $(0.0500\ L)(0.095\ mol\cdot L^{-1}) = 4.7 \times 10^{-3}$ mol NaOH added (a strong base)
Produces 4.7×10^{-3} mol $CH_3CO_2^-$ from CH_3COOH after adding NaOH:

$$[CH_3COOH] = \frac{(0.100\ L)(0.150\ mol\cdot L^{-1})}{0.1500\ L} - \frac{4.7 \times 10^{-3}\ mol}{0.150\ L}$$
$$= 6.9 \times 10^{-2}\ mol\cdot L^{-1}$$
$$[CH_3CO_2^-] = \frac{(0.100\ L)(0.50\ mol\cdot L^{-1})}{0.150\ L} + \frac{4.7 \times 10^{-3}\ mol}{0.150\ L}$$
$$= 3.6 \times 10^{-1}\ mol\cdot L^{-1}$$
$$pH = 4.75 + \log\left(\frac{3.6 \times 10^{-1}}{6.9 \times 10^{-2}}\right) = 4.75 + 0.72 = 5.47 \quad \text{(after adding base)}$$

11.97 Equilibrium equation: $HCO_3^-(aq) + H_2O(l) \rightleftharpoons H_3O^+(aq) + CO_3^{2-}(aq)$

$$K_a = \frac{[H_3O^+][CO_3^{2-}]}{[HCO_3^-]}$$
$$pH = pK_a + \log\left(\frac{[CO_3^{2-}]}{[HCO_3^-]}\right)$$
$$10 = -\log(5.6 \times 10^{-11}) + \log\left(\frac{[CO_3^{2-}]}{[HCO_3^-]}\right)$$
$$\log\left(\frac{[CO_3^{2-}]}{[HCO_3^-]}\right) = 10 - 10.25 = -0.25$$
$$\frac{[CO_3^{2-}]}{[HCO_3^-]} = 0.56$$

Then prepare a solution containing Na_2CO_3 and $NaHCO_3$ in a molar ratio of 0.56:1.

11.99 $CaF_2(s) \rightleftharpoons Ca^{2+}(aq) + 2\ F^-(aq)$

$[Ca^{2+}][F^-]^2 = 4.0 \times 10^{-11}$

$[Ca^{2+}](5 \times 10^{-5})^2 = 4.0 \times 10^{-11} \quad [Ca^{2+}] = 0.016\ mol\cdot L^{-1}$

The maximum concentration of $[Ca^{2+}]$ allowed will be $0.016\ mol\cdot L^{-1}$.

11.101 moles of Ag^+ = (0.100 L)(1.0×10^{-4} mol·L^{-1}) = 1.0×10^{-5} mol Ag^+

moles of CO_3^{2-} = (0.100 L)(1.0×10^{-4} mol·L^{-1}) = 1.0×10^{-5} mol CO_3^{2-}

$Ag_2CO_3(s) \rightleftharpoons 2\,Ag^+(aq) + CO_3^{2-}(aq)$

$[Ag^+]^2[CO_3^{2-}] = Q_{sp}$

$$\left(\frac{1.0 \times 10^{-5}}{0.200}\right)^2\left(\frac{1.0 \times 10^{-5}}{0.200}\right) = 1.3 \times 10^{-13} = Q_{sp}$$

Because Q_{sp} (1.3×10^{-13}) is less than K_{sp} (6.2×10^{-12}), no precipitate will form.

11.103 The K_{sp} values from Tables 11.5 and 11.7 are Cu^{2+}, 1.3×10^{-36}; Co^{2+}, 5×10^{-22}; Cd^{2+}, 4×10^{-29}. All the salts have the same expression for K_{sp}, $K_{sp} = [M^{2+}][S^{2-}]$, so the compound with the smallest K_{sp} will precipitate first, in this case CuS.

11.105 (a) First, calculate the stoichiometry of the reaction of the strong base with the weak acid, according to the equation:

$H_2SO_3(aq) + OH^-(aq) \longrightarrow HSO_3^-(aq) + H_2O(l)$

mol OH^- = (0.0250 L)(0.150 mol·L^{-1}) = 0.003 75 mol

mol H_2SO_3 = (0.0250 L)(0.125 mol·L^{-1}) = 0.003 12 mol

Because there is more OH^- than H_2SO_3, all of the H_2SO_3 will react to form HSO_3^-, with excess OH^- leftover.

0.003 12 mol of H_2SO_3 will form 0.003 12 mol HSO_3^-

mol OH^- remaining = 0.003 75 mol − 0.003 12 mol =
0.000 63 mol remaining

This excess OH^- will then react with HSO_3^- to form SO_3^{2-}:

$HSO_3^-(aq) + OH^-(aq) \longrightarrow SO_3^{2-}(aq) + H_2O(l)$

We then allow this reaction to go to completion. All the OH^- will be consumed. 0.003 12 mol HSO_3^- will react with 0.000 63 mol OH^- to produce 0.000 63 mol of SO_3^{2-} and leave 0.003 12 − 0.000 63 mol = 0.002 49 mol HSO_3^-. The concentrations of these species will be

$$[HSO_3^-] = \frac{0.002\ 49\text{ mol}}{0.0500\text{ L}} = 0.0498\text{ mol·L}^{-1}$$

$$[SO_3^{2-}] = \frac{0.000\ 63\text{ mol}}{0.0500\text{ L}} = 0.013\text{ mol·L}^{-1}$$

This places us in the second buffer region of H_2SO_3, so the following equation will dominate the equilibrium (for which $K_{a2} = 1.2 \times 10^{-7}$)

Concentration $(\text{mol}\cdot\text{L}^{-1})$	$HSO_3^-(aq)$	$+ H_2O(l) \rightleftharpoons$	$SO_3^{2-}(aq)$	$+ H_3O^+(aq)$
initial	0.0498	—	0.013	0
change	$-x$	—	$+x$	$+x$
final	$0.0498 - x$	—	$0.013 + x$	$+x$

$$K_a = \frac{[SO_3^{2-}][OH^-]}{[HSO_3^-]} = \frac{(0.013 + x)(x)}{(0.0498 - x)} = 1.2 \times 10^{-7}$$

If $x \ll 0.013$, this simplifies to

$$\frac{0.013x}{0.0498} = 1.2 \times 10^{-7}$$

$x = 4.6 \times 10^{-7}$

$[H_3O^+] = 4.6 \times 10^{-7}$, pH = 6.34

This number is in the region in which the autoprotolysis of water may influence the value, but a more rigorous calculation taking this into account gives the same value.

(b) The problem is worked as in (a) but with additional OH^-, more SO_3^{2-} will be formed:

total mol $OH^- = (0.0450\ \text{L})(0.150\ \text{mol}\cdot\text{L}^{-1}) = 0.006\ 75$ mol

mol $H_2SO_3 = (0.0250\ \text{L})(0.125\ \text{mol}\cdot\text{L}^{-1}) = 0.003\ 12$ mol

Again, all the H_2SO_3 will react to form HSO_3^-, with excess OH^- leftover.

0.003 12 mol H_2SO_3 will form 0.003 12 mol HSO_3^-

mol OH^- remaining = 0.006 75 mol − 0.003 12 mol = 0.003 63 mol remaining

This excess OH^- will then react with HSO_3^- to form SO_3^{2-}:

$$HSO_3^-(aq) + OH^-(aq) \longrightarrow SO_3^{2-}(aq) + H_2O(l)$$

We then allow this reaction to go to completion. All the HSO_3^- will be consumed. 0.003 12 mol HSO_3^- will react to form 0.003 12 mol SO_3^{2-} and 0.003 63 − 0.003 12 mol = 0.000 51 OH^- will remain. The concentrations of these species will be

$$[SO_3^{2-}] = \frac{0.003\ 12\ \text{mol}}{0.0700\ \text{L}} = 0.0446\ \text{mol}\cdot\text{L}^{-1}$$

$$[OH^-] = \frac{0.000\ 51\ \text{mol}}{0.0700\ \text{L}} = 0.0073\ \text{mol}\cdot\text{L}^{-1}$$

This places us in first base hydrolysis region of SO_3^{2-}:

Concentration $(mol \cdot L^{-1})$	$SO_3^{2-}(aq)$	+ $H_2O(l) \rightleftharpoons$	$HSO_3^-(aq)$	+ $OH^-(aq)$
initial	0.0446	—	0	0.0073
change	$-x$	—	$+x$	$+x$
final	$0.0446 - x$	—	$+x$	$0.0073 + x$

$$K_{b1} = \frac{K_w}{K_{a2}} = \frac{1.00 \times 10^{-14}}{1.2 \times 10^{-7}} = 8.3 \times 10^{-8}$$

$$K_a = \frac{[HSO_3^-][OH^-]}{[SO_3^{2-}]} = \frac{(x)(0.0073 + x)}{(0.0446 - x)} = 8.3 \times 10^{-8}$$

If $x \ll 0.0073$, this simplifies to

$$\frac{0.0073x}{0.0446} = 8.3 \times 10^{-8}$$

$x = 5.1 \times 10^{-7}$

$[OH^-] = 0.0073 + 5.1 \times 10^{-7} = 0.0073$, pOH = 2.14

pH = 14.00 − 2.14 = 11.86

x is small compared to 0.0073 and so the assumption was valid. What this means here is that the pH is controlled by the amount of excess base added to the solution.

11.107 (1) first stoichiometric point: $OH^-(aq) + H_2SO_4(aq) \longrightarrow H_2O(l) + HSO_4^-(aq)$
Then, because the volume of the solution has doubled, $[HSO_4^-] = 0.10$ M,

Concentration $(mol \cdot L^{-1})$	$HSO_4^-(aq)$	+ $H_2O(l) \rightleftharpoons$	$SO_4^{2-}(aq)$	+ $H_3O^+(aq)$
initial	0.10	—	0	0
change	$-x$	—	$+x$	$+x$
equilibrium	$0.10 - x$	—	x	x

$$K_{a2} = 0.012 = \frac{x^2}{0.10 - x}$$

$0.0012 - 0.012x = x^2$

$x^2 + 0.012x - 0.0012 = 0$

$$x = \frac{-0.012 + \sqrt{(0.012)^2 + (4)(0.0012)}}{2}$$

$x = [H_3O^+] = 0.029\ mol \cdot L^{-1}$

pH = −log(0.029) = 1.54

(2) second stoichiometric point:

$HSO_4^-(aq) + OH^-(aq) \longrightarrow SO_4^{2-}(aq) + H_2O(l)$. Because the volume of the solution has increased by an equal amount,

Concentration ($mol\cdot L^{-1}$)	$SO_4^{2-}(aq)$	+ $H_2O(l)$ ⇌	$HSO_4^-(aq)$	+ $OH^-(aq)$
initial	0.067	—	0	1×10^{-7}
change	$-x$	—	$+x$	$+x$
equilibrium	$0.067 - x$	—	x	$1 \times 10^{-7} + x$

$$K_b = 8.3 \times 10^{-3} = \frac{(x)(1 \times 10^{-7} + x)}{0.067 - x}$$

$$(5.6 \times 10^{-14}) - (8.3 \times 10^{-13}\,x) = 1 \times 10^{-7}\,x + x^2$$

$$x^2 + (1 \times 10^{-7}x) - (5.6 \times 10^{-14}) = 0$$

$$x = \frac{-1 \times 10^{-7} + \sqrt{(1 \times 10^{-7})^2 + (4)(5.6 \times 10^{-14})}}{2}$$

$$x = [HSO_4^-] = 1.9 \times 10^{-7}\ mol\cdot L^{-1}$$

$$[OH^-] = (1.0 \times 10^{-7}) + (1.9 \times 10^{-7}) = 2.9 \times 10^{-7}\ mol\cdot L^{-1}$$

$$pOH = -\log(2.9 \times 10^{-7}) = 6.54, \quad pH = 14.00 - 6.54 = 7.46$$

11.109 (a) $PbCl_2(s) \rightleftharpoons Pb^{2+}(aq) + 2\,Cl^-(aq)$

$[Pb^{2+}][Cl^-]^2 = K_{sp} = 1.6 \times 10^{-5}$

$[0.010][Cl^-]^2 = 1.6 \times 10^{-5}$

$[Cl^-]^2 = 1.6 \times 10^{-3}$

$[Cl^-] = 4.0 \times 10^{-2}\ mol \cdot L^{-1}$; will precipitate lead (II) ion

$Ag^+(aq) + Cl^-(aq) \rightleftharpoons AgCl(s)$

$[Ag^+][Cl^-] = K_{sp} = 1.6 \times 10^{-10}$

$[0.010][Cl^-] = 1.6 \times 10^{-10}$

$[Cl^-] = 1.6 \times 10^{-8}\ mol \cdot L^{-1}$; will precipitate Ag^+

(b) from part (a), Ag^+ will precipitate first.

(c) $[Ag^+](4.0 \times 10^{-2}) = 1.6 \times 10^{-10}$

$$[Ag^+] = \frac{1.6 \times 10^{-10}}{4.0 \times 10^{-2}} = 4.0 \times 10^{-9}\ mol \cdot L^{-1}$$

(d) percentage Ag^+ remaining $= \dfrac{4.0 \times 10^{-9}}{0.010} \times 10^2 = 4.0 \times 10^{-5}$ % unprecipitated; virtually 100% of the first cation (Ag^+) is precipitated

11.111 The K_{sp} value for PbF_2 obtained from Table 11.5 is 3.7×10^{-8}. Using this value, the $\Delta G°$ of the dissolution reaction can be obtained from

$\Delta G° = -RT \ln K.$

$\Delta G° = -(8.314\ J\cdot K^{-1}\cdot mol^{-1})(298.2\ K)\ln(3.7 \times 10^{-8})$

$\Delta G° = +42.43\ kJ\cdot mol^{-1}$

From the Appendices we find that $\Delta G^\circ_f(F^-, aq) = -278.79\ kJ\cdot mol^{-1}$ and $\Delta G^\circ_f(Pb^{2+}, aq) = -24.43\ kJ\cdot mol^{-1}$.

$$\Delta G^\circ = +42.43\ kJ\cdot mol^{-1} = \Delta G^\circ_f(Pb^{2+}, aq) + \Delta G^\circ_f(F^-, aq) - \Delta G^\circ_f(PbF_2, s)$$

$$+42.43\ kJ\cdot mol^{-1} = (-24.43\ kJ\cdot mol^{-1}) + (-278.79\ kJ\cdot mol^{-1}) - \Delta G^\circ_f(PbF_2, s)$$

$$\Delta G^\circ_f(PbF_2, s) = -345.65\ kJ\cdot mol^{-1}$$

11.113 (a) The K_{sp} values are:

Cu_2S 2.0×10^{-47} $\quad$ CuS 1.3×10^{-36}

Although the K_{sp} values seem to indicate that the Cu_2S is less soluble, we really need to calculate the molar solubilities because there are differing numbers of ions involved.

For Cu_2S: $\quad Cu_2S(s) \rightleftharpoons 2\ Cu^+(aq) + S^{2-}(aq)$

conc. change $\quad +2x \quad +x$

$K_{sp} = [Cu^+]^2[S^{2-}] = (2x)^2(x) = 4\,x^3 = 2.0 \times 10^{-47}$

$x = 1.7 \times 10^{-16}$

molar solubility of $Cu_2S = 1.7 \times 10^{-16}\ mol\cdot L^{-1}$

For CuS: $\quad CuS(s) \rightleftharpoons Cu^{2+}(aq) + S^{2-}(aq)$

$K_{sp} = [Cu^{2+}][S^{2-}] = (x)(x) = x^2 = 1.3 \times 10^{-36}$

$x = 1.1 \times 10^{-18}$

molar solubility of $CuS = 1.1 \times 10^{-18}\ mol\cdot L^{-1}$

The molar solubility is greater for Cu_2S than for CuS.

(b) One might be tempted to take the ratios of the concentrations found in (a) as the solution to this question ($3.4 \times 10^{-16} \div 1.1 \times 10^{-18} = 3.1 \times 10^2$), however, the common ion effect will suppress the dissolution of CuS. Because Cu_2S is more soluble, we will assume that equilibrium will establish the concentration of S^{2-} in solution. From (a) we find that $[S^{2-}] = 1.7 \times 10^{-16}\ mol\cdot L^{-1}$ and $[Cu^+] = 2 \times 1.7 \times 10^{-16}\ mol\cdot L^{-1} = 3.4 \times 10^{-16}\ mol\cdot L^{-1}$. If we use the $[S^{2-}]$ from the first equilibrium to calculate the Cu^{2+} concentration, we obtain:

$1.3 \times 10^{-36} = [Cu^{2+}][1.7 \times 10^{-16}]$

$[Cu^{2+}] = 7.6 \times 10^{-21}\ mol\cdot L^{-1}$

From this value we can see that the concentratin of S^{2-} will not be significantly changed by the dissolution of CuS: $1.7 \times 10^{-16} + 7.6 \times 10^{-18} \cong 1.7 \times 10^{-16}$

The ratio of Cu^+ to Cu^{2+} will be $\dfrac{3.4 \times 10^{-16}}{7.6 \times 10^{-21}} = 4.5 \times 10^4$

CHAPTER 12
ELECTROCHEMISTRY

12.1 In each case, first obtain the balanced half-reactions. Multiply the oxidation and reduction half-reactions by appropriate factors that will result in the same number of electrons being present in both half-reactions. Then add the half-reactions, canceling electrons in the process, to obtain the balanced equation for the whole reaction. Check to see that the final equation is balanced.

(a) $4[Cl_2(g) + 2\,e^- \longrightarrow 2\,Cl^-(aq)]$

$1[S_2O_3^{2-}(aq) + 5\,H_2O(l) \longrightarrow 2\,SO_4^{2-}(aq) + 10\,H^+(aq) + 8\,e^-]$

$4\,Cl_2(g) + S_2O_3^{2-}(aq) + 5\,H_2O(l) + 8\,e^- \longrightarrow$

$8\,Cl^-(aq) + 2\,SO_4^{2-}(aq) + 10\,H^+(aq) + 8\,e^-$

$4\,Cl_2(g) + S_2O_3^{2-}(aq) + 5\,H_2O(l) \longrightarrow 8\,Cl^-(aq) + 2\,SO_4^{2-}(aq) + 10\,H^+(aq)$

Cl_2 is the oxidizing agent and $S_2O_3^{2-}$ is the reducing agent.

(b) $2[MnO_4^-(aq) + 8\,H^+(aq) + 5\,e^- \longrightarrow Mn^{2+}(aq) + 4\,H_2O(l)]$

$5[H_2SO_3(aq) + H_2O(l) \longrightarrow HSO_4^-(aq) + 3\,H^+(aq) + 2\,e^-]$

$2\,MnO_4^-(aq) + 16\,H^+(aq) + 5\,H_2SO_3(aq) + 5\,H_2O(l) + 10\,e^- \longrightarrow$

$2\,Mn^{2+}(aq) + 8\,H_2O(l) + 5\,HSO_4^-(aq) + 15\,H^+(aq) + 10\,e^-$

$2\,MnO_4^-(aq) + H^+(aq) + 5\,H_2SO_3(aq) \longrightarrow$

$2\,Mn^{2+}(aq) + 3\,H_2O(l) + 5\,HSO_4^-(aq)$

MnO_4^- is the oxidizing agent and H_2SO_3 is the reducing agent.

(c) $Cl_2(g) + 2e^- \longrightarrow 2\,Cl^-(aq)$

$H_2S(aq) \longrightarrow S(s) + 2\,H^+(aq) + 2e^-$

$Cl_2(g) + H_2S(aq) + 2\,e^- \longrightarrow 2\,Cl^-(aq) + S(s) + 2\,H^+(aq) + 2\,e^-$

$Cl_2(g) + H_2S(aq) \longrightarrow 2\,Cl^-(aq) + S(s) + 2\,H^+(aq)$

Cl_2 is the oxidizing agent and H_2S is the reducing agent.

(d) $Cl_2(g) + 2\,e^- \longrightarrow 2\,Cl^-(aq)$

$2\,H_2O(l) + Cl_2(g) \longrightarrow 2\,HOCl(aq) + 2\,H^+(aq) + 2\,e^-$

$2\,H_2O(l) + Cl_2(g) + 2\,e^- \longrightarrow 2\,HOCl(aq) + 2\,H^+(aq) + 2\,Cl^-(aq) + 2\,e^-$

or $H_2O(l) + Cl_2(g) \longrightarrow HOCl(aq) + H^+(aq) + Cl^-(aq)$

Cl_2 is both the oxidizing and the reducing agent.

12.3 (a) $O_3(g) \longrightarrow O_2(g)$

$O_3(g) \longrightarrow O_2(g) + H_2O(l)$ (balances O's)

$2\ H_2O(l) + O_3(g) \longrightarrow O_2(g) + H_2O(l) + 2\ OH^-(aq)$ (balances H's)
$H_2O(l) + O_3(g) \longrightarrow O_2(g) + 2\ OH^-(aq)$ (cancels H_2O)
$H_2O(l) + O_3(g) + 2\ e^- \longrightarrow O_2(g) + 2\ OH^-(aq)$ (balances charge);
$Br^-(aq) \longrightarrow BrO_3^-(aq)$
$3\ H_2O(l) + Br^-(aq) \longrightarrow BrO_3^-(aq)$ (balances O's)
$6\ OH^-(aq) + 3\ H_2O(l) + Br^-(aq) \longrightarrow BrO_3^-(aq) + 6\ H_2O(l)$ (balances H's)
$6\ OH^-(aq) + 3\ H_2O(l) + Br^-(aq) \longrightarrow$
$BrO_3^-(aq) + 6\ H_2O(l) + 6\ e^-$ (balances charge)

Combining half-reactions yields

$3[H_2O(l) + O_3(g) + 2\ e^- \longrightarrow O_2(g) + 2OH^-(aq)]$
$6\ OH^-(aq) + 3\ H_2O(l) + Br^-(aq) \longrightarrow BrO_3^-(aq) + 6\ H_2O(l) + 6\ e^-$

$6\ H_2O(l) + 3\ O_3(g) + 6\ OH^-(aq) + Br^-(aq) + 6\ e^- \longrightarrow$
$3\ O_2(g) + 6\ OH^-(aq) + BrO_3^-(aq) + 6\ H_2O(l) + 6\ e^-$

and, $3\ O_3(g) + Br^-(aq) \longrightarrow 3\ O_2(g) + BrO_3^-(aq)$

O_3 is the oxidizing agent and Br^- is the reducing agent.

(b) $Br_2(l) + 2\ e^- \longrightarrow 2\ Br^-(aq)$ (balanced reduction half-reaction)
$Br_2(l) + 6\ H_2O(l) \longrightarrow 2\ BrO_3^-(aq)$ (O's balanced); then,
$Br_2(l) + 6\ H_2O(l) + 12\ OH^-(aq) \longrightarrow$
$2\ BrO_3^-(aq) + 12\ H_2O(l)$ (H's balanced); and,
$Br_2(l) + 12\ OH^-(aq) \longrightarrow$
$2\ BrO_3^-(aq) + 6\ H_2O(l) + 10\ e^-$ (electrons balanced)

Combining half-reactions yields

$5[Br_2(l) + 2\ e^- \longrightarrow 2\ Br^-(aq)]$
$1[Br_2(l) + 12\ OH^-(aq) \longrightarrow 2\ BrO_3^-(aq) + 6\ H_2O(l) + 10\ e^-]$

$6\ Br_2(l) + 12\ OH^-(aq) + 10\ e^- \longrightarrow$
$10\ Br^-(aq) + 2\ BrO_3^-(aq) + 6\ H_2O(l) + 10\ e^-$
$6\ Br_2(l) + 12\ OH^-(aq) \longrightarrow 10\ Br^-(aq) + 2\ BrO_3^-(aq) + 6\ H_2O(l)$

dividing by 2 gives:

$3\ Br_2(l) + 6\ OH^-(aq) \longrightarrow 5\ Br^-(aq) + BrO_3^-(aq) + 3\ H_2O(l)$

Br_2 is both the oxidizing agent and the reducing agent.

(c) $Cr^{3+}(aq) + 4\ H_2O(l) \longrightarrow CrO_4^{2-}(aq)$ (O's balanced); then,
$Cr^{3+}(aq) + 4\ H_2O(l) + 8\ OH^-(aq) \longrightarrow CrO_4^{2-}(aq) + 8\ H_2O(l)$ (H's balanced); and,
$Cr^{3+}(aq) + 8\ OH^-(aq) \longrightarrow CrO_4^{2-}(aq) + 4\ H_2O(l) + 3\ e^-$ (charge balanced)
$MnO_2(s) \longrightarrow Mn^{2+}(aq) + 2\ H_2O(l)$; then,
$MnO_2(s) + 4\ H_2O(l) \longrightarrow Mn^{2+}(aq) + 2\ H_2O(l) + 4\ OH^-(aq)$ (H's balanced); and,
$MnO_2(s) + 2\ H_2O(l) + 2\ e^- \longrightarrow Mn^{2+}(aq) + 4\ OH^-(aq)$ (charge balanced)

Combining half-reactions yields

$2[Cr^{3+}(aq) + 8\ OH^-(aq) \longrightarrow CrO_4^{2-}(aq) + 4\ H_2O(l) + 3\ e^-]$

$3[MnO_2(s) + 2\ H_2O(l) + 2\ e^- \longrightarrow Mn^{2+}(aq) + 4\ OH^-(aq)]$

$2\ Cr^{3+}(aq) + 16\ OH^-(aq) + 3\ MnO_2(s) + 6\ H_2O(l) + 6\ e^- \longrightarrow$
$2\ CrO_4^{2-}(aq) + 8\ H_2O(l) + 3\ Mn^{2+}(aq) + 12\ OH^-(aq) + 6\ e^-$

$2\ Cr^{3+}(aq) + 4\ OH^-(aq) + 3\ MnO_2(s) \longrightarrow$
$2\ CrO_4^{2-}(aq) + 2\ H_2O(l) + 3\ Mn^{2+}(aq)$

Cr^{3+} is the reducing agent and MnO_2 is the oxidizing agent.

(d) $3[P_4(s) + 8\ OH^-(aq) \longrightarrow 4\ H_2PO_2^-(aq) + 4\ e^-]$

$P_4(s) + 12\ H_2O(l) + 12\ e^- \longrightarrow 4\ PH_3(g) + 12\ OH^-(aq)$

$4\ P_4(s) + 12\ H_2O(l) + 24\ OH^-(aq) + 12\ e^- \longrightarrow$
$12\ H_2PO_2^-(aq) + 4\ PH_3(g) + 12\ OH^-(aq) + 12\ e^-$

$4\ P_4(s) + 12\ H_2O(l) + 12\ OH^-(aq) \longrightarrow 12\ H_2PO_2^-(aq) + 4\ PH_3(g)$

or $P_4(s) + 3\ H_2O(l) + 3\ OH^-(aq) \longrightarrow 3\ H_2PO_2^-(aq) + PH_3(g)$

$P_4(s)$ is both the oxidizing and the reducing agent.

12.5 The half-reactions are

$MnO_4^-(aq) \longrightarrow Mn^{2+}(aq) + 4\ H_2O(l)$ (O's balanced)

$MnO_4^-(aq) + 8\ H^+(aq) \longrightarrow Mn^{2+}(aq) + 4\ H_2O(l)$ (H's balanced)

$MnO_4^-(aq) + 8\ H^+(aq) + 5\ e^- \longrightarrow Mn^{2+}(aq) + 4\ H_2O(l)$ (charge balanced)

$C_6H_{12}O_6(aq) + 6\ H_2O(l) \longrightarrow 6\ CO_2(g)$ (O's balanced)

$C_6H_{12}O_6(aq) + 6\ H_2O(l) \longrightarrow 6\ CO_2(g) + 24\ H^+(aq)$ (H's balanced)

$C_6H_{12}O_6(aq) + 6\ H_2O(l) \longrightarrow$
$6\ CO_2(g) + 24\ H^+(aq) + 24\ e^-$ (charge balanced)

Adding half-reactions yields

$24[MnO_4^-(aq) + 8\ H^+(aq) + 5\ e^- \longrightarrow Mn^{2+}(aq) + 4\ H_2O(l)]$

$5[C_6H_{12}O_6(aq) + 6\ H_2O(l) \longrightarrow 6\ CO_2(g) + 24\ H^+(aq) + 24\ e^-]$

$24\ MnO_4^-(aq) + 192\ H^+(aq) + 5\ C_6H_{12}O_6(aq) + 30\ H_2O(l) + 120\ e^- \longrightarrow$
$24\ Mn^{2+}(aq) + 96\ H_2O(l) + 30\ CO_2(g) + 120\ H^+(aq) + 120\ e^-$

$24\ MnO_4^-(aq) + 72\ H^+(aq) + 5\ C_6H_{12}O_6(aq) \longrightarrow$
$24\ Mn^{2+}(aq) + 66\ H_2O(l) + 30\ CO_2(g)$

12.7 See Figs. 12.1 and 12.5

(a) anode (b) positive

12.9 See Figs. 12.1 and 12.5

(a) electrons (b) ions (cations and anions)

12.11 (a) $Ag^{+}(aq) + e^{-} \longrightarrow Ag(s)$ $E°(\text{cathode}) = +0.80$ V

$Fe^{3+}(aq) + e^{-} \longrightarrow Fe^{2+}(aq)$ $E°(\text{anode}) = +0.77$ V

Reversing the anode half-reaction yields

$Fe^{2+}(aq) \longrightarrow Fe^{3+}(aq) + e^{-}$

and the cell reaction is, upon addition of the half-reactions,

$Ag^{+}(aq) + Fe^{2+}(aq) \longrightarrow Ag(s) + Fe^{3+}(aq)$ $E°_{cell} = +0.80\text{ V} - 0.77\text{ V}$

$= +0.03$ V

(b) $2\,H^{+}(aq) + 2\,e^{-} \longrightarrow H_2(g)$ $E°(\text{anode}) = 0.00$ V

$Cl_2(g) + 2\,e^{-} \longrightarrow 2\,Cl^{-}(aq)$ $E°(\text{cathode}) = +1.36$ V

Therefore, at the anode, after reversal,

$H_2(g) \longrightarrow 2\,H^{+}(aq) + 2\,e^{-}$

and, the cell reaction is, upon addition of the half-reactions,

$Cl_2(g) + H_2(g) \longrightarrow 2\,H^{+}(aq) + 2\,Cl^{-}(aq)$ $E°_{cell} = +1.36\text{ V} - 0.00\text{ V}$

$= +1.36$ V

(c) $2[U^{3+}(aq) + 3\,e^{-} \longrightarrow U(s)]$ $E°(\text{anode}) = -1.79$ V

$3[V^{2+}(aq) + 2\,e^{-} \longrightarrow V(s)]$ $E°(\text{cathode}) = -1.19$ V

Therefore, at the anode, after reversal,

$2[U(s) \longrightarrow U^{3+}(aq) + 3\,e^{-}]$

and, the cell reaction is, upon addition of the half-reactions,

$3\,V^{2+}(aq) + 2\,U(s) \longrightarrow 2\,U^{3+}(aq) + 3\,V(s)$ $E°_{cell} = -1.19\text{ V} - (-1.79\text{ V})$

$= +0.60$ V

(d) $O_2(g) + 2\,H_2O(l) + 4\,e^{-} \longrightarrow 4\,OH^{-}(aq)$ $E°(\text{cathode}) = 0.40$ V

$O_2(g) + 4\,H^{+}(aq) + 4\,e^{-} \longrightarrow 2\,H_2O(l)$ $E°(\text{anode}) = 1.23$ V

Reversing the anode half-reaction yields

$2\,H_2O(l) \longrightarrow O_2(g) + 4\,H^{+}(aq) + 4\,e^{-}$

and the cell reaction is, upon addition of the half-reactions,

$4\,H_2O(l) \longrightarrow 4\,H^{+}(aq) + 4\,OH^{-}(aq)$ $E°_{cell} = 0.40\text{ V} - 1.23\text{ V} = -0.83$ V

or, $H_2O(l) \longrightarrow H^{+}(aq) + OH^{-}(aq)$

Note: This balanced equation corresponds to the cell notation given. The spontaneous process is the reverse of this reaction.

(e) $Sn^{4+}(aq) + 2\,e^{-} \longrightarrow Sn^{2+}(aq)$ $E°(\text{anode}) = +0.15$ V

$Hg_2Cl_2(s) + 2\,e^{-} \longrightarrow 2\,Hg(l) + 2\,Cl^{-}(aq)$ $E°(\text{cathode}) = +0.27$ V

Therefore, at the anode, after reversal,

$Sn^{2+}(aq) \longrightarrow Sn^{4+}(aq) + 2\,e^{-}$

and, the cell reaction is, upon addition of the half-reactions,

$Sn^{2+}(aq) + Hg_2Cl_2(s) \longrightarrow 2\,Hg(l) + 2\,Cl^{-}(aq) + Sn^{4+}(aq)$

$E°_{cell} = 0.27\text{ V} - 0.15\text{ V} = 0.12$ V

12.13 (a) $Ni^{2+}(aq) + 2\ e^- \longrightarrow Ni(s)$ $E°(\text{cathode}) = -0.23$ V

$Zn^{2+}(aq) + 2\ e^- \longrightarrow Zn(s)$ $E°(\text{anode}) = -0.76$ V

Reversing the anode reaction yields

$Zn(s) \longrightarrow Zn^{2+}(aq) + 2\ e^-$ (at anode), then, upon addition,

$Ni^{2+}(aq) + Zn(s) \longrightarrow Ni(s) + Zn^{2+}(aq)$ (overall cell)

$E°_{cell} = -0.23\text{ V} - (-0.76\text{ V}) = +0.53\text{ V}$

and $Zn(s)\,|\,Zn^{2+}(aq)\,||\,Ni^{2+}(aq)\,|\,Ni(s)$

(b) $2[Ce^{4+}(aq) + e^- \longrightarrow Ce^{3+}(aq)]$ $E°(\text{cathode}) = +1.61$ V

$I_2(s) + 2\ e^- \longrightarrow 2\ I^-(aq)$ $E°(\text{anode}) = +0.54$ V

Reversing the anode reaction yields

$2\ I^-(aq) \longrightarrow 2\ e^- + I_2(s)$ (at anode), then, upon addition,

$2\ I^-(aq) + 2\ Ce^{4+}(aq) \longrightarrow 2\ Ce^{3+}(aq) + I_2(s)$ (overall cell)

$E°_{cell} = +1.61\text{ V} - 0.54\text{ V} = +1.07\text{ V}$

and $Pt(s)\,|\,I^-(aq)\,|\,I_2(s)\,||\,Ce^{4+}(aq), Ce^{3+}(aq)\,|\,Pt(s)$

An inert electrode such as Pt is necessary when both oxidized and reduced species are in the same solution.

(c) $Cl_2(g) + 2\ e^- \longrightarrow 2\ Cl^-(aq)$ $E°(\text{cathode}) = +1.36$ V

$2\ H^+(aq) + 2\ e^- \longrightarrow H_2(g)$ $E°(\text{anode}) = 0.00$ V

Reversing the anode reaction yields

$H_2(g) \longrightarrow 2\ H^+(aq) + 2\ e^-$ (at anode), then, upon addition,

$H_2(g) + Cl_2(g) \longrightarrow 2\ HCl(aq)$ (overall cell) $E°_{cell} = +1.36\text{ V} - 0.00\text{ V}$
$= +1.36\text{ V}$

and $Pt(s)\,|\,H_2(g)\,|\,H^+(aq)\,||\,Cl^-(aq)\,|\,Cl_2(g)\,|\,Pt(s)$

An inert electrode such as Pt is necessary for gas/ion electrode reactions.

(d) $3[Au^+(aq) + e^- \longrightarrow Au(s)]$ $E°(\text{cathode}) = +1.69$ V

$Au^{3+}(aq) + 3\ e^- \longrightarrow Au(s)$ $E°(\text{anode}) = +1.40$ V

Reversing the anode reaction yields

$Au(s) \longrightarrow Au^{3+}(aq) + 3\ e^-$ then, upon addition (anode),

$3\ Au^+(aq) \longrightarrow 2\ Au(s) + Au^{3+}(aq)$ (overall cell) $E°_{cell} = +1.69\text{ V} - 1.40\text{ V}$
$= +0.29\text{ V}$

and $Au(s)\,|\,Au^{3+}(aq)\,||\,Au^+(aq)\,|\,Au(s)$

12.15 (a) $Ag^+(aq) + e^- \longrightarrow Ag(s)$ $E°(\text{cathode}) = +0.80$ V

$AgBr(s) + e^- \longrightarrow Ag(s) + Br^-(aq)$ $E°(\text{anode}) = +0.07$ V

Reversing the anode reaction yields

$Ag(s) + Br^-(aq) \longrightarrow AgBr(s) + e^-$ then, upon addition,

$Ag^+(aq) + Br^-(aq) \longrightarrow AgBr(s)$ (overall cell) $E°_{cell} = +0.80\text{ V} - 0.07\text{ V}$
$= +0.73\text{ V}$

This is the direction of the spontaneous cell reaction that could be used to study the given solubility equilibrium. The reverse of this cell reaction corresponds to the reaction as given. It is not spontaneous. Thus,

$AgBr(s) \longrightarrow Ag^+(aq) + Br^-(aq) \quad E°_{cell} = -0.73$ V

For this, the cathode and anode reactions are reversed relative to those above. A cell diagram for the nonspontaneous process is

$Ag(s) | Ag^+(aq) || Br^-(aq) | AgBr(s) | Ag(s)$

(b) To conform to the notation of this chapter, the neutralization is rewritten as

$H^+(aq) + OH^- \longrightarrow H_2O(l)$

$O_2(g) + 4\ H^+(aq) + 4\ e^- \longrightarrow 2\ H_2O(l) \quad E°(\text{cathode}) = +1.23$ V

$O_2(g) + 2\ H_2O(l) + 4\ e^- \longrightarrow 4\ OH^-(aq) \quad E°(\text{anode}) = +0.40$ V

Reversing the anode reaction yields

$4\ OH^-(aq) \longrightarrow O_2(g) + 2\ H_2O(l) + 4\ e^-$, then, upon addition,

$4\ H^+(aq) + 4\ OH^-(aq) \longrightarrow 4\ H_2O(l)$

or $H^+(aq) + OH^-(aq) \longrightarrow H_2O(l)$ (overall cell) $E° = +1.23\text{ V} - 0.40\text{ V}$
$= +0.83$ V

and $Pt(s) | O_2(g) | OH^-(aq) || H^+(aq) | O_2(g) | Pt(s)$

(c) $Cd(OH)_2(s) + 2\ e^- \longrightarrow Cd(s) + 2\ OH^-(aq) \quad E°(\text{anode}) = -0.81$ V

$Ni(OH)_3(s) + e^- \longrightarrow Ni(OH)_2(s) + OH^-(aq) \quad E°(\text{cathode}) = +0.49$ V

Reversing the anode reaction and multiplying the cathode reaction by 2 yields

$Cd(s) + 2\ OH^-(aq) \longrightarrow Cd(OH)_2(s) + 2\ e^-$

$2\ Ni(OH)_3 + 2\ e^- \longrightarrow 2\ Ni(OH)_2(s) + 2\ OH^-(aq)$ then, upon addition,

$2\ Ni(OH)_2(s) + Cd(s) \longrightarrow Cd(OH)_2(s) + 2\ Ni(OH)_2(s)$

and $Cd(s) | Cd(OH)_2(s) | KOH(aq) || Ni(OH)_3(s) | Ni(OH)_2(s) | Ni(s)$

12.17 (a) $MnO_4^-(aq) + 8\ H^+(aq) + 5\ e^- \longrightarrow Mn^{2+}(aq) + 4\ H_2O(l)$ (cathode half-reaction)

$5[Fe^{3+}(aq) + e^- \longrightarrow Fe^{2+}(aq)]$ (anode half-reaction)

(b) Reversing the anode reaction and adding the two equations yields

$MnO_4^-(aq) + 5\ Fe^{2+}(aq) + 8\ H^+(aq) \longrightarrow Mn^{2+}(aq) + 5\ Fe^{3+}(aq) + 4\ H_2O(l)$

The cell diagram is

$Pt(s) | Fe^{2+}(aq), Fe^{3+}(aq) || MnO_4^-(aq), Mn^{2+}(aq), H^+(aq) | Pt(s)$

12.19 A galvanic cell has a positive potential difference; therefore, identify as cathode and anode the electrodes that make $E°$(cell) positive upon calculating

$E°(\text{cell}) = E°(\text{cathode}) - E°(\text{anode})$

There are only two possibilities: If your first guess gives a negative $E°$(cell), switch your identification.

(a) $Cu^{2+}(aq) + 2\,e^- \longrightarrow Cu(s)$ $E°$(cathode) = +0.34 V

$Cr^{3+}(aq) + 1\,e^- \longrightarrow Cr^{2+}(aq)$ $E°$(anode) = −0.41 V

$E°$(cell) = +0.34 V − (−0.41 V) = +0.75 V

(b) $AgCl(s) + e^- \longrightarrow Ag(s) + Cl^-(aq)$ $E°$(cathode) = +0.22 V

$AgI(s) + e^- \longrightarrow Ag(s) + I^-(aq)$ $E°$(anode) = −0.15 V

$E°$(cell) = +0.22V − (−0.15V) = +0.37 V

(c) $Hg_2{}^{2+}(aq) + 2\,e^- \longrightarrow 2\,Hg(l)$ $E°$(cathode) = +0.79 V

$Hg_2Cl_2(s) + 2\,e^- \longrightarrow 2\,Hg(l) + 2\,Cl^-(aq)$ $E°$(anode) = +0.27 V

$E°$(cell) = +0.79 V − (+0.27 V) = +0.52 V

(d) $Pb^{4+}(aq) + 2\,e^- \longrightarrow Pb^{2+}(aq)$ $E°$(cathode) = +1.67 V

$Sn^{4+}(aq) + 2\,e^- \longrightarrow Sn^{2+}(aq)$ $E°$(anode) = +0.15 V

$E°$(cell) = +1.67 V − (+0.15 V) = +1.52 V

12.21 See Exercise 12.19 solutions for $E°$ (cell) values. In each case, $\Delta G°_r = -nFE°$. $1\ V = 1\ J\cdot C^{-1}$. n is determined by balancing the equation for the cell reaction constructed from the half-reactions given in Exercise 12.19.

(a) $Cu^{2+}(aq) + 2\,Cr^{2+}(aq) \longrightarrow Cu(s) + 2\,Cr^{3+}(aq)$, $n = 2$

$E°_{cell} = +0.75\ V$ and $\Delta G°_r = -nFE° = -(2)(9.6485 \times 10^4\ C\cdot mol^{-1})(0.75\ J\cdot C^{-1})$
$= -145\ kJ\cdot mol^{-1}$

(b) $AgCl(s) + I^-(aq) \longrightarrow AgI(s) + Cl^-(aq)$, $n = 1$

$E°_{cell} = +0.37\ V$ and $\Delta G°_r = -nFE°$
$= -1 \times 9.6485 \times 10^4\ C\cdot mol^{-1} \times 0.37\ J\cdot C^{-1} = -36\ kJ\cdot mol^{-1}$

(c) $Hg_2{}^{2+}(aq) + 2\,Cl^-(aq) \longrightarrow Hg_2Cl_2(s)$, $n = 2$

$E°_{cell} = +0.52\ V$ and $\Delta G°_r = -nFE° = -(2)(9.6485 \times 10^4\ C\cdot mol^{-1})(0.52\ J\cdot C^{-1})$
$= -100\ kJ\cdot mol^{-1}$

(d) $Pb^{4+}(aq) + Sn^{2+}(aq) \longrightarrow Pb^{2+}(aq) + Sn^{4+}(aq)$, $n = 2$

$E°_{cell} = +1.52\ V$ and $\Delta G°_r = -nFE° = -(2)(9.6485 \times 10^4\ C\cdot mol^{-1})(1.52\ J\cdot C^{-1})$
$= -293\ kJ\cdot mol^{-1}$

12.23 The cell, as written $Cu(s)|Cu^{2+}(aq)||Pb^{2+}(aq)|Pb(s)$, makes the Cu/Cu^{2+} electrode the anode, because this is where oxidation is occurring; the Pb^{2+}/Pb electrode is the cathode. The calculation is

$E° = E°$(cathode) − $E°$(anode)

−0.47 V = $E°$(cathode) − (+0.34 V)

$E°$(cathode) = −0.13 V

12.25 Refer to Appendix 2B. The more negative (less positive) the standard reduction potential, the stronger is the metal as a reducing agent.

(a) Cu < Fe < Zn < Cr

(b) Mg < Na < K < Li

(c) V < Ti < Al < U

(d) Au < Ag < Sn < Ni

12.27 In each case, identify the couple with the more positive reduction potential. This will be the couple at which reduction occurs, and therefore contains the oxidizing agent. The other couple contains the reducing agent.

(a) Co^{2+}/Co $E° = -0.28$ V, Co^{2+} is the oxidizing agent (cathode)

Ti^{3+}/Ti^{2+} $E° = -0.37$ V, Ti^{2+} is the reducing agent (anode)

$Pt(s) | Ti^{2+}(aq), Ti^{3+}(aq) || Co^{2+}(aq) | Co(s)$

$E°_{cell} = E°(\text{cathode}) - E°(\text{anode}) = -0.28\text{ V} - (-0.37\text{ V}) = +0.09\text{ V}$

(b) U^{3+}/U $E° = -1.79$ V, U^{3+} is the oxidizing agent (cathode)

La^{3+}/La $E° = -2.52$ V, La is the reducing agent (anode)

$La(s) | La^{3+}(aq) || U^{3+}(aq) | U(s)$

$E°_{cell} = -1.79\text{ V} - (-2.52\text{ V}) = +0.73\text{ V}$

(c) Fe^{3+}/Fe^{2+} $E° = +0.77$ V, Fe^{3+} is the oxidizing agent (cathode)

H^+/H_2 $E° = 0.00$ V, H_2 is the reducing agent (anode)

$Pt(s) | H_2(g) | H^+(aq) || Fe^{2+}(aq), Fe^{3+}(aq) | Pt(s)$

$E°_{cell} = +0.77\text{ V} - 0.00\text{ V} = +0.77\text{ V}$

(d) $O_3/O_2,OH^-$ $E° = +1.24$ V, O_3 is the oxidizing agent (cathode)

Ag^+/Ag $E° = +0.80$ V, Ag is the reducing agent (anode)

$Ag(s) | Ag^+(aq) || OH^-(aq) | O_3(g), O_2(g) | Pt(s)$

$E°_{cell} = +1.24\text{ V} - 0.80\text{ V} = +0.44\text{ V}$

12.29 (a) H_2/H^+ $E° = 0.00$ V; Ni^{2+}/Ni $E° = -0.23$ V

No, Ni is the better reducing agent; its couple has a more negative $E°$.

(b) Cr^{3+}/Cr $E° = -0.74$ V; Pb^{2+}/Pb $E° = -0.13$ V

Yes, Cr is the better reducing agent; its couple has a more negative $E°$.

$2[Cr(s) \longrightarrow Cr^{3+}(aq) + 3\,e^-]$ (anode)

$3[Pb^{2+}(aq) + 2\,e^- \longrightarrow Pb(s)]$ (cathode)

$3\,Pb^{2+}(aq) + 2\,Cr(s) \longrightarrow 3\,Pb(s) + 2\,Cr^{3+}(aq)$

$E°_{cell} = E°(\text{cathode}) - E°(\text{anode}) = -0.13\text{ V} - (-0.74\text{ V}) = +0.61\text{ V}$

(c) $MnO_4^-, H^+/Mn^{2+}$ $E° = +1.51$ V; Cu^{2+}/Cu $E° = +0.34$ V

Yes, MnO_4^- is the better oxidizing agent; its couple has a more positive $E°$.

$5[Cu(s) \longrightarrow Cu^{2+}(aq) + 2\ e^-]$ (anode)

$2[MnO_4^-(aq) + 8\ H^+(aq) + 5\ e^- \longrightarrow Mn^{2+}(aq) + 4\ H_2O(l)]$ (cathode)

$5\ Cu(s) + 2\ MnO_4^-(aq) + 16\ H^+(aq) \longrightarrow 5\ Cu^{2+}(aq) + 2\ Mn^{2+}(aq) + 8\ H_2O(l)$

$E°_{cell} = 1.51\ V - 0.34\ V = 1.17\ V$

(d) Fe^{3+}/Fe^{2+} $E° = +0.77\ V$; Hg_2^{2+}/Hg $E° = +0.79\ V$

No, Hg_2^{2+} is the better oxidizing agent; its couple has the more positive $E°$.

12.31 (a) $Mg^{2+} + Cu \longrightarrow$ no reaction; $E°$ for $Cu^{2+}/Cu > E°$ for Mg^{2+}/Mg

(b) $2\ Al + 3\ Pb^{2+} \longrightarrow 2\ Al^{3+} + 3\ Pb$

$Pb^{2+} + 2\ e^- \longrightarrow Pb$ reduction (cathode) $E° = -0.13\ V$

$Al \longrightarrow Al^{3+} + 3\ e^-$ oxidation (anode) $E° = -1.66\ V$

$E°_{cell} = E°(\text{cathode}) - E°(\text{anode}) = -0.13\ V - (-1.66\ V) = +1.53\ V$

$\Delta G°_r = -nFE° = -(6)(9.6485 \times 10^4\ C \cdot mol^{-1})(+1.53\ J \cdot C^{-1})$

$= -886\ kJ \cdot mol^{-1}$; therefore, spontaneous

12.33 $E°\ (Br_2, Br^-) = +1.09\ V$ $E°\ (Cl_2, Cl^-) = +1.36\ V$ $E°\ (O_2, H^+, H_2O)$

$= +1.23\ V$

O_2 could be used because $E°\ (O_2, H^+, H_2O) > E°\ (Br_2, Br^-)$. It is not used because that reaction is so much slower than the one with Cl_2.

12.35 (a) $3\ Au^+(aq) \rightleftharpoons 2\ Au(s) + Au^{3+}(aq)$

(b) $Au^+(aq) + e^- \longrightarrow Au(s)$ $E° = +1.69\ V$

$Au^{3+}(aq) + 3\ e^- \longrightarrow Au(s)$ $E° = +1.40\ V$

Multiplying the first equation by three and subtracting the second equation gives the net equation desired. The potential is given simply by subtracting the second from the first:

$E° = 1.69\ V - 1.40\ V = +0.29\ V$

Because $E°$ is positive, the process should be spontaneous.

12.37 The appropriate half-reactions are:

$U^{4+} + e^- \longrightarrow U^{3+}$ $E° = -0.61$ (A)

$U^{3+} + 3\ e^- \longrightarrow U$ $E° = -1.79$ (B)

(A) and (B) add to give the desired half-reaction (C):

$U^{4+} + 4\ e^- \longrightarrow U$ $E° = ?$ (C)

In order to calculate the potential of a *half-reaction* we need to convert the $E°$ values into $\Delta G°$ values:

$\Delta G°(A) = -nFE°(A) = -1F(-0.61\ V)$

$\Delta G°(B) = -nFE°(B) = -3F(-1.79\ V)$

$\Delta G°(C) = -nFE°(C) = -4FE°(C)$

$\Delta G°(C) = \Delta G°(A) + \Delta G°(B)$

$-4FE°(C) = -1F(-0.61\text{ V}) + [-3F(-1.79\text{ V})]$

The constant F will cancel from both sides leaving:

$-4E°(C) = -1(-0.61\text{ V}) - 3(-1.79\text{ V})$

$E°(C) = -[0.61\text{ V} + 5.37\text{ V}]/4 = -1.50\text{ V}$

12.39 (a) $E°_{cell} = E°(\text{cathode}) - E°(\text{anode}) = 0.34\text{ V} - (-0.41\text{ V}) = 0.75\text{ V}$, and

$$\ln K = \frac{nFE°}{RT} \text{ at } 25°\text{C} = \frac{nE°}{0.025\,69\text{ V}} \quad \ln K = \frac{(2)(0.75\text{ V})}{0.025\,69\text{ V}} = 58 \quad \text{and}$$

$K = 10^{25}$

(b) $Ti^{2+}(aq) + 2\,e^- \longrightarrow Ti(s)$ $\quad E°(\text{cathode}) = -1.63\text{ V}$

$Mn^{2+}(aq) + 2\,e^- \longrightarrow Mn(s)$ $\quad E°(\text{anode}) = -1.18\text{ V}$

Note: These equations represent the cathode and anode half-reactions for the overall reaction as written. The spontaneous direction of this reaction under standard conditions is the opposite of that given.

$E°_{cell} = E°(\text{cathode}) - E°(\text{anode}) = -1.63\text{ V} - (-1.18\text{ V}) = -0.45\text{ V}$, and

$$\ln K = \frac{nFE°}{RT} \text{ at } 25°\text{C} = \frac{nE°}{0.025\,69\text{ V}} \quad \ln K = \frac{(2)(-0.45\text{ V})}{0.025\,69\text{ V}} = -35 \quad \text{and}$$

$K = 10^{-15}$

(c) $E°$ for $Hg_2^{2+}/Hg = +0.79\text{ V} > E°$ for $Pb^{2+}/Pb = -0.13\text{ V}$

Therefore, Hg_2^{2+} is reduced and Pb is oxidized.

$Hg_2^{2+}(aq) + Pb(s) \longrightarrow Pb^{2+}(aq) + 2\,Hg(l)$

$Hg_2^{2+} + 2\,e^- \longrightarrow 2\,Hg(l)$ $\quad E°(\text{cathode}) = 0.79\text{ V}$

$Pb^{2+} + 2\,e^- \longrightarrow Pb(s)$ $\quad E°(\text{anode}) = -0.13\text{ V}$

$E°_{cell} = E°(\text{cathode}) - E°(\text{anode}) = 0.79\text{ V} - (-0.13\text{ V}) = 0.92\text{ V}$, and

$$\ln K = \frac{nFE°}{RT} \text{ at } 25°\text{C} = \frac{nE°}{0.025\,69\text{ V}} \quad \ln K = \frac{(2)(0.92\text{ V})}{0.025\,69\text{ V}} = +72 \quad \text{and}$$

$K = 10^{31}$

(d) $In^{3+}(aq) + e^- \longrightarrow In^{2+}(aq)$ $\quad E°(\text{cathode}) = -0.49\text{ V}$

$U^{4+}(aq) + e^- \longrightarrow U^{3+}(aq)$ $\quad E°(\text{anode}) = -0.61\text{ V}$

$E°_{cell} = E°(\text{cathode}) - E°(\text{anode}) = -0.49\text{ V} - (-0.61\text{ V}) = +0.12\text{ V}$

$$\ln K = \frac{nFE°}{RT} = \frac{nE°}{0.025\,69\text{ V}} \text{ at } 25°\text{C} \quad \ln K = \frac{1 \times (+0.12\text{ V})}{0.025\,69\text{ V}} = 4.7 \quad K = 1 \times 10^2$$

12.41 Consider the half-reactions involved. Construct a cell reaction from them and calculate its standard potential. If positive, the oxidation will occur.

$S_2O_8^{2-}(aq) + 2\,e^- \longrightarrow 2\,SO_4^{2-}(aq)$ $\quad E°(\text{cathode}) = +2.05\text{ V}$

$Ag^{2+}(aq) + e^- \longrightarrow Ag^+(aq) \quad E°(\text{anode}) = +1.98\text{ V}$

Reverse the anode reaction and multiply by 2.

$S_2O_8^{2-} + 2\,e^- \longrightarrow 2\,SO_4^{2-}$

$2[Ag^+ \longrightarrow Ag^{2+} + e^-]$

$S_2O_8^{2-}(aq) + Ag^+(aq) \longrightarrow 2\,Ag^{2+}(aq) + 2\,SO_4^{2-}(aq) \quad E°_{cell} = 2.05\text{ V} - (+1.98\text{ V}) = +0.07\text{ V}$

Yes, the oxidation will work.

$$\ln K = \frac{nFE°}{RT}, \text{ at } 25°C = \frac{nE°}{0.025\,69\text{ V}} = \frac{(2)(0.07\text{ V})}{0.025\,69\text{ V}} = 5 \quad \text{and } K = 10^2$$

12.43 (a) $Pb^{4+}(aq) + 2e^- \longrightarrow Pb^{2+}(aq) \quad E°(\text{cathode}) = +1.67\text{ V}$

$Sn^{2+}(aq) \longrightarrow Sn^{4+}(aq) + 2e^- \quad E°(\text{anode}) = +0.15\text{ V}$

$Pb^{4+}(aq) + Sn^{2+}(aq) \longrightarrow Pb^{2+}(aq) + Sn^{4+}(aq) \quad E°_{cell} = 1.67\text{ V} - (0.15\text{ V}) = +1.52\text{ V}$

$$\text{Then, } E = E° - \left(\frac{0.025\,693\text{ V}}{n}\right)\ln Q; \quad 1.33\text{ V} = 1.52\text{ V} - \left(\frac{0.025\,693\text{ V}}{2}\right)\ln Q$$

$$\ln Q = \frac{1.52\text{ V} - 1.33\text{ V}}{0.0129\text{ V}} = \frac{0.19\text{ V}}{0.0129\text{ V}} = 15 \quad Q = 10^6$$

(b) $2[Cr_2O_7^{2-}(aq) + 14\,H^+(aq) + 6\,e^- \longrightarrow 2\,Cr^{3+}(aq) + 7\,H_2O(l)] \quad E°(\text{cathode}) = 1.33\text{ V}$

$3[2\,H_2O(l) \longrightarrow O_2(g) + 4\,H^+(aq) + 4\,e^-] \quad E°(\text{anode}) = +1.23\text{ V}$

$2\,Cr_2O_7^{2-}(aq) + 16\,H^+(aq) \longrightarrow 4\,Cr^{3+}(aq) + 8\,H_2O(l) + 3\,O_2(g) \quad E°_{cell} = 0.10\text{ V}$

$$\text{Then, } E = E° - \left(\frac{0.0257\text{ V}}{n}\right)\ln Q; \quad 0.10\text{ V} = +0.10\text{ V} - \left(\frac{0.0257\text{ V}}{12}\right)\ln Q$$

$\ln Q = 0.00 \quad Q = 1.0$

12.45 (a) $Cu^{2+}(aq, 0.010\text{ M}) + 2\,e^- \longrightarrow Cu(s)$ (cathode)

$Cu^{2+}(aq, 0.0010\text{ M}) + 2\,e^- \longrightarrow Cu(s)$ (anode)

$Cu^{2+}(aq, 0.010\text{ M}) \longrightarrow Cu^{2+}(aq, 0.0010\text{ M}),\ n = 2$

$E°_{cell} = E°(\text{cathode}) - E°(\text{anode}) = 0\text{ V}$

$$E_{cell} = E°_{cell} - \left(\frac{RT}{nF}\right)\ln Q = -\left(\frac{0.025\,693\text{ V}}{2}\right)\ln Q \text{ at } 25°C$$

$$E_{cell} = -\left(\frac{0.025\,693\text{ V}}{2}\right)\ln\left(\frac{0.0010\text{ M}}{0.010\text{ M}}\right) = +0.030\text{ V}$$

(b) at pH = 3.0, $[H^+] = 1 \times 10^{-3}$ M

at pH = 4.0, $[H^+] = 1 \times 10^{-4}$ M

Cell reaction is $H^+(aq, 1 \times 10^{-3}\ \text{M}) \longrightarrow H^+(aq, 1 \times 10^{-4}\ \text{M}),\ n = 1$

$$E°_{cell} = 0\ \text{V} \quad E_{cell} = E°_{cell} - \left(\frac{RT}{nF}\right) \ln Q = -\left(\frac{0.025\ 693\ \text{V}}{1}\right) \ln \left(\frac{1 \times 10^{-4}}{1 \times 10^{-3}}\right)$$

$$= +6 \times 10^{-2}\ \text{V}$$

12.47 In each case, $E°_{cell} = E°(\text{cathode}) - E°(\text{anode})$. Recall that the values for $E°$ at the electrodes refer to the electrode potential for the half-reaction written as a reduction reaction. In balancing the cell reaction, the half-reaction at the anode is reversed. However, this does not reverse the sign of electrode potential used at the anode, because the value always refers to the reduction potential.

(a) $2\ H^+(aq, 1.0\ \text{M}) + 2\ e^- \longrightarrow H_2(g, 1\ \text{atm}) \quad E°(\text{cathode}) = 0.00\ \text{V}$

$H_2(g, 1\ \text{atm}) \longrightarrow 2\ H^+(aq, 0.0010\ \text{M}) + 2\ e^- \quad E°(\text{anode}) = 0.00\ \text{V}$

$2\ H^+(aq, 1.0\ \text{M}) + H_2(g, 1\ \text{atm}) \longrightarrow 2\ H^+(aq, 0.0010\ \text{M}) + H_2(g, 1\ \text{atm})$

$E°_{cell} = 0.00\ \text{V}$

$$\text{Then, } E = E° - \left(\frac{0.025\ 693\ \text{V}}{n}\right) \ln \left(\frac{[H^+, 0.0010\ \text{M}]^2 P_{H_2}}{[H^+, 1.0\ \text{M}]^2 P_{H_2}}\right)$$

$$E = 0.00\ \text{V} - \left(\frac{0.025\ 693\ \text{V}}{2}\right) \ln \left(\frac{(0.0010\ \text{M})^2 \times 1\ \text{atm}}{(1.0\ \text{M})^2 \times 1\ \text{atm}}\right)$$

$$E = -0.0129\ \text{V} \ln (1.0 \times 10^{-3})^2 = +0.18\ \text{V}$$

(b) $Ni^{2+}(aq) + 2\ e^- \longrightarrow Ni(s) \quad E°(\text{cathode}) = -0.23\ \text{V}$

$Zn(s) \longrightarrow Zn^{2+}(aq) + 2\ e^- \quad E°(\text{anode}) = -0.76\ \text{V}$

$Ni^{2+}(aq) + Zn(s) \longrightarrow Ni(s) + Zn^{2+}(aq) \quad E°_{cell} = +0.53\ \text{V}$

$$\text{Then, } E = E° - \left(\frac{0.025\ 693\ \text{V}}{n}\right) \ln \left(\frac{[Zn^{2+}]}{[Ni^{2+}]}\right)$$

$$E = 0.53\ \text{V} - \left(\frac{0.025\ 693\ \text{V}}{2}\right) \ln \left(\frac{0.10}{0.001}\right) = 0.53\ \text{V} - 0.06\ \text{V} = 0.47\ \text{V}$$

(c) $2\ H^+(aq) + 2\ e^- \longrightarrow H_2(g) \quad E°(\text{cathode}) = 0.00\ \text{V}$

$2\ Cl^-(aq) \longrightarrow Cl_2(g) + 2\ e^- \quad E°(\text{anode}) = +1.36\ \text{V}$

$2\ H^+(aq) + 2\ Cl^-(aq) \longrightarrow H_2(g) + Cl_2(g) \quad E°_{cell} = -1.36\ \text{V}$

Then,

$$E = E° - \left(\frac{0.025\ 693\ \text{V}}{n}\right) \ln \left(\frac{P_{H_2} P_{Cl_2}}{[H^+]^2 [Cl^-]^2}\right)$$

$$E = -1.36\ \text{V} - \left(\frac{0.025\ 693\ \text{V}}{2}\right) \ln \left(\frac{\left(\frac{450}{760}\right)\left(\frac{100}{760}\right)}{(0.010)^2 (1.0)^2}\right)$$

$$E = -1.36\ \text{V} - (0.0129\ \text{V}) \ln \left(\frac{(0.592)(0.132)}{(1.0 \times 10^{-4})(1)}\right) = -1.36\ \text{V} - 0.09\ \text{V}$$

$$= -1.45\ \text{V}$$

(d) $Sn^{4+}(aq, 0.060\ \text{M}) + 2\ e^- \longrightarrow Sn^{2+}(aq, 1.0\ \text{M})$ $E°(\text{cathode}) = +0.15$ V

$Sn(s) \longrightarrow Sn^{2+}(aq, 0.020\ \text{M}) + 2\ e^-$ $E°(\text{anode}) = -0.14$ V

$Sn^{4+}(aq, 0.060\ \text{M}) + Sn(s) \longrightarrow Sn^{2+}(aq, 1.0\ ) + Sn^{2+}(aq, 0.020\ \text{M})$

$E°_{cell} = 0.29$ V

$$E = E° - \left(\frac{0.025\ 693\ \text{V}}{n}\right) \ln \left(\frac{(1.0)(0.020)}{(0.060)}\right)$$

$$E = 0.29\ \text{V} - \left(\frac{0.025\ 693\ \text{V}}{2}\right) \ln \left(\frac{(1)(0.020)}{(0.060)}\right) = 0.30\ \text{V}$$

12.49 In each case, obtain the balanced equation for the cell reaction from the half-cell reactions at the electrodes, by reversing the reduction equation for the half-reaction at the anode, multiplying the half-reaction equations by an appropriate factor to balance the number of electrons, and then adding the half-reactions. Calculate $E°_{cell} = E°(\text{cathode}) - E°(\text{anode})$. Then write the Nernst equation for the cell reaction and solve for the unknown.

(a) $Hg_2Cl_2(s) + 2\ e^- \longrightarrow 2\ Hg(l) + 2\ Cl^-(aq)$ $E°(\text{cathode}) = +0.27$ V

$H_2(g) \longrightarrow 2\ H^+(aq) + 2\ e^-$ $E°(\text{anode}) = 0.00$ V

$H_2(g) + Hg_2Cl_2(s) \longrightarrow 2\ H^+(aq) + 2\ Hg(l) + 2\ Cl^-(aq)$ $E°_{cell} = +0.27$ V

$$E = E° - \left(\frac{0.025\ 693\ \text{V}}{n}\right) \ln \left(\frac{[H^+]^2[Cl^-]^2}{[H_2]}\right)$$

$$0.33\ \text{V} = 0.27\ \text{V} - \left(\frac{0.025\ 693\ \text{V}}{2}\right) \ln \left(\frac{[H^+]^2(1)^2}{(1)}\right)$$

$$= 0.27\ \text{V} - (0.0129\ \text{V})\ln[H^+]^2$$

$$0.06\ \text{V} = -0.0257\ \text{V} \ln[H^+] = -0.0257\ \text{V} \times (2.303 \log [H^+])$$

$$\text{pH} = \frac{0.06\ \text{V}}{(2.303)(0.025\ 693\ \text{V})} = 1$$

(b) $2[MnO_4^-(aq) + 8\ H^+(aq) + 5\ e^- \longrightarrow Mn^{2+}(aq) + 4\ H_2O(l)]$

$E°(\text{cathode}) = +1.51$ V

$5[2\ Cl^-(aq) \longrightarrow Cl_2(g) + 2\ e^-]$ $E°(\text{anode}) = +1.36$ V

$2\ MnO_4^-(aq) + 16\ H^+(aq) + 10\ Cl^-(aq) \longrightarrow 5\ Cl_2(g) + 2\ Mn^{2+}(aq) + 8\ H_2O(l)$ $E°_{cell} = +0.15$ V

$$E = E° - \left(\frac{0.0257\ \text{V}}{n}\right) \ln \left(\frac{[Cl_2]^5[Mn^{2+}]^2}{[MnO_4]^2[H^+]^{16}[Cl^-]^{10}}\right)$$

$$-0.30\ \text{V} = +0.15\ \text{V} - \left(\frac{0.0257\ \text{V}}{10}\right) \ln \left(\frac{(1)^5(0.10)^2}{(0.010)^2(1 \times 10^{-4})^{16}(Cl^-)^{10}}\right)$$

$$-0.45\ \text{V} = -(0.002\ 5693\ \text{V}) \log\left(\frac{1 \times 10^{-2}}{(1 \times 10^{-4})(1 \times 10^{-64})[\text{Cl}^-]^{10}}\right)$$

$$= -0.002\ 5693\ \text{V}\left[\ln(1 \times 10^{66}) + \ln\left(\frac{1}{[\text{Cl}^-]^{10}}\right)\right]$$

$$= -0.390\ \text{V} + (0.0025\ 693\text{V}) \ln[\text{Cl}^-]^{10}$$

$$-0.0594\ \text{V} = 0.002\ 5693\ \text{V} \ln[\text{Cl}^-]^{10}$$

$$= (0.025\ 693\ \text{V}) \ln[\text{Cl}^-]$$

$$\ln[\text{Cl}^-] = \frac{-0.06\ \text{V}}{0.025\ 693\ \text{V}} = -2$$

$$[\text{Cl}^-] = 10^{-1}\ \text{mol} \cdot \text{L}^{-1}$$

12.51 To calculate this value, we need to determine the $E°$ value for the solubility reaction:

$AgI(s) \rightleftharpoons Ag^+(aq) + I^-(aq) \quad E° = ?$

The relation $\Delta G° = -nRT \ln K = -nFE°$ can be used to calculate the value of K_{sp}. The equations that will add to give the net equation we want are

$AgI(s) + e^- \longrightarrow Ag(s) + I^-(aq) \quad E° = -0.15\ \text{V}$

$Ag(s) \longrightarrow Ag^+(aq) + e^- \quad E° = -0.80\ \text{V}$

Notice that the second equation is reversed from the reduction reaction given in the Appendix, and consequently the $E°$ value is changed in sign.

Adding these two equations together gives the desired net reaction, and summing the $E°$ values will give the $E°$ value for that process:

$E° = (-0.15\ \text{V}) + (-0.80\ \text{V}) = -0.95\ \text{V}$

$$\ln K_{sp} = \frac{nFE°}{RT} = \frac{(1)(9.65 \times 10^4\ \text{C}\cdot\text{mol}^{-1})(-0.95\ \text{V})}{(8.314\ \text{J}\cdot\text{K}^{-1}\cdot\text{mol}^{-1})(298\ \text{K})} = -37.00$$

$K_{sp} = 8.5 \times 10^{-17}$

(b) This value is of the same order of magnitude as the value reported in Table 11.5 (1.5×10^{-16}).

12.53 For the standard calomel electrode, $E° = +0.27$ V. If this were set equal to 0, all other potentials would also be decreased by 0.27 V. (a) Therefore, the standard hydrogen electrode's standard reduction potential would be 0.00 V − 0.27 V or −0.27 V. (b) The standard reduction potential for Cu^{2+}/Cu would be 0.34 V − 0.27 V or +0.07 V.

12.55 $Fe^{3+}(aq) + 3\ e^- \longrightarrow Fe(s) \quad E° = -0.04\ \text{V}$

$Cr^{3+}(aq) + 3\ e^- \longrightarrow Cr(s) \quad E° = -0.74\ \text{V}$

$Fe^{2+}(aq) + 2\ e^- \longrightarrow Fe(s) \quad E° = -0.44\ \text{V}$

$Cr^{2+}(aq) + 2\ e^- \longrightarrow Cr(s) \quad E° = -0.91\ V$

Comparison of the reduction potentials shows that Cr is more easily oxidized than Fe, so the presence of Cr retards the rusting of Fe. At the position of the scratch, the gap is filled with oxidation products of Cr, thereby preventing contact of air and water with the iron.

12.57 (a) $Fe_2O_3 \cdot H_2O$ (b) H_2O and O_2 jointly oxidize iron. (c) Water is more highly conducting if it contains dissolved ions, so the rate of rusting is increased.

12.59 (a) aluminum or magnesium; both are below titanium in the electrochemical series.

(b) cost, availability, and toxicity of products in the environment

(c) $Cu^{2+} + 2\ e^- \longrightarrow Cu(s) \quad E° = +0.34\ V$

$Cu^+ + e^- \longrightarrow Cu(s) \quad E° = +0.52\ V$

$Fe^{3+} + 3\ e^- \longrightarrow Fe(s) \quad E° = -0.04\ V$

$Fe^{2+} + 2\ e^- \longrightarrow Fe(s) \quad E° = -0.44\ V$

Fe could act as the anode of an electrochemical cell if Cu^{2+} or Cu^+ were present; therefore, it could be oxidized at the point of contact. Water with dissolved ions would act as the electrolyte.

12.61 See Figs. 12.19 and 12.21. (a) anode (same as galvanic cell) (b) positive (opposite of galvanic cell)

12.63 The strategy is to consider the possible competing cathode and anode reactions. At the cathode, choose the reduction reaction with the most positive (least negative) standard reduction potential ($E°$ value). At the anode, choose the oxidation reaction with the least positive (most negative) standard reduction potential ($E°$ value, as given in the table). Then calculate $E°_{cell} = E°(\text{cathode}) - E°(\text{anode})$. The negative of this value is the minimum potential that must be supplied.

(a) cathode: $Co^{2+}(aq) + 2\ e^- \longrightarrow Co(s) \quad E° = -0.28\ V$

(rather than $2\ H_2O(l) + 2\ e^- \longrightarrow H_2(g) + 2\ OH^-(aq) \quad E° = -0.83\ V$)

(b) anode: $2\ H_2O(l) \longrightarrow O_2(g) + 4\ H^+(aq) + 4\ e^- \quad E° = +1.23\ V$

(the SO_4^{2-} ion will not oxidize)

(c) $E°_{cell} = E°(\text{cathode}) - E°(\text{anode}) = -0.28\ V - (+0.81\ V) = -1.09\ V$

Therefore E (supplied) must be $> +1.09\ V$ (1.09 V is the minimum).

12.65 In each case, compare the reduction potential of the ion to the reduction potential of water ($E° = -0.42$ V) and choose the process with the least negative $E°$ value.

(a) $Mn^{2+}(aq) + 2\,e^- \longrightarrow Mn(s)$ $E° = -1.18$ V

(b) $Al^{3+}(aq) + 3\,e^- \longrightarrow Al(s)$ $E° = -1.66$ V

The reactions in (a) and (b) evolve hydrogen rather than yield a metallic deposit because water is reduced, according to $2\,H_2O(l) + 2\,e^- \longrightarrow H_2(g) + 2\,OH^-(aq)$ ($E° = -0.42$ V, at pH = 7)

(c) $Ni^{2+}(aq) + 2\,e^- \longrightarrow Ni(s)$ $E° = -0.23$ V

(d) $Au^{3+}(aq) + 3\,e^- \longrightarrow Au(s)$ $E° = +1.69$ V

In (c) and (d) the metal ion will be reduced.

12.67 (a) $Cu^{2+} + 2\,e^- \longrightarrow Cu$

$$\text{amount (moles) of } e^- = (5.12\text{ g Cu})\left(\frac{1\text{ mol Cu}}{63.54\text{ g Cu}}\right)\left(\frac{2\text{ mol } e^-}{1\text{ mol Cu}}\right) = 0.161\text{ mol } e^-$$

(b) $Al^{3+} + 3\,e^- \longrightarrow Al$

$$\text{amount (moles) of } e^- = (200\text{ g Al})\left(\frac{1\text{ mol Al}}{26.98\text{ g Al}}\right)\left(\frac{3\text{ mol } e^-}{1\text{ mol Al}}\right) = 22.2\text{ mol } e^-$$

(c) $2\,H_2O \longrightarrow O_2 + 4\,H^+ + 4\,e^-$

$$\text{amount (moles) of } e^- = (200\text{ L } O_2)\left(\frac{1\text{ mol } O_2}{22.4\text{ L } O_2}\right)\left(\frac{4\text{ mol } e^-}{1\text{ mol } O_2}\right) = 35.7\text{ mol } e^-$$

Note: 22.4 L is the volume of 1 mole of an ideal gas at 273 K and 1 atm.

12.69 (a) $Ag^+(aq) + e^- \longrightarrow Ag(s)$

$$\text{time} = (4.4\text{ mg Ag})\left(\frac{10^{-3}\text{ g}}{1\text{ mg}}\right)\left(\frac{1\text{ mol Ag}}{107.98\text{ g Ag}}\right)\left(\frac{1\text{ mol } e^-}{1\text{ mol Ag}}\right)\left(\frac{9.65\times10^4\text{ C}}{1\text{ mol } e^-}\right)\left(\frac{1\text{ A}\cdot\text{s}}{1\text{ C}}\right)\left(\frac{1}{0.50\text{ A}}\right) = 7.9\text{ s}$$

(b) $Cu^{2+}(aq) + 2\,e^- \longrightarrow Cu(s)$

$$\text{mass Cu} = (7.9\text{ s})(0.50\text{ A})\left(\frac{1\text{ C}}{1\text{ A}\cdot\text{s}}\right)\left(\frac{1\text{ mol } e^-}{9.65\times10^4\text{ C}}\right)\left(\frac{0.50\text{ mol Cu}}{1\text{ mol } e^-}\right)\left(\frac{63.5\text{ g Cu}}{1\text{ mol Cu}}\right) = 1.3\text{ mg Cu}$$

12.71 (a) $Cr(VI) + 6\,e^- \longrightarrow Cr(s)$

$$\text{current} = \frac{\text{charge}}{\text{time}}$$

$$= \frac{4.0\text{ g Cr}\left(\frac{1\text{ mol Cr}}{52.00\text{ g Cr}}\right)\left(\frac{6\text{ mol e}^-}{1\text{ mol Cr}}\right)\left(\frac{9.65\times10^4\text{ C}}{1\text{ mol e}^-}\right)}{24\text{ h}\times3600\text{ s}\cdot\text{h}^{-1}}$$

$$= 0.52\text{ C}\cdot\text{s}^{-1} = 0.52\text{ A}$$

(b) $Na^+ + e^- \longrightarrow Na(s)$

$$\text{current} = \frac{4.0\text{ g Na}\left(\frac{1\text{ mol Na}}{22.99\text{ g Na}}\right)\left(\frac{1\text{ mol e}^-}{1\text{ mol Na}}\right)\left(\frac{9.65\times10^4\text{ C}}{1\text{ mol e}^-}\right)}{24\text{ h}\times3600\text{ s}\cdot\text{h}^{-1}}$$

$$= 0.19\text{ C}\cdot\text{s}^{-1} = 0.19\text{ A}$$

12.73 $Ru^{n+}(aq) + n\,e^- \longrightarrow Ru(s)$; solve for n

$$\text{moles of Ru} = (0.0310\text{ g Ru})\left(\frac{1\text{ mol}}{101.07\text{ g Ru}}\right) = 3.07\times10^{-4}\text{ mol}$$

$$\text{total charge} = (500\text{ s})(120\text{ mA})\left(\frac{10^{-3}\text{ A}}{1\text{ mA}}\right)\left(\frac{1\text{ C}\cdot\text{s}^{-1}}{1\text{ A}}\right) = 60\text{ C}$$

$$\text{moles of e}^- = (60\text{ C})\left(\frac{1\text{ mol e}^-}{96\,500\text{ C}}\right) = 6.2\times10^{-4}\text{ mol e}^-$$

$$n = \frac{6.2\times10^{-4}\text{ mol e}^-}{3.07\times10^{-4}\text{ mol}} = \frac{2\text{ mol charge}}{1\text{ mol}}$$

Therefore, oxidation number of Ru^{2+} is +2.

12.75 $Zn^{2+} + 2\,e^- \longrightarrow Zn$ (2 mol e^-/1 mol Zn)

$$\text{charge used} = (1.0\text{ mA})\left(\frac{10^{-3}\text{ A}}{1\text{ mA}}\right)(31\text{ d})\left(\frac{24\text{ h}}{1\text{ d}}\right)\left(\frac{3600\text{ s}}{1\text{ h}}\right)$$

$$= 2.7\times10^3\text{ C}$$

$$\text{moles of e}^-\text{ used} = (2.7\times10^3\text{ C})\left(\frac{1\text{ mol e}^-}{96\,500\text{ C}}\right) = 2.8\times10^{-2}\text{ mol e}^-$$

$$\text{moles of Zn} = (2.8\times10^{-2}\text{ mol e}^-)\left(\frac{1\text{ mol Zn}}{2\text{ mol e}^-}\right) = 1.4\times10^{-2}\text{ mol Zn}$$

$$\text{mass of Zn} = (1.4\times10^{-2}\text{ mol Zn})\left(\frac{65.37\text{ g Zn}}{1\text{ mol Zn}}\right) = 0.92\text{ g Zn}$$

12.77 A primary cell is the primary source of the electrical energy produced by its operation. A secondary cell is an energy storage device that stores electrical energy produced elsewhere and releases it upon operation (discharge). It is the secondary source of the electrical energy. Most primary cells produce electricity from chemicals that were sealed into them when they were made. They are not normally rechargeable. A secondary cell is one that must be charged from some other electrical supply before use. It is normally rechargeable.

12.79 (a) The electrolyte is KOH(aq)/HgO(s), which will have the consistency of a moist paste.
(b) The oxidizing agent is HgO(s).
(c) $HgO(s) + Zn(s) \longrightarrow Hg(l) + ZnO(s)$

12.81 See Box 12.1.
The anode reaction is $Zn(s) \longrightarrow Zn^{2+}(aq) + 2\ e^-$; this reaction supplies the electrons to the external circuit. The cathode reaction is $MnO_2(s) + H_2O(l) + e^- \longrightarrow MnO(OH)_2(s) + OH^-(aq)$. The $OH^-(aq)$ produced reacts with $NH_4^+(aq)$ from the $NH_4Cl(aq)$ present: $NH_4^+(aq) + OH^-(aq) \longrightarrow H_2O(l) + NH_3(g)$. The $NH_3(g)$ produced complexes with the $Zn^{2+}(aq)$ produced in the anode reaction: $Zn^{2+}(aq) + 4\ NH_3(g) \longrightarrow [Zn(NH_3)_4]^{2+}(aq)$. The overall reaction is complicated.

12.83 See Table 12.2. (a) KOH(aq) (b) In the charging process, the cell reaction is the reverse of what occurs in discharge. Therefore, at the anode, $2\ Ni(OH)_2(s) + 2\ OH^-(aq) \longrightarrow 2\ Ni(OH)_3 + 2\ e^-$.

12.85

$$2[Zn^{2+}(aq) + 2\ e^- \longrightarrow Zn(s)] \qquad E°(\text{cathode}) = -0.76\ V$$
$$M(s) \longrightarrow M^{4+}(aq) + 4\ e^- \qquad E°(\text{anode}) = x$$
$$M(s) + 2\ Zn^{2+}(aq) \longrightarrow 2\ Zn(s) + M^{4+}(aq) \qquad E°_{\text{cell}} = 0.16\ V$$

$E°_{\text{cell}} = E°(\text{cathode}) - E°(\text{anode})$

$+0.16\ V = -0.76\ V - (x)$

$x = -0.92\ V = E°(M^{4+}/M)$

12.87 The strategy is to find the $E°$ value for the solubility reaction and then find appropriate half reactions which add to give that solubility reaction. One of these half reactions is our unknown, the other is obtained from Appendix 2B:

$$Cu(IO_3)_2(s) + 2\ e^- \longrightarrow Cu(s) + 2\ IO_3^-(aq) \qquad E° = ? \qquad (A)$$
$$Cu(s) \longrightarrow Cu^{2+}(aq) + 2\ e^- \qquad E° = -0.34\ V \qquad (B)$$
$$Cu(IO_3)_2(s) + \longrightarrow Cu^{2+}(aq) + 2\ IO_3^-(aq) \qquad (C)$$

$$E° = \frac{RT \ln K_{sp}}{nF}$$

$$= \frac{(8.314\ J\cdot K^{-1}\cdot mol^{-1})(298.2\ K) \ln(1.4 \times 10^{-7})}{2(9.65 \times 10^4\ C\cdot mol^{-1})}$$

$$= -0.20\ V$$

$-0.20\text{ V} = E°(\text{A}) + (-0.34\text{ V})$

$E°(\text{A}) = +0.14\text{ V}$

12.89 $\text{Au} \longrightarrow \text{Au}^{3+} + 3\,e^-$ $E°(\text{anode}) = +1.40\text{ V}$

$MnO_4^- + 8\,H^+ + 5\,e^- \longrightarrow Mn^{2+} + 4\,H_2O$ $E°(\text{cathode}) = +1.51\text{ V}$

$E°_{\text{cell}} = E°(\text{cathode}) - E°(\text{anode}) = 1.51\text{ V} - (+1.40\text{ V}) = +0.11\text{ V}$

Therefore, spontaneous; so gold will be oxidized under standard conditions.

However, $\text{Au} \longrightarrow \text{Au}^{3+} + 3\,e^-$ $E°(\text{anode}) = +1.40\text{ V}$

$Cr_2O_7^{2-} + 14\,H^+ + 6\,e^- \longrightarrow 2\,Cr^{3+} + 7\,H_2O$ $E°(\text{cathode}) = 1.33\text{ V}$

$E°_{\text{cell}} = E°(\text{cathode}) - E°(\text{anode}) = 1.33\text{ V} - (+1.40\text{ V}) = -0.07\text{ V}$

Therefore, not spontaneous; so gold will not be oxidized under standard conditions.

12.91 In each case, determine the cathode and anode half-reactions corresponding to the reaction *as written*. Look up the standard reduction potentials for these half-reactions and then calculate $E°_{\text{cell}} = E°(\text{cathode}) - E°(\text{anode})$. If $E°_{\text{cell}}$ is positive, the reaction is spontaneous under standard conditions.

(a) $E°_{\text{cell}} = E°(\text{cathode}) - E°(\text{anode}) = +0.96\text{ V} - (+0.79\text{ V}) = +0.17\text{ V}$

Therefore, spontaneous galvanic cell:

$\text{Hg(l)} \mid Hg_2^{2+}\text{(aq)} \parallel NO_3^-\text{(aq)}, H^+\text{(aq)} \mid \text{NO(g)} \mid \text{Pt(s)}$

$\Delta G°_r = -nFE° = -(6)(9.65 \times 10^4\text{ C}\cdot\text{mol}^{-1})(+0.17\text{ J}\cdot\text{C}^{-1}) = -98\text{ kJ}\cdot\text{mol}^{-1}$

(b) $E°_{\text{cell}} = E°(\text{cathode}) - E°(\text{anode}) = +0.92\text{ V} - (+1.09\text{ V}) = -0.17\text{ V}$

Therefore, not spontaneous.

(c) $E°_{\text{cell}} = E°(\text{cathode}) - E°(\text{anode}) = +1.33\text{ V} - (+0.97\text{ V}) = +0.36\text{ V}$

Therefore, spontaneous galvanic cell.

$\text{Pt(s)} \mid Pu^{3+}\text{(aq)}, Pu^{4+}\text{(aq)} \parallel Cr_2O_7^{2-}\text{(aq)}, Cr^{3+}\text{(aq)}, H^+\text{(aq)} \mid \text{Pt(s)}$

$\Delta G°_r = -nFE° = -(6)(9.65 \times 10^4\text{ C}\cdot\text{mol}^{-1})(0.36\text{ J}\cdot\text{C}^{-1}) = -208\text{ kJ}\cdot\text{mol}^{-1}$

12.93 In each case, break down the solution equilibrium reaction into two half-reactions and identify the cathode and anode half-reaction corresponding to the dissolution reaction. Then calculate $E°_{\text{cell}} = E°(\text{cathode}) - E°(\text{anode})$. From this, calculate K_{sp} and, finally, the solubility.

(a) $\text{AgCl(s)} + e^- \longrightarrow \text{Ag(s)} + Cl^-\text{(aq)}$ $E°(\text{cathode}) = +0.22\text{ V}$

$\text{Ag(s)} \longrightarrow Ag^+\text{(aq)} + e^-$ $E°(\text{anode}) = +0.80\text{ V}$

$\text{AgCl(s)} \longrightarrow Ag^+\text{(aq)} + Cl^-\text{(aq)}$ $E°_{\text{cell}} = -0.58\text{ V}$

$$\ln K_{sp} = \frac{n(E°_{\text{cell}})}{0.0257\text{ V}} = \frac{(1)(-0.58)}{0.0257} = -23$$

$K_{sp} = 10^{-10}$

Because the solubility of AgCl $= \sqrt{K_{sp}}$, the solubility of AgCl $= 10^{-5}\ \text{mol}\cdot\text{L}^{-1}$.

(b) $Hg_2Cl_2(s) + 2\ e^- \longrightarrow 2\ Hg(l) + 2\ Cl^-(aq)$ $E°(\text{cathode}) = +0.27$ V

$Hg^0(l) \longrightarrow Hg_2^{2+}(aq) + 2\ e^-$ $E°(\text{anode}) = +0.79$ V

$Hg_2Cl_2(s) \longrightarrow Hg_2^{2+}(aq) + 2\ Cl^-(aq)$ $E°_{cell} = -0.52$ V

$$\ln K_{sp} = \frac{n(E°_{cell})}{0.0257\ \text{V}} = \frac{(2)(-0.52)}{0.0257} = -40$$

$K_{sp} = 10^{-18}$

For Hg_2Cl_2, $K_{sp} = 10^{-18} = (S)(2\ S)^2 = 4\ S^3$ $S = 10^{-6}\ \text{mol}\cdot\text{L}^{-1}$

(c) $PbSO_4(s) + 2\ e^- \longrightarrow Pb + SO_4^{2-}(aq)$ $E°(\text{cathode}) = -0.36$ V

$Pb \longrightarrow Pb^{2+}(aq) + 2\ e^-$ $E°(\text{anode}) = -0.13$ V

$PbSO_4(s) \longrightarrow Pb^{2+}(aq) + SO_4^{2-}(aq)$ $E°_{cell} = -0.23$ V

$$\ln K_{sp} = \frac{n(E°_{cell})}{0.0257\ \text{V}} = \frac{(2)(-0.23)}{0.0257\ \text{V}} = -18$$

$K_{sp} = 10^{-8}$

Solving the K_{sp} expression for $PbSO_4$ yields the solubility of $PbSO_4 = \sqrt{10^{-8}} = 10^{-4}\ \text{mol}\cdot\text{L}^{-1}$

12.95 $F_2(g) + 2\ e^- \longrightarrow 2\ F^-(aq)$ $E°(\text{cathode}) = +2.87$ V

$2\ HF(aq) \longrightarrow F_2(g) + 2\ H^+(aq) + 2\ e^-$ $E°(\text{anode}) = +3.03$ V

$2\ HF(aq) \longrightarrow 2\ H^+(aq) + 2\ F^-(aq)$ $E°_{cell} = -0.16$ V

For the above reaction, $K = \dfrac{[H^+]^2[F^-]^2}{[HF]^2}$ and $\ln K = \dfrac{nFE°}{RT}$

$$\text{at } 25°\text{C} = \frac{nE°}{0.025\ 69\ \text{V}} = \frac{(2)(-0.16\ \text{V})}{0.025\ 69\ \text{V}} = -12$$

$K = 10^{-5}$

$K_a = \sqrt{K} = \sqrt{10^{-5}} = 10^{-3}$

12.97 $Ag^+(aq) + e^- \longrightarrow Ag$ $E°(\text{cathode}) = +0.80$ V

$Fe^{2+}(aq) \longrightarrow Fe^{3+}(aq) + e^-$ $E°(\text{anode}) = +0.77$ V

$Ag^+(aq) + Fe^{2+}(aq) \longrightarrow Fe^{3+}(aq) + Ag(s)$ $E°_{cell} = +0.03$ V

$$E_{cell} = E°_{cell} - \left(\frac{0.0257\ \text{V}}{n}\right) \ln\left(\frac{[Fe^{3+}]}{[Ag^+][Fe^{2+}]}\right)$$

$$= 0.03\ \text{V} - (0.0257\ \text{V}) \ln\left(\frac{1}{(0.010)(0.0010)}\right) = 0.03\ \text{V} - 0.30\ \text{V}$$

$$= -0.27\ \text{V}$$

Comment: The cell changes from spontaneous to nonspontaneous as a function of concentration.

12.99 The wording of this exercise suggests that K^+ ions participate in an electrolyte concentration cell reaction. Therefore, $E°_{cell} = 0.00$ V, because the two half cells would be identical under standard conditions.

Then,

$$E = E° - \left(\frac{0.0257\ \text{V}}{n}\right) \ln\left(\frac{[\text{K}_{out}^+]}{[\text{K}_{in}^+]}\right) = 0.00\ \text{V} - \left(\frac{0.0257\ \text{V}}{1}\right) \ln\left(\frac{1}{30}\right)$$

$$= +0.09\ \text{V}$$

and $$E = 0.00\ \text{V} - \left(\frac{0.0257\ \text{V}}{1}\right) \ln\left(\frac{1}{20}\right) = +0.08\ \text{V}$$

The range of potentials is 0.08 V to 0.09 V.

12.101 (a) Closed: energy is exchanged, but not matter.

(b) Open: matter, in the form of fuel, continuously enters the system, and reaction products are continuously expelled.

(c) Closed: energy, but not matter, is exchanged with the surroundings.

12.103 $Ti^{n+} + ne^- \longrightarrow Ti(s)$; solve for n.

Charge consumed $= 4.70\ \text{C}\cdot\text{s}^{-1} \times 6.00\ \text{h} \times 3600\ \text{s}\cdot\text{h}^{-1} = 1.02 \times 10^5\ \text{C}$

$$\text{Moles of charge consumed} = \frac{1.02 \times 10^5\ \text{C}}{9.6485 \times 10^4\ \text{C}\cdot\text{mol}^{-1}} = 1.06\ \text{mol e}^-$$

$$\text{Moles of Ti lost} = \frac{12.57\ \text{g Ti}}{47.88\ \text{g Ti/mol Ti}} = 0.2625\ \text{mol Ti}$$

$$n = \frac{1.06\ \text{mol e}^-}{0.2625\ \text{mol Ti}} = \frac{4.04\ \text{mol e}^-}{1.0\ \text{mol Ti}} = (\text{necessarily})\ \frac{4\ \text{mol e}^-}{1\ \text{mol Ti}}$$

Therefore, the oxidation number is 4, i.e., the species is Ti^{4+}.

12.105 First reaction:

$Au(s) + OH^-(aq) + CN^-(aq) + O_2(aq) \longrightarrow Au(CN)_2^-(aq)$

at anode: $Au(s) + 2\ CN^-(aq) \longrightarrow Au(CN)_2^-(aq) + e^-$

$O_2(aq) \longrightarrow 2\ H_2O(l)$ (O's balanced)

$O_2(aq) + 4\ H_2O(l) \longrightarrow 2\ H_2O(l) + 4\ OH^-(aq)$ (H's balanced)

$O_2(aq) + 4\ H_2O(l) + 4\ e^- \longrightarrow 2\ H_2O(l) + 4\ OH^-(aq)$ (charge balanced)

at cathode: $O_2(aq) + 2\ H_2O(l) + 4\ e^- \longrightarrow 4\ OH^-(aq)$

Combining the two half-reactions:

$4[2\ CN^-(aq) + Au(s) \longrightarrow Au(CN)_2^-(aq) + e^-]$

$1[O_2(aq) + 2\ H_2O(l) + 4\ e^- \longrightarrow 4\ OH^-(aq)]$

$8\ CN^-(aq) + 4\ Au(s) + O_2(aq) + 2\ H_2O(l) \longrightarrow 4\ Au(CN)_2^-(aq) + 4\ OH^-(aq)$

Second reaction:

$Au(CN)_2^-(aq) + Zn(s) \longrightarrow Au(s) + Zn(CN)_4^{2-}(aq)$

The first half-reaction is the reverse of the precipitation reaction above:

$Au(CN)_2^-(aq) + e^- \longrightarrow Au(s) + 2\ CN^-(aq)$

The second half-reaction:

$Zn(s) + 4\ CN^-(aq) \longrightarrow Zn(CN)_4^{2-}(aq) + 2\ e^-$

Combining the two half-reactions:

$2[Au(CN)_2^-(aq) + e^- \longrightarrow Au(s) + 2\ CN^-(aq)]$

$1[Zn(s) + 4\ CN^-(aq) \longrightarrow Zn(CN)_4^{2-}(aq) + 2\ e^-]$

$2\ Au(CN)_2^-(aq) + Zn(s) \longrightarrow 2\ Au(s) + Zn(CN)_4^{2-}(aq)$

12.107 (a) $O_2(g) + 4\ H^+(aq) + 4\ e^- \longrightarrow 2\ H_2O(l)$ $E°(\text{cathode}) = +1.23\ V$

$2[Sn^{2+}(aq) \longrightarrow Sn^{4+}(aq) + 2\ e^-]$ $E°(\text{anode}) = +0.15\ V$

$E°_{cell} = +1.08\ V$

(b) $O_2(g) + 4\ H^+(aq) + 2\ Sn^{2+}(aq) \longrightarrow 2\ Sn^{4+}(aq) + 2\ H_2O(l)$

(c) $$E_{cell} = E°_{cell} - \frac{0.025\ 693\ V}{n} \ln \frac{[Sn^{4+}]^2}{[Sn^{2+}]^2[H^+]^4[O_2]}$$

$$E_{cell} = 1.08\ V - \frac{0.025\ 693\ V}{4} \ln \frac{(0.010)^2}{(0.10)^2(1.0 \times 10^{-4})^4(1)}$$

$E_{cell} = 1.08\ V - 0.21\ V = 0.87\ V$

(d) $$\ln K = \frac{nFE°}{RT} = \frac{nE°}{0.025\ 693\ V} = \frac{(4)(0.87)}{0.025\ 693} = 1.4 \times 10^2 \text{ and } K = 6 \times 10^{60}$$

(e) $\Delta G_r = -nFE = -(4)(9.6485) \times 10^4\ C \cdot mol^{-1})(0.87\ J \cdot C^{-1})$

$= -3.4 \times 10^2\ kJ \cdot mol^{-1}$

(f) $$E_{cell} = E°_{cell} - \frac{0.025\ 693\ V}{n} \ln \frac{[Sn^{4+}]^2}{[Sn^{2+}]^2[H^+]^4[O_2]}$$

$$0.89\ V = 1.08\ V - \frac{0.025\ 693\ V}{4} \ln \frac{(0.010)^2}{(0.10)^2(H^+)^4(1)}$$

Solving, $[H^+] = 1.9 \times 10^{-4}\ mol \cdot L^{-1}$

and $pH = -\log[H^+] = -\log(1.9 \times 10^{-4}) = 3.72$

12.109 buffer system = $HA \rightleftharpoons H^+ + A^-$

$$Q = \frac{(H^+)(A^-)}{(HA)}$$

Note: (H^+), as opposed to $[H^+]$, indicates a nonequilibrium molarity.

Because in a buffer system $(A) \approx (HA)$, we can write

$Q = (\mathrm{H}^+)$

$$E_{\text{cell}} = E°_{\text{cell}} - \frac{RT}{nF}\ln(\mathrm{H}^+)$$

$$0.060\ \mathrm{V} = E°_{\text{cell}} - \left(\frac{0.025\ 693}{1}\right)(2.303)(\log(\mathrm{H}^+))$$

Because $\log(\mathrm{H}^+) = -[-\log(\mathrm{H}^+)] = -\mathrm{pH}$, we have

$0.060\ \mathrm{V} = E°_{\text{cell}} - 0.0592 \times (-\mathrm{pH})$

$0.060\ \mathrm{V} = E°_{\text{cell}} + 0.0592 \times \mathrm{pH}$

$0.060\ \mathrm{V} = E°_{\text{cell}} + 0.0592 \times 9.40$

$0.060\ \mathrm{V} = E°_{\text{cell}} + 0.556\ \mathrm{V}$

$E° = 0.060\ \mathrm{V} - 0.556\ \mathrm{V} = -0.496\ \mathrm{V}$

Similarly, $0.22\ \mathrm{V} = -0.496\ \mathrm{V} + 0.0592\ \mathrm{V} \times \mathrm{pH}$

$$\mathrm{pH} = \frac{0.22\ \mathrm{V} + 0.496\ \mathrm{V}}{0.0592\ \mathrm{V}} = 12$$

12.111 A simplified electrochemical series for the relevant couples can be written:

Reducing agent	Oxidizing agents	$E°$(V)
	Au^{3+}	+1.40
	Ag^+	+0.80
Cu		+0.34
	Ni^{2+}	−0.23
	Co^{2+}	−0.28
	Fe^{2+}	−0.44

It is evident that, at least under standard conditions (by no means guaranteed in an industrial process!), Cu cannot reduce Ni^{2+}, Co^{2+}, and Fe^{2+}, leaving Au and Ag as "anode mud." (Chemical engineers will adjust conditions, which will be non-standard, to secure this effect.)

12.113 (A) $ClO_4^- + 2\ H^+ + 2\ e^- \longrightarrow ClO_3^- + H_2O \qquad E° = +1.23\ \mathrm{V}$
(B) $ClO_4^- + H_2O + 2\ e^- \longrightarrow ClO_3^- + 2\ OH^- \qquad E° = +0.36\ \mathrm{V}$

(a) The Nernst equation can be used to derive the potential as a function of pH:

$$E' = E° - \frac{RT}{nF}\ln Q$$

For (A), $E'(\mathrm{A}) = 1.23\ \mathrm{V} - \dfrac{0.059\ 16}{2}\log\dfrac{[\mathrm{ClO_3^-}]}{[\mathrm{ClO_4^-}][\mathrm{H^+}]^2}$

We are only interested in varying $[\mathrm{H^+}]$, so the $[\mathrm{ClO_3^-}]$ and $[\mathrm{ClO_4^-}]$ will be left at the standard values of 1 M.

$$E'(A) = +1.23\text{ V} - \frac{0.059\ 16}{2}\log\frac{1}{[H^+]^2}$$
$$= +1.23\text{ V} - \frac{0.059\ 16}{2} \times 2\log\frac{1}{[H^+]}$$
$$= +1.23\text{ V} - 0.059\ 16\,(-\log\,[H^+])$$
$$= +1.23\text{ V} - 0.059\ 16\text{ pH}$$

Similarly for (B):

$$E'(B) = +0.36\text{ V} - \frac{0.059\ 16}{2}\log\frac{[ClO_3^-][OH^-]^2}{[ClO_4^-]}$$

As above, we are only interested in varying $[OH^-]$, so the $[ClO_3^-]$ and $[ClO_4^-]$ will be left at the standard values of 1 M.

$$E'(B) = +0.36\text{ V} - \frac{0.059\ 16}{2}\log\,[OH^-]^2$$
$$= +0.36\text{ V} - \frac{0.059\ 16}{2} \times 2\log\,[OH^-]$$
$$= +0.36\text{ V} - 0.059\ 16 \times \log\,[OH^-]$$
$$= +0.36\text{ V} + 0.059\ 16\text{ pOH}$$

Because pOH + pH = pK_w = 14.00, we can write:

$$\text{pOH} = 14.00 - \text{pH}$$
$$E'(B) = +0.36\text{ V} + 0.059\ 16[14.00 - \text{pH}]$$
$$= +0.36\text{ V} + 0.83\text{ V} - 0.059\ 16\text{ pH}$$
$$= +1.19\text{ V} - 0.059\ 16\text{ pH}$$

If we compare this to $E°(A)$, we find that the equations are essentially the same. They should be identical, the difference being due to the limitation of the number of significant figures available for the calculations.

(b) From the discussion above, we can see that the potential in neutral solution should be the same, regardless of which half-reaction we use to calculate the value.

Using $E°(A)$ $= +1.23\text{ V} - 0.059\ 16\text{ pH} = +1.23\text{ V} - 0.059\ 16\,(7.00) = 0.82$
Using $E'(B)$ $= +0.36\text{ V} + 0.059\ 16\text{ pOH} = +0.36 + 0.059\ 16\,(7.00) = 0.77$

Although these numbers differ slightly, they should be identical; again the difference lies in the limitation of the number of significant figures.

12.115 (a) $Fe^{2+} + 2\,e^- \longrightarrow Fe$ $\quad E° = -0.44$ V
$Mn^{2+} + 2\,e^- \longrightarrow Mn$ $\quad E° = -1.18$

Because these are reduction reactions, we need a corresponding oxidation. The nitrate ion contains N in its highest oxidation state so it cannot be oxidized further. The logical choice is the oxidation of water. The appropriate reduction potential is

$$O_2 + 4\,H^+ + 4\,e^- \longrightarrow 2\,H_2O \qquad E° = +1.23\text{ V}$$

The two overall reactions will be:

$$2\,Fe^{2+} + 2\,H_2O \longrightarrow 2\,Fe + O_2 + 4\,H^+ \qquad E° = -0.44\text{ V} - 1.23\text{ V} = -1.67\text{ V}$$

$$2\,Mn^{2+} + 2\,H_2O \longrightarrow 2\,Mn + O_2 + 4\,H^+ \qquad E° = -1.18\text{ V} - 1.23\text{ V} = -2.41\text{ V}$$

(b) The actual potentials, however, will differ from these standard potentials because the concentrations of the metal ions and hydrogen ions are not 1 M, and the pressure of O_2 is not 1 bar. To calculate the actual values, the Nernst equation is used.

For the Fe reaction: $E = -1.67\text{ V} - \dfrac{0.059\ 16}{4}\log\dfrac{P_{O_2}[H^+]^4}{[Fe^{2+}]^2}$

In an open beaker, with the metal ions dissolved in water with pH = 5.00, the pressure of O_2 will be $0.2095 \times 1.00\text{ atm} \times 0.987\text{ bar}\cdot\text{atm}^{-1} = 0.207$ bar. Substituting the specific values will give:

$$E = -1.67\text{ V} - \frac{0.059\ 16}{4}\log\frac{(0.207)(1.00 \times 10^{-5})^4}{(0.100)^2}$$

$$= -1.67\text{ V} + 0.28\text{ V} = -1.39\text{ V}$$

For the Mn reaction:

$$E = -2.41\text{ V} - \frac{0.059\ 16}{4}\log\frac{(0.207)(1.00 \times 10^{-5})^4}{(0.150)^2}$$

$$= -2.41\text{ V} + 0.28\text{ V} = -2.13\text{ V}$$

In order to plate out iron from this mixture, 1.39 V must be applied, and 2.13 V must be applied to cause the reduction of Mn^{2+}.

(c) Because the potential for reducing iron(II) is more positive than the potential for reducing manganese(II), the iron will plate out first.

(d) The answer to this question is obtained from the Nernst equation by determining the concentration of Fe^{2+} when the applied potential reaches 2.13 V:

$$-2.13\text{ V} = -1.67\text{ V} - \frac{0.059\ 16}{4}\log\frac{(0.207)[H^+]^4}{[Fe^{2+}]^2}$$

$$-0.46\text{ V} = -\frac{0.059\ 16}{4}\log\frac{(0.207)[H^+]}{[Fe^{2+}]^2}$$

$$31.10 = \log 0.207 + \log\frac{[H^+]^4}{[Fe^{2+}]^2}$$

$$31.78 = \log\frac{[H^+]^4}{[Fe^{2+}]^2}$$

$$\frac{[H^+]^4}{[Fe^{2+}]^2} = 6.0 \times 10^{31}$$

$$\frac{[H^+]^2}{[Fe^{2+}]} = 7.7 \times 10^{15}$$

For the last ratio to be 7.7×10^{15}, essentially all of the Fe^{2+} must be converted to Fe(s). This means that $[H^+]$ will essentially be $0.200\ mol \cdot L^{-1}$. Substituting this number gives $[Fe^{2+}] = 5 \times 10^{-18}\ mol \cdot L^{-1}$. We can say that the iron is quantitatively precipitated by this point.

We might note, however, that the potential of 2.13 V is now no longer the potential at which Mn^{2+} will begin to be reduced. Because the reduction of Fe^{2+} has produced a considerable amount of acid, the original reduction of Mn^{2+} should be recalculated:

$$E = -2.41\ V - \frac{0.059\ 16}{4} \log \frac{(0.207)(0.200)^4}{(0.150)^2} = -2.38\ V$$

Thus, even less iron will remain in solution.

CHAPTER 13
CHEMICAL KINETICS

13.1 (a) $\text{rate}(N_2) = \text{rate}(H_2) \times \left(\dfrac{1\text{ mol } N_2}{3\text{ mol } H_2}\right) = \dfrac{1}{3} \times \text{rate}(H_2)$

(b) $\text{rate}\ (NH_3) = \text{rate}(H_2) \times \left(\dfrac{2\text{ mol } NH_3}{3\text{ mol } H_2}\right) = \dfrac{2}{3} \times \text{rate}(H_2)$

(c) $\text{rate}(NH_3) = \text{rate}(N_2) \times \left(\dfrac{2\text{ mol } NH_3}{1\text{ mol } N_2}\right) = 2 \times \text{rate}(N_2)$

13.3 (a) rate of decomposition of ozone $= \left(1.5 \times 10^{-3}\ \dfrac{\text{mol } O_2}{\text{L}\cdot\text{s}}\right)\left(\dfrac{2\text{ mol } O_3}{3\text{ mol } O_2}\right)$

$= 1.0 \times 10^{-3}\ (\text{mol } O_3)\cdot\text{L}^{-1}\cdot\text{s}^{-1}$

(b) (a) $1.5 \times 10^{-3}\ \text{mol}\cdot\text{L}^{-1}\cdot\text{s}^{-1} \div 2 = 7.5 \times 10^{-4}\ \text{mol}\cdot\text{L}^{-1}\cdot\text{s}^{-1}$

13.5 (a) rate of formation of dichromate ions $= \left(\dfrac{0.14\text{ mol } Cr_2O_7^{\ 2-}}{\text{L}\cdot\text{s}}\right)\left(\dfrac{2\text{ mol } CrO_4^{\ 2-}}{1\text{ mol } Cr_2O_7^{\ 2-}}\right)$

$= 0.28\ \text{mol}\cdot\text{L}^{-1}\cdot\text{s}^{-1}$

(b) $0.14\ \text{mol}\cdot\text{L}^{-1}\cdot\text{s}^{-1} \div 1 = 0.14\ \ \text{mol}\cdot\text{L}^{-1}\cdot\text{s}^{-1}$

13.7 For A $\longrightarrow$ products, rate $= (\text{mol A})\cdot\text{L}^{-1}\cdot\text{s}^{-1}$

(a) rate $[(\text{mol A})\cdot\text{L}^{-1}\cdot\text{s}^{-1}] = k_0[\text{A}]^0 = k_0$, so units of k_0 are $(\text{mol A})\cdot\text{L}^{-1}\cdot\text{s}^{-1}$ (same as the units for the rate, in this case)

(b) rate $[(\text{mol A})\cdot\text{L}^{-1}\cdot\text{s}^{-1}] = k_1[\text{A}]$, so units of k_1 are $\dfrac{(\text{mol A})\cdot\text{L}^{-1}\cdot\text{s}^{-1}}{(\text{mol A})\cdot\text{L}^{-1}} = \text{s}^{-1}$

(c) rate $[(\text{mol A})\cdot\text{L}^{-1}\cdot\text{s}^{-1}] = k_2[\text{A}]^2$, so units of k_2 are $\dfrac{(\text{mol A})\cdot\text{L}^{-1}\cdot\text{s}^{-1}}{[(\text{mol A})\cdot\text{L}^{-1}]^2} =$

$\text{L}\cdot(\text{mol A})^{-1}\cdot\text{s}^{-1}$

13.9 From the units of the rate constant, k, it follows that the reaction is first order, thus rate $= k[N_2O_5]$.

$[N_2O_5] = \left(\dfrac{2.00\text{ g } N_2O_5}{1.00\text{ L}}\right)\left(\dfrac{1\text{ mol } N_2O_5}{108.02\text{ g } N_2O_5}\right) = 0.0185\ \text{mol}\cdot\text{L}^{-1}$

rate $= 5.2 \times 10^{-3}\ \text{s}^{-1} \times 0.0185\ \text{mol}\cdot\text{L}^{-1} = 9.6 \times 10^{-5}\ (\text{mol } N_2O_5)\cdot\text{L}^{-1}\cdot\text{s}^{-1}$

13.11 From the units of the rate constant, it follows that the reaction is second order, therefore:

$$\text{rate} = k[\text{H}_2][\text{I}_2]$$

$$= (0.063\ \text{L}\cdot\text{mol}^{-1}\cdot\text{s}^{-1})\left(\frac{0.15\ \text{g H}_2}{0.500\ \text{L}}\right)\left(\frac{1\ \text{mol H}_2}{2.016\ \text{g H}_2}\right)\left(\frac{0.32\ \text{g I}_2}{0.500\ \text{L}}\right)\left(\frac{1\ \text{mol I}_2}{253.8\ \text{g I}_2}\right)$$

$$= 2.4 \times 10^{-5}\ \text{mol}\cdot\text{L}^{-1}\cdot\text{s}^{-1}$$

(b) rate(new) $= k \times 2 \times [\text{H}_2]_{\text{initial}}[\text{I}_2] = 2 \times$ rate(initial), so, by a factor of 2

13.13 Because the rate increased in direct proportion to the concentrations of both reactants, the rate is first order in both reactants. rate $= k[\text{CH}_3\text{Br}][\text{OH}^-]$

13.15 (a) rate $= k[\text{A}]^a[\text{B}]^b$, in which the orders a and b are to be determined. When the concentration of A was decreased by a factor of 3, the rate decreased by a factor of 9; thus $(\frac{1}{3})^a = (\frac{1}{9})$, giving $a = 2$, and the reaction is second order in A. When the concentration of B was decreased by a factor of 3, the rate decreased by a factor of 3; thus $(\frac{1}{3})^b = (\frac{1}{3})$, giving $b = 1$, and the reaction is first order in B. The overall order $= 2 + 1 = 3$.

(b) rate $= k[\text{A}]^2[\text{B}]$

(c) $$k = \frac{\text{rate}}{[\text{A}]^2[\text{B}]} = \left(\frac{12.6\ \text{mol}}{\text{L}\cdot\text{s}}\right)\left(\frac{\text{L}}{0.60\ \text{mol}}\right)^2\left(\frac{\text{L}}{0.30\ \text{mol}}\right)$$

$$= 1.2 \times 10^2\ \text{L}^2\cdot\text{mol}^{-2}\cdot\text{s}^{-1}$$

The data from any of the experiments 1, 2, or 3 could have been used.

(d) $$\text{rate} = \left(\frac{1.2 \times 10^2\ \text{L}^2}{\text{mol}^2\cdot\text{s}}\right)\left(\frac{0.17\ \text{mol}}{\text{L}}\right)^2\left(\frac{0.25\ \text{mol}}{\text{L}}\right)$$

$$= 0.87\ \text{mol}\cdot\text{L}^{-1}\cdot\text{s}^{-1}$$

13.17 When the concentration of ICl is doubled, the rate doubles (experiments 1 and 2). Therefore, the reaction is first order in ICl. When the concentration of H_2 is tripled, the rate triples (experiments 2 and 3); thus, the reaction is first order in H_2.

(a) rate $= k[\text{ICl}][\text{H}_2]$

(b) $$k = \left(\frac{22 \times 10^{-7}\ \text{mol}}{\text{L}\cdot\text{s}}\right)\left(\frac{\text{L}}{3.0 \times 10^{-3}\ \text{mol}}\right)\left(\frac{\text{L}}{4.5 \times 10^{-3}\ \text{mol}}\right)$$

$$= 0.16\ \text{L}\cdot\text{mol}^{-1}\cdot\text{s}^{-1}$$

(c) $$\text{rate} = \left(\frac{0.16\ \text{L}}{\text{mol}\cdot\text{s}}\right)\left(\frac{4.7 \times 10^{-3}\ \text{mol}}{\text{L}}\right)\left(\frac{2.7 \times 10^{-3}\ \text{mol}}{\text{L}}\right)$$

$$= 2.0 \times 10^{-6}\ \text{mol}\cdot\text{L}^{-1}\cdot\text{s}^{-1}$$

13.19 (a) Doubling the concentration of A (experiments 1 and 2) doubles the rate;

therefore, the reaction is first order in A. Increasing the concentration of B by the ratio 3.02/1.25 (experiments 2 and 3) increases the rate by $(3.02/1.25)^2$; hence, the reaction is second order in B. Tripling the concentration of C (experiment 3 and 4) increases the rate by $3^2 = 9$; thus, the reaction is second order in C. Therefore, rate $= k[A][B]^2[C]^2$.

(b) overall order = 5

(c) $k = \dfrac{\text{rate}}{[A][B]^2[C]^2}$

Using the data from experiment 4, we get

$$k = \left(\frac{0.457\ \text{mol}}{\text{L}\cdot\text{s}}\right)\left(\frac{\text{L}}{1.25 \times 10^{-3}\ \text{mol}}\right)\left(\frac{\text{L}}{3.02 \times 10^{-3}\ \text{mol}}\right)^2\left(\frac{\text{L}}{3.75 \times 10^{-3}\ \text{mol}}\right)^2$$
$$= 2.85 \times 10^{12}\ \text{L}^4\cdot\text{mol}^{-4}\cdot\text{s}^{-1}$$

From experiment 3, we get

$$k = \left(\frac{5.08 \times 10^{-2}\ \text{mol}}{\text{L}\cdot\text{s}}\right)\left(\frac{\text{L}}{1.25 \times 10^{-3}\ \text{mol}}\right)\left(\frac{\text{L}}{3.02 \times 10^{-3}\ \text{mol}}\right)^2\left(\frac{\text{L}}{1.25 \times 10^{-3}\ \text{mol}}\right)^2$$
$$= 2.85 \times 10^{12}\ \text{L}^4\cdot\text{mol}^{-4}\cdot\text{s}^{-1} \quad \text{(Checks!)}$$

(d) $$\text{rate} = \left(\frac{2.85 \times 10^{12}\ \text{L}^4}{\text{mol}^4\cdot\text{s}}\right)\left(\frac{3.01 \times 10^{-3}\ \text{mol}}{\text{L}}\right)\left(\frac{1.00 \times 10^{-3}\ \text{mol}}{\text{L}}\right)^2\left(\frac{1.15 \times 10^{-3}\ \text{mol}}{\text{L}}\right)^2$$
$$= 1.13 \times 10^{-2}\ \text{mol}\cdot\text{L}^{-1}\cdot\text{s}^{-1}$$

13.21 (a) $k = \dfrac{0.693}{t_{1/2}} = \dfrac{0.693}{1000\ \text{s}} = 6.93 \times 10^{-4}\ \text{s}^{-1}$

(b) We use $\ln\left(\dfrac{[A]_0}{[A]_t}\right) = kt$ and solve for k.

$$k = \frac{\ln([A]_0/[A]_t)}{t} = \frac{\ln\left(\dfrac{0.33\ \text{mol}\cdot\text{L}^{-1}}{0.14\ \text{mol}\cdot\text{L}^{-1}}\right)}{47\ \text{s}} = 1.8 \times 10^{-2}\ \text{s}^{-1}$$

(c) $$[A]_t = \left(\frac{0.050\ \text{mol A}}{\text{L}}\right) - \left[\left(\frac{2\ \text{mol A}}{1\ \text{mol B}}\right)\left(\frac{0.015\ \text{mol B}}{\text{L}}\right)\right]$$
$$= 0.020\ (\text{mol A})\cdot\text{L}^{-1}$$

$$k = \frac{\ln\left(\dfrac{0.050\ \text{mol}\cdot\text{L}^{-1}}{0.020\ \text{mol}\cdot\text{L}^{-1}}\right)}{120\ \text{s}} = 7.6 \times 10^{-3}\ \text{s}^{-1}$$

13.23 (a) $t_{1/2} = \dfrac{0.693}{k} = \left(\dfrac{0.693\ \text{s}}{3.7 \times 10^{-5}}\right)\left(\dfrac{1\ \text{min}}{60\ \text{s}}\right)\left(\dfrac{1\ \text{h}}{60\ \text{min}}\right) = 5.2\ \text{h}$

(b) $[A]_t = [A]_0\, e^{-kt}$

$t = 2.0\ \text{h} = 2.0\ \text{h} \times 3600\ \text{s}\cdot\text{h}^{-1} = 7.2 \times 10^3\ \text{s}$

$[N_2O_5] = 2.33 \times 10^{-2}\ \text{mol}\cdot\text{L}^{-1} \times e^{-(3.7\times10^{-5}\ \text{s}^{-1})(7.2\times10^3\ \text{s})} = 1.79 \times 10^{-2}\ \text{mol}\cdot\text{L}^{-1}$

(c) Solve for t from $\ln\left(\frac{[A]_0}{[A]_t}\right) = kt$, which gives

$$t = \frac{\ln\left(\frac{[A]_0}{[A]_t}\right)}{k} = \frac{\ln\left(\frac{[N_2O_5]_0}{[N_2O_5]_t}\right)}{k} = \frac{\ln\left(\frac{2.33}{1.76}\right)}{3.7 \times 10^{-5}\ s^{-1}} = 7.6 \times 10^3\ s$$

$$= (7.6 \times 10^3\ s)\left(\frac{1\ min}{60\ s}\right) = 1.3 \times 10^2\ min$$

13.25 (a) This is the half-life itself, 200 s.

(b) $\frac{[A]}{[A]_0} = \frac{1}{16} = \left(\frac{1}{2}\right)^4$; so the time elapsed is 4 half-lives.

$t = 4 \times 200\ s = 800\ s$

(c) Because $\frac{1}{9}$ is not a multiple of $\frac{1}{2}$, we cannot work directly from the half-life. But $k = 0.693/t_{1/2}$

$$\text{so } k = \frac{0.693}{200\ s} = 3.46 \times 10^{-3}\ s^{-1}$$

Then [see the solution to Exercise 13.23(c)],

$$t = \frac{\ln\left(\frac{[A]_0}{[A]_t}\right)}{k} = \frac{\ln\left(\frac{9}{1}\right)}{3.46 \times 10^{-3}\ s^{-1}} = 635\ s$$

13.27 (a) $t_{1/2} = \frac{0.693}{k} = \frac{0.693}{2.81 \times 10^{-3}\ min^{-1}} = 247\ min$

(b) See the solutions to Exercises 13.31(c) and 13.33(c).

$$t = \frac{\ln\left(\frac{[SO_2Cl_2]_0}{[SO_2Cl_2]_t}\right)}{k} = \frac{\ln 10}{2.81 \times 10^{-3}\ min^{-1}} = 819\ min$$

(c) $[A]_t = [A]_0\, e^{-kt}$

Because the vessel is sealed, masses and concentrations are proportional, and we write

$$\begin{aligned}(\text{mass left})_t &= (\text{mass})_0\, e^{-kt} \\ &= 14.0\ g \times e^{-(2.81\times10^{-3}\ min^{-1}\times 60\ min\cdot h^{-1}\times 1.5\ h)} \\ &= 10.9\ g\end{aligned}$$

Note: Knowledge of the volume of the vessel is not required. However, we could have converted mass to concentration, solved for the new concentration at 1.5 h, and finally converted back to the new (remaining) mass. But this is not necessary.

13.29 (a) We first calculate the concentration of A at 3.0 min.

$$[A]_t = [A]_0 - \left(\frac{1 \text{ mol A}}{3 \text{ mol B}}\right) \times [B]_t$$

$$= 0.015 \text{ mol}\cdot\text{L}^{-1} - \left(\frac{1 \text{ mol A}}{3 \text{ mol B}}\right) \times 0.020 \text{ (mol B)}\cdot\text{L}^{-1}$$

$$= 0.008 \text{ mol}\cdot\text{L}^{-1}$$

The rate constant is then determined from the first-order integrated rate law.

$$k = \frac{\ln\left(\frac{[A]_0}{[A]_t}\right)}{t} = \frac{\ln\left(\frac{0.015}{0.008}\right)}{3.0 \text{ min}} = 0.21 \text{ min}^{-1}$$

(b) $[A]_t = 0.015 \text{ mol}\cdot\text{L}^{-1} - \left(\frac{1 \text{ mol A}}{3 \text{ mol B}}\right) \times 0.040 \text{ (mol B)}\cdot\text{L}^{-1}$

$$= 0.002 \text{ mol}\cdot\text{L}^{-1}$$

$$t = \frac{\ln\left(\frac{[A]_0}{[A]_t}\right)}{k} = \frac{\ln\left(\frac{0.015}{0.002}\right)}{0.21 \text{ min}^{-1}} = 10 \text{ min}$$

additional time = 10 min − 3.0 min = 7 min

13.31 We draw up the following table:

Time, s	$[N_2O_5]$, 10^{-3} mol·L^{-1}	$[A]_0/[A]$	$\ln([A]_0/[A])$
0	2.15	1.00	0.000
4000	1.88	1.14	0.134
8000	1.64	1.31	0.270
12 000	1.43	1.50	0.405
16 000	1.25	1.72	0.542

(a) The integrated rate equation for a first-order reaction is

$$\ln[A]_t = \ln[A]_0 - kt \quad \text{or} \quad \ln\left(\frac{[A]_0}{[A]_t}\right) = kt$$

We then plot $\ln([A]_0/[A]_t)$ versus t and see whether a straight line through the origin results.

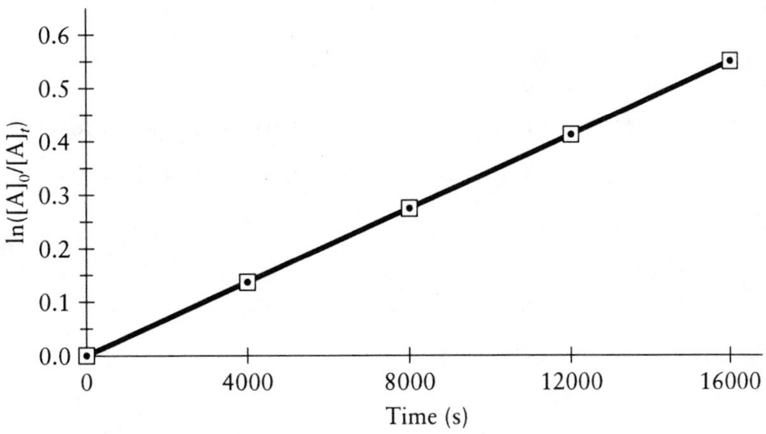

The data fit the first-order equation well.

(b) The rate constant is the slope of this line. It is easily determined with plotting software, with a graphing calculator, or by the methods of Appendix 1E. Using first and last points in the graph yields

$$\text{slope} = \frac{0.542 - 0.000}{(16\,000 - 0)\text{ s}} = 3.39 \times 10^{-5}\text{ s}^{-1} = k$$

13.33 (a) Draw up the following table and plot 1/[HI] against time.

Time, s	[HI], 10^{-3} mol·L^{-1}	1/[HI], 10^{3} L·mol^{-1}
0	1000	0.001 00
1000	112	0.008 93
2000	61	0.016
3000	41	0.024
4000	31	0.032

Equation 17.b in the text can be rearranged as

$$\frac{1}{[\text{A}]_t} = \frac{1 + [\text{A}]_0\, kt}{[\text{A}]_0} = \frac{1}{[\text{A}]_0} + kt$$

Thus, if the reaction is second order, a plot of 1/[HI] against time should give a straight line of slope k.

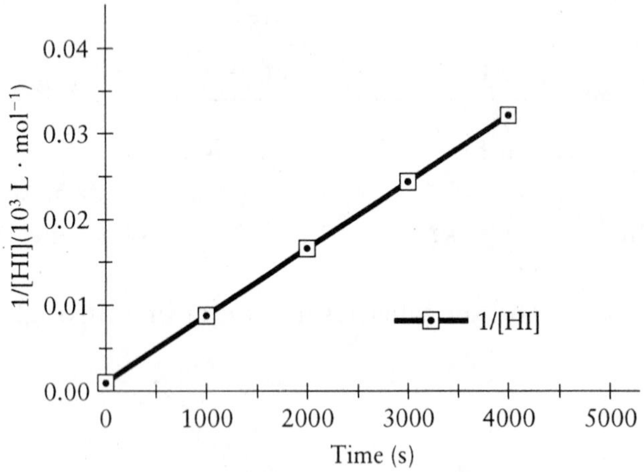

As can be seen from the graph, the data fit the equation for a second-order reaction quite well. The slope can be determined from any pair of data points, but we choose the points at 4000 and 1000 s.

$$\text{(b) slope} = \frac{(0.032 - 0.008\,93) \times 10^{3}\text{ L·mol}^{-1}}{(4000 - 1000)\text{ s}}$$

$$= 7.7 \times 10^{-3}\text{ L·mol}^{-1}\text{·s}^{-1} = k$$

13.35 It is convenient to obtain an expression for the half-life of a second-order reaction. We work with Eq. 17b.

$$[A]_t = \frac{[A]_0}{1 + [A]_0 kt} \qquad (17.b)$$

$$\frac{[A]_{t1/2}}{[A]_0} = \frac{1}{2} = \frac{1}{1 + [A]_0 kt_{1/2}}$$

Therefore, $1 + [A]_0 kt_{1/2} = 2$, or $[A]_0 kt_{1/2} = 1$, or

$$t_{1/2} = \frac{1}{k[A]_0} \quad \text{and} \quad k = \frac{1}{t_{1/2}[A]_0}$$

It is also convenient to rewrite Eq. 17.b to solve for t. We take reciprocals:

$$\frac{1}{[A]_t} = \frac{1}{[A]_0} + kt$$

giving

$$t = \frac{\dfrac{1}{[A]_t} - \dfrac{1}{[A]_0}}{k}$$

(a) $$k = \frac{1}{t_{1/2}[A]_0} = \frac{1}{(50.5\ \text{s})(0.84\ \text{mol}\cdot\text{L}^{-1})} = 0.024\ \text{L}\cdot\text{mol}^{-1}\cdot\text{s}^{-1}$$

$$t = \frac{\dfrac{1}{[A]} - \dfrac{1}{[A]_0}}{k} = \frac{\dfrac{16}{[A]_0} - \dfrac{1}{[A]_0}}{k} = \frac{15}{k[A]_0}$$

$$= \frac{15}{(0.024\ \text{L}\cdot\text{mol}^{-1}\cdot\text{s}^{-1})(0.84\ \text{mol}\cdot\text{L}^{-1})} = 7.46 \times 10^2\ \text{s}$$

(b) $$t = \frac{\dfrac{4}{[A]_0} - \dfrac{1}{[A]_0}}{k} = \frac{3}{k[A]_0}$$

$$= \frac{3}{(0.024\ \text{L}\cdot\text{mol}^{-1}\cdot\text{s}^{-1})(0.84\ \text{mol}\cdot\text{L}^{-1})} = 1.5 \times 10^2\ \text{s}$$

(c) $$t = \frac{\dfrac{5}{[A]_0} - \dfrac{1}{[A]_0}}{k} = \frac{4}{k[A]_0}$$

$$= \frac{4}{(0.024\ \text{L}\cdot\text{mol}^{-1}\cdot\text{s}^{-1})(0.84\ \text{mol}\cdot\text{L}^{-1})} = 2.0 \times 10^2\ \text{s}$$

13.37 See the solution to Exercise 13.35 for the derivation of the formulas needed here.

(a) $$t = \frac{\dfrac{1}{[A]} - \dfrac{1}{[A]_0}}{k} = \frac{\dfrac{1\ \text{L}}{0.080\ \text{mol}} - \dfrac{1\ \text{L}}{0.10\ \text{mol}}}{0.010\ \text{L}\cdot\text{mol}^{-1}\cdot\text{min}^{-1}} = 2.5 \times 10^2\ \text{min}$$

$$\text{(b) } [\mathrm{A}] = \frac{0.45 \text{ mol A}}{\mathrm{L}} - \left[\left(\frac{0.45 \text{ mol B}}{\mathrm{L}}\right)\left(\frac{1 \text{ mol A}}{2 \text{ mol B}}\right)\right]$$

$$= 0.23(\text{mol A}) \cdot \mathrm{L}^{-1} = \tfrac{1}{2}[\mathrm{A}]_0$$

$$t = t_{1/2} = \frac{1}{k[\mathrm{A}]_0} = \frac{1}{(0.0045 \text{ L}\cdot\text{mol}^{-1}\cdot\text{min}^{-1})(0.45 \text{ mol}\cdot\text{L}^{-1})}$$

$$= 4.9 \times 10^2 \text{ min}$$

13.39 (a) Make the following table and graph:

T, K	$1/T$, K^{-1}	k, s^{-1}	ln k
750	0.001 33	1.8×10^{-4}	−8.62
800	0.001 25	2.7×10^{-3}	−5.91
850	0.001 18	3.0×10^{-2}	−3.51
900	0.001 11	2.6×10^{-1}	−1.35

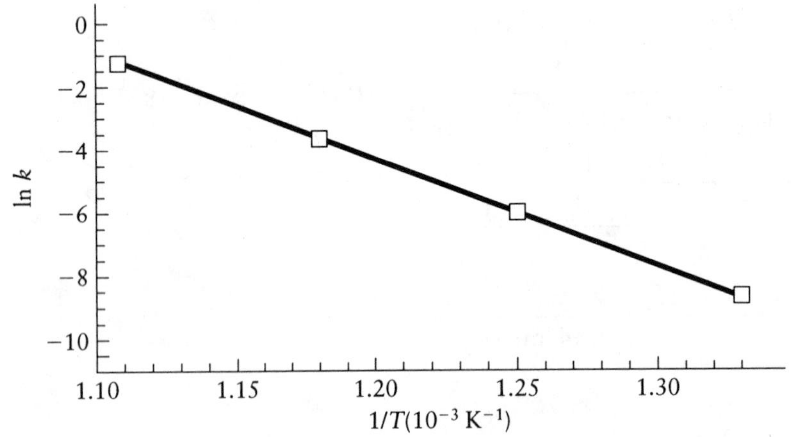

$$\ln k = \ln A - E_a/RT$$

$$\text{slope of plot} = -E_a/R = \frac{[-1.35 - (-8.62)] \text{ K}}{0.001\,11 - 0.001\,33} = 3.3 \times 10^4 \text{ K}$$

$$E_a = (3.3 \times 10^4 \text{ K})(8.31 \times 10^{-3} \text{ kJ}\cdot\text{mol}^{-1}\cdot\text{K}^{-1}) = 2.74 \times 10^2 \text{ kJ}\cdot\text{mol}^{-1}$$

$$\text{(b) } \ln\left(\frac{k'}{k}\right) = \frac{E_a}{R}\left(\frac{1}{T} - \frac{1}{T'}\right)$$

$$= \left(\frac{2.74 \times 10^2 \text{ kJ}\cdot\text{mol}^{-1}}{8.31 \times 10^{-3} \text{ kJ}\cdot\text{K}^{-1}\cdot\text{mol}^{-1}}\right)\left(\frac{1}{750 \text{ K}} - \frac{1}{873 \text{ K}}\right)$$

$$= 6.19$$

$$\frac{k'}{k} = 4.9 \times 10^2$$

$$k' = (4.9 \times 10^2)(1.8 \times 10^{-4} \text{ s}^{-1}) = 8.8 \times 10^{-2} \text{ s}^{-1}$$

An approximate value of k' can also be obtained from the plot itself and yields the same value.

13.41 We use $\ln\left(\frac{k'}{k}\right) = \frac{E_a}{R}\left(\frac{1}{T} - \frac{1}{T'}\right) = \frac{E_a}{R}\left(\frac{T' - T}{T'T}\right)$

$$\ln\left(\frac{k'}{k}\right) = \ln\left(\frac{0.87\ \text{s}^{-1}}{0.38\ \text{s}^{-1}}\right)$$

$$= \left(\frac{E_a}{8.31 \times 10^{-3}\ \text{kJ·K}^{-1}\text{·mol}^{-1}}\right)\left(\frac{1030\ \text{K} - 1000\ \text{K}}{1030\ \text{K} \times 1000\ \text{K}}\right)$$

$$E_a = \frac{(8.31 \times 10^{-3}\ \text{kJ·K}^{-1}\text{·mol}^{-1})(1000\ \text{K})(1030\ \text{K})(0.83)}{30\ \text{K}} = 2.4 \times 10^{2}\ \text{kJ·mol}^{-1}$$

13.43 We use $\ln\left(\frac{k'}{k}\right) = \frac{E_a}{R}\left(\frac{1}{T} - \frac{1}{T'}\right) = \frac{E_a}{R}\left(\frac{T' - T}{TT'}\right)$

k' = rate constant at 700°C, $T' = (700 + 273)\ \text{K} = 973\ \text{K}$

$$\ln\left(\frac{k'}{k}\right) = \left(\frac{315\ \text{kJ · mol}^{-1}}{8.314 \times 10^{-3}\ \text{kJ·K}^{-1}\text{·mol}^{-1}}\right)\left(\frac{973\ \text{K} - 1073\ \text{K}}{973\ \text{K} \times 1073\ \text{K}}\right)$$

$$= -3.63;\quad \frac{k'}{k} = 0.026$$

$$k' = 0.026 \times 9.7 \times 10^{10}\ \text{L·mol}^{-1}\text{·s}^{-1} = 2.5 \times 10^{9}\ \text{L·mol}^{-1}\text{·s}^{-1}$$

13.45 $\ln\left(\frac{k'}{k}\right) = \frac{E_a}{R}\left(\frac{1}{T} - \frac{1}{T'}\right) = \frac{E_a}{R}\left(\frac{T' - T}{TT'}\right)$

$$= \left(\frac{103\ \text{kJ·mol}^{-1}}{8.314 \times 10^{-3}\ \text{kJ·K}^{-1}\text{·mol}^{-1}}\right)\left(\frac{323\ \text{K} - 318\ \text{K}}{318\ \text{K} \times 323\ \text{K}}\right) = 0.60$$

$$\frac{k'}{k} = 1.8$$

$$k' = 1.8 \times 5.1 \times 10^{-4}\ \text{s}^{-1} = 9.2 \times 10^{-4}\ \text{s}^{-1}$$

13.47 $2\ \text{ICl} + H_2 \longrightarrow 2\ \text{HCl} + I_2$

HI is the only intermediate. It is the only species that does not appear as either a reactant or a product.

13.49 The first elementary reaction is the rate-controlling step, because it is the slow step. The second elementary reaction is fast and does not affect the overall reaction order, which is second order as a result of the fact that the rate-controlling step is bimolecular.

rate = $k[\text{NO}][\text{Br}_2]$

13.51 The overall rate is determined by the slow step. rate = $k_3[\text{COCl}][\text{Cl}_2]$. But COCl is an intermediate and its concentration has to be eliminated.

$$k_2[\text{Cl}][\text{CO}] = k'_2[\text{COCl}]$$

$$[\text{COCl}] = \frac{k_2}{k'_2}[\text{Cl}][\text{CO}]$$

but we cannot leave [Cl] in the expression either:

$$k_1[\text{Cl}_2] = k'_1[\text{Cl}]^2, \text{ giving } [\text{Cl}] = \sqrt{\frac{k_1}{k'_1}}[\text{Cl}_2]^{1/2}$$

$$[\text{COCl}] = \left(\frac{k_2}{k'_2}\right)\sqrt{\frac{k_1}{k'_1}}[\text{Cl}_2]^{1/2}[\text{CO}], \text{ therefore}$$

$$\text{rate} = k_3(k_2/k'_2)(k_1/k'_1)^{1/2}[\text{CO}][\text{Cl}_2]^{3/2} = k[\text{CO}][\text{Cl}_2]^{3/2}$$

13.53 If mechanism (a) were correct, the rate law would be rate = $k_2[NO_2][CO]$. But this expression does not agree with the experimental result and can be eliminated as a possibility. Mechanism (b) has rate = $k_2[NO_2]^2$ from the slow step. Step 2 does not influence the overall rate, but it is necessary to achieve the correct overall reaction; thus this mechanism agrees with the experimental data $k = k_2$. Mechanism (c) is not correct, which can be seen from the rate expression for the slow step, rate = $k_2[NO_3][CO]$. [CO] cannot be eliminated from this expression to yield the experimental result, which does not contain [CO].

13.55 (a) All chemical reactions involve a change in enthalpy, ΔH. As a consequence, the products are not at the same energy level as the reactants, and activation energies for the forward and backward reactions are different. Therefore, the rate constants will be different, because $k = Ae^{-E_a/RT}$, and they depend on E_a. However, at equilibrium, the *rates* of the forward and reverse reactions are equal.

(b) For a reversible reaction,

$$K_c = \frac{k(\text{forward})}{k(\text{reverse})} = \frac{k}{k'} \qquad \text{Eq. 29}$$

Therefore, if K_c is very large, $k(\text{forward}) \gg k(\text{reverse})$.

13.57 (a) The equilibrium constant will be given by the ratio of the rate constant of the forward reaction to the rate constant of the reverse reaction:

$$K = \frac{k}{k'} = \frac{265\ \text{L}\cdot\text{mol}^{-1}\cdot\text{min}^{-1}}{392\ \text{L}\cdot\text{mol}^{-1}\cdot\text{min}^{-1}} = 0.676$$

(b) The reaction profile corresponds to a plot similar to that shown in Fig. 13.28a. The reaction is endothermic—the reverse reaction has a lower activation barrier than the forward reaction.

(c) Raising the temperature will increase the rate constant of the reaction with the higher activation barrier more than it will the rate constant of the reaction with the lower energy barrier. We expect the rate of the forward reaction to go up substantially more than for the reverse reaction in this case. k will increase more than k' and consequently the equilibrium constant K will increase. This is consistent with Le Chatelier's principle.

13.59 cat = catalyzed, uncat = uncatalyzed $E_{a,cat} = \frac{1}{2} E_a = \frac{1}{2} E_{a,uncat}$

$$\frac{\text{rate(cat)}}{\text{rate(uncat)}} = \frac{k_{cat}}{k_{uncat}} = \frac{Ae^{-E_{a,cat}/RT}}{Ae^{-E_a/RT}} = \frac{e^{-(1/2)E_a/RT}}{e^{-E_a/RT}} = e^{(1/2)E_a/RT}$$

$$e^{[(1/2)(100\ \text{kJ}\cdot\text{mol}^{-1})/(8.314\times10^{-3}\ \text{kJ}\cdot\text{K}^{-1}\cdot\text{mol}^{-1}\times 400\ \text{K})]} = 3 \times 10^6$$

13.61 cat = catalyzed, uncat = uncatalyzed

$$\frac{\text{rate(cat)}}{\text{rate(uncat)}} = \frac{k_{cat}}{k_{uncat}} = 1000 = \frac{Ae^{-E_{a,cat}/RT}}{Ae^{-E_a/RT}} = \frac{e^{-E_{a,cat}/RT}}{e^{-E_a/RT}}$$

$$\ln 1000 = \frac{-E_{a,cat}}{RT} + \frac{E_a}{RT}$$

$$\begin{aligned} E_{a,cat} &= E_a - RT \ln 1000 \\ &= 98\ \text{kJ}\cdot\text{mol}^{-1} - (8.31 \times 10^{-3}\ \text{kJ}\cdot\text{K}^{-1}\cdot\text{mol}^{-1})(298\ \text{K})(\ln 1000) \\ &= 80.9\ \text{kJ}\cdot\text{mol}^{-1} \end{aligned}$$

13.63 (a) rate of formation of O_2 $= \left(6.5 \times 10^{-3}\ \frac{\text{mol NO}_2}{\text{L}\cdot\text{s}}\right) \times \left(\frac{1\ \text{mol O}_2}{2\ \text{mol NO}_2}\right)$

$$= 3.3 \times 10^{-3}\ (\text{mol O}_2)\cdot\text{L}^{-1}\cdot\text{s}^{-1}$$

(b) $6.5 \times 10^{-3}\ \text{mol}\cdot\text{L}^{-1}\cdot\text{s}^{-1} \div 2 = 3.3 \times 10^{-3}\ \text{mol}\cdot\text{L}^{-1}\cdot\text{s}^{-1}$

13.65 rate $= k[C_4H_9Br]^a[OH^-]^b$

Because the rate increases in direct proportion to $[C_4H_9Br]$, a must be 1. Because $[OH^-]$ has no effect on the rate, b must be 0. Therefore,

rate $= k[C_4H_9Br]^1[OH^-]^0 = k[C_4H_9Br]$

The reaction is first order in $[C_4H_9Br]$, zero order in $[OH^-]$, and first order overall.

13.67 rate $= k[H_2SeO_3][I^-]^3[H^+]^2$ $k = 5.0 \times 10^5\ \text{L}^5\cdot\text{mol}^{-5}\cdot\text{s}^{-1}$

$$\text{rate}_0 = \left(\frac{5.0 \times 10^5\ \text{L}^5}{\text{mol}^5\cdot\text{s}}\right)\left(\frac{0.020\ \text{mol}}{\text{L}}\right)\left(\frac{0.020\ \text{mol}}{\text{L}}\right)^3\left(\frac{0.010\ \text{mol}}{\text{L}}\right)^2$$

$$= 8.0 \times 10^{-6}\ \text{mol}\cdot\text{L}^{-1}\cdot\text{s}^{-1}$$

13.69 (a) $k = \dfrac{\dfrac{1}{[A]_t} - \dfrac{1}{[A]_0}}{t} = \dfrac{\dfrac{1}{0.0050\ \text{mol}\cdot\text{L}^{-1}} - \dfrac{1}{0.040\ \text{mol}\cdot\text{L}^{-1}}}{12\ \text{h}} = 15\ \text{L}\cdot\text{mol}^{-1}\cdot\text{h}^{-1}$

(b) $[CX_2] = 0.040\ \dfrac{\text{mol } CX_2}{\text{L}} - \left(0.070\ \dfrac{\text{mol X}}{\text{L}}\right)\left(\dfrac{1\text{ mol } CX_2}{2\text{ mol X}}\right)$

$= 0.005\ \text{mol}\cdot\text{L}^{-1}$

$$k = \frac{\dfrac{1}{[CX_2]_t} - \dfrac{1}{[CX_2]_0}}{t} = \frac{\dfrac{1}{0.005\ \text{mol}\cdot\text{L}^{-1}} - \dfrac{1}{0.040\ \text{mol}\cdot\text{L}^{-1}}}{15\ \text{h}}$$

$= 10\ \text{L}\cdot\text{mol}^{-1}\cdot\text{h}^{-1}$(1 sf)

13.71

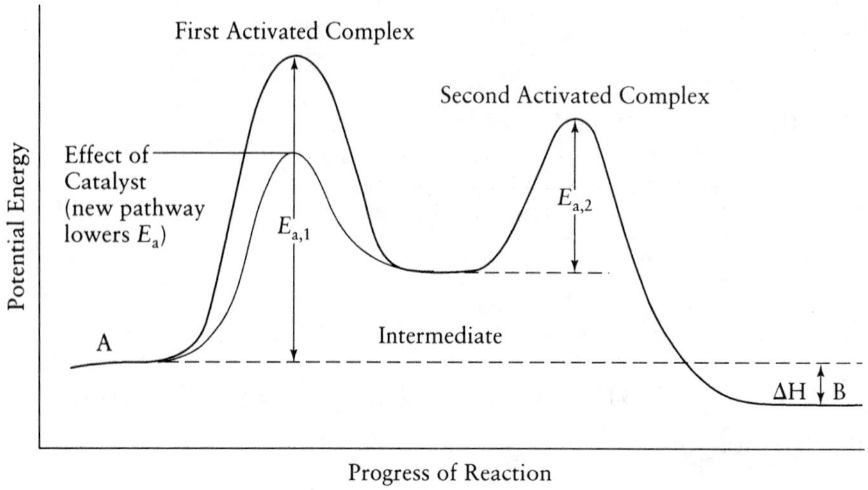

13.73 x = amount of original sample = 1.0 mg

n = number of half-lives

$\left(\dfrac{1}{2}\right)^n \times x$ = amount remaining

$\dfrac{5.2}{12.3} = 0.42$ half-lives

$\left(\dfrac{1}{2}\right)^{0.42} \times 1.0\ \text{mg} = 0.75\ \text{mg}$

13.75 (a) The objective is to reproduce the observed rate law. If step 2 is the slow step, if step 1 is a rapid equilibrium, and if step 3 is fast also, then our proposed rate law will be rate = $k_2[N_2O_2][H_2]$. Consider the equilibrium of Step 1: $k_1[NO]^2 = k_1'[N_2O_2]$

$[N_2O_2] = \dfrac{k}{k_1'}[NO]^2$ Substituting in our proposed rate law, we have

Rate $= k_2\left(\dfrac{k_1}{k_1'}\right)[NO]^2[H_2] = k[NO]^2[H_2]$ where $k = k_2\left(\dfrac{k_1}{k_1'}\right)$

The assumptions made above reproduce the observed rate law; therefore, step 2 is the slow step.

(b)

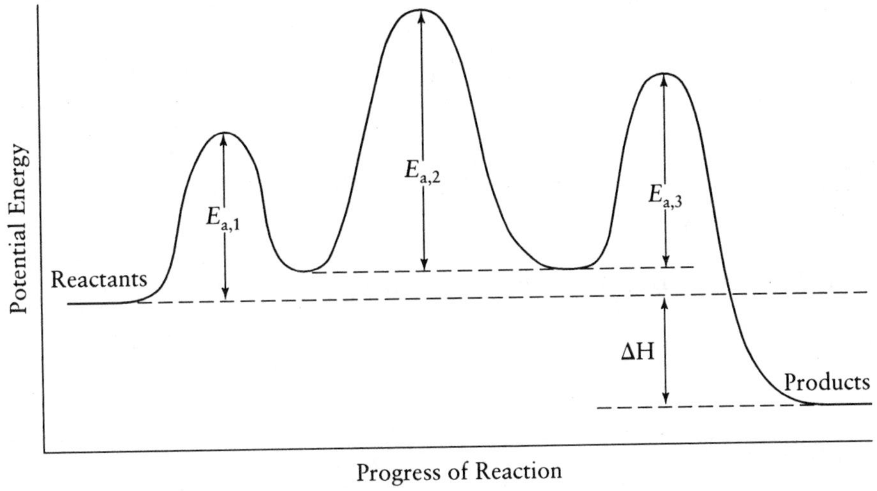

Note: The dips that represent the formation of the intermediate N_2O_2 and N_2O will not be at the same energy, but we have no information to determine which should be lower.

13.77 (a) Make the following table and graph:

T, K	$1/T$, K^{-1}	$(k \times 10^3)$ $L \cdot mol^{-1} \cdot s^{-1}$	$\ln(k \times 10^3)$
297	0.003 37	1.3	0.26
301	0.003 32	2.0	0.69
305	0.003 28	3.0	1.1
309	0.003 24	4.4	1.5
313	0.003 19	6.4	1.9

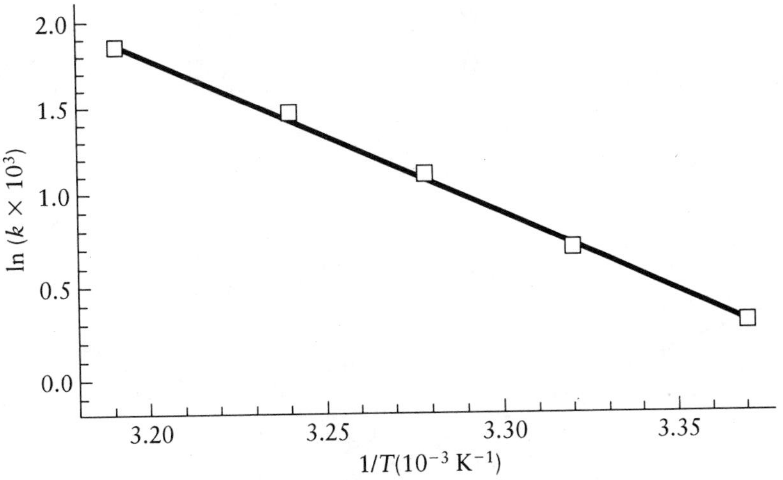

$$\text{slope of straight line} = -\left(\frac{E_a}{R}\right) = \frac{(1.9 - 0.26)\ K}{0.003\ 19 - 0.003\ 37} = -8.9 \times 10^3\ K$$

$$E_a = (8.9 \times 10^3\ K)(8.314 \times 10^{-3}\ kJ \cdot K^{-1} \cdot mol^{-1}) = 74\ kJ \cdot mol^{-1}$$

(b) $T = 298$ K, $1/T = 3.36 \times 10^{-3}$ K^{-1}

From the plot, $\ln(k \times 10^3) \approx 0.37$, $k = 1.4 \times 10^{-3}$ L·mol^{-1}·s^{-1}

13.79 (a) ClO is the reaction intermediate; Cl is the catalyst.

(b) Cl, ClO, O, O_2

(c) Step 1 is initiating; step 2 is propagating.

(d) $Cl + Cl \longrightarrow Cl_2$

13.81

Concentration (mol·L^{-1})	$2\,N_2O_5$ $\rightleftharpoons$	$4\,NO_2$ +	O_2
initial	P_0	0	0
change	$-x$	$+2x$	$+0.5x$
At time t	$P_0 - x$	$2x$	$0.5x$

Therefore, P_{total} at time $t = P_0 + 1.5x$. This allows calculation of x at each time, which in turn allows calculation of $P_{N_2O_5}(= P_0 - x)$ at these times. Converting the units to atmospheres by dividing by 101.325 kPa·atm^{-1} and to $[N_2O_5]$ by dividing by RT allows us to make the following table:

t, min	x, kPa	$P_{N_2O_5}$, kPa	$P_{N_2O_5}$, atm	$[N_2O_5]$, mol·L^{-1}	$\ln[N_2O_5]$
0	0	27.3	0.269	0.0100	−4.605
5	10.9	16.4	0.162	6.01×10^{-3}	−5.114
10	17.5	9.85	0.0972	3.61×10^{-3}	−5.624
15	21.4	5.9	0.058	2.2×10^{-3}	−6.12
20	23.8	3.5	0.035	1.3×10^{-3}	−6.65
30	26.0	1.3	0.013	4.8×10^{-4}	−7.64

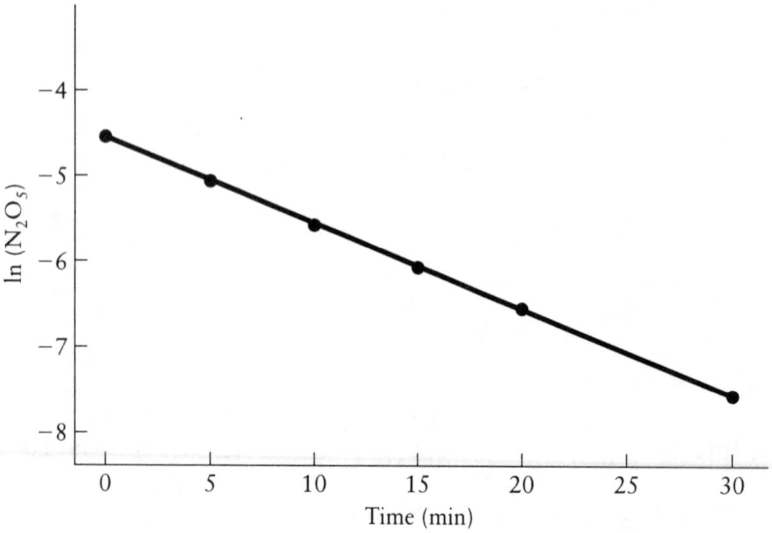

The data fit closely to a straight line; therefore, this is a first-order reaction. The rate constant can be obtained from the slope, which is

$$\frac{-4.605 - (-7.64)}{30 \text{ min}} = 0.101 \text{ min}^{-1} = k$$

Rate $= k[N_2O_5] = 0.101 \text{ min}^{-1}[N_2O_5]$, which gives the results in the table below.

t, min	Rate, mol·L^{-1}·min^{-1}
0	1.01×10^{-3}
5	6.07×10^{-4}
10	3.65×10^{-4}
15	2.2×10^{-4}
20	1.3×10^{-4}
30	4.8×10^{-5}

13.83 For a third-order reaction,

$$t_{1/2} \propto \frac{1}{[A_0]^2} \quad \text{or} \quad t_{1/2} = \frac{\text{constant}}{[A_0]^2}$$

(a) The time necessary for the concentration to fall to one-half of the initial concentration is one half-life:

$$\text{first half-life} = t_1 = t_{1/2} = \frac{\text{constant}}{[A_0]^2}$$

(b) This time, $t_{1/4}$, is two half-lives, but because of different starting concentrations, the half-lives are not the same:

$$\text{second half-life} = t_2 = \frac{\text{constant}}{(\frac{1}{2}[A_0])^2} = \frac{4(\text{constant})}{[A_0]^2} = 4t_1$$

$$\text{total time} = t_1 + t_2 = t_1 + 4t_1 = 5t_1 = t_{1/4}$$

(c) This time, $t_{1/16}$, is four half-lives; again, the half-lives are not the same:

$$\text{third half-life} = t_3 = \frac{\text{constant}}{(\frac{1}{4}[A_0])^2} = \frac{16(\text{constant})}{[A_0]^2} = 16t_1$$

$$\text{fourth half-life} = t_4 = \frac{\text{constant}}{(\frac{1}{8}[A_0])^2} = \frac{64(\text{constant})}{[A_0]^2} = 64t_1$$

$$\text{total time} = t_1 + t_2 + t_3 + t_4 = t_1 + 4t_1 + 16t_1 + 64t_1 = 85t_1 = t_{1/16}$$

If t_1 is known, the times $t_{1/4}$ and $t_{1/16}$ can be calculated easily.

13.85 The reaction reaches equilibrium. Consider, for example, $A \rightleftharpoons B$. Assume that the reaction is first order in both directions:

rate(forward) $= k[A]$

rate(reverse) $= k'[B]$

At equilibrium, rate(forward) = rate(reverse), so

rate = 0 = rate(forward) − rate(reverse)

or $k[\mathrm{A}] - k'[\mathrm{B}] = 0$

or $k[\mathrm{A}] - k'([\mathrm{A}]_0 - [\mathrm{A}]) = 0$

or $(k + k')[\mathrm{A}] = k'[\mathrm{A}]_0$

or $[\mathrm{A}] = \dfrac{k'}{(k + k')}[\mathrm{A}]_0$, which is not zero.

13.87 By analogy with the reaction in Exercise 13.80, the overall reaction here is

$CH_4(g) + Cl_2(g) \longrightarrow CH_3Cl + HCl$

(a) Initiation: $Cl_2 \longrightarrow 2\ Cl$

Propagation: $Cl + CH_4 \longrightarrow CH_3Cl + H$

$H + Cl_2 \longrightarrow HCl + Cl$

Termination: $Cl + Cl \longrightarrow Cl_2$

$H + H \longrightarrow H_2$

$H + Cl \longrightarrow HCl$

(b) CH_3Cl and HCl

13.89 The strategy for working this problem is to obtain the equilibrium constants for the reaction at two or more temperatures and then use those values to calculate ΔH°_r and ΔS°_r. From Table 13.1 we can obtain K values at 4 temperatures:

$$K = \frac{k}{k'}$$

$$K_{500} = \frac{6.4 \times 10^{-9}\ \mathrm{L{\cdot}mol^{-1}{\cdot}s^{-1}}}{4.3 \times 10^{-7}\ \mathrm{L{\cdot}mol^{-1}{\cdot}s^{-1}}} = 0.015$$

$$K_{600} = \frac{9.7 \times 10^{-6}\ \mathrm{L{\cdot}mol^{-1}{\cdot}s^{-1}}}{4.4 \times 10^{-4}\ \mathrm{L{\cdot}mol^{-1}{\cdot}s^{-1}}} = 0.022$$

$$K_{700} = \frac{1.8 \times 10^{-3}\ \mathrm{L{\cdot}mol^{-1}{\cdot}s^{-1}}}{6.3 \times 10^{-2}\ \mathrm{L{\cdot}mol^{-1}{\cdot}s^{-1}}} = 0.028$$

$$K_{700} = \frac{9.7 \times 10^{-2}\ \mathrm{L{\cdot}mol^{-1}{\cdot}s^{-1}}}{2.6\ \mathrm{L{\cdot}mol^{-1}{\cdot}s^{-1}}} = 0.037$$

We can choose to calculate the desired quantities from any two of these points, or we can plot the data and determine the values from the slope and intercept of the graph:

$$\ln K = -\frac{\Delta H^\circ_r}{R}\left(\frac{1}{T_1}\right) + \frac{\Delta S^\circ_r}{R}$$

The plot should be ln K versus $\dfrac{1}{T}$. The slope will be $-\dfrac{\Delta H^\circ_r}{R}$ and the intercept will be $\dfrac{\Delta S^\circ_r}{R}$.

T(K)	$\frac{1}{T}$(K^{-1})	K	ln K
500	0.0200	0.015	−4.20
600	0.001 67	0.022	−3.82
700	0.001 43	0.028	−3.58
800	0.001 25	0.037	−3.30

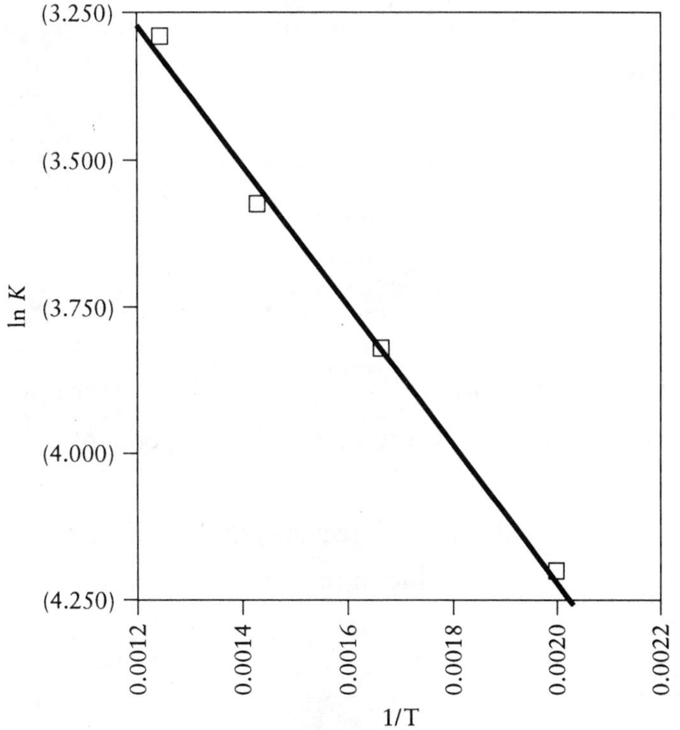

ln $K = -1177\ T^{-1} - 1.86$

$$-\frac{\Delta H^\circ_r}{R} = -1177,\ \Delta H^\circ_r = 9.8\ \text{kJ}\cdot\text{mol}^{-1}$$

$$\frac{\Delta S^\circ_r}{R} = -1.86,\ \Delta S^\circ_r = -15\ \text{J}\cdot\text{K}^{-1}\cdot\text{mol}^{-1}$$

13.91 In order for the reaction to be catalyzed heterogeneously, the reacting species must attach themselves to the surface of the catalyst. The concentration of the reactants is usually much greater than the number of active sites available on the catalyst so that the rate is determined by the surface area of the catalyst and not by the concentrations or pressures of reactants.

13.93 (a) $OCl^- + H_2O \underset{k_1'}{\overset{k_1}{\rightleftharpoons}} HOCl + OH^-$ fast equilibrium

$HOCl + I^- \xrightarrow{k_2} HOI + Cl^-$ very slow

$$HOI + OH^- \underset{k_3'}{\overset{k_3}{\rightleftharpoons}} OI^- + H_2O \quad \text{fast equilibrium}$$

The overall reaction is $OCl^- + I^- \longrightarrow OI^- + Cl^-$

(b) The rate law will be based upon the slow step of the reaction:

rate $= k_2[HOCl][I^-]$

Even though HOCl is a stable species, because it is an intermediate in the reaction as written, we technically should not leave the rate law in this form. The concentration of HOCl can be expressed in terms of the reactants and products using the fast equilibrium approach:

$$K = \frac{k_1}{k_1'} = \frac{[HOCl][OH^-]}{[OCl^-]}$$

$$[HOCl] = \frac{k_1}{k_1'}\frac{[OCl^-]}{[OH^-]}$$

$$\text{rate} = \frac{k_2k_1}{k_1'}\frac{[OCl^-][I^-]}{[OH^-]}$$

(c) An examination of the rate law shows that the rate is dependent upon the concentration of OH^-, which means that the rate will be dependent upon the pH of the solution.

(d) If the reaction is carried out in an organic solvent, then H_2O is no longer the solvent and its concentration must be included in calculating the equilibrium concentration of HOCl:

$$K = \frac{k_1}{k_1'} = \frac{[HOCl][OH^-]}{[OCl^-][H_2O]}$$

$$[HOCl] = \frac{k_1}{k_1'}\frac{[OCl^-][H_2O]}{[OH^-]}$$

$$\text{rate} = \frac{k_2k_1}{k_1'}\frac{[OCl^-][I^-][H_2O]}{[OH^-]}$$

The rate of reaction will then show a dependence upon the concentration of water, which will be obscured when the reaction is done with water as the solvent.

CHAPTER 14
THE ELEMENTS: THE FIRST FOUR MAIN GROUPS

14.1 (a) ruthenium (b) lithium (c) indium (d) iodine

14.3 (a) cobalt (b) cobalt (c) sodium (d) oxygen

14.5 iodine < bromine < chlorine

14.7 chlorine

14.9 tellurium

14.11 (a) K^+; (b) Na^+; (c) Tl^+

14.13 $2\,K(s) + H_2(g) \longrightarrow 2\,KH(s)$

14.15 (a) saline (b) molecular (c) molecular (d) metallic

14.17 (a) acidic (b) amphoteric (c) acidic (d) basic

14.19 (a) CO_2 (b) SO_3 (c) B_2O_3

14.21 (a) $C_2H_2(g) + H_2(g) \longrightarrow H_2C{=}CH_2(g)$

Oxidation number of C in $C_2H_2 = -1$; of C in $H_2C{=}CH_2 = -2$, carbon has been reduced.

(b) $CO(g) + H_2O(g) \rightleftharpoons CO_2(g) + H_2(g)$

(c) $BaH_2(s) + 2\,H_2O(l) \longrightarrow Ba(OH)_2 + 2\,H_2(g)$

14.23 (a) $CH_4(g) + H_2O(g) \longrightarrow CO(g) + 3\,H_2(g)$

$$\begin{aligned}\Delta H^\circ_r &= \Delta H^\circ_f(CO, g) - [\Delta H^\circ_f(CH_4, g) + \Delta H^\circ_f(H_2O, g)]\\ &= (-110.53\ kJ\cdot mol^{-1}) - [(-74.81\ kJ\cdot mol^{-1}) + (-241.82\ kJ\cdot mol^{-1})]\\ &= +206.10\ kJ\cdot mol^{-1}\end{aligned}$$

(b) $\Delta S°_r = S°(CO, g) + 3S°(H_2, g) - [S°(CH_4, g) - S°(H_2O, g)]$

$$= 197.67\ J\cdot K^{-1}\cdot mol^{-1} + 3(130.68\ J\cdot K^{-1}\cdot mol^{-1}) - [186.26\ J\cdot K^{-1}\cdot mol^{-1} + 188.83\ J\cdot K^{-1}\cdot mol^{-1}]$$

$$= +214.62\ J\cdot K^{-1}\cdot mol^{-1}$$

(c) $\Delta G°_r = \Delta G°_f(CO, g) - [\Delta G°_f(CH_4, g) + \Delta G°_f(H_2O, g)]$

$$= (-137.17\ kJ\cdot mol^{-1}) - [(-50.72\ kJ\cdot mol^{-1}) + (-228.57\ kJ\cdot mol^{-1})]$$

$$= +142.12\ kJ\cdot mol^{-1}$$

$\Delta G°_r$ can also be calculated from $\Delta H°_r$ and $\Delta S°_r$:

$$\Delta G°_r = \Delta H°_r - T\Delta S°_r$$

$$= +206.10\ kJ\cdot mol^{-1} - (298\ K)(+214.62\ J\cdot K^{-1}\cdot mol^{-1})/(1000\ J\cdot kJ^{-1})$$

$$= +142.14\ kJ\cdot mol^{-1}$$

14.25 (a) At STP (273 K and 1.00 atm), 1.00 mol of $H_2(g)$, assumed to be ideal, occupies 22.4 L. Then, for

$$CaH_2(s) + 2\ H_2O(l) \longrightarrow Ca(OH)_2(s) + H_2(g)$$

$$\text{volume of } H_2(g) = (500\ g\ CaH_2)\left(\frac{1\ mol\ CaH_2}{42.10\ g\ CaH_2}\right)\left(\frac{2\ mol\ H_2}{1\ mol\ CaH_2}\right)\left(\frac{22.4\ L\ H_2}{1\ mol\ H_2}\right)$$

$$= 532\ L\ H_2$$

(b) $$\text{volume of } H_2O(l) = (500\ g\ CaH_2)\left(\frac{1\ mol\ CaH_2}{42.10\ g\ CaH_2}\right)\left(\frac{2\ mol\ H_2O}{1\ mol\ CaH_2}\right)$$

$$\left(\frac{18.02\ g\ H_2O}{1\ mol\ H_2O}\right)\left(\frac{1.0\ mL\ H_2O}{1.0\ g\ H_2O}\right) = 428\ mL\ H_2O$$

14.27 (a) $H_2(g) + Cl_2(g) \xrightarrow{\text{light}} 2\ HCl(g)$

(b) $H_2(g) + 2\ Na(l) \xrightarrow{\Delta} 2\ NaH(s)$

(c) $P_4(s) + 6\ H_2(g) \longrightarrow 4\ PH_3(g)$

(d) $Cu(s) + H_2(g) \longrightarrow$ N.R.

14.29 Lithium is the only Group I element that reacts directly with nitrogen to form lithium nitride:

$$6\ Li(s) + N_2(g) \xrightarrow{\Delta} 2\ Li_3N(s)$$

Lithium reacts with oxygen to form mainly the oxide:

$$4\ Li(s) + O_2(g) \longrightarrow 2\ Li_2O(s)$$

The other members of the group form mainly the peroxide or superoxide. Lithium exhibits the diagonal relationship that is common to many first members of a group. Li is similar in many of its compounds to the compounds of Mg. This is related to the small ionic radius of Li^+, 58 pm, which is closer to the ionic radius of Mg^{2+}, 72 pm, but substantially less than that of Na^+, 102 pm.

14.31 (a) $4\ Li(s) + O_2(g) \longrightarrow 2\ Li_2O(s)$

(b) $6\ Li(s) + N_2(g) \xrightarrow{\Delta} 2\ Li_3N(s)$

(c) $2\ Na(s) + 2\ H_2O(l) \longrightarrow 2\ NaOH(aq) + H_2(g)$

(d) $4\ KO_2(s) + 2\ H_2O(g) \longrightarrow 4\ KOH(s) + 3\ O_2(g)$

14.33 1 mol $Na_2CO_3 \cdot 10\ H_2O$ yields 1 mol Na_2CO_3 in water.

$$\begin{aligned}\text{mass of } Na_2CO_3 \cdot 10\ H_2O &= 0.250\ L \times 0.100\ mol \cdot L^{-1} \\ &\quad \times 286.15\ g\ Na_2CO_3 \cdot 10\ H_2O \cdot mol^{-1} \\ &= 7.15\ g\ Na_2CO_3 \cdot 10\ H_2O\end{aligned}$$

14.35 $Mg(s) + 2\ H_2O(l) \longrightarrow Mg(OH)_2 + H_2(g)$

14.37 (a) $CaO(s) + H_2O(l) \longrightarrow Ca(OH)_2(s)$

(b)
$$\begin{aligned}\Delta G^\circ_r &= \Delta G^\circ_f(Ca(OH)_2, s) - [\Delta G^\circ_f(CaO, s) + \Delta G^\circ_f(H_2O, l)] \\ &= -898.49\ kJ \cdot mol^{-1} - [(-604.03\ kJ \cdot mol^{-1}) + (-237.13\ kJ \cdot mol^{-1})] \\ &= -57.33\ kJ \cdot mol^{-1}\end{aligned}$$

14.39 Be is the weakest reducing agent; Mg is stronger, but weaker than the remaining members of the group, all of which have approximately the same reducing strength. This effect is related to the very small radius of the Be^{2+} ion, 27 pm; its strong polarizing power introduces much covalent character into its compounds. Thus, Be attracts electrons more strongly and does not release them as readily as other members of the group. Mg^{2+} is also a small ion, 58 pm, so the same reasoning applies to it also, but to a lesser extent. The remaining ions of the group are considerably larger, release electrons more readily, and are better reducing agents.

14.41 (a) $2\ Al(s) + 2\ OH^-(aq) + 6\ H_2O(l) \longrightarrow 2[Al(OH)_4]^-(aq) + 3\ H_2(g)$

(b) $Be(s) + 2\ OH^-(aq) + 2\ H_2O(l) \longrightarrow [Be(OH)_4]^{2-}(aq) + H_2(g)$

The similarity of Be and Al in chemical reactions is an example of the diagonal relationship in the periodic table; namely, the similar chemical behavior of elements that are diagonal neighbors, such as Be and Al.

14.43 (a) $Mg(OH)_2(s) + 2\ HCl(aq) \longrightarrow MgCl_2(aq) + 2\ H_2O(l)$

(b) $Ca(s) + 2\ H_2O(l) \longrightarrow Ca(OH)_2(aq) + H_2(g)$

(c) $BaCO_3(s) \xrightarrow{\Delta} BaO(s) + CO_2(g)$

14.45 (a) $:\ddot{\underset{..}{Cl}}—Be—\ddot{\underset{..}{Cl}}:$ (b) 180° (c) *sp*

14.47 (a) $CaCO_3(s) \xrightarrow{\Delta} CaO(s) + CO_2(g)$

$$\Delta H°_r = (-635.09\ kJ\cdot mol^{-1}) + (-393.51\ kJ\cdot mol^{-1}) - (-1206.9\ kJ\cdot mol^{-1})$$
$$= +178.3\ kJ\cdot mol^{-1}$$
$$\Delta S°_r = (39.75\ J\cdot K^{-1}\cdot mol^{-1}) + (213.74\ J\cdot K^{-1}\cdot mol^{-1}) - (92.9\ J\cdot K^{-1}\cdot mol^{-1})$$
$$= +160.6\ J\cdot K^{-1}\cdot mol^{-1}$$

(b) The temperature at which the equilibrium constant becomes greater than 1 is the temperature at which $\Delta G°_r$ crosses over between positive and negative values. Therefore, it is the temperature at which $\Delta G°_r = 0$.

$$\Delta G°_r = 0 = \Delta H°_r - T\Delta S°_r = -RT \ln K \text{ (for } K = 1, \ln K = 0)$$
$$T = \frac{\Delta H°_r}{\Delta S°_r} = \frac{178\ 300\ J\cdot mol^{-1}}{160.6\ J\cdot K^{-1}\cdot mol^{-1}} = 1110\ K = 837°C$$

Consequently, at temperatures above 1110 K, $\ln K > 0$ and K becomes greater than 1.

14.49 The overall equation for the electrolytic reduction in the Hall process is

$$4\ Al^{3+}(melt) + 6\ O^{2-}(melt) + 3\ C(s, gr) \longrightarrow 4\ Al(s) + 3\ CO_2(g)$$

14.51 (a) $B_2O_3(s) + 3\ Mg(l) \xrightarrow{\Delta} 2\ B(s) + 3\ MgO(s)$

(b) $2\ Al(s) + 3\ Cl_2(g) \longrightarrow 2\ AlCl_3(s)$

(c) $4\ Al(s) + 3\ O_2(g) \longrightarrow 2\ Al_2O_3(s)$

14.53 (a) The hydrate of $AlCl_3$, that is, $AlCl_3\cdot 6\ H_2O$, functions as a deodorant and antiperspirant.

(b) α-Alumina is corundum. It is used as an abrasive in sandpaper.

(c) $B(OH)_3$ is an antiseptic and insecticide.

14.55 The cathode reaction is $Al^{3+}(melt) + 3\ e^- \longrightarrow Al(l)$

$$\text{charge consumed} = (8.0\ h)\left(\frac{3600\ s}{1\ h}\right)(1.0 \times 10^5\ C\cdot s^{-1}) = 2.88 \times 10^9 C$$

$$\text{mass of Al produced} = (2.88 \times 10^9\ C)\left(\frac{1\ mol\ e^-}{9.65 \times 10^4\ C}\right)\left(\frac{1\ mol\ Al}{3\ mol\ e^-}\right)\left(\frac{26.98\ g\ Al}{1\ mol\ Al}\right)$$
$$= 2.7 \times 10^5\ g\ Al$$

14.57 (a) We want $E°$ for $Tl^{3+}(aq) + 3\ e^- \longrightarrow Tl(s)$, $n = 3$.
This reaction is the reverse of the formation reaction

$Tl(s) \longrightarrow Tl^{3+}(aq) + 3\ e^-$

Therefore, for the Tl^{3+}/Tl couple, $\Delta G°_r = -215\ kJ \cdot mol^{-1}$.

$$\Delta G°_r = -nFE° = -215\ kJ \cdot mol^{-1}$$
$$E° = \frac{\Delta G°_r}{-nF} = \frac{-2.15 \times 10^5\ J \cdot mol^{-1}}{-3 \times 9.65 \times 10^4\ C \cdot mol^{-1}} = +0.743\ V$$

(b) Using the potential from part (a) and the potential from Appendix 2B for the reduction of Tl^+, we can determine if Tl^+ will spontaneously disproportionate in solution. The equation of interest is

$3\ Tl^+(aq) \longrightarrow 2\ Tl(s) + Tl^{3+}(aq)$

The half-reactions to combine are:

$Tl^+ \longrightarrow Tl^{3+} + 2\ e^-$ (1)
$Tl^+ + e^- \longrightarrow Tl$ (2)

The potential for reaction (1) must be obtained using the $\Delta G°$ values of the two known half-reactions:

$Tl \longrightarrow Tl^{3+} + 3\ e^-$ $\Delta G° = +215\ kJ \cdot mol^{-1}$
$Tl^+ + e^- \longrightarrow Tl$ $\Delta G° = -nFE°$
$= -1(9.65 \times 10^4\ J \cdot V^{-1} \cdot mol^{-1})(-0.34\ V)/(1000\ J \cdot kJ^{-1})$
$= +33\ kJ \cdot mol^{-1}$

$\Delta G°$ for the combined half-reaction $Tl^+ \longrightarrow Tl^{3+} + 2\ e^-$ is the sum of these two numbers:
$+215\ kJ \cdot mol^{-1} + 33\ kJ \cdot mol^{-1} = +248\ kJ \cdot mol^{-1}$.
We can now combine this with the reduction process for Tl^+ to get the desired equation:

$Tl^+ \longrightarrow Tl^{3+} + 2\ e^-$ $\Delta G° = +248\ kJ \cdot mol^{-1}$
$+2(Tl^+ + e^- \longrightarrow Tl)$ $\Delta G° = 2(+33\ kJ \cdot mol^{-1})$

$3\ Tl^+ \longrightarrow 2\ Tl + Tl^{3+}$ $\Delta G° = +248\ kJ \cdot mol^{-1} + 2(+33\ kJ \cdot mol^{-1})$
$= +314\ kJ \cdot mol^{-1}$

This is not spontaneous.

14.59 Silicon occurs widely in the Earth's crust in the form of silicates in rocks and as silicon dioxide in sand. It is obtained from quartzite, a form of quartz (SiO_2), by the following processes:

(1) reduction in an electric arc furnace

$$SiO_2(s) + 2\ C(s) \longrightarrow Si(s, \text{crude}) + 2\ CO(g)$$

(2) purification of the crude product in two steps

$$Si(s, \text{crude}) + 2\ Cl_2(g) \longrightarrow SiCl_4(l)$$

followed by reduction with hydrogen to the pure element

$$SiCl_4(l) + 2\ H_2(g) \longrightarrow Si(s, \text{pure}) + 4\ HCl(g)$$

14.61 In diamond, carbon is sp^3 hybridized and forms a tetrahedral, three-dimensional network structure, which is extremely rigid. Graphite carbon is sp^2 hybridized and planar, and its application as a lubricant results from the fact that the two-dimensional sheets can "slide" across one another, thereby reducing friction. In graphite, the unhybridized *p*-electrons are free to move from one carbon atom to another, which results in its high electrical conductivity. In diamond, all electrons are localized in sp^3 hybridized C—C σ-bonds, so diamond is a poor conductor of electricity.

14.63 (a) $SiCl_4(l) + 2\ H_2(g) \longrightarrow Si(s) + 4\ HCl(g)$

(b) $SiO_2(s) + 3\ C(s) \xrightarrow{2000°C} SiC(s) + 2\ CO(g)$

(c) $Ge(s) + 2\ F_2(g) \longrightarrow GeF_4(s)$

(d) $CaC_2(s) + 2\ H_2O(l) \longrightarrow Ca(OH)_2(s) + C_2H_2(g)$

14.65 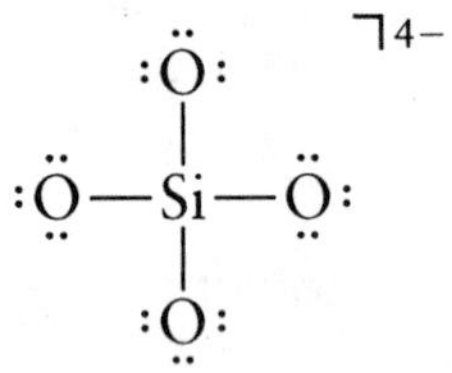

Formal charges: Si = 0, O = −1

oxidation numbers: Si = +4, O = −2

This is an AX_4 VSEPR structure; therefore, the shape is tetrahedral.

14.67 $SiO_2(s) + 2\ C(s) \longrightarrow Si(s) + 2\ CO(g)$

$$\begin{aligned}\Delta H°_r &= \Delta H°_f(\text{products}) - \Delta H°_f(\text{reactants})\\ &= [(2)(-110.53\ \text{kJ}\cdot\text{mol}^{-1})] - [-910.94\ \text{kJ}\cdot\text{mol}^{-1}]\\ &= +689.88\ \text{kJ}\cdot\text{mol}^{-1}\end{aligned}$$

$$\begin{aligned}\Delta S^\circ_r &= S^\circ(\text{products}) - S^\circ(\text{reactants})\\ &= [18.83\ \text{J}\cdot\text{K}^{-1}\cdot\text{mol}^{-1} + (2)(197.67\ \text{J}\cdot\text{K}^{-1}\cdot\text{mol}^{-1})]\\ &\quad - [41.84\ \text{J}\cdot\text{K}^{-1}\cdot\text{mol}^{-1} + (2)(5.740\ \text{J}\cdot\text{K}^{-1}\cdot\text{mol}^{-1})]\\ &= +360.85\ \text{J}\cdot\text{K}^{-1}\cdot\text{mol}^{-1}\end{aligned}$$

$$\begin{aligned}\Delta G^\circ_r &= \Delta H^\circ_r - T\Delta S^\circ_r\\ &= 689.88 \times \text{kJ} - (298.15\ \text{K})(360.85\ \text{J}\cdot\text{K}^{-1}\cdot\text{mol}^{-1})/1000\ \text{J}\cdot\text{kJ}^{-1}\\ &= +5.8229 \times 10^2\ \text{kJ}\cdot\text{mol}^{-1}\end{aligned}$$

The temperature at which the equilibrium constant becomes greater than 1 is the temperature at which $\Delta G^\circ_r = -RT \ln K = 0$, because $\ln 1 = 0$. Above this temperature, the equilibrium constant is greater than 1. $\Delta G^\circ_r = 0$ when $T\Delta S^\circ_r = \Delta H^\circ_r$, or

$$T = \frac{\Delta H^\circ_r}{\Delta S^\circ_r} = \frac{+689.88 \times 10^3\ \text{J}\cdot\text{mol}^{-1}}{+360.85\ \text{J}\cdot\text{K}^{-1}\cdot\text{mol}^{-1}} = 1912\ \text{K}$$

14.69 mass of HF $= (2.00 \times 10^{-3}\ \text{g})\left(\frac{1\ \text{mol SiO}_2}{60.09\ \text{g SiO}_2}\right)\left(\frac{6\ \text{mol HF}}{1\ \text{mol SiO}_2}\right)\left(\frac{20.01\ \text{g HF}}{1\ \text{mol HF}}\right)$

$= 4.00 \times 10^{-3}$ g HF = 4.00 mg HF

14.71 (a) The $Si_2O_7^{6-}$ ion is built from two SiO_4^{4-} tetrahedral ions in which the silicate tetrahedra share one O atom. This is the only case in which one O is shared.
(b) The pyroxenes, for example, jade, $NaAl(SiO_3)_2$, consist of chains of SiO_4 units in which two O atoms are shared by neighboring units. The repeating unit has the formula SiO_3^{2-}. See Fig. 14.42.

14.73 SiF_6^{2-}

14.75 Ionic fluorides react with water to liberate HF, which then reacts with the glass. Glass bottles used to store metal fluorides become brittle and may disintegrate upon standing on the shelf.

14.77 The iron ions impart a deep red color to the clay, which is not desirable for manufacture of fine china; a white base is aesthetically more pleasing.

14.79 In the majority of its reactions, hydrogen acts as a reducing agent. Examples are $2\ H_2(g) + O_2(g) \longrightarrow 2\ H_2O(l)$ and various ore reduction processes, such as

$NiO(s) + H_2(g) \xrightarrow{\Delta} Ni(s) + H_2O(g)$. With highly electropositive elements, such as the alkali metals, $H_2(g)$ acts as an oxidizing agent and forms metal hydrides, for example, $2\ K(s) + H_2(g) \longrightarrow 2\ KH(s)$.

14.81 (a)

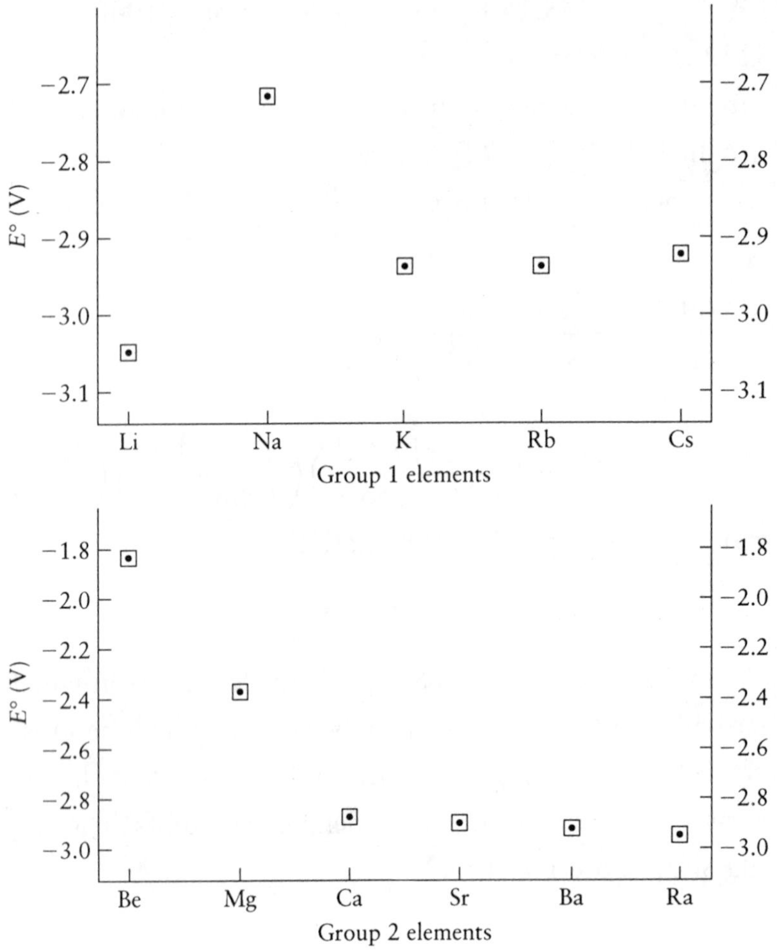

(b) For both groups, the trend in standard potentials with increasing atomic number is overall downward (they become more negative), but lithium is anomalous. This overall downward trend makes sense, because we expect that it is easier to remove electrons that are farther away from the nuclei. However, because there are several factors that influence ease of removal, the trend is not smooth. The potentials are a net composite of the free energies of sublimation of solids, dissociation of gaseous molecules, ionization enthalpies, and enthalpies of hydration of gaseous ions. The origin of the anomalously strong reducing power of Li is the strongly exothermic energy of hydration of the very small Li^+ ion, which favors the ionization of the element in aqueous solution.

14.83 (a, b)

$H_2(g) + F_2(g) \longrightarrow 2\ HF(g)$, explosive $\Delta G°_f = -273.2\ kJ \cdot mol^{-1}$

$H_2(g) + Cl_2(g) \longrightarrow 2\ HCl(g)$, explosive $\Delta G°_f = -95.30\ kJ \cdot mol^{-1}$

$H_2(g) + Br_2(g) \longrightarrow 2\ HBr(g)$, vigorous $\Delta G°_f = -53.45\ kJ \cdot mol^{-1}$

$H_2(g) + I_2(g) \longrightarrow 2\ HI(g)$, less vigorous $\Delta G°_f = +1.70\ kJ \cdot mol^{-1}$

The word *vigor* as used here has two components, a thermodynamic one and a kinetic one. $\Delta G°_f$ is most negative for HF and becomes slightly positive for HI. So, the formation of HF, HCl, and HBr are all thermodynamically spontaneous. The kinetics parallel, in this case, the thermodynamic spontaneity, although this parallel behavior is not necessarily true in other systems.

(c) hydrofluoric acid
hydrochloric acid
hydrobromic acid
hydroiodic acid

14.85 (a) The oxide ion, O^{2-}, in CaO acts as a Lewis base and reacts with the Lewis acid, SiO_2, in a Lewis acid-base reaction:

$$CaO(s) + SiO_2(s) \longrightarrow CaSiO_3(l)$$

SiO_2, which is an impurity in iron ore, is removed by this reaction. The calcium oxide in this reaction can be obtained from limestone [$CaCO_3(s) \xrightarrow{\Delta} CaO(s) + CO_2(g)$]. This is the reason that limestone is important in the iron industry.

(b) $CaO(s) + CO_2(g) \longrightarrow CaCO_3(s)$ (not an efficient preparation of $CaCO_3$, because of the weak Lewis acidity of CO_2)

14.87 (a) The "hardness" of water is due to the presence of calcium and magnesium salts (particularly their hydrogen carbonates). In laundering and bathing, Ca^{2+} and Mg^{2+} cations convert soluble Na^+ soaps to insoluble Ca^{2+} and Mg^{2+} soaps (thereby reducing the detergent efficiency).

(b) Softening can be achieved by removing the Ca^{2+} and Mg^{2+} before they form these insoluble compounds, with the use of lime.

$Ca(HCO_3)_2(aq) + Ca(OH)_2(aq) \longrightarrow 2\ CaCO_3(s) + 2\ H_2O(l)$

$Mg(HCO_3)_2(aq) + Ca(OH_2)(aq) \longrightarrow Mg(OH)_2(s) + Ca(HCO_3)_2(aq)$

14.89 (a) H_3BO_3, acid; $B(OH)_4^-$, conjugate base

(b) $B(OH)_3(aq) + 2\ H_2O(l) \rightleftharpoons H_3O^+(aq) + B(OH)_4^-$

14.91 (a)

Element		Ionization energy, $kJ \cdot mol^{-1}$		Atomic radius, pm	
Group 13	B	799	Gen. decreasing ↓	88	Increasing ↓
	Al	577		143	
	Ga	577		153	
	In	556		167	
	Tl	590		171	
Group 14	C	1090	Gen. decreasing ↓	77	Increasing ↓
	Si	786		118	
	Ge	784		122	
	Sn	707		158	
	Pb	716		175	

(b) The ionization energies generally decrease down a group. As the atomic number of an element increases, atomic shells and subshells that are farther from the nucleus are filled. The outermost valence electrons are consequently easier to remove. The radii increase down a group for the same reason. The radii are primarily determined by the outer shell electrons, which are farther from the nucleus in the heavy elements.

(c) The trends correlate well with elemental properties, for example, the greater ease of outermost electron removal correlates with increased metallic character, that is, ability to form positive ions by losing one or more electrons.

14.93 (a) $2\,H_2O(l) + 2\,e^- \longrightarrow H_2(g) + 2\,OH^-(aq)$

(b) cathode, because this process is reduction

(c) At STP (273 K and 1 atm), 1 mol H_2 occupies 22.4 L.

$$V_{H_2(g)} = (10.0\text{ A})(30\text{ min})\left(\frac{60\text{ s}}{1\text{ min}}\right)\left(\frac{1\text{ mol e}^-}{9.65\times 10^4\text{ A}\cdot\text{s}}\right)\left(\frac{1\text{ mol H}_2}{2\text{ mol e}^-}\right)\left(\frac{22.4\text{ L H}_2}{1\text{ mol H}_2}\right)$$

$$= 2.1\text{ L H}_2$$

14.95 In the majority of its reactions, hydrogen acts as a reducing agent, that is, $H_2(g) \longrightarrow 2\,H^+(aq) + 2\,e^-$, $E° = 0$ V. In these reactions, hydrogen resembles Group 1 elements, such as Na and K. However, as described in the text and in the answer to Exercise 14.79, it may also act as an oxidizing agent; that is, $H_2(g) + 2\,e^- \longrightarrow 2\,H^-(aq)$, $E° = -2.25$ V. In these reactions, hydrogen resembles Group 17 elements, such as Cl and Br. Consequently, H_2 will oxidize elements with standard reduction potentials more negative than -2.25 V, such as the alkali and alkaline earth metals (except Be). The compounds formed are hydrides and contain the H^- ion; the singly charged negative ion is reminiscent of the halide ions. Hydrogen also forms diatomic molecules and covalent bonds like the halogens.

The atomic radius of H is 78 pm, which compares rather well to that of F (64 pm) but not as well to that of Li (157 pm). The ionization energy of H is $1310\ kJ\cdot mol^{-1}$, which is similar to that of F ($1680\ kJ\cdot mol^{-1}$) but not similar to that of Li ($519\ kJ\cdot mol^{-1}$). The electron affinity of H is $+73\ kJ\cdot mol^{-1}$, that of F is $+328\ kJ\cdot mol^{-1}$, and that of Li is $60\ kJ\cdot mol^{-1}$. So in its atomic radius and ionization energy, H more closely resembles the Period 2 halogen, fluorine, in Group 17, than the Period 2 alkali metal, lithium, in Group 1; whereas in electron affinity, it more closely resembles lithium, Group 1. In electronegativity, H does not resemble elements in either Group 1 or Group 17, although its electronegativity is somewhat closer to those of Group 1. Consequently, hydrogen could be placed in either Group 1 or Group 17. But it is best to think of hydrogen as a unique element that has properties in common with both metals and nonmetals; therefore, it should probably be centered in the periodic table, as it is shown in the table in the text.

14.97 (a)

Ion	Radius, pm	Polarizing ability (×1000)	Ion	Radius, pm	Polarizing ability (×1000)
Li^+	58	17	Be^{2+}	27	74
Na^+	102	9.80	Mg^{2+}	72	28
K^+	138	7.25	Ca^{2+}	100	20.0
Rb^+	149	6.71	Sr^{2+}	116	17.2
Cs^+	170	5.88	Ba^{2+}	136	14.7

(b) These data roughly support the diagonal relationship. Li^+ is more like Mg^{2+} than Be^{2+}, and Na^+ is more like Ca^{2+} than Mg^{2+}; but further down the group, the correlation fails. Charge divided by r^3 would be a better measure of polarizing ability.

14.99 $H_2(g) + Br_2(l) \longrightarrow 2\ HBr(g)$

$$\text{number of moles of HBr} = (0.120\ L\ H_2)\left(\frac{1\ mol\ H_2}{22.4\ L\ H_2}\right)\left(\frac{2\ mol\ HBr}{1\ mol\ H_2}\right)$$

$$= 0.0107\ mol$$

$$\text{molar concentration of HBr} = \frac{0.0107\ mol}{0.150\ L} = 0.0713\ mol\cdot L^{-1}$$

14.101 The smaller the cation, the greater is the ability of the cation to polarize and weaken the carbonate ion, CO_3^{2-}. On that basis, we would predict that within a group the carbonates of the first members of the group are less stable than those of the later members. Thus, $Li_2CO_3 < Na_2CO_3 < K_2CO_3 < Rb_2CO_3 < Cs_2CO_3$ and $BeCO_3 < MgCO_3 < CaCO_3 < SrCO_3 < BaCO_3$. Between groups, we would expect the stability of the carbonates in one period to decrease from Group 1 to Group 13 because of the smaller size of Group 13 ions. Thus, $Al_2(CO_3)_2 < MgCO_3 < Na_2CO_3$. Carbonates of Group 13 M^{3+} ions are, in fact, so unstable that they do not exist. Tl_2CO_3 where Tl has an oxidation number of +1 is known, however.

14.103 NaH, like other saline hydrides, reacts with O—H bonds to liberate H_2. For example, reaction of NaH with H_2O gives NaOH(aq) and H_2(g). The corresponding reaction with methanol is

$NaH(s) + CH_3OH(l) \longrightarrow NaOCH_3 + H_2(g)$

Sodium methoxide (also known as sodium methylate), $NaOCH_3$, is a strong base which is used in organic chemistry.

14.105 (a) The unit cell described will contain a total of 4 B atoms and 4 N atoms. The volume of the cell is $(361.5\ \text{pm})^3 = (3.615 \times 10^{-8}\ \text{cm})^3 =$ or $4.724 \times 10^{-23}\ \text{cm}^{-3}$. The mass in the unit cell will be $(4 \times 10.81\ \text{g}\cdot\text{mol}^{-1} + 4 \times 14.01\ \text{g}\cdot\text{mol}^{-1}) \div 6.022 \times 10^{23}\ \text{mol}^{-1} = 1.649 \times 10^{-22}\ \text{g}$.

$$d = \frac{1.649 \times 10^{-22}\ \text{g}}{4.724 \times 10^{-23}\ \text{cm}^3} = 3.491\ \text{g}\cdot\text{cm}^{-3}$$

(b) Because the density of cubic boron nitride is greater than that of hexagonal BN, we would expect the cubic form to be favored at high pressures, exactly as found for the cubic (diamond) and hexagonal (graphite) forms of carbon.

CHAPTER 15
THE ELEMENTS: THE LAST FOUR MAIN GROUPS

15.1 The first step is the liquefaction of air, which is 76% by mass nitrogen. Air is cooled to below its boiling point by a series of expansion and compression steps in a kind of refrigerator. Nitrogen gas is then obtained by distillation of liquid air. The nitrogen boils off at $-196°C$, but gases with higher boiling points, principally O_2, remain as a liquid. The pure nitrogen gas is then liquefied by repeating the process.

15.3 $CO(NH_2)_2 + 2\ H_2O \longrightarrow (NH_4)_2CO_3$

$$\text{mass of } (NH_4)_2CO_3 = (5.0\text{ kg urea})\left(\frac{10^3\text{ g}}{1\text{ kg}}\right)\left(\frac{1\text{ mol}}{60.06\text{ g}}\right)$$
$$\left(\frac{1\text{ mol }(NH_4)_2CO_3}{1\text{ mol urea}}\right)\left(\frac{96.09\text{ g }(NH_4)_2CO_3}{1\text{ mol }(NH_4)_2CO_3}\right)$$
$$= 8.0 \times 10^3\text{ g (or 8.0 kg)}(NH_4)_2CO_3$$

15.5 (a) 1 mole of $N_2(g)$ occupies 22.4 L at STP. For the reaction $Pb(N_3)_2 \longrightarrow Pb + 3\ N_2$, the volume of $N_2(g)$ produced is

$$(1.0\text{ g }Pb(N_3)_2)\left(\frac{1\text{ mol }Pb(N_3)_2}{291.25\text{ g }Pb(N_3)_2}\right)\left(\frac{3\text{ mol }N_2}{1\text{ mol }Pb(N_3)_2}\right)\left(\frac{22.4\text{ L }N_2}{1\text{ mol }N_2}\right)$$
$$= 0.23\text{ L }N_2(g)$$

(b) $Hg(N_3)_2$ would produce a larger volume, because its molar mass is less. Note that molar mass occurs in the denominator in this calculation.

15.7 N_2O: $H_2N_2O_2$; $N_2O(g) + H_2O(l) \longrightarrow H_2N_2O_2(aq)$
N_2O_3: HNO_2; $N_2O_3(g) + H_2O(l) \longrightarrow 2\ HNO_2(aq)$
N_2O_5: HNO_3; $N_2O_5(g) + H_2O(l) \longrightarrow 2\ HNO_3(aq)$

15.9

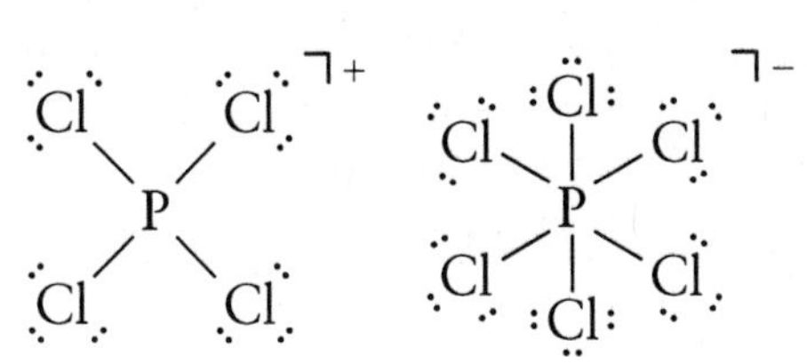

PCl_4^+, AX_4 tetrahedral; PCl_6^-, AX_6 octahedral

15.11 superphosphate, $2\ CaSO_4 + Ca(H_2PO_4)_2$

$$\%P = \frac{2(30.97\ g\cdot mol^{-1})}{506.33\ g\cdot mol^{-1}} \times 100 = 12.23\%$$

15.13 To answer this question, we can use the thermodynamic quantities from Appendix 2A to calculate whether this reaction is spontaneous:

$$Cu(s) + H_2SO_4(aq) \longrightarrow Cu^{2+}(aq) + SO_4^{\ 2-}(aq) + H_2(g)$$

$$\Delta G^\circ_r = 65.49\ kJ\cdot mol^{-1} + (-744.53\ kJ\cdot mol^{-1}) - (-755.91\ kJ\cdot mol^{-1})$$

$$= +76.87\ kJ\cdot mol^{-1}$$

Because ΔG°_r is positive, we do not expect this reaction to be spontaneous. Note: $H_2SO_4(aq)$ is approximated using $H^+(aq) + HSO_4^{\ -}(aq)$ since it is a strong acid.

15.15 (a) $4\ Li(s) + O_2(g) \xrightarrow{\Delta} 2\ Li_2O(s)$

(b) $2\ Na(s) + 2\ H_2O(l) \longrightarrow 2\ NaOH(aq) + H_2(g)$

(c) $2\ F_2(g) + 2\ H_2O(l) \longrightarrow 4\ HF(aq) + O_2(g)$

(d) $2\ H_2O(l) \longrightarrow O_2(g) + 4\ H^+(aq) + 4\ e^-$

15.17 (a) $2\ H_2S(g) + 3\ O_2(g) \longrightarrow 2\ SO_2(g) + 2\ H_2O(g)$

(b) $CaO(s) + H_2O(l) \longrightarrow Ca(OH)_2(aq)$

(c) $PCl_5(s) + 4\ H_2O(l) \longrightarrow H_3PO_4(aq) + 5\ HCl(g)$

15.19

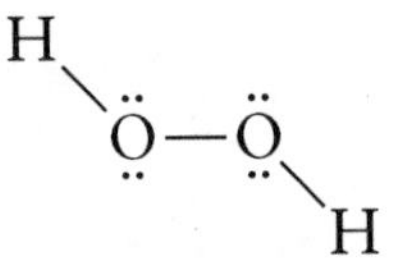

Each O in H_2O_2 is an AX_2E_2 structure; therefore, the bond angle is predicted to be $<109.5°$. In actuality, it is 97°.

15.21 $2\ H_2O_2(l) \longrightarrow 2\ H_2O(l) + O_2(g)$

Assume 3% by mass. At 273 K and 1.00 atm, 1 mol O_2 has a volume of 22.4 L.

$$\text{volume of } O_2 = (500\ mL)\left(\frac{1.0\ g}{1.0\ mL}\right)\left(\frac{0.03\ g\ H_2O_2}{1.0\ g\ soln}\right)$$

$$\left(\frac{1\ mol\ H_2O_2}{34.02\ g\ H_2O_2}\right)\left(\frac{1\ mol\ O_2}{2\ mol\ H_2O_2}\right)\left(\frac{22.4\ L\ O_2}{1\ mol\ O_2}\right) = 5\ L\ O_2$$

15.23 $O_2^{\ 2-} + H_2O \rightleftharpoons HO_2^{\ -} + OH^-$ essentially complete

$$HO_2^{\ -} + H_2O \rightleftharpoons H_2O_2 + OH^- \quad K_b = \frac{K_w}{K_{a1}}$$

$K_{a1} = 1.8 \times 10^{-12} \quad K_b = \dfrac{1.00 \times 10^{-14}}{1.8 \times 10^{-12}} = 5.6 \times 10^{-3}$

Because this K_b is relatively small, we can assume that essentially all the OH^- is formed in the first ionization; therefore,

$$[OH^-] = \left(\frac{2.00 \text{ g } Na_2O_2}{0.200 \text{ L}}\right)\left(\frac{1 \text{ mol } Na_2O_2}{77.98 \text{ g } Na_2O_2}\right)\left(\frac{1 \text{ mol } OH^-}{1 \text{ mol } Na_2O_2}\right)$$
$$= 0.128 \text{ mol}\cdot\text{L}^{-1}$$

$pOH = -\log(0.128) = 0.893 \quad pH = 14.00 - 0.893 = 13.11$

If we do not ignore the second ionization, then the additional contribution to $[OH^-]$ can be approximately calculated as follows:

$$K_b = \frac{[H_2O_2][OH^-]}{[HO_2^-]} = \frac{x(0.128 + x)}{(0.128 - x)} = 5.6 \times 10^{-3}$$

To a first approximation, $x = 5.6 \times 10^{-3} \text{ mol}\cdot\text{L}^{-1}$

To a second approximation, $x = \dfrac{K_b(0.128 - 0.0056)}{(0.128 + 0.0056)} = 0.005$

Then $[OH^-] = 0.128 + 0.005 = 0.133$; $pOH = -\log(0.133) = 0.876$; and $pH = 13.12$. The difference between calculations is slight.

15.25 The weaker the H—X bond, the stronger the acid. H_2Te has the weakest bond; H_2O, the strongest. Therefore, the acid strengths are

$H_2Te > H_2Se > H_2S > H_2O$

15.27 $\frac{3}{2} O_2(g) \longrightarrow O_3(g)$

(a) $\Delta H^\circ_f(O_3, g) = 142.7 \text{ kJ}\cdot\text{mol}^{-1}$

$\Delta S^\circ_f(O_3, g) = 238.93 \text{ J}\cdot\text{K}^{-1}\cdot\text{mol}^{-1} - \frac{3}{2} \times 205.14 \text{ J}\cdot\text{K}^{-1}\cdot\text{mol}^{-1}$
$= -68.78 \text{ J}\cdot\text{K}^{-1}\cdot\text{mol}^{-1}$

(b) $\Delta G^\circ_f(O_3, g) \approx \Delta H^\circ_f(O_3, g, 25°C) - T\Delta S^\circ_f(O_3, g, 25°C)$

Because $\Delta G^\circ_f(O_3, g)$ is positive at all temperatures, the reaction is not spontaneous at any temperature. It is less favored at high temperatures.

(c) Because the reaction entropy is negative, the $-T\Delta S^\circ_f$ term is always positive; so the entropy contribution to ΔG°_f is always positive, and the entropy does not favor the spontaneous formation of ozone.

15.29 (a) The reaction is: $H_2SO_4(l) \longrightarrow H_2SO_4(aq)$

where $H_2SO_4(aq)$ is $H^+(aq) + HSO_4^-(aq)$ since H_2SO_4 is a strong acid. (Note: $\Delta H^\circ_f(H^+)$is defined as 0).

$\Delta H^\circ_r = -887.34 \text{ kJ}\cdot\text{mol}^{-1} - (-813.99 \text{ kJ}\cdot\text{mol}^{-1}) = -73.35 \text{ kJ}\cdot\text{mol}^{-1}$

(b) The number of moles of H_2SO_4 is $0.981 \div 98.07\ g \cdot mol^{-1} = 0.0100$ mol
The amount of heat generated should be

$$0.0100\ mol \times -73.35\ kJ \cdot mol^{-1} = -0.734\ kJ.$$

The heat capacity of water is $4.18\ J \cdot (°C)^{-1} \cdot g^{-1}$. Adding 0.734 kJ of heat to the water should raise the temperature by

$$\Delta t = \frac{0.734\ kJ \times 1000\ J \cdot kJ^{-1}}{4.18\ J \cdot (°C)^{-1} \cdot g^{-1} \times 10.0\ g} = 17.5°$$

The final temperature should be 25.0°C + 17.5°C = 42.5°C.

15.31 Fluorine comes from the minerals fluorspar, CaF_2; cryolite Na_3AlF_6; and the fluorapatites, $Ca_5F(PO_4)_3$. The free element is prepared from HF and KF by electrolysis, but the HF and the KF needed for the electrolysis are prepared in the laboratory. Chlorine primarily comes from the mineral rock salt, NaCl. The pure element is obtained by electrolysis of liquid NaCl.

15.33 (a) HIO(aq) $H = +1, O = -2$; therefore, $I = +1$
(b) ClO_2 $O = -2$; therefore, $Cl = +4$
(c) Cl_2O_7 $O = -2$; therefore, $Cl = +14/2 = +7$
(d) $NaIO_3$ $Na = +1, O = -2$; therefore, $I = +5$

15.35 (a) $4\ KClO_3(l) \xrightarrow{\Delta} 3\ KClO_4(s) + KCl(s)$
(b) $Br_2(l) + H_2O(l) \longrightarrow HBrO(aq) + HBr(aq)$
(c) $NaCl(s) + H_2SO_4(aq) \longrightarrow NaHSO_4(aq) + HCl(g)$
(d) (a) and (b) are redox reactions. In (a), Cl is both oxidized and reduced. In (b), Br is both oxidized and reduced. (c) is a Brønsted acid-base reaction; H_2SO_4 is the acid, and Cl^- the base.

15.37 (a) $HClO < HClO_2 < HClO_3 < HClO_4$ ($HClO_4$ is strongest; HClO, weakest)
(b) The oxidation number of Cl increases from HClO to $HClO_4$. In $HClO_4$, chlorine has its highest oxidation number of +7, so $HClO_4$ will be the strongest oxidizing agent.

15.39

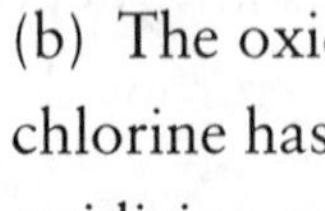

AX_2E_2, angular, slightly less than 109°

15.41

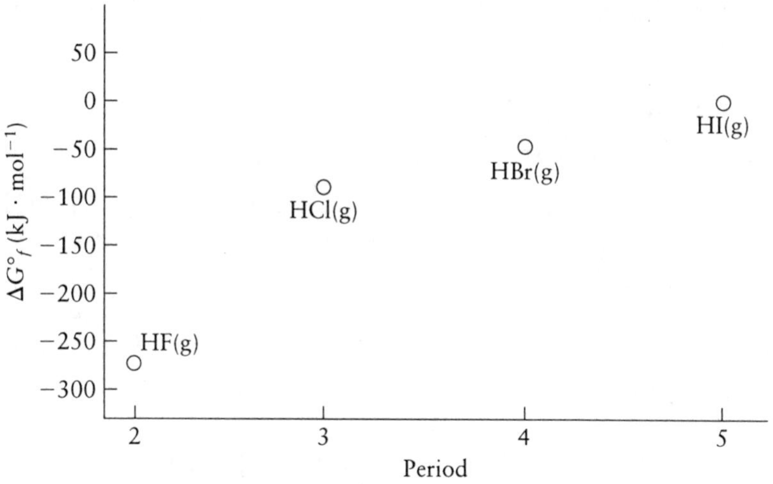

The thermodynamic stability of the hydrogen halides decreases down the group. The $\Delta G°_f$ values of HCl, HBr, and HI fit nicely on a straight line; whereas HF is anomalous. In other properties, HF is also the anomalous member of the group, in particular, its acidity. Also see Exercise 15.42.

15.43 $Cl_2(g) + 2\,e^- \longrightarrow 2\,Cl^-(aq) \quad E° = +1.36\text{ V}$

$MnO_4^-(aq) + 8\,H^+(aq) + 5\,e^- \longrightarrow Mn^{2+}(aq) + 4\,H_2O(l) \quad E° = +1.51\text{ V}$

$E°_{cell} = (1.36 - 1.51)\text{ V} = -0.15\text{ V}$

Because $E°_{cell}$ is negative, $Cl_2(g)$ will not oxidize Mn^{2+} to form the permanganate ion in an acidic solution.

15.45 $\frac{1}{2}\,H_2(g) + \frac{1}{2}\,I_2(s) \rightleftharpoons HI(g)$

$\Delta G°_r = \Delta G°_f(HI, g) = 1.70\text{ kJ}\cdot\text{mol}^{-1}$

$\Delta G°_r = -RT\ln K_p$

$$\ln K_p = \frac{-1.70 \times 10^3\text{ J}\cdot\text{mol}^{-1}}{8.314\text{ J}\cdot\text{mol}^{-1}\cdot\text{K}^{-1} \times 298\text{ K}} = -0.686$$

$K_p = 0.504$

15.47 $F^-(aq) + PbCl_2(s) \longrightarrow Cl^-(aq) + PbClF(s)$

$$\text{molarity of } F^- \text{ ions} = \left(\frac{0.765\text{ g PbClF}}{0.0250\text{ L}}\right)\left(\frac{1\text{ mol PbClF}}{261.64\text{ g PbClF}}\right)\left(\frac{1\text{ mol }F^-}{1\text{ mol PbClF}}\right)$$
$$= 0.117\text{ mol}\cdot\text{L}^{-1}$$

15.49 $3\,NH_4ClO_4(s) + 3\,Al(s) \longrightarrow Al_2O_3(s) + AlCl_3(s) + 6\,H_2O(g) + 3\,NO(g)$

The number of moles of NH_4ClO_4 is

$1.00\text{ kg} \times 1000\text{ g}\cdot\text{kg}^{-1} \div 117.49\text{ g}\cdot\text{mol}^{-1} = 8.51\text{ mol}$

The number of moles of Al is

$1.00\ \text{kg} \times 1000\ \text{g}\cdot\text{kg}^{-1} \div 26.98\ \text{g}\cdot\text{mol}^{-1} = 37.06\ \text{mol}$

The limiting reagent is the NH_4ClO_4.

The standard enthalpy for the reaction is given by

$$\begin{aligned}\Delta H^\circ_r &= (-1675.7\ \text{kJ}\cdot\text{mol}^{-1}) + (-704.2\ \text{kJ}\cdot\text{mol}^{-1}) + 6(-241.82\ \text{kJ}\cdot\text{mol}^{-1}) \\ &\quad + 3(90.25\ \text{kJ}\cdot\text{mol}^{-1}) - [3(-295.31\ \text{kJ}\cdot\text{mol}^{-1})] \\ &= -2674.1\ \text{kJ}\cdot\text{mol}^{-1}\end{aligned}$$

This value is the amount of heat released for 3 mol NH_4ClO_4. The amount released for 8.51 mol will be

$$8.51\ \text{mol}\ NH_4ClO_4 \times \frac{-2674.1\ \text{kJ}\cdot\text{mol}^{-1}}{3\ \text{mol}\ NH_4ClO_4} = -7.59 \times 10^3\ \text{kJ}$$

There will be 7.59×10^3 kJ of heat released.

15.51 Helium occurs as a component of natural gases found under rock formations in certain locations, especially some in Texas. Argon is obtained by distillation of liquid air.

15.53 (a) KrF_2: F = -1; therefore, Kr = $+2$

(b) XeF_6: F = -1; therefore, Xe = $+6$

(c) KrF_4: F = -1; therefore, Kr = $+4$

(d) XeO_4^{2-}: O = -2, $N_{ox}(\text{Xe}) - 8 = -2$; therefore, $N_{ox}(\text{Xe}) = +6$

15.55 $XeF_4 + 4\ H^+ + 4\ e^- \longrightarrow Xe + 4\ HF$

15.57 Because H_4XeO_6 has more highly electronegative O atoms bonded to Xe, we predict that H_4XeO_6 is more acidic than H_2XeO_4.

15.59 A sol is colloid comprised of solid particles suspended in a liquid. Muddy water is a type of sol. A foam is a suspension of a gas in a solid or liquid. Styrofoam, foam rubber, soapsuds, and aerogels are all types of foams.

15.61 (a) both a sol and an emulsion (b) a foam (c) a sol

15.63 In fluorescence, the light that is emitted is of lower energy than the light that is absorbed, and the fluorescence stops as soon as the exciting radiation is stopped. In phosphorescence, the phosphorescing molecules remain excited for a period of time after the stimulus has stopped.

15.65 The mercury atoms are an energy transfer agent. They absorb the energy from a high-voltage discharge and emit ultraviolet light in the region of 254 and 185 nm. This emitted light then excites a fluorescent material that emits radiation in the visible region of the spectrum, which is the light observed when the lamp is turned on.

15.67 The larger the value of $E°$ for the reduction $X_2 + 2\,e^- \longrightarrow 2\,X^-$, the greater the oxidizing strength of the halogen X_2. From Appendix 2B,

$F_2 + 2\,e^- \longrightarrow 2\,F^- \quad E° = +2.87\ V$

$Cl_2 + 2\,e^- \longrightarrow 2\,Cl^- \quad E° = +1.36\ V$

$Br_2 + 2\,e^- \longrightarrow 2\,Br^- \quad E° = +1.09\ V$

$I_2 + 2\,e^- \longrightarrow 2\,I^- \quad E° = +0.54\ V$

Thus, $I_2 < Br_2 < Cl_2 < F_2$.

15.69 Ionization energies increase; electron affinities also increase (in magnitude). Large ionization energies and electron affinities are characteristic of nonmetals. Electronegativities and standard reduction potentials increase as well; large values of these properties are also characteristic of nonmetals.

15.71 (a)

NO_2^- 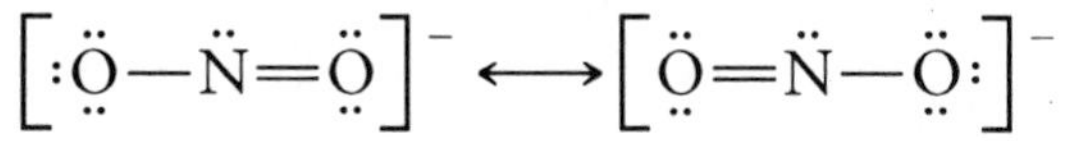

NO_3^- 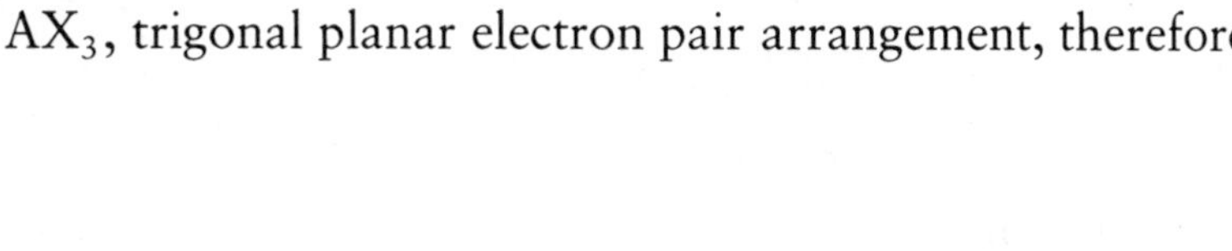

(b) NO_2^-: AX_2E, trigonal planar electron pair arrangement, therefore sp^2 hybridization

NO_3^-: AX_3, trigonal planar electron pair arrangement, therefore sp^2 hybridization

15.73 oxidation:

$As_2S_3(s) + 8\,H_2O(l) \longrightarrow 2\,AsO_4^{3-}(aq) + 3\,S^{2-}(aq) + 16\,H^+(aq) + 4\,e^-$

reduction: $H_2O_2(aq) + 2\,H^+(aq) + 2\,e^- \longrightarrow 2\,H_2O(l)$

Multiply the reduction reaction by 2, cancel electrons, and add.

overall:

$As_2S_3(s) + 2\,H_2O_2(aq) + 4\,H_2O(l) \longrightarrow 2\,AsO_4^{3-}(aq) + 3\,S^{2-}(aq) + 12\,H^+(aq)$

or $As_2S_3(s) + 2\,H_2O_2(aq) + 4\,H_2O(l) \longrightarrow 2\,H_3AsO_4(aq) + 3\,H_2S(aq)$

15.75 This ratio, $\Delta H_{vap}/T_b$, is the entropy of vaporization. Hydrogen bonding is much stronger in $H_2O(l)$ than in $H_2S(l)$. Thus $H_2O(l)$ has a more ordered arrangement than $H_2S(l)$. Consequently, the change in entropy upon transformation to the gaseous state is greater for H_2O than for H_2S.

15.77 $Ca_3(PO_4)_2(s) + 3\ H_2SO_4(l) \longrightarrow 2\ H_3PO_4(l) + 3\ CaSO_4(s)$

$$V = (1000\text{ kg } H_3PO_4)\left(\frac{10^3\text{ g}}{1\text{ kg}}\right)\left(\frac{1\text{ mol } H_3PO_4}{97.99\text{ g } H_3PO_4}\right)$$
$$\left(\frac{3\text{ mol } H_2SO_4}{2\text{ mol } H_3PO_4}\right)\left(\frac{98.08\text{ g } H_2SO_4}{1\text{ mol } H_2SO_4}\right)\left(\frac{1\text{ mL}}{1.84\text{ g}}\right)\left(\frac{10^{-3}\text{ L}}{1\text{ mL}}\right)$$
$$= 8.16 \times 10^2\text{ L conc. } H_2SO_4$$

15.79 (a) $SO_2(g) + H_2O(l) \longrightarrow H_2SO_3(l)$ This is a Lewis acid-base reaction. SO_2 is the acid and H_2O is the base.
(b) $2\ F_2(g) + 2\ NaOH(aq) \longrightarrow OF_2(g) + 2\ NaF(aq) + H_2O(l)$ This is a redox reaction illustrating the oxidizing ability of F_2 in basic solution and is used for the preparation of $OF_2(g)$. O is oxidized and F is reduced.
(c) $S_2O_3^{2-}(aq) + 4\ Cl_2(g) + 13\ H_2O(l) \longrightarrow 2\ HSO_4^-(aq) + 8\ H_3O^+(aq) + 8\ Cl^-(aq)$ This is a redox reaction illustrating the oxidizing power of $Cl_2(g)$ in acidic solution. S is oxidized and Cl is reduced.
(d) $2\ XeF_6(s) + 16\ OH^-(aq) \longrightarrow XeO_6^{4-}(aq) + Xe(g) + 12\ F^-(aq) + 8\ H_2O(l) + O_2(g)$ This is a redox reaction that is also a disproportionation reaction in that Xe goes from oxidation number +6 to +8 and to 0. Xe is both oxidized and reduced.

15.81 (a) $I_2(s) + 3\ F_2(g) \longrightarrow 2\ IF_3(s)$; $I_2(s) + 5\ F_2(g) \longrightarrow 2\ IF_5(s)$, etc.
(b) $I_2(aq) + I^-(aq) \longrightarrow I_3^-(aq)$
(c) $Cl_2(g) + H_2O(l) \longrightarrow HCl(aq) + HOCl(aq)$
But there are competing reactions, such as $Cl_2(g) + H_2O(l) \longrightarrow 2\ HCl(aq) + \frac{1}{2}\ O_2(g)$. The predominant reaction is determined by the temperature and pH.
(d) $2\ F_2(g) + 2\ H_2O(l) \longrightarrow 4\ HF(aq) + O_2(g)$

15.83 (a) $CaCl_2(s) + 2\ H_2SO_4(aq, conc) \longrightarrow Ca(HSO_4)_2(aq) + 2\ HCl(g)$
(b) $KBr(s) + H_3PO_4(aq) \xrightarrow{\Delta} KH_2PO_4(aq) + HBr(g)$
(c) $KI(s) + H_3PO_4(aq) \xrightarrow{\Delta} KH_2PO_4(aq) + HI(g)$

15.85 (a) +3: N_2O_3, dinitrogen trioxide
HNO_2, nitrous acid
KNO_2, potassium nitrite
+5: N_2O_5, dinitrogen pentoxide
HNO_3, nitric acid
KNO_3, potassium nitrate

(b) +3: P_4O_6, phosphorus(III) oxide
H_3PO_3, phosphorous acid
K_2HPO_3, potassium hydrogen phosphite
+5: P_4O_{10}, phosphorous(V) oxide
H_3PO_4, phosphoric acid
$Na_4P_2O_7$, sodium pyrophosphate

15.87 The heads of matches consist of a paste of potassium chlorate ($KClO_3$), antimony sulfide, (Sb_2S_3), sulfur, and powdered glass. The striking strip contains red phosphorus. When the match is struck against the red phosphorus surface, a reaction of the red phosphorus and potassium chlorate causes the match to ignite. The Sb_2S_3 and sulfur are the fuels that are consumed by combustion after the ignition. The powdered glass helps to produce the friction required for ignition.

15.89 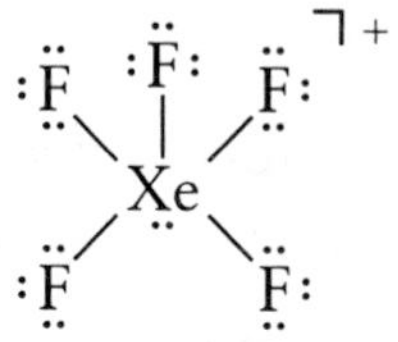

AX_5E, square pyramidal

15.91 Orpiment is As_2S_3 and realgar is As_4S_4. Orpiment is yellow and realgar is orange-red. They are both used as pigments.

15.93 (a) $[\ddot{N}=N=\ddot{N}]^-$
AX_2, linear, 180°

(b) F^-, 133 pm; N_3^-, 148 pm; Cl^-, 181 pm; therefore, between fluorine and chlorine

(c) HCl, HBr, and HI are all strong acids. For HF, $K_a = 3.5 \times 10^{-4}$, so HF is slightly more acidic than HN_3. The small size of the azide ion suggests that the H—N bond in HN_3 is similar in strength to that of the H—F bond, so it is expected to be a weak acid.

(d) ionic: NaN_3, $Pb(N_3)_2$, AgN_3, etc.
covalent: HN_3, $B(N_3)_3$, FN_3, etc.

15.95 $$[ClO^-] = (0.028\ 34\ L)\left(\frac{0.110\ mol\ S_2O_3^{\ 2-}}{1\ L\ Na_2S_2O_3}\right)\left(\frac{1\ mol\ I_2}{2\ mol\ S_2O_3^{\ 2-}}\right)\left(\frac{1\ mol\ ClO^-}{1\ mol\ I_2}\right)\left(\frac{1}{0.010\ 00\ L\ ClO^-}\right)$$
$$= 0.156\ mol\cdot L^{-1}$$

15.97 The solubility of the ionic halides is determined by a variety of factors, especially the lattice enthalpy and enthalpy of hydration. There is a delicate balance between the two factors, with the lattice enthalpy usually being the determining one. Lattice enthalpies decrease from chloride to iodide, so water molecules can more readily separate the ions in the latter. Less ionic halides, such as the silver halides, generally have a much lower solubility, and the trend in solubility is the reverse of the more ionic halides. For the less ionic halides, the covalent character of the bond allows the ion pairs to persist in water. The ions are not easily hydrated, making them less soluble. The polarizability of the halide ions, and thus, the covalency of their bonding, increases down the group.

15.99 To answer this question, we can compare equilibrium "vapor pressures" of water over each of these reagents. The one with the lowest equilibrium vapor pressure will be the better drying agent. The two reactions of interest are:
$$CaO(s) + H_2O(g) \longrightarrow Ca(OH)_2(s)$$
$$P_4O_{10}(s) + 6\ H_2O(g) \longrightarrow 4\ H_3PO_4(s)$$
First, the free energies of the reactions are calculated and from these the equilibrium pressure of water can be obtained.
For CaO:
$$\Delta G°_r = -898.49\ kJ\cdot mol^{-1} - [(-604.03\ kJ\cdot mol^{-1}) + (-228.57\ kJ\cdot mol^{-1})]$$
$$= -65.89\ kJ\cdot mol^{-1}$$
$$\Delta G°_r = -RT \ln K$$
$$K = e^{-\frac{\Delta G°}{RT}} = e^{-\frac{-65\ 890\ J\cdot mol^{-1}}{(8.314\ J\cdot K^{-1}\cdot mol^{-1})(298\ K)}} = 3.5 \times 10^{11}$$
$$K = \frac{1}{P_{H_2O}} = 3.5 \times 10^{11}$$
$$P_{H_2O} = 2.8 \times 10^{-12}\ bar$$

For P_4O_{10}:

$$\Delta G^\circ_r = 4(-1119.2\ \text{kJ}\cdot\text{mol}^{-1}) - [(-2697.0\ \text{kJ}\cdot\text{mol}^{-1}) + 6(-228.57\ \text{kJ}\cdot\text{mol}^{-1})]$$
$$= -408.4\ \text{kJ}\cdot\text{mol}^{-1}$$

$$K = e^{-\frac{\Delta G^\circ}{RT}} = e^{-\frac{-408\,400\ \text{J}\cdot\text{mol}^{-1}}{(8.314\ \text{J}\cdot\text{K}^{-1}\cdot\text{mol}^{-1})(298\ \text{K})}} = 3.9 \times 10^{71}$$

$$K = \frac{1}{(P_{H_2O})^6} = 3.9 \times 10^{71}$$

$$P_{H_2O} = 1.2 \times 10^{-12}\ \text{bar}$$

Because the pressure of water possible above CaO is greater than that above P_4O_{10}, CaO will be a poorer drying agent.

15.101 (a) $Ag^+ + e^- \longrightarrow Ag \qquad E^\circ = +0.80\ \text{V}$

$I_2 + 2\,e^- \longrightarrow 2\,I^- \qquad E^\circ = +0.54\ \text{V}$

For $2\,Ag^+ + 2\,I^- \longrightarrow Ag + I_2$, $E^\circ = +0.80\ \text{V} - 0.54\ \text{V} = +0.26\ \text{V}$.

The process should be spontaneous.

(b) The formation of AgI precipitate means that the concentration of Ag^+ ions is never high enough to achieve the conditions necessary for the redox reaction to take place. The solubility product K_{sp} limits the concentrations in solution, so that the actual redox potential is not the value calculated, which represents the values when $[Ag^+] = 1\ \text{M}$ and $[I^-] = 1\ \text{M}$. If we use the concentrations established by the solubility equilibrium and the Nernst equation, we can calculate the actual redox potential:

$$E = E^\circ - \frac{RT}{nF} \ln Q$$

where, in this case, $Q = \dfrac{1}{K_{sp}^{\ 2}}$ for the reaction as written.

$K_{sp} = 1.5 \times 10^{-16}$ for AgI.

$$E = +0.26\ \text{V} - \frac{(8.314\ \text{J}\cdot\text{K}^{-1}\cdot\text{mol}^{-1})(298\ \text{K})}{(2)(96\,485\ \text{J}\cdot\text{V}^{-1}\cdot\text{mol}^{-1})} \ln \frac{1}{(1.5 \times 10^{-16})^2}$$
$$= +0.26\ \text{V} - \frac{(8.314\ \text{J}\cdot\text{K}^{-1}\cdot\text{mol}^{-1})(298\ \text{K})}{(96\,485\ \text{J}\cdot\text{V}^{-1}\cdot\text{mol}^{-1})} \ln \frac{1}{(1.5 \times 10^{-16})}$$
$$= +0.26\ \text{V} - 0.94\ \text{V}$$
$$= -0.68\ \text{V}$$

The fact that the concentrations of Ag^+ and I^- are limited in solution means that the redox potential for a spontaneous reaction is never achieved.

CHAPTER 16
THE *d*-BLOCK: METALS IN TRANSITION

16.1 (a) Sc, $[Ar]3d^14s^2$: one unpaired 3*d*-electron
(b) V, $[Ar]3d^34s^2$: three unpaired 3*d*-electrons
(c) Cu, $[Ar]3d^{10}4s^1$: one unpaired 4*s*-electron
(d) Au, $[Xe]4f^{14}5d^{10}6s^1$: one unpaired 6*s*-electron

16.3 Elements at the left of the *d*-block tend to have strongly negative standard potentials.

16.5 See Figs. 16.2 and 16.4.
(a) Sc (b) Au (c) Nb

16.7 See Fig. 1.44 and Appendix 2D.
(a) Ti (b) Cu (c) Zn

16.9 The lanthanide contraction accounts for the failure of the third-row (Period 6) metallic radii to increase as expected, relative to the radii of the second row (Period 5). This results from the presence of the *f*-block orbitals. The *f*-electrons present in the lanthanides are even poorer as nuclear shields than *d*-electrons, and a marked decrease in metallic radius occurs along the *f*-block elements as a result of the increased effective nuclear charge, which pulls the electrons inward. When the *d*-block resumes (at lutetium), the metallic radius has "contracted" from 188 to 157 pm. All the elements following lutetium have smaller-than-expected radii. Examples of this effect include the high density of the Period 6 elements and lack of reactivity of gold and platinum.

16.11 Hg is much more dense than Cd, because the shrinkage in atomic radius that occurs between $Z = 58$ and $Z = 71$ (the lanthanide contraction) causes the atoms following the rare earths to be smaller than might have been expected for their atomic masses and atomic numbers. Zn and Cd have densities that are not too dissimilar, because the radius of Cd is subject to only a smaller *d*-block contraction.

16.13 Proceeding down a group in the *d*-block (for example, from Cr to Mo to W), there is an increasing probability of finding the elements in higher oxidation states. That is, higher oxidation states become more stable on going down a group.

16.15 In MO_3, M has an oxidation number of +6. Of these three elements, the +6 oxidation state is most stable for Cr. See Fig. 16.7.

16.17 (a) $TiO_2(s) + 2\ C(s) + Cl_2(g) \xrightarrow{1000°C} TiCl_4(g) + 2\ CO(g)$

followed by $TiCl_4(g) + 2\ Mg(l) \xrightarrow{700°C} Ti(s) + 2\ MgCl_2(s)$

(b) $V_2O_5(g) + 5\ Ca(l) \longrightarrow 5\ CaO(s) + 2\ V(s)$

16.19 (a) Ti(s), $MgCl_2(s)$

$TiCl_4(g) + 2\ Mg(l) \longrightarrow Ti(s) + 2\ MgCl_2(s)$

(b) $Co^{2+}(aq)$, $HCO_3^-(aq)$, $NO_3^-(aq)$

$CoCO_3(s) + HNO_3\ (aq) \longrightarrow Co^{2+}(aq) + HCO_3^-(aq) + NO_3^-(aq)$

(c) V(s), CaO(s)

$V_2O_5(s) + 5\ Ca(l) \xrightarrow{\Delta} 2\ V(s) + 5\ CaO(s)$

16.21 (a) titanium(IV) oxide, TiO_2

(b) iron(III) oxide, Fe_2O_3

(c) manganese(IV) oxide, MnO_2

16.23 (a) $V^{2+} + 2\ e^- \longrightarrow V(s) \quad E° = -1.19\ V$

$V^{3+} + e^- \longrightarrow V^{2+} \quad E° = -0.26\ V$

$2\ H^+ + 2\ e^- \longrightarrow H_2(g) \quad E° = 0.00\ V$

Therefore, V(s) will be oxidized to V^{3+}. The products are V^{3+}, H_2, and Cl^-.

(b) $Hg_2^{2+} + 2\ e^- \longrightarrow 2\ Hg \quad E° = +0.79\ V$

$Hg^{2+} + 2\ e^- \longrightarrow Hg \quad E° = +1.62\ V$

$2\ H^+ + 2\ e^- \longrightarrow H_2(g) \quad E° = 0.00\ V$

Therefore, no reaction.

(c) $Co^{2+} + 2\ e^- \longrightarrow Co(s) \quad E° = -0.28\ V$

$Co^{3+} + e^- \longrightarrow Co^{2+} \quad E° = +1.81\ V$

$2\ H^+ + 2\ e^- \longrightarrow H_2(g) \quad E° = 0.00\ V$

Therefore, Co(s) will be oxidized to Co^{2+}. The products are Co^{2+}, H_2, and Cl^-. The further oxidation to Co^{3+} is not favorable by reaction of H^+ with Co^{2+}. However, Co^{2+} is oxidized in air to Co^{3+}.

16.25 All three elements in this group—Cu, Ag, and Au—are chemically rather inert, Ag more so than Cu, and Au more so than Ag. The reduction potentials of their ions are all positive, in the order Au > Ag > Cu, so they are not readily oxidized. They have a common electron configuration, $(n-1)d^{10}ns^1$.

16.27 (a) $2\,ZnS(s) + 3\,O_2(g) \xrightarrow{\Delta} 2\,ZnO(s) + 2\,SO_2(g)$

followed by $ZnO(s) + C(s) \xrightarrow{\Delta} Zn(l) + CO(g)$

(b) $HgS(s) + O_2(g) \xrightarrow{\Delta} Hg(g) + SO_2(g)$

16.29

$$Cu^+ + e^- \longrightarrow Cu \qquad E° = +0.52\ V$$

$$Cu^+ \longrightarrow Cu^{2+} + e^- \qquad E° = +0.15\ V$$

$$2\,Cu^+ \longrightarrow Cu^{2+} + Cu \qquad E° = +0.37\ V$$

$$\Delta G°_r = -nFE° = -RT \ln K \quad (n = 1)$$

$$nFE° = RT \ln K$$

$$\ln K = \frac{(1)(9.65 \times 10^4\ C\cdot mol^{-1})(0.37\ V)}{(8.314\ J\cdot mol^{-1}\cdot K^{-1})(298\ K)} = 14$$

$$K = 10^6$$

16.31 Let x = the oxidation number to be determined.

(a) $x(Fe) + 6 \times (-1) = -4$

$x(Fe) = -4 - (-6) = +2$

(b) $x(Co) + 6 \times (0) = +3$

$x(Co) = +3$

(c) $x(Co) + 5 \times (-1) + 1 \times (0) = -2$

$x(Co) = -2 - (-5) = +3$

(d) $x(Co) + 1 \times (-2) + 5 \times (0) = +1$

$x(Co) = +1 - (-2) = +3$

16.33 (a) hexacyanoferrate(II) ion

(b) hexaamminecobalt(III) ion

(c) aquapentacyanocobaltate(III) ion

(d) pentaamminesulfatocobalt(III) ion

16.35 (a) $K_3[Cr(CN)_6]$

(b) $[Co(NH_3)_5(SO_4)]Cl$

(c) $[Co(NH_3)_4(H_2O)_2]Br_3$

(d) $Na[Fe(H_2O)_2(C_2O_4)_2]$

16.37 The molecule $HN(CH_2CH_2NH_2)_2$ has three nitrogen atoms, each with a lone pair of electrons that may be used for bonding to a metal center. The molecule can thus function as a tridentate ligand. The CO_3^{2-} ion can bind to a metal ion through either one or two oxygen atoms. It may, therefore, serve as a mono- or bidentate ligand. H_2O is always a monodenate ligand. The oxalate ion can bind through two oxygen atoms and is usually a bidentate ligand.

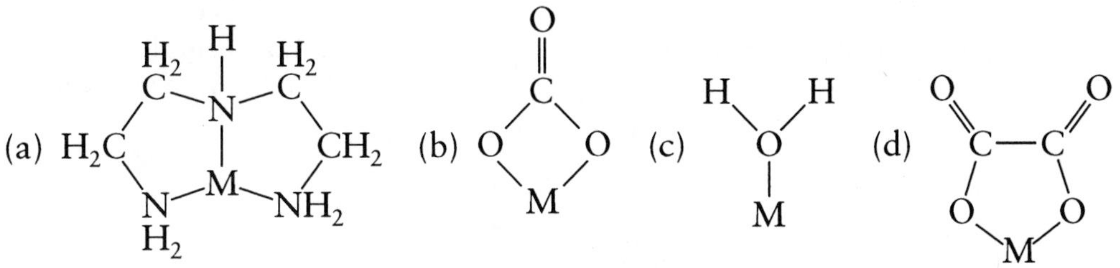

16.39 As shown below, only the molecule (b) can function as a chelating ligand. The two amine groups in (a) and (c) are arranged so that they would not be able to coordinate simultaneously to the same metal center. It is possible for each of the amine groups in (a) and (c) to coordinate to two different metal centers, however. This is not classified as chelating. When a single ligand binds to two different metal centers, it is known as a *bridging* ligand.

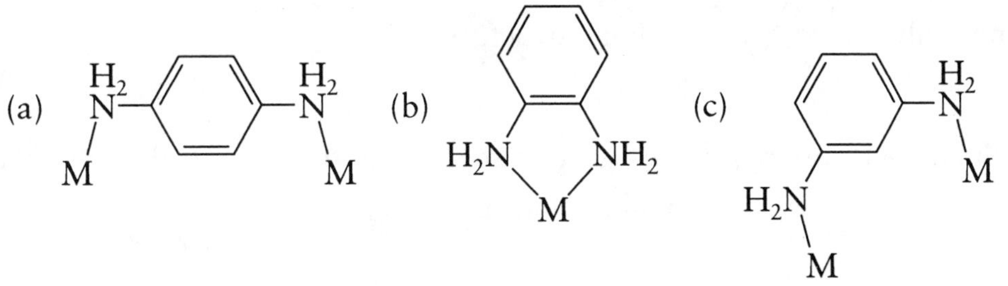

16.41 (a) 4 (b) 2 (c) 6 (en is bidentate) (d) 6 (EDTA is hexadentate)

16.43 (a) structural isomers, linkage isomers

(b) structural isomers, ionization isomers

(c) structural isomers, linkage isomers

(d) structural isomers, ionization isomers

16.45 $[Co(H_2O)_6]Cl_3$, $[CoCl(H_2O)_5]Cl_2 \cdot H_2O$, $[CoCl_2(H_2O)_4]Cl \cdot 2\ H_2O$, and $[CoCl_3(H_2O)_3] \cdot 3\ H_2O$

16.47 $[CoCl(NO_2)(en)_2]Cl$ and $[CoCl(ONO)(en)_2]Cl$

16.49 (a) yes

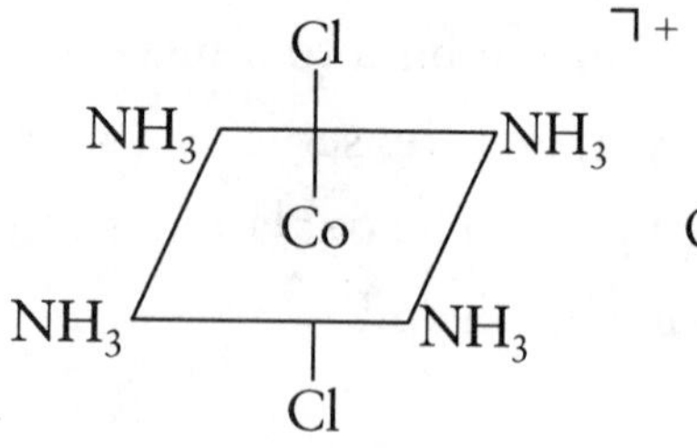

$Cl^- \cdot H_2O$ and

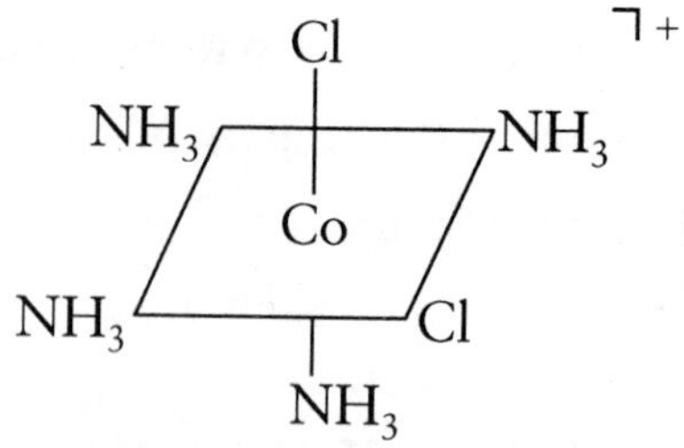

$Cl^- \cdot H_2O$

trans-Tetraamminedichlorocobalt(III) chloride monohydrate

cis-Tetraamminedichlorocobalt(III) chloride monohydrate

(b) no

(c) yes

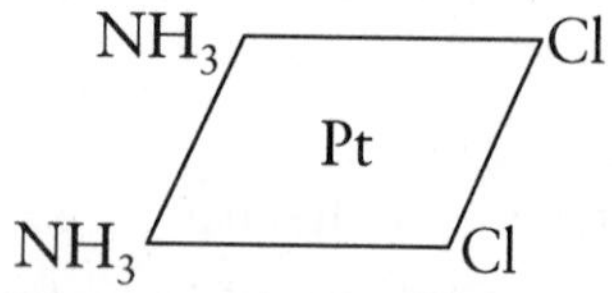

and

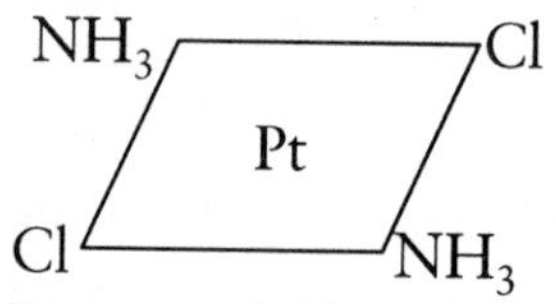

cis-Diamminedichloroplatinum(II)

trans-Diamminedichloroplatinum(II)

16.51 (a) yes, in the form of optical isomerism; see part (c)

(b) no, only

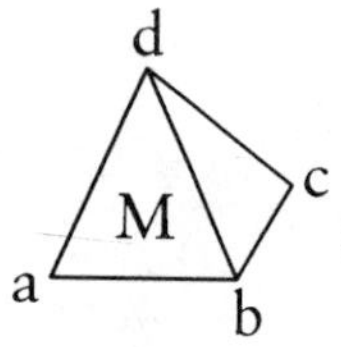

; a single structure

(c) Yes; if four different ligand groups are bonded to the central atom, then the central atom is chiral and exhibits optical activity.

16.53

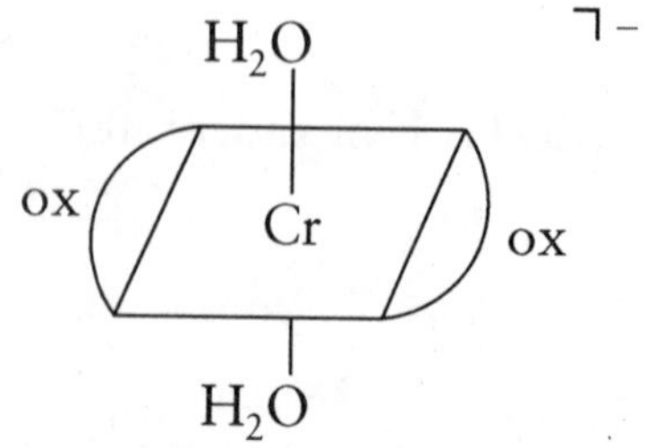

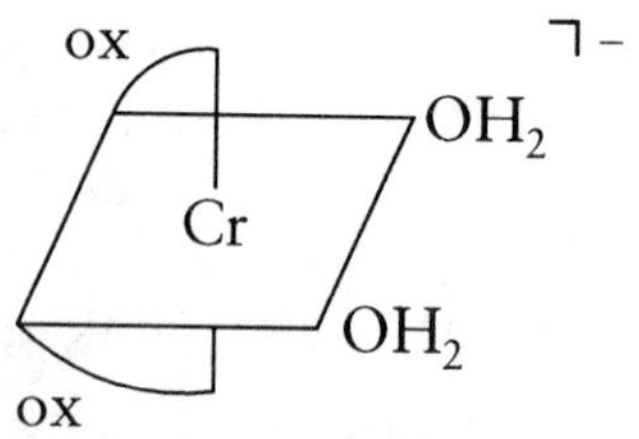

trans-Diaquabis(oxalato)chromate(III) ion

cis-Diaquabis(oxalato)chromate(III) ion

A second *cis* isomer with the same name exists.

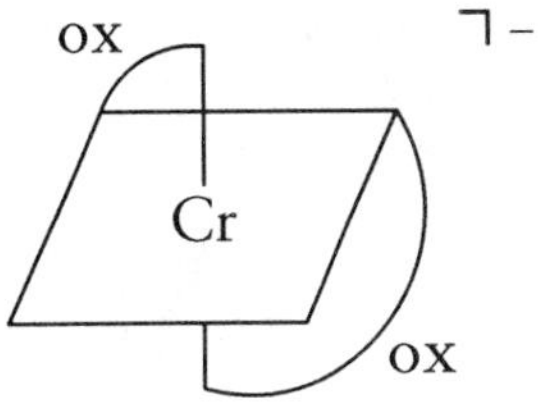

It is the mirror image of the first *cis* complex.

16.55 First complex:

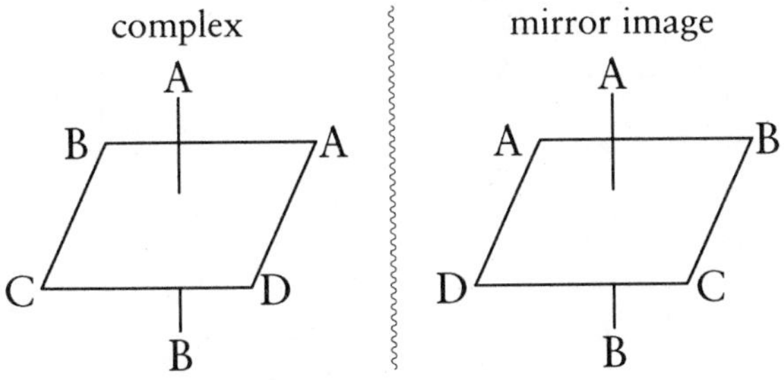

No rotation will make the complex and its mirror image match; therefore, it is chiral.

Second complex:

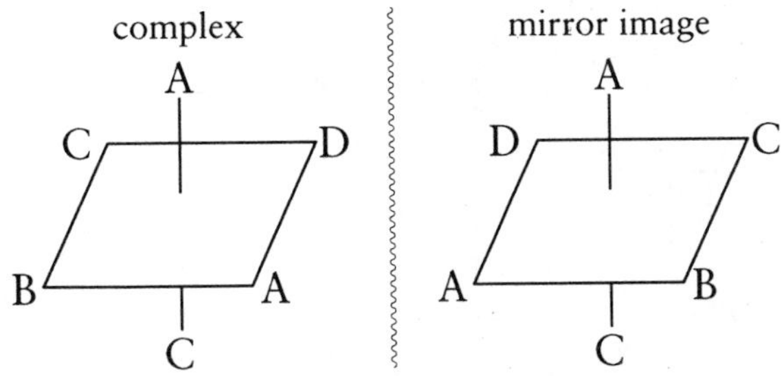

A double rotation shows that the complex and its mirror image are superimposable; therefore, it is not chiral.

The two complexes are not enantiomers; they are not even isomers.

16.57

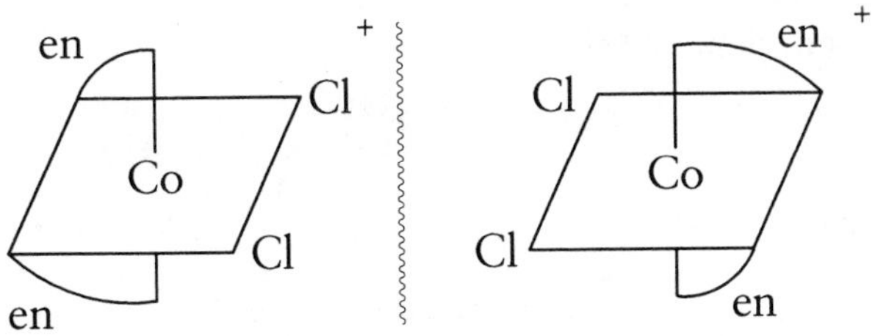

16.59 (a) octahedral: strong-field ligand, 6 e^-

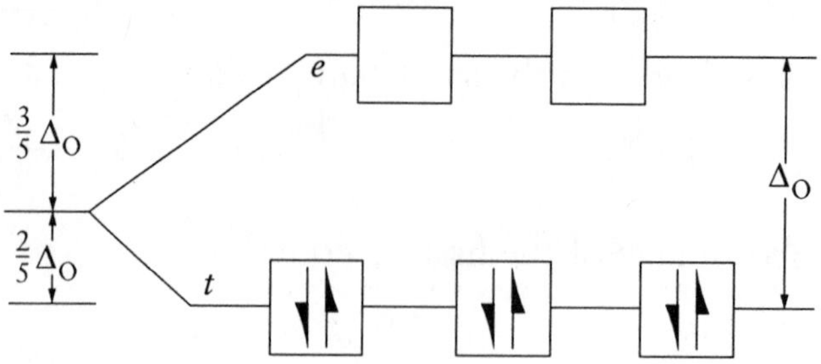

diamagnetic
no unpaired electrons

(b) tetrahedral: weak-field ligand, 8 e^-

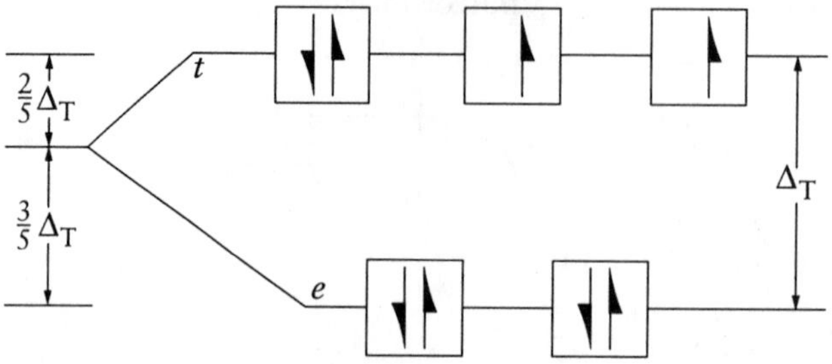

2 unpaired electrons

(c) octahedral: weak-field ligand, 5 e^-

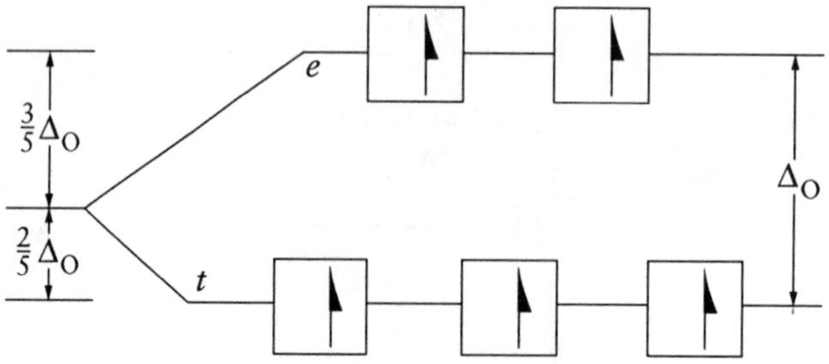

5 unpaired electrons

(d) octahedral: strong-field ligand, 5 e^-

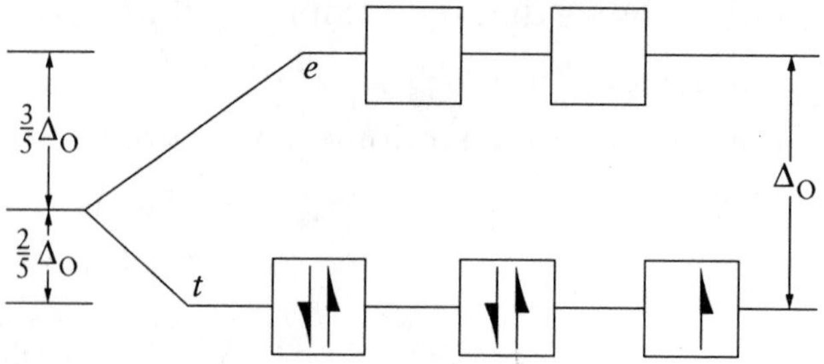

one unpaired electron

16.61 (a) $[Co(en)_3]^{3+}$ 6 e^-  0 unpaired electrons

(b) $[Mn(CN)_6]^{3-}$ 4 e^- 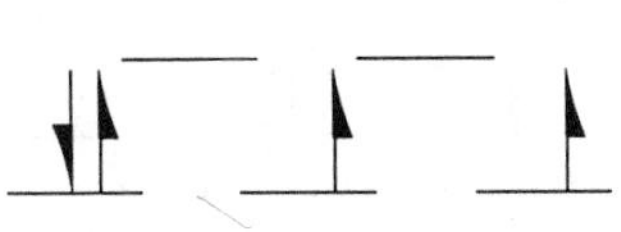 2 unpaired electrons

16.63 Weak-field ligands do not interact strongly with the *d*-electrons in the metal ion, so they produce only a small crystal field splitting of the *d*-electron energy states. The opposite is true of strong-field ligands. With weak-field ligands, unpaired electrons remain unpaired if there are unfilled orbitals; hence, a weak-field ligand is likely to lead to a high-spin complex. Strong-field ligands cause electrons to pair up with electrons in lower energy orbitals. A strong-field ligand is likely to lead to a low-spin complex. Ligands arranged in the spectrochemical series help to distinguish strong-field and weak-field ligands. Measurement of magnetic susceptibility (paramagnetism) can be used to determine the number of unpaired electrons, which, in turn, establishes whether the associated ligand is weak-field or strong-field in nature.

16.65 (a) $[CoF_6]^{3-}$ 6 e^- F^- is a weak-field ligand; therefore,

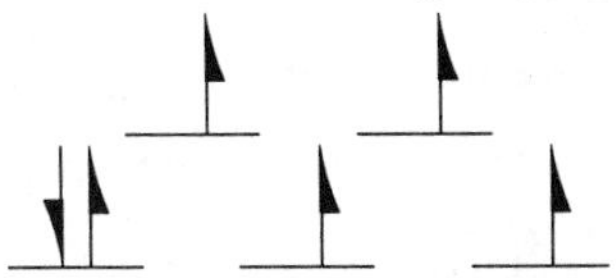

4 unpaired electrons

(b) $[Co(en)_3]^{3+}$ 6 e^- en is a strong-field ligand; therefore,

0 unpaired electrons

Because F^- is a weak-field ligand and en a strong-field ligand, the splitting between levels is less in (a) than in (b). Therefore, (a) will absorb light of longer wavelength than will (b) and consequently will display a shorter wavelength color. Blue light is shorter in wavelength than yellow light, so (a) $[CoF_6]^{3-}$ is blue and (b) $[Co(en)_3]^{3+}$ is yellow.

16.67 In Zn^{2+}, the 3*d*-orbitals are filled (d^{10}). Therefore, there can be no electronic transitions between the *t* and *e* levels; hence, no visible light is absorbed and the aqueous ion is colorless. The d^{10} configuration has no unpaired electrons, so Zn compounds would not be paramagnetic.

16.69 (a) $\Delta_O = \dfrac{hc}{\lambda} = \dfrac{(6.63 \times 10^{-34}\ \text{J·s}^{-1})(3.00 \times 10^{8}\ \text{m·s}^{-1})}{740 \times 10^{-9}\ \text{m}} = 2.69 \times 10^{-19}\ \text{J}$

(b) $\Delta_O = \frac{hc}{\lambda} = \frac{(6.63 \times 10^{-34}\ \text{J}\cdot\text{s}^{-1})(3.00 \times 10^{8}\ \text{m}\cdot\text{s}^{-1})}{460 \times 10^{-9}\ \text{m}} = 4.32 \times 10^{-19}\ \text{J}$

(c) $\Delta_O = \frac{hc}{\lambda} = \frac{(6.63 \times 10^{-34}\ \text{J}\cdot\text{s}^{-1})(3.00 \times 10^{8}\ \text{m}\cdot\text{s}^{-1})}{575 \times 10^{-9}\ \text{m}} = 3.46 \times 10^{-19}\ \text{J}$

These numbers can be multiplied by 6.02×10^{23} to obtain $\text{kJ}\cdot\text{mol}^{-1}$.

(a) $2.69 \times 10^{-19}\ \text{J} \times 6.02 \times 10^{23}\ \text{mol}^{-1} = 162\ \text{kJ}\cdot\text{mol}^{-1}$

(b) $4.32 \times 10^{-19}\ \text{J} \times 6.02 \times 10^{23}\ \text{mol}^{-1} = 260\ \text{kJ}\cdot\text{mol}^{-1}$

(c) $3.46 \times 10^{-19}\ \text{J} \times 6.02 \times 10^{23}\ \text{mol}^{-1} = 208\ \text{kJ}\cdot\text{mol}^{-1}$

$Cl < H_2O < NH_3$ (spectrochemical series)

16.71 The e_g set, which is comprised of the $d_{x^2-y^2}$ and d_{z^2} orbitals.

16.73 (a) The CN^- ion is a π-acid ligand accepting electrons into the empty π^* orbital created by the C—N multiple bond. (b) The Cl^- ion has extra lone pairs in addition to the one that is used to form the σ-bond to the metal, and so it can act as a π-base, donating electrons in a *p*-orbital to an empty *d*-orbital on the metal. (c) H_2O, like Cl^-, also has an "extra" lone pair of electrons that can be donated to a metal center, making it a weak π-base; (d) en is neither a π-acid nor a π-base, because it does not have any empty π-type antibonding orbitals nor does it have any extra lone pairs of electrons to donate. $Cl^- < H_2O <$ en $< CN^-$. Note that the spectrochemical series orders the ligands as π-bases $<$ σ-bond only ligands $<$ π-acceptors.

16.75 Nonbinding or slightly antibonding. In a complex that forms only σ-bonds, the t_{2g} set of orbitals is nonbonding. If the ligands can function as weak π-donors (those close to the middle of the spectrochemical series, such as H_2O), the t_{2g} set becomes slightly antibonding by interacting with the filled *p*-orbitals on the ligands.

16.77 Antibonding. The e_g set of orbitals on an octahedral metal ion are always antibonding because of interactions with ligand orbitals that form the σ-bonds. This is true regardless of whether the ligands are π-acceptors, π-donors, or neither.

16.79 Water has two lone pairs of electrons. Once one of these is used to form the σ-bond to the metal ion, the second may be used to form a π-bond. This causes the t_{2g} set of orbitals to move up in energy, making Δ smaller; therefore, water is a weak field ligand. Ammonia does not have this extra lone pair of electrons and consequently cannot function as a π-donor ligand.

16.81 (a) CO

(b) $Fe_3O_4(s) + CO(g) \longrightarrow 3\ FeO(s) + CO_2(g)$ (Zone D)

$Fe_2O_3(s) + CO(g) \longrightarrow 2\ FeO(s) + CO_2(g)$ (Zone C)

followed by $FeO(s) + CO(g) \longrightarrow Fe(s) + CO_2(g)$ (Zone C)

$Fe_2O_3(s) + 3\ CO(g) \longrightarrow 2\ Fe(s) + 3\ CO_2(g)$ (Zone D)

(c) carbon

16.83 The major impurity is carbon; it is removed by oxidation of the carbon to CO_2, followed by capture of the CO_2 by base to form a slag.

16.85 Copper and zinc

16.87 Alloys are usually (1) harder and more brittle, and (2) poorer conductors of electricity than the metals from which they are made.

16.89 (a) A ferrofluid is a mixture containing a viscous, nonpolar liquid (oil), a detergent that can form micelles, and a finely powdered magnetic material, such as Fe_3O_4. When placed in a magnetic field, the magnetic particles in the fluid tend to align with the field, which means that the position and motion of the liquid can be controlled by the applied magnetic field. (b) At first glance, the oxidation state in Fe_3O_4 appears to be the nonintegral value of $+8/3$. Another way to view this is that the compound is composed of both Fe^{2+} and Fe^{3+} in a 1:2 ratio. The compound could also be written as $FeO \cdot Fe_2O_3$.

16.91 $[Sc(H_2O)_6]^{3+}(aq) + H_2O(l) \longrightarrow [Sc(H_2O)_5OH]^{2+}(aq) + H_3O^+(aq)$. Here the hydrated ion acts as a proton donor, and therefore it is a Brønsted acid.

16.93 (a) More than one kind of reduction occurs.

In Zone C,

$Fe_2O_3(s) + 3\ CO(g) \longrightarrow 2\ Fe(s) + 3\ CO_2(g)$

In Zone D,

$3\ Fe_2O_3(s) + CO(g) \longrightarrow 2\ Fe_3O_4(s) + CO_2(g)$

$Fe_3O_4(s) + CO(g) \longrightarrow 3\ FeO(s) + CO_2(g)$

These reactions combine to give

$Fe_2O_3(s) + CO(g) \longrightarrow 2\ FeO(s) + CO_2(g)$

In Zone C,

$FeO(s) + CO(g) \longrightarrow Fe(s) + CO_2(g)$

(b) $TiCl_4(g) + 2\ Mg(l) \xrightarrow{\Delta} Ti(s) + 2\ MgCl_2(s)$

(c) $CaO(s) + SiO_2(s) \xrightarrow{\Delta} CaSiO_3(l)$

16.95 (a) Cr^{3+} ions in water form the complex $[Cr(H_2O)_6]^{3+}(aq)$, which behaves as a Brønsted acid:

$[Cr(H_2O)_6]^{3+}(aq) + H_2O(l) \rightleftharpoons [Cr(H_2O)_5OH]^{2+}(aq) + H_3O^+(aq)$

(b) The gelatinous precipitate is the hydroxide $Cr(OH)_3$. The precipitate dissolves as the $Cr(OH)_4^-$ complex ion is formed:

$Cr^{3+}(aq) + 3\ OH^-(aq) \longrightarrow Cr(OH)_3(s)$

$Cr(OH)_3(s) + OH^-(aq) \longrightarrow Cr(OH)_4^-(aq)$

16.97 (a) tetrakis(oxalato)zirconate(IV) ion; CN = 8, Zn = +4

(b) diaquatetrachlorocuprate(II) ion; CN = 6, Cu = +2

(c) amminetrichloroplatinate(II) ion; CN = 4, Pt = +2

(d) tetracyanotetrahydroxomolybdate(IV) ion; CN = 8, Mo = +4

16.99 $[PtBrCl(NH_3)_2]$

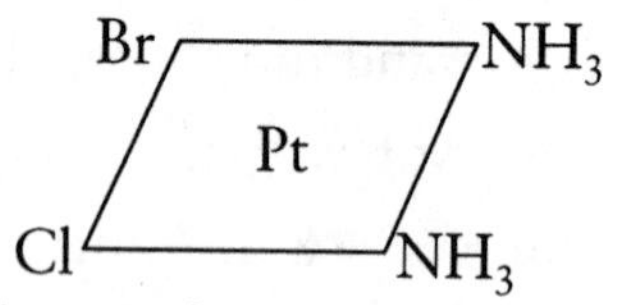

cis-Diamminebromochloroplatinum(II)

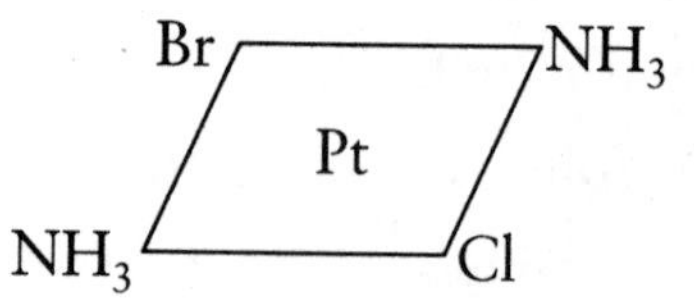

trans-Diamminebromochloroplatinum(II)

16.101 (a) The first, $[Ni(SO_4)(en)_2]Cl_2$, will give a precipitate of AgCl when $AgNO_3$ is added; the second will not.

(b) The second, $[NiCl_2(en)_2]I_2$, will show free I_2 when mildly oxidized with, for example, Br_2, but the first will not.

16.103 (a), (b)

$[MnCl_6]^{4-}$ 5 e^- Cl^- is a weak-field ligand

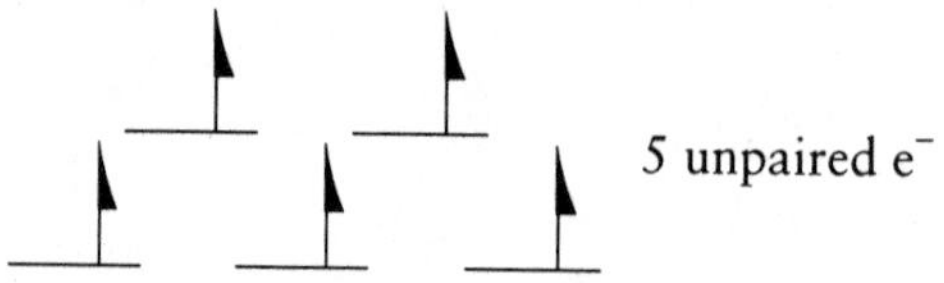

$[Mn(CN)_6]^{4-}$ 5 e^- CN^- is a strong-field ligand

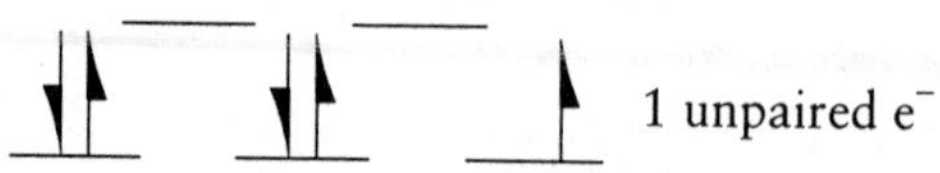

(c) A weak-field ligand absorbs long-wavelength light; therefore, $[MnCl_6]^{4-}$ transmits shorter wavelengths and $[Mn(CN)_6]^{4-}$ transmits longer wavelengths.

16.105 4, 5, 6, and 7

16.107 High spin Mn^{2+} ions have a d^5 configuration with 5 unpaired electrons, as shown below.

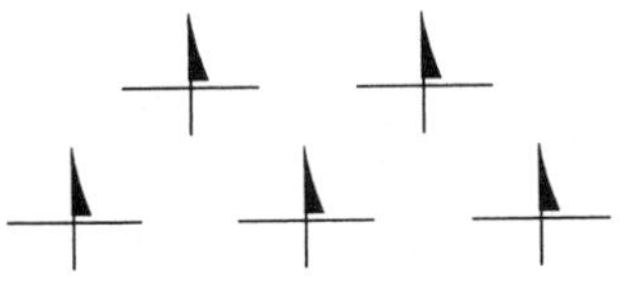

In order for light to be absorbed in the visible region of the spectrum, an electron from the t_{2g} set has to be moved into the e_g set of orbitals. Because these orbitals already each have an electron in them, it would be necessary to pair the electrons, which requires additional energy. This extra energy requirement makes the absorption of the visible light less probable, and so the complexes usually are only faintly colored.

16.109 Consider the formation reaction of a metal complex from the metal ion and its ligands:

$M^{x+} + nL \longrightarrow ML_n^{\;x+}$

It can be seen that there is a considerable loss in entropy upon creating the complex. Polydentate ligands that have two or more places on a ligand that can bind to a metal, do not have nearly so unfavorable an entropy change—the coordination of the second site on the ligand occurs intramolecularly, once the first part of the ligand is attached. Thus, in general, metal complexes with chelating ligands have much higher formation constants than complexes with all monodentate ligands. This known as the *chelation effect;* it is so strong that once one end of a polydentate ligand binds to a metal center, the other end usually binds extremely rapidly.

16.111 The correct structure for $[Co(NH_3)_6]Cl_3$ consists of four ions, $Co(NH_3)_6^{\;3+}$ and 3 Cl^- in aqueous solution. The chloride ions can be easily precipitated as AgCl. This would not be possible if they were bonded to the other (NH_3) ligands. If the structure were $Co(NH_3—NH_3—Cl)_3$, VSEPR theory would predict that the Co^{3+} ion would have a trigonal planar ligand arrangement. The splitting of the d-orbital energies would not be the same as the octahedral arrangement and would lead to different spectroscopic and magnetic properties inconsistent with the experimental evidence. In addition, neither optical nor geometrical isomers would be observed.

16.113 The ligand field strength of the halide ions shows them to lie in the order $I^- < Br^- < Cl^- < F^-$, which shows that the less electronegative the ligand is, the smaller the value of Δ for the complex. The value of Δ correlates with the ability of the ligand's extra lone pairs of electrons to interact with the t_{2g} set of the octahedral metal ion. This means that the less electronegative the ligand is, the easier it is for the ligand to donate electrons to the metal ion.

16.115 In order to determine these relationships, we need to consider the types of interactions that the ligands on the ends of the spectrochemical series will have with the metal ions. Those that are weak-field (form high spin complexes, π-bases) have extra lone pairs of electrons that can be donated to a metal ion in a π fashion. The strong-field ligands (form low spin complexes, π-acids) accept electrons from the metals. The complexes that will be more stable will be produced in general by the match between ligand and metal. Thus, the early transition metals in high oxidation states will have few or no electrons in the *d*-orbitals. These metal ions will become stabilized by ligands that can donate more electrons to the metal—the *d*-orbitals that are empty can readily accept electrons. The more stable complexes will be formed with the weak-field ligands. The opposite is true for metals with many electrons, which are the ones at the right side of the periodic table, in low oxidation states. These metals generally have most of the *d*-orbitals filled, so they would, in fact, be destabilized by π-donation. They form instead more stable complexes with the π-acceptor ligands that can remove some of the electron density from the metal ions.

CHAPTER 17
NUCLEAR CHEMISTRY

17.1 $\lambda = \dfrac{c}{\nu}$, $E = N_A h\nu$, $1\ \text{Hz} = 1\ \text{s}^{-1}$

(a) $\lambda = \dfrac{3.00 \times 10^{8}\ \text{m}\cdot\text{s}^{-1}}{9.4 \times 10^{19}\ \text{s}^{-1}} = 3.2 \times 10^{-12}\ \text{m}$

$E = 6.02 \times 10^{23}\ \text{mol}^{-1} \times 6.63 \times 10^{-34}\ \text{J}\cdot\text{s} \times 9.4 \times 10^{19}\ \text{s}^{-1}$
$= 3.8 \times 10^{10}\ \text{J}\cdot\text{mol}^{-1}$

(b) $\lambda = \dfrac{3.00 \times 10^{8}\ \text{m}\cdot\text{s}^{-1}}{5.7 \times 10^{21}\ \text{s}^{-1}} = 5.3 \times 10^{-14}\ \text{m}$

$E = 6.02 \times 10^{23}\ \text{mol}^{-1} \times 6.63 \times 10^{-34}\ \text{J}\cdot\text{s} \times 5.7 \times 10^{21}\ \text{s}^{-1}$
$= 2.3 \times 10^{12}\ \text{J}\cdot\text{mol}^{-1}$

(c) $\lambda = \dfrac{3.00 \times 10^{8}\ \text{m}\cdot\text{s}^{-1}}{3.7 \times 10^{20}\ \text{s}^{-1}} = 8.1 \times 10^{-13}\ \text{m}$

$E = 6.02 \times 10^{23}\ \text{mol}^{-1} \times 6.63 \times 10^{-34}\ \text{J}\cdot\text{s} \times 3.7 \times 10^{20}\ \text{s}^{-1}$
$= 1.5 \times 10^{11}\ \text{J}\cdot\text{mol}^{-1}$

(d) $\lambda = \dfrac{3.00 \times 10^{8}\ \text{m}\cdot\text{s}^{-1}}{7.3 \times 10^{22}\ \text{s}^{-1}} = 4.1 \times 10^{-15}\ \text{m}$

$E = 6.02 \times 10^{23}\ \text{mol}^{-1} \times 6.63 \times 10^{-34}\ \text{J}\cdot\text{s} \times 7.3 \times 10^{22}\ \text{s}^{-1}$
$= 2.9 \times 10^{13}\ \text{J}\cdot\text{mol}^{-1}$

17.3 We assume that all the change in energy goes into the energy of the γ ray emitted. Then, in each case,

$$\nu = \frac{\Delta E}{h}, \qquad \lambda = \frac{c}{\nu}$$

$$\text{energy of 1 MeV} = \left(\frac{10^{6}\ \text{eV}}{1\ \text{MeV}}\right)\left(\frac{1.602 \times 10^{-19}\ \text{J}}{1\ \text{eV}}\right) = 1.602 \times 10^{-13}\ \text{J}\cdot\text{MeV}^{-1}$$

(a) $\Delta E = (1.33\ \text{MeV})\left(\dfrac{1.602 \times 10^{-13}\ \text{J}}{1\ \text{MeV}}\right) = 2.13 \times 10^{-13}\ \text{J}$

$$\nu = \frac{\Delta E}{h} = \frac{2.13 \times 10^{-13}\ \text{J}}{6.63 \times 10^{-34}\ \text{J}\cdot\text{s}} = 3.21 \times 10^{20}\ \text{s}^{-1} = 3.21 \times 10^{20}\ \text{Hz}$$

$$\lambda = \frac{c}{\nu} = \frac{3.00 \times 10^{8}\ \text{m}\cdot\text{s}^{-1}}{3.21 \times 10^{20}\ \text{s}^{-1}} = 9.35 \times 10^{-13}\ \text{m}$$

(b) $\Delta E = (1.64 \text{ MeV})\left(\dfrac{1.602 \times 10^{-13} \text{ J}}{1 \text{ MeV}}\right) = 2.63 \times 10^{-13} \text{ J}$

$$\nu = \frac{\Delta E}{h} = \frac{2.63 \times 10^{-13} \text{ J}}{6.63 \times 10^{-34} \text{ J}\cdot\text{s}} = 3.97 \times 10^{20} \text{ s}^{-1} = 3.97 \times 10^{20} \text{ Hz}$$

$$\lambda = \frac{3.00 \times 10^{8} \text{ m}\cdot\text{s}^{-1}}{3.95 \times 10^{20} \text{ s}^{-1}} = 7.59 \times 10^{-13} \text{ m}$$

(c) $\Delta E = (1.10 \text{ MeV})\left(\dfrac{1.602 \times 10^{-13} \text{ J}}{1 \text{ MeV}}\right) = 1.76 \times 10^{-13} \text{ J}$

$$\nu = \frac{\Delta E}{h} = \frac{1.76 \times 10^{-13} \text{ J}}{6.63 \times 10^{-34} \text{ J}\cdot\text{s}} = 2.65 \times 10^{20} \text{ s}^{-1} = 2.65 \times 10^{20} \text{ Hz}$$

$$\lambda = \frac{c}{\nu} = \frac{3.00 \times 10^{8} \text{ m}\cdot\text{s}^{-1}}{2.65 \times 10^{20} \text{ s}^{-1}} = 1.13 \times 10^{-12} \text{ m}$$

17.5 (a) ${}^{3}_{1}\text{T} \longrightarrow {}^{0}_{-1}\text{e} + {}^{A}_{Z}\text{E}$ $A = 3 - 0 = 3, Z = 1 - (-1) = 2$, E = He

so, ${}^{3}_{1}\text{T} \longrightarrow {}^{0}_{-1}\text{e} + {}^{3}_{2}\text{He}$

(b) ${}^{83}_{39}\text{Y} \longrightarrow {}^{0}_{1}\text{e} + {}^{A}_{Z}\text{E}$ $A = 83 - 0 = 83, Z = 39 - 1 = 38$, E = Sr

so, ${}^{83}_{39}\text{Y} \longrightarrow {}^{0}_{1}\text{e} + {}^{83}_{38}\text{Sr}$

(c) ${}^{87}_{36}\text{Kr} \longrightarrow {}^{0}_{-1}\text{e} + {}^{A}_{Z}\text{E}$ $A = 87 - 0 = 87, Z = 36 - (-1) = 37$, E = Rb

so, ${}^{87}_{36}\text{Kr} \longrightarrow {}^{0}_{-1}\text{e} + {}^{87}_{37}\text{Rb}$

(d) ${}^{225}_{91}\text{Pa} \longrightarrow {}^{4}_{2}\alpha + {}^{A}_{Z}\text{E}$ $A = 225 - 4 = 221, Z = 91 - 2 = 89$, E = Ac

so, ${}^{225}_{91}\text{Pa} \longrightarrow {}^{4}_{2}\alpha + {}^{221}_{89}\text{Ac}$

17.7 (a) ${}^{8}_{5}\text{B} \longrightarrow {}^{0}_{1}\text{e} + {}^{A}_{Z}\text{E}$ $A = 8 - 0 = 8, Z = 5 - 1 = 4$, E = Be

so, ${}^{8}_{5}\text{B} \longrightarrow {}^{0}_{1}\text{e} + {}^{8}_{4}\text{Be}$

(b) ${}^{63}_{28}\text{Ni} \longrightarrow {}^{0}_{-1}\text{e} + {}^{A}_{Z}\text{E}$ $A = 63 - 0 = 63, Z = 28 - (-1) = 29$, E = Cu

so, ${}^{63}_{28}\text{Ni} \longrightarrow {}^{0}_{-1}\text{e} + {}^{63}_{29}\text{Cu}$

(c) ${}^{185}_{79}\text{Au} \longrightarrow {}^{4}_{2}\alpha + {}^{A}_{Z}\text{E}$ $A = 185 - 4 = 181, Z = 79 - 2 = 77$, E = Ir

so, ${}^{185}_{79}\text{Au} \longrightarrow {}^{4}_{2}\alpha + {}^{181}_{77}\text{Ir}$

(d) ${}^{7}_{4}\text{Be} + {}^{0}_{-1}\text{e} \longrightarrow {}^{A}_{Z}\text{E}$ $A = 7 + 0 = 7, Z = 4 - 1 = 3$, E = Li

so, ${}^{7}_{4}\text{Be} + {}^{0}_{-1}\text{e} \longrightarrow {}^{7}_{3}\text{Li}$

17.9 (a) ${}^{24}_{11}\text{Na} \longrightarrow {}^{24}_{12}\text{Mg} + {}^{0}_{-1}\text{e}$; a β particle is emitted.

(b) ${}^{128}_{50}\text{Sn} \longrightarrow {}^{128}_{51}\text{Sb} + {}^{0}_{-1}\text{e}$; a β particle is emitted.

(c) ${}^{140}_{57}\text{La} \longrightarrow {}^{140}_{56}\text{Ba} + {}^{0}_{1}\text{e}$; a positron ($\beta^+$) is emitted.

(d) ${}^{228}_{90}\text{Th} \longrightarrow {}^{224}_{88}\text{Ra} + {}^{4}_{2}\alpha$; an α particle is emitted.

17.11 (a) $A/Z = 68/29 = 2.34 > (A/Z)_{\text{based}}$; hence, ${}^{68}_{29}\text{Cu}$ is neutron rich, and β decay is most likely.

$${}^{68}_{29}\text{Cu} \longrightarrow {}^{0}_{-1}\text{e} + {}^{68}_{30}\text{Zn}$$

(b) $A/Z = 103/48 = 2.15 < (A/Z)_{\text{based}}$; therefore, $^{103}_{48}\text{Cd}$ is proton rich, and β^+ decay is most likely.

$$^{103}_{48}\text{Cd} \longrightarrow {}^{0}_{1}\text{e} + {}^{103}_{47}\text{Ag}$$

(c) $^{243}_{97}\text{Bk}$ has $Z > 83$ and is proton rich; therefore, α decay is most likely.

$$^{243}_{97}\text{Bk} \longrightarrow {}^{4}_{2}\alpha + {}^{239}_{95}\text{Am}$$

(d) $^{260}_{105}\text{Db}$ has $Z > 83$; therefore, α decay is most likely.

$$^{260}_{105}\text{Db} \longrightarrow {}^{4}_{2}\alpha + {}^{256}_{103}\text{Lr}$$

17.13

α	$^{235}_{92}\text{U} \longrightarrow {}^{4}_{2}\alpha + {}^{231}_{90}\text{Th}$
β	$^{231}_{90}\text{Th} \longrightarrow {}^{0}_{-1}\text{e} + {}^{231}_{91}\text{Pa}$
α	$^{231}_{91}\text{Pa} \longrightarrow {}^{4}_{2}\alpha + {}^{227}_{89}\text{Ac}$
β	$^{227}_{89}\text{Ac} \longrightarrow {}^{0}_{-1}\text{e} + {}^{227}_{90}\text{Th}$
α	$^{227}_{90}\text{Th} \longrightarrow {}^{4}_{2}\alpha + {}^{223}_{88}\text{Ra}$
α	$^{223}_{88}\text{Ra} \longrightarrow {}^{4}_{2}\alpha + {}^{219}_{86}\text{Rn}$
α	$^{219}_{86}\text{Rn} \longrightarrow {}^{4}_{2}\alpha + {}^{215}_{84}\text{Po}$
β	$^{215}_{84}\text{Po} \longrightarrow {}^{0}_{-1}\text{e} + {}^{215}_{85}\text{At}$
α	$^{215}_{85}\text{At} \longrightarrow {}^{4}_{2}\alpha + {}^{211}_{83}\text{Bi}$
β	$^{211}_{83}\text{Bi} \longrightarrow {}^{0}_{-1}\text{e} + {}^{211}_{84}\text{Po}$
α	$^{211}_{84}\text{Po} \longrightarrow {}^{4}_{2}\alpha + {}^{207}_{82}\text{Pb}$

17.15 To determine the charge and mass of the unknown particle, it helps to write $^{1}_{1}\text{p}$ and $^{1}_{0}\text{n}$ for the proton and neutron, respectively; and $^{0}_{-1}\text{e}$ and $^{0}_{1}\text{e}$ for the β particle and positron, respectively.

(a) $^{14}_{7}\text{N} + {}^{4}_{2}\alpha \longrightarrow {}^{17}_{8}\text{O} + {}^{1}_{1}\text{p}$

(b) $^{248}_{96}\text{Cm} + {}^{1}_{0}\text{n} \longrightarrow {}^{249}_{97}\text{Bk} + {}^{0}_{-1}\text{e}$

(c) $^{243}_{95}\text{Am} + {}^{1}_{0}\text{n} \longrightarrow {}^{244}_{96}\text{Cm} + {}^{0}_{-1}\text{e} + \gamma$

(d) $^{13}_{6}\text{C} + {}^{1}_{0}\text{n} \longrightarrow {}^{14}_{6}\text{C} + \gamma$

17.17 (a) $^{20}_{10}\text{Ne} + {}^{4}_{2}\alpha \longrightarrow {}^{8}_{4}\text{Be} + {}^{16}_{8}\text{O}$

(b) $^{20}_{10}\text{Ne} + {}^{20}_{10}\text{Ne} \longrightarrow {}^{16}_{8}\text{O} + {}^{24}_{12}\text{Mg}$

(c) $^{44}_{20}\text{Ca} + {}^{4}_{2}\alpha \longrightarrow \gamma + {}^{48}_{22}\text{Ti}$

(d) $^{27}_{13}\text{Al} + {}^{2}_{1}\text{H} \longrightarrow {}^{1}_{1}\text{p} + {}^{28}_{13}\text{Al}$

17.19 In each case, identify the unknown particle by performing a mass and charge balance as you did in the solutions to Exercises 17.5 and 17.7. Then write the complete nuclear equation.

(a) $^{14}_{7}N + ^{4}_{2}\alpha \longrightarrow ^{17}_{8}O + ^{1}_{1}p$

(b) $^{239}_{94}Pu + ^{1}_{0}n \longrightarrow ^{240}_{95}Am + ^{0}_{-1}e$

17.21 (a) untriquadium, Utq (b) unquadpentium, Uqp (c) binilunium, Bnu

17.23 $\text{activity} = (3.7 \times 10^{6}\ \text{Bq})\left(\frac{1\ \text{Ci}}{3.7 \times 10^{10}\ \text{Bq}}\right) = 1.0 \times 10^{-4}\ \text{Ci}$

17.25 1 Bq = 1 disintegration per second (dps)

(a) $1.0\ \text{Ci} = 3.7 \times 10^{10}\ \text{dps} = 3.7 \times 10^{10}\ \text{Bq}$

(b) $$(82\ \text{mCi})\left(\frac{10^{-3}\ \text{Ci}}{1\ \text{mCi}}\right)\left(\frac{3.7 \times 10^{10}\ \text{dps}}{1\ \text{Ci}}\right) = 3.0 \times 10^{9}\ \text{dps} = 3.0 \times 10^{9}\ \text{Bq}$$

(c) $$(1.0\ \mu\text{Ci})\left(\frac{10^{-6}\ \text{Ci}}{1\ \mu\text{Ci}}\right)\left(\frac{3.7 \times 10^{10}\ \text{dps}}{1\ \text{Ci}}\right) = 3.7 \times 10^{4}\ \text{dps} = 3.7 \times 10^{4}\ \text{Bq}$$

17.27 $\text{dose in rads} = 1.0\ \text{J}\cdot\text{kg}^{-1} \times \left(\frac{1\ \text{rad}}{10^{-2}\ \text{J}\cdot\text{kg}^{-1}}\right) = 1.0 \times 10^{2}\ \text{rad}$

dose equivalent in rems = Q × dose in rads

$$= \left(\frac{1\ \text{rem}}{1\ \text{rad}}\right)(1.0 \times 10^{2}\ \text{rad}) = 1.0 \times 10^{2}\ \text{rem}$$

$1.0 \times 10^{2}\ \text{rem} \div 100\ \text{rem/Sv} = 1.0\ \text{Sv}$

17.29 $1.0\ \text{rad}\cdot\text{day}^{-1} = (1.0\ \text{rad}\cdot\text{day}^{-1})\left(\frac{1\ \text{rem}}{1\ \text{rad}}\right) = 1\ \text{rem}\cdot\text{day}^{-1}$

$100\ \text{rem} = 1\ \text{rem}\cdot\text{day}^{-1} \times \text{time}$

time = 100 day

17.31 $k = \frac{0.693}{t_{1/2}}$

(a) $k = \frac{0.693}{12.3\ \text{y}} = 5.63 \times 10^{-2}\ \text{y}^{-1}$

(b) $k = \frac{0.693}{0.84\ \text{s}} = 0.83\ \text{s}^{-1}$

(c) $k = \frac{0.693}{10.0\ \text{min}} = 0.0693\ \text{min}^{-1}$

17.33 In each case, $k = \dfrac{0.693}{t_{1/2}}$, initial activity $\propto N_0$, final activity $\propto N$, and $N = N_0 e^{-kt}$

Therefore, $\dfrac{\text{initial activity}}{\text{final activity}} = \dfrac{N_0}{N} = e^{kt}$ and $\ln\left(\dfrac{N_0}{N}\right) = kt$

Solving for t, $t = \left(\dfrac{1}{k}\right)\ln\left(\dfrac{N_0}{N}\right) = \left(\dfrac{1}{k}\right)\ln\left(\dfrac{\text{initial activity}}{\text{final activity}}\right)$

(a) $k = \dfrac{0.693}{1.60 \times 10^3\ \text{y}} = 4.33 \times 10^{-4}\ \text{y}^{-1}$

$t = \left(\dfrac{1}{4.33 \times 10^{-4}\ \text{y}^{-1}}\right)\ln\left(\dfrac{1.0\ \text{Ci}}{0.10\ \text{Ci}}\right) = 5.3 \times 10^3\ \text{y}$

(b) $k = \dfrac{0.693}{1.26 \times 10^9\ \text{y}} = 5.50 \times 10^{-10}\ \text{y}^{-1}$

$t = \left(\dfrac{1}{5.50 \times 10^{-10}\ \text{y}^{-1}}\right)\ln\left(\dfrac{1.0 \times 10^{-6}\ \text{Ci}}{10 \times 10^{-9}\ \text{Ci}}\right) = 8.4 \times 10^9\ \text{y}$

(c) $k = \dfrac{0.693}{5.26\ \text{y}} = 0.132\ \text{y}^{-1}$

$t = \left(\dfrac{1}{0.132\ \text{y}^{-1}}\right)\ln\left(\dfrac{10\ \text{Ci}}{8\ \text{Ci}}\right) = 2\text{y}$

17.35 We know that initial activity $\propto N_0$, and final activity $\propto N$. Therefore,

$$\frac{\text{final activity}}{\text{initial activity}} = \frac{N}{N_0} = e^{-kt}$$

$$k = \frac{0.693}{t_{1/2}} = \frac{0.693}{5.26\ \text{y}} = 0.132\ \text{y}^{-1}$$

final activity $=$ initial activity $\times e^{-kt}$

$= 4.4\ \text{Ci} \times e^{-(0.132\ \text{y}^{-1} \times 50\ \text{y})}$

$= 6.0 \times 10^{-3}\ \text{Ci}$

17.37 In each case, $k = \dfrac{0.693}{t_{1/2}}$, $N = N_0 e^{-kt}$, $\dfrac{N}{N_0} = e^{-kt}$, and the percentage remaining $= 100\% \times (N/N_0)$

(a) $k = \dfrac{0.693}{5.73 \times 10^3\ \text{y}} = 1.21 \times 10^{-4}\ \text{y}^{-1}$

percentage remaining $= 100\% \times e^{-(1.21 \times 10^{-4}\ \text{y}^{-1} \times 1000\ \text{y})} = 88.6\%$

(b) $k = \dfrac{0.693}{12.3\ \text{y}} = 0.0563\ \text{y}^{-1}$

percentage remaining $= 100\% \times e^{-(0.0563\ \text{y}^{-1} \times 20.0\ \text{y})} = 32.4\%$

17.39 (a) $t_{1/2} = 4.5 \times 10^9$ y, $k = \dfrac{0.693}{t_{1/2}} = \dfrac{0.693}{4.5 \times 10^9 \text{ y}} = 1.54 \times 10^{-10} \text{ y}^{-1}$

$$\text{fraction remaining} = \frac{N}{N_0} = e^{-kt}$$
$$= e^{-(1.54 \times 10^{-10} \text{ y}^{-1} \times 9.0 \times 10^9 \text{ y})}$$
$$= e^{-1.4} = 0.25$$

(b) fraction remaining $= \dfrac{N}{N_0} = \dfrac{1}{2}$; hence age $= t_{1/2} = 1.26 \times 10^9$ y

17.41 Let dis = disintegrations

$$\text{activity from "old" sample} = \frac{1500 \text{ dis}/0.250 \text{ g}}{10.0 \text{ h}} = 600 \text{ dis}\cdot\text{g}^{-1}\cdot\text{h}^{-1}$$

activity from current sample $= 920 \text{ dis}\cdot\text{g}^{-1}\cdot\text{h}^{-1}$

$$k = \frac{0.693}{t_{1/2}} = \frac{0.693}{5.73 \times 10^3 \text{ y}} = 1.21 \times 10^{-4} \text{ y}^{-1}$$

"old" activity $\propto N$, current activity $\propto N_0$

$$\frac{\text{"old" activity}}{\text{current activity}} = \frac{N}{N_0} = e^{-kt}, \frac{N_0}{N} = e^{kt}, \ln\left(\frac{N_0}{N}\right) = kt$$

Solve for t (=age):

$$t = \frac{\ln\left(\frac{N_0}{N}\right)}{k} = \frac{\ln\left(\frac{920}{600}\right)}{1.21 \times 10^{-4} \text{ y}^{-1}} = 3.53 \times 10^3 \text{ y}$$

17.43 In each case, $k = \dfrac{0.693}{t_{1/2} \text{ (in s)}}$, activity in $Bq = k \times N$

$$\text{activity in Ci} = \frac{\text{activity in } Bq}{3.7 \times 10^{10} \; Bq\cdot\text{Ci}^{-1}}$$

Note: Bq(= disintegrating nuclei per second) has the units of nuclei·s^{-1}

(a) $k = \left(\dfrac{0.693}{1.60 \times 10^3 \text{ y}}\right)\left(\dfrac{1 \text{ y}}{3.17 \times 10^7 \text{ s}}\right) = 1.37 \times 10^{-11} \text{ s}^{-1}$

$$N = (1.0 \times 10^{-3} \text{ g})\left(\frac{1 \text{ mol}}{226 \text{ g}}\right)\left(\frac{6.02 \times 10^{23} \text{ nuclei}}{1 \text{ mol}}\right) = 2.7 \times 10^{18} \text{ nuclei}$$

$$\text{activity} = 1.37 \times 10^{-11} \text{ s}^{-1} \times 2.7 \times 10^{18} \text{ nuclei} \times \left(\frac{1 \text{ Ci}}{3.7 \times 10^{10} \text{ Bq}}\right)$$
$$= 1.0 \times 10^{-3} \text{ Ci}$$

(b) $k = \left(\dfrac{0.693}{28.1 \text{ y}}\right)\left(\dfrac{1 \text{ y}}{3.17 \times 10^7 \text{ s}}\right) = 7.80 \times 10^{-10} \text{ s}^{-1}$

$$N = (2.0 \times 10^{-6} \text{ g})\left(\frac{1 \text{ mol}}{90 \text{ g}}\right)\left(\frac{6.02 \times 10^{23} \text{ nuclei}}{1 \text{ mol}}\right) = 1.3 \times 10^{16} \text{ nuclei}$$

$$\text{activity} = (7.80 \times 10^{-10}\ \text{s}^{-1})(1.3 \times 10^{16}\ \text{nuclei})\left(\frac{1\ \text{Ci}}{3.7 \times 10^{10}\ \text{Bq}}\right)$$

$$= 2.7 \times 10^{-4}\ \text{Ci}$$

$$\text{(c)}\ k = \left(\frac{0.693}{2.6\ \text{y}}\right)\left(\frac{1\ \text{y}}{3.17 \times 10^{7}\ \text{s}}\right) = 8.4 \times 10^{-9}\ \text{s}^{-1}$$

$$N = (0.43 \times 10^{-3}\ \text{g})\left(\frac{1\ \text{mol}}{147\ \text{g}}\right)\left(\frac{6.02 \times 10^{23}\ \text{nuclei}}{1\ \text{mol}}\right) = 1.8 \times 10^{18}\ \text{nuclei}$$

$$\text{activity} = (8.4 \times 10^{-9}\ \text{s}^{-1})(1.8 \times 10^{18}\ \text{nuclei})\left(\frac{1\ \text{Ci}}{3.7 \times 10^{10}\ \text{Bq}}\right) = 0.41\ \text{Ci}$$

17.45 If isotopically enriched water, such as $H_2{}^{18}O$, is used in the reaction, the label can be followed. Once the products are separated, a suitable technique, such as vibrational spectroscopy of mass spectrometry, can be used to determine whether the product has incorporated the ^{18}O. For example, if the methanol ends up with the O atom from the water molecules, then its molar mass would be 34 $\text{g}\cdot\text{mol}^{-1}$, rather than 32 $\text{g}\cdot\text{mol}^{-1}$ found for methanol with elements present at their natural isotopic abundance.

17.47 The vibrational frequency is proportional to the reduced mass of the two atoms that form the bond according the equation:

$$\nu = \frac{1}{2\pi}\sqrt{\frac{k}{\mu}}$$

$$\text{where } \mu = \frac{m_A m_B}{m_A + m_B}$$

Because we are not given ν, it is easiest to make a relative comparison by taking the ratio of ν for the C—D molecule versus ν for the C—H molecule:

$$\frac{\nu_{C-D}}{\nu_{C-H}} = \frac{\dfrac{1}{2\pi}\sqrt{\dfrac{k}{\mu_{C-D}}}}{\dfrac{1}{2\pi}\sqrt{\dfrac{k}{\mu_{C-H}}}} = \sqrt{\frac{\mu_{C-H}}{\mu_{C-D}}} = \sqrt{\frac{\dfrac{m_C m_H}{m_C + m_H}}{\dfrac{m_C m_D}{m_C + m_D}}} = \sqrt{\frac{\dfrac{(12.01)(1.0079)}{12.01 + 1.0079}}{\dfrac{(12.01)(2.00)}{12.01 + 2.00}}}$$

$$= \sqrt{\frac{\left(\dfrac{12.02}{13.02}\right)}{\left(\dfrac{24.02}{14.01}\right)}} = 0.7338$$

We would thus expect the vibrational frequency for the C—D bond to be approximately 0.73 times the value for the C—H bond (lower in energy).

17.49 This term refers to self-sustaining nuclear chain reactions. For a chain reaction to be self-sustaining, each nucleus that splits must provide an average of at least one new neutron that results in the fission of another nucleus. If the mass of the fissionable material is too small, the neutrons will escape before they can produce fission. The critical mass is the smallest mass that can sustain a nuclear chain reaction.

17.51 In each case, first calculate ΔE for the process described. Note whether ΔE is positive or negative, corresponding to energy added or removed from the system. Then calculate the change in mass from the change in energy using $\Delta E = (\Delta m)c^2$ or

$$\Delta m = \frac{\Delta E}{c^2}$$

(a) $\Delta E = 250\ \text{g} \times 0.39\ \text{J}\cdot(°\text{C})^{-1}\cdot\text{g}^{-1} \times (250°\text{C} - 35°\text{C}) = 2.1 \times 10^4\ \text{J}$

$$\Delta m = \frac{2.1 \times 10^4\ \text{J}}{(3.00 \times 10^8\ \text{m}\cdot\text{s}^{-1})^2} = 2.3 \times 10^{-13}\ \text{kg} = 2.3 \times 10^{-10}\ \text{g} = \text{mass gained}$$

(b) $\Delta E = -n\Delta H°_{\text{melt}}$, where n = number of moles

$$\Delta E = -(6.01\ \text{kJ}\cdot\text{mol}^{-1})\left(\frac{50.0\ \text{g}}{18.02\ \text{g}\cdot\text{mol}^{-1}}\right) = -16.7\ \text{kJ} = -1.67 \times 10^4\ \text{J}$$

$$\Delta m = \frac{-1.67 \times 10^4\ \text{J}}{(3.00 \times 10^8\ \text{m}\cdot\text{s}^{-1})^2} = -1.86 \times 10^{-13}\ \text{kg} = -1.86 \times 10^{-10}\ \text{g}$$

$= \text{mass lost}$

(c) $\Delta E = 2\ \text{mol} \times \Delta H°_f(\text{PCl}_5, \text{g}) = 2\ \text{mol} \times (-374.9\ \text{kJ}\cdot\text{mol}^{-1}) = -749.8\ \text{kJ}$

$= -7.498 \times 10^5\ \text{J}$

$$\Delta m = \frac{-7.498 \times 10^5\ \text{J}}{(3.00 \times 10^8\ \text{m}\cdot\text{s}^{-1})^2} = -8.33 \times 10^{-12}\ \text{kg} = -8.33 \times 10^{-9}\ \text{g} = \text{mass lost}$$

17.53 Remember to convert g to kg.

(a) $E = mc^2 = (1.0 \times 10^{-3}\ \text{kg})(3.00 \times 10^8\ \text{m}\cdot\text{s}^{-1})^2$

$= 9.0 \times 10^{13}\ \text{kg}\cdot\text{m}^2\cdot\text{s}^{-2} = 9.0 \times 10^{13}\ \text{J}$

(b) $E = mc^2 = (9.109 \times 10^{-31}\ \text{kg})(3.00 \times 10^8\ \text{m}\cdot\text{s}^{-1})^2$

$= 8.20 \times 10^{-14}\ \text{kg}\cdot\text{m}^2\cdot\text{s}^{-2} = 8.20 \times 10^{-14}\ \text{J}$

17.55 $$\Delta m = \frac{\Delta E}{c^2} = \frac{-3.9 \times 10^{26}\ \text{J}\cdot\text{s}^{-1}}{(3.00 \times 10^8\ \text{m}\cdot\text{s}^{-1})^2} = -4.3 \times 10^9\ \text{kg}\cdot\text{s}^{-1}$$

17.57 $1\ \text{u} = 1.6605 \times 10^{-27}\ \text{kg}$

In each case, calculate the difference in mass between the nucleus and the free particles from which it may be considered to have been formed. Then obtain the binding energy from the relation $E_{\text{bind}} = \Delta mc^2$.

(a) ${}^{4}_{2}\text{He}$: $2\,{}^{1}\text{H} + 2\,\text{n} \longrightarrow {}^{4}_{2}\text{He}$

$\Delta m = 4.0026\ \text{u} - (2 \times 1.0078\ \text{u} + 2 \times 1.0087\ \text{u}) = -0.0304\ \text{u}$

$$\Delta m = (-0.0304\ \text{u})\left(\frac{1.6605 \times 10^{-27}\ \text{kg}}{1\ \text{u}}\right) = -5.05 \times 10^{-29}\ \text{kg}$$

$$E_{\text{bind}} = -(5.05 \times 10^{-29}\ \text{kg})(3.00 \times 10^{8}\ \text{m}\cdot\text{s}^{-1})^2$$
$$= -4.55 \times 10^{-12}\ \text{kg}\cdot\text{m}^2\cdot\text{s}^{-2} = -4.55 \times 10^{-12}\ \text{J}$$

$$E_{\text{bind}}/\text{nucleon} = \frac{-4.55 \times 10^{-12}\ \text{J}}{4\ \text{nucleons}} = -1.14 \times 10^{-12}\ \text{J}\cdot\text{nucleon}^{-1}$$

(b) ${}^{239}_{94}\text{Pu}$: $94\,{}^{1}\text{H} + 145\,\text{n} \longrightarrow {}^{239}_{94}\text{Pu}$

$\Delta m = 239.0522\ \text{u} - (94 \times 1.0078\ \text{u} + 145 \times 1.0087\ \text{u}) = -1.9425\ \text{u}$

$$\Delta m = -1.9425\ \text{u} \times \left(\frac{1.6605 \times 10^{-27}\ \text{kg}}{1\ \text{u}}\right) = -3.2255 \times 10^{-27}\ \text{kg}$$

$$E_{\text{bind}} = -3.2255 \times 10^{-27}\ \text{kg} \times (2.997 \times 10^{8}\ \text{m}\cdot\text{s}^{-1})^2 = -2.897 \times 10^{-10}\ \text{J}$$

$$E_{\text{bind}}/\text{nucleon} = \frac{-2.897 \times 10^{-10}\ \text{J}}{239\ \text{nucleons}} = -1.212 \times 10^{-12}\ \text{J}\cdot\text{nucleon}^{-1}$$

(c) ${}^{2}_{1}\text{H}$: ${}^{1}\text{H} + \text{n} \longrightarrow {}^{2}_{1}\text{H}$

$\Delta m = 2.0141\ \text{u} - (1.0078\ \text{u} + 1.0087\ \text{u}) = -0.0024\ \text{u}$

$$\Delta m = -0.0024\ \text{u} \times \left(\frac{1.6605 \times 10^{-27}\ \text{kg}}{1\ \text{u}}\right) = -4.0 \times 10^{-30}\ \text{kg}$$

$$E_{\text{bind}} = -4.0 \times 10^{-30}\ \text{kg} \times (3.00 \times 10^{8}\ \text{m}\cdot\text{s}^{-1})^2 = -3.6 \times 10^{-13}\ \text{J}$$

$$E_{\text{bind}}/\text{nucleon} = \frac{-3.6 \times 10^{-13}\ \text{J}}{2\ \text{nucleons}} = -1.8 \times 10^{-13}\ \text{J}\cdot\text{nucleon}^{-1}$$

(d) ${}^{56}_{26}\text{Fe}$: $26\,{}^{1}\text{H} + 30\,\text{n} \longrightarrow {}^{56}_{26}\text{Fe}$

$\Delta m = 55.9349\ \text{u} - (26 \times 1.0078\ \text{u} + 30 \times 1.0087\ \text{u}) = -0.5289\ \text{u}$

$$\Delta m = -0.5289\ \text{u} \times \left(\frac{1.661 \times 10^{-27}\ \text{kg}}{1\ \text{u}}\right) = -8.785 \times 10^{-28}\ \text{kg}$$

$$E_{\text{bind}} = -8.785 \times 10^{-28}\ \text{kg} \times (2.998 \times 10^{8}\ \text{m}\cdot\text{s}^{-1})^{-2} = -7.896 \times 10^{-11}\ \text{J}$$

$$E_{\text{bind}}/\text{nucleon} = \frac{-7.896 \times 10^{-11}\ \text{J}}{56\ \text{nucleons}} = -1.410 \times 10^{-12}\ \text{J}\cdot\text{nucleon}^{-1}$$

${}^{56}\text{Fe}$ is the most stable, because it has the largest binding energy per nucleon.

17.59 In each case, we first determine the change in mass, Δm = (mass of products) − (mass of reactants). We then calculate the energy released from $\Delta E = (\Delta m)c^2$.

(a) $\text{D} + \text{D} \longrightarrow {}^{3}\text{He} + \text{n}$

$2.0141\ \text{u} + 2.0141\ \text{u} \longrightarrow 3.0160\ \text{u} + 1.0087\ \text{u}$

$4.0282\ \text{u} \longrightarrow 4.0247\ \text{u}$

$\Delta m = -0.0035\ \text{u}$

$$\Delta m = (-0.0035\ \text{u})\left(\frac{1.661 \times 10^{-27}\ \text{kg}}{1\ \text{u}}\right) = -5.8 \times 10^{-30}\ \text{kg}$$

$$\Delta E = \Delta mc^2 = (-5.8 \times 10^{-30}\ \text{kg})(3.00 \times 10^8\ \text{m·s}^{-1})^2 = -5.2 \times 10^{-13}\ \text{J}$$

$$\left(\frac{-5.2 \times 10^{-13}\ \text{J}}{4.0282\ \text{u}}\right)\left(\frac{1\ \text{u}}{1.661 \times 10^{-24}\ \text{g}}\right) = -7.8 \times 10^{10}\ \text{J·g}^{-1}$$

(b) $^{3}\text{He} + \text{D} \longrightarrow {}^{4}\text{He} + {}^{1}_{1}\text{H}$

$$3.0160\ \text{u} + 2.0141\ \text{u} \longrightarrow 4.0026\ \text{u} + 1.0078\ \text{u}$$

$$5.0301\ \text{u} \longrightarrow 5.0104\ \text{u}$$

$$\Delta m = -0.0197\ \text{u}$$

$$\Delta m = -0.0197\ \text{u} \times \left(\frac{1.661 \times 10^{-27}\ \text{kg}}{1\ \text{u}}\right) = -3.27 \times 10^{-29}\ \text{kg}$$

$$\Delta E = \Delta mc^2 = -(3.27 \times 10^{-29}\ \text{kg})(3.00 \times 10^8\ \text{m·s}^{-1})^2 = -2.94 \times 10^{-12}\ \text{J}$$

$$\left(\frac{-2.94 \times 10^{-12}\ \text{J}}{5.0301\ \text{u}}\right)\left(\frac{1\ \text{u}}{1.661 \times 10^{-24}\ \text{g}}\right) = -3.52 \times 10^{11}\ \text{J·g}^{-1}$$

(c) $^{7}\text{Li} + {}^{1}_{1}\text{H} \longrightarrow 2\ {}^{4}\text{He}$

$$7.0160\ \text{u} + 1.0078\ \text{u} \longrightarrow 2(4.0026\ \text{u})$$

$$8.0238\ \text{u} \longrightarrow 8.0052\ \text{u}$$

$$\Delta m = -0.0186\ \text{u}$$

$$\Delta m = (-0.0186\ \text{u})\left(\frac{1.661 \times 10^{-27}\ \text{kg}}{1\ \text{u}}\right) = -3.09 \times 10^{-29}\ \text{kg}$$

$$\Delta E = \Delta mc^2 = (-3.09 \times 10^{-29}\ \text{kg})(3.00 \times 10^8\ \text{m·s}^{-1})^2 = -2.78 \times 10^{-12}\ \text{J}$$

$$\left(\frac{-2.78 \times 10^{-12}\ \text{J}}{8.0238\ \text{u}}\right)\left(\frac{1\ \text{u}}{1.661 \times 10^{-24}\ \text{g}}\right) = -2.09 \times 10^{11}\ \text{J·g}^{-1}$$

(d) $\text{D} + \text{T} \longrightarrow {}^{4}\text{He} + {}^{1}_{1}\text{H}$

$$2.0141\ \text{u} + 3.0160\ \text{u} \longrightarrow 4.0026\ \text{u} + 1.0078\ \text{u}$$

$$5.0301\ \text{u} \longrightarrow 5.00104\ \text{u}$$

$$\Delta m = -0.0197\ \text{u}$$

$$\Delta m = (-0.0197\ \text{u})\left(\frac{1.661 \times 10^{-27}\ \text{kg}}{1\ \text{u}}\right) = -3.27 \times 10^{-29}\ \text{kg}$$

$$\Delta E = \Delta mc^2 = (-3.27 \times 10^{-29}\ \text{kg})(3.00 \times 10^8\ \text{m·s}^{-1})^2 = -2.94 \times 10^{-12}\ \text{J}$$

$$\left(\frac{-2.94 \times 10^{-12}\ \text{J}}{5.0301\ \text{u}}\right)\left(\frac{1\ \text{u}}{1.661 \times 10^{-24}\ \text{g}}\right) = -3.52 \times 10^{11}\ \text{J·g}^{-1}$$

17.61 In each case, solve for the unknown particle by doing a mass and charge balance as you did in the solutions to Exercises 17.5 and 17.7.

(a) $^{244}_{95}\text{Am} \longrightarrow {}^{134}_{53}\text{I} + {}^{107}_{42}\text{Mo} + 3\ {}^{1}_{0}\text{n}$

(b) $^{235}_{92}\text{U} + {}^{1}_{0}\text{n} \longrightarrow {}^{96}_{40}\text{Zr} + {}^{138}_{52}\text{Te} + 2\ {}^{1}_{0}\text{n}$

(c) $^{235}_{92}\text{U} + {}^{1}_{0}\text{n} \longrightarrow {}^{101}_{42}\text{Mo} + {}^{132}_{50}\text{Sn} + 3\ {}^{1}_{0}\text{n}$

17.63 The stabilities of nuclei vary and the greatest stabilities are associated with certain numbers of nucleons. These numbers are referred to as magic numbers; they are

2, 8, 20, 50, 82, 126

The existence of this series of numbers reminds us of the magic numbers of electrons in the electronic configurations of the noble gases:

2, 10, 18, 36, 54, 86

Because this series of numbers corresponds to a shell model for electrons in atoms, the analogous series of numbers for nucleons suggests a nuclear shell model.

17.65 (a) ${}^{11}_{5}\text{B} + {}^{4}_{2}\alpha \longrightarrow 2\ {}^{1}_{0}\text{n} + {}^{13}_{7}\text{N}$

(b) ${}^{35}_{17}\text{Cl} + {}^{2}_{1}\text{D} \longrightarrow {}^{1}_{0}\text{n} + {}^{36}_{18}\text{Ar}$

(c) ${}^{96}_{42}\text{Mo} + {}^{2}_{1}\text{D} \longrightarrow {}^{1}_{0}\text{n} + {}^{97}_{43}\text{Tc}$

(d) ${}^{45}_{21}\text{Sc} + {}^{1}_{0}\text{n} \longrightarrow {}^{4}_{2}\alpha + {}^{42}_{19}\text{K}$

17.67 (a) 1 Ci = 3.7×10^{10} decays per second (dps)

$$\text{decays per minute (dpm) for 4 pCi} = 4 \times 10^{-12}\ \text{Ci} \times 3.7 \times 10^{10}\ \text{dps} \times \left(\frac{60\ \text{s}}{1\ \text{min}}\right)$$
$$= 9\ \text{dpm}$$

(b) $$\text{volume(L)} = (2.0 \times 3.0 \times 2.5)\ \text{m}^3 \times \left(\frac{10^3\ \text{L}}{1\ \text{m}^3}\right) = 1.5 \times 10^4\ \text{L}$$

$$\text{number of decays} = (1.5 \times 10^4\ \text{L})\left(\frac{4\ \text{pCi}}{1\ \text{L}}\right)\left(\frac{9\ \text{decays}\cdot\text{min}^{-1}}{4\ \text{pCi}}\right)(5.0\ \text{min})$$
$$= 7 \times 10^5\ \text{decays}$$

17.69 $$N_0 = \text{number of } {}^{222}\text{Rn atoms} = 1.0 \times 10^{-5}\ \text{mol} \times 6.0 \times 10^{23}\ \text{atoms}\cdot\text{mol}^{-1}$$
$$= 6.0 \times 10^{18}\ \text{atoms}$$

$$k = \frac{\ln 2}{t_{1/2}} = \frac{0.693}{3.82\ \text{d}} = 0.181\ \text{d}^{-1}$$

(a) $$\text{rate of decay} = k \times N = \left(\frac{0.181}{\text{d}}\right)\left(\frac{1\ \text{d}}{8.64 \times 10^4\ \text{s}}\right)(6.0 \times 10^{18}\ \text{atoms})$$
$$= 1.26 \times 10^{13}\ \text{atoms}\cdot\text{s}^{-1}\ \text{(dps or Bq)}$$

$$\text{initial activity} = (1.26 \times 10^{13}\ \text{Bq})\left(\frac{1\ \text{Ci}}{3.7 \times 10^{10}\ \text{Bq}}\right)\left(\frac{1\ \text{pCi}}{10^{-12}\ \text{Ci}}\right)\left(\frac{1}{2000\ \text{m}^3}\right)\left(\frac{1\ \text{m}^3}{10^3\ \text{L}}\right)$$
$$= 1.7 \times 10^8\ \text{pCi}\cdot\text{L}^{-1}$$

(b) $$N = N_0 e^{-kt} = 6.0 \times 10^{18}\ \text{atoms} \times e^{-0.181\ \text{d}^{-1} \times 1\ \text{d}} = 5.0 \times 10^{18}\ \text{atoms}$$

(c) $$\ln\left(\frac{\text{activity}}{\text{initial activity}}\right) = -kt$$

$$t = -\left(\frac{1}{k}\right)\ln\left(\frac{\text{activity}}{\text{initial activity}}\right) = -\left(\frac{1}{0.181\ \text{d}^{-1}}\right)\ln\left(\frac{4}{1.70 \times 10^8}\right)$$
$$= 1 \times 10^2\ \text{days}$$

17.71 $k = \dfrac{0.693}{t_{1/2}} = \dfrac{0.693}{8.05\ \text{d}} = 0.0861\ \text{d}^{-1}$

$N = N_0 e^{-kt}$ and $\dfrac{N}{N_0} = e^{-kt}$

Taking the natural log of both sides gives

$$\ln\left(\frac{N}{N_0}\right) = -kt$$

Because activity is proportional to N (Eq. 2), we can write

$$\ln\left(\frac{\text{final activity}}{\text{initial activity}}\right) = -kt$$

Solving for t gives

$$t = -\left(\frac{1}{k}\right)\ln\left(\frac{\text{final activity}}{\text{initial activity}}\right) = -\left(\frac{1}{0.0861\ \text{d}^{-1}}\right)\ln\left(\frac{10}{500}\right) = 45\ \text{d}$$

17.73 (a) activity $\propto N$; and, because $\ln\left(\dfrac{N}{N_0}\right) = -kt$

$$\ln\left(\frac{\text{final activity}}{\text{initial activity}}\right) = -kt$$

$$\ln\left(\frac{32}{58}\right) = -k \times 12.3\ \text{d}$$

$k = 0.048\ \text{d}^{-1}$

$$t_{1/2} = \frac{0.693}{k} = \frac{0.693}{0.048\ \text{d}^{-1}} = 14\ \text{d}$$

(b) $\ln\left(\dfrac{N}{N_0}\right) = -0.048\ \text{d}^{-1} \times 30\ \text{d} = -1.4$

$\dfrac{N}{N_0} = \text{fraction remaining} = 0.25$

17.75 $10.0\ \text{Ci} = 10.0 \times 3.7 \times 10^{10}\ \text{nuclei}\cdot\text{s}^{-1} \times \dfrac{8.64 \times 10^4\ \text{s}}{1\ \text{d}}$

$= 3.2 \times 10^{16}\ \text{nuclei}\cdot\text{d}^{-1}$

$k = \dfrac{0.693}{t_{1/2}} = \dfrac{0.693}{88\ \text{d}} = 7.9 \times 10^{-3}\ \text{d}^{-1}$

activity = rate = $3.2 \times 10^{16}\ \text{nuclei}\cdot\text{d}^{-1} = 7.9 \times 10^{-3}\ \text{d}^{-1} \times N$

$$N = \frac{3.2 \times 10^{16}\ \text{nuclei}\cdot\text{d}^{-1}}{7.9 \times 10^{-3}\ \text{s}^{-1}} = 4.1 \times 10^{18}\ \text{nuclei}$$

mass of $^{35}S = (35\ \text{u})\left(\dfrac{1.661 \times 10^{-24}\ \text{g}}{1\ \text{u}}\right)(4.1 \times 10^{18}\ \text{nuclei})$

$= 2.4 \times 10^{-4}\ \text{g}$

17.77 $^{24}_{11}\text{Na} \longrightarrow {}^{24}_{12}\text{Mg} + {}^{0}_{-1}\text{e}$

mass ($^{24}_{11}$Na) = 23.990 96 u

mass ($^{24}_{12}$Mg) = 23.985 04 u

The mass of the electron does not need to be explicitly included in the calculation because it is already included in the mass of Mg.

$$\Delta m = \text{mass}\,(^{24}_{12}\text{Mg}) - \text{mass}\,(^{24}_{11}\text{Na}) = 23.985\,04\text{ u} - 23.990\,96\text{ u}$$

$$= -5.92 \times 10^{-3}\text{ u}$$

$$\Delta m\text{ (in kg)} = -5.92 \times 10^{-3}\text{ u} \times 1.661 \times 10^{-27}\text{ kg u}^{-1} = -9.83 \times 10^{-30}\text{ kg}$$

It is convenient to work part (b) first.

(b) $\Delta E = \Delta mc^2 = -(9.83 \times 10^{-30}\text{ kg})(3.00 \times 10^8\text{ m}\cdot\text{s}^{-1}) = -8.85 \times 10^{-13}\text{ J}$

(a) $\Delta E\text{ (per nucleon)} = \dfrac{-8.85 \times 10^{-13}\text{ J}}{24\text{ nucleons}} = -3.69 \times 10^{-14}\text{ J}\cdot\text{nucleon}^{-1}$

17.79 (a) At first thought, it might seem that a fusion bomb would be more suitable for excavation work, because the fusion process itself does not generate harmful radioactive waste products. However, in practice, fusion cannot be initiated in a bomb in the absence of the high temperatures that can only be generated by a fission bomb. So, there is no environmental advantage to the use of a fusion bomb. The fission bomb has the advantage that its destructive power can be more carefully controlled. It is possible to make small fission bombs whose destructive effect can be contained within a small area.

(b) The principal argument for the use of bombs in excavation is speed, and therefore cost-effectiveness, of the process. The principal argument against their use is environmental damage.

17.81 $k = \dfrac{0.693}{4.5 \times 10^9\text{ y}} = 1.5 \times 10^{-10}\text{ y}^{-1}$

$t(=\text{age}) = -\left(\dfrac{1}{k}\right)\ln\left(\dfrac{N}{N_0}\right)$ [Example 17.2]

$$\frac{N}{N_0} = \frac{\text{mass of }^{238}\text{U}}{\text{initial mass of }^{238}\text{U}} = \frac{1}{1 + \dfrac{\text{mass of }^{206}\text{Pb}}{\text{mass of }^{238}\text{U}}}$$

(a) $\dfrac{N}{N_0} = \dfrac{1}{1 + 1.00} = \dfrac{1}{2.00}$, therefore age $= t_{1/2} = 4.5 \times 10^9$ y

(b) $\dfrac{N}{N_0} = \dfrac{1}{1 + \dfrac{1}{1.25}} = 0.556$

$$t(=\text{age}) = -\left(\frac{1}{1.5 \times 10^{-10}\text{ y}^{-1}}\right)\ln(0.556) = 3.9 \times 10^9\text{ y}$$

17.83 (a) activity $= (17.3\ \text{Ci})\left(\dfrac{3.7 \times 10^{10}\ \text{Bq}}{1\ \text{Ci}}\right)$

$$= 6.4 \times 10^{11}\ \text{Bq} = 6.4 \times 10^{11}\ \text{nuclei}\cdot\text{s}^{-1}$$

$$N = (2.0 \times 10^{-6}\ \text{g})\left(\frac{1\ \text{u}}{1.661 \times 10^{-24}\ \text{g}}\right)\left(\frac{1\ \text{nucleus}}{24\ \text{u}}\right) = 5.0 \times 10^{16}\ \text{nuclei}$$

$$k = \frac{\text{activity}}{N} = \frac{6.4 \times 10^{11}\ \text{nuclei}\cdot\text{s}^{-1}}{5.0 \times 10^{16}\ \text{nuclei}} = 1.3 \times 10^{-5}\ \text{s}^{-1} = 1.1\ \text{d}^{-1}$$

$$t_{1/2} = \frac{0.693}{k} = \frac{0.693}{1.3 \times 10^{-5}\ \text{s}^{-1}} = 5.3 \times 10^{4}\ \text{s} = 15\ \text{h} = 0.63\ \text{d}$$

(b) $m = m_0 e^{-kt} = 2.0\ \mu\text{g} \times e^{-1.11\ \text{d}^{-1} \times 2.0\ \text{d}} = 0.22\ \mu\text{g}$

CHAPTER 18
ORGANIC CHEMISTRY I: THE HYDROCARBONS

18.1 (a) alkene (b) alkyne (c) alkane (d) both alkene and alkyne (e) alkene

18.3 (a) $(CH_3)_3CH$ or C_4H_{10}, alkane (b) $C_6H_7CH_3$ or C_7H_{10}, alkene (c) C_6H_{12}, alkane (d) C_6H_{12}, alkane

18.5 (a) C_5H_{10}, alkane (b) C_5H_6, alkene (c) C_8H_{18}, alkane (d) $C_{13}H_{16}$, aromatic hydrocarbon

18.7 (a) ethane (b) hexane (c) octane (d) methane

18.9 (a) methyl (b) pentyl (c) propyl

18.11 (a) propane (b) ethane (c) pentane (d) 2,3-dimethylbutane

18.13 (a) 4-methyl-2-pentene (b) 2,3-dimethyl-2-phenylpentane

18.15 (a) $CH_2{=}CHCH(CH_3)CH_2CH_3$ (b) $CH_3CH_2C(CH_3)_2CH(CH_2CH_3)(CH_2)_2CH_3$ (c) $CH{\equiv}C(CH_2)_2C(CH_3)_3$ (d) $CH_3CH(CH_3)CH(CH_2CH_3)CH(CH_3)_2$

18.17 (a)

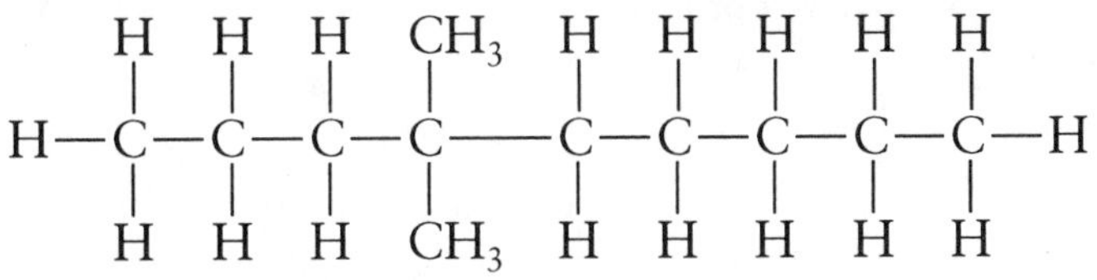

(b)

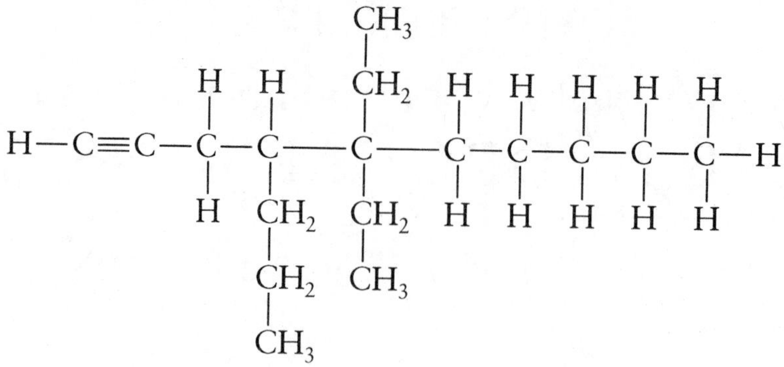

(c)

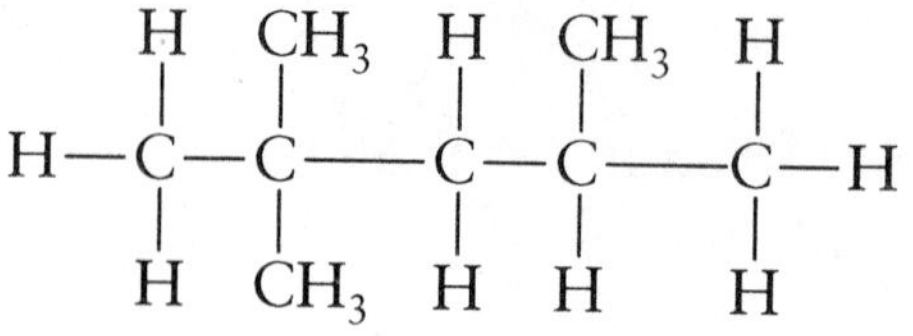

(d)

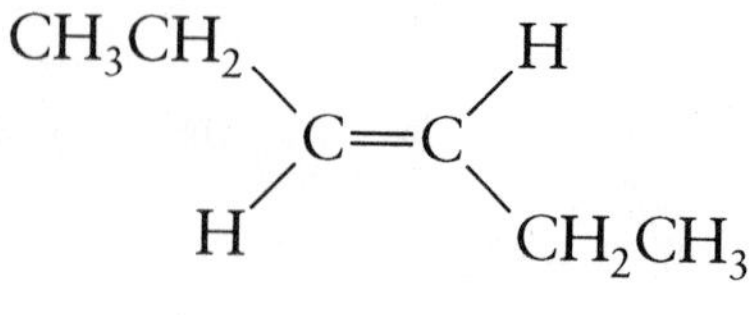

18.19 (a)

(b)

(c)

18.21 (a) C_4H_8

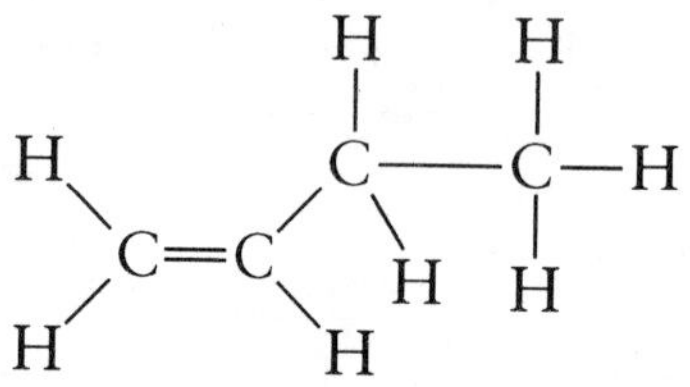

1-Butene

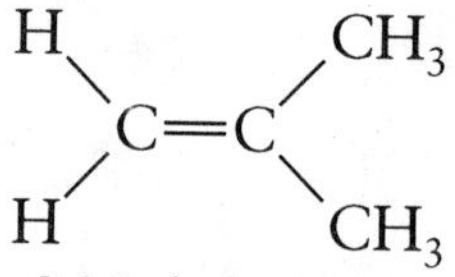

2-Methylpropene

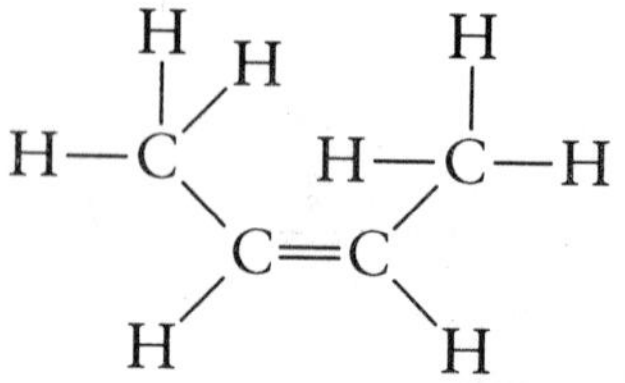

cis-2-Butene

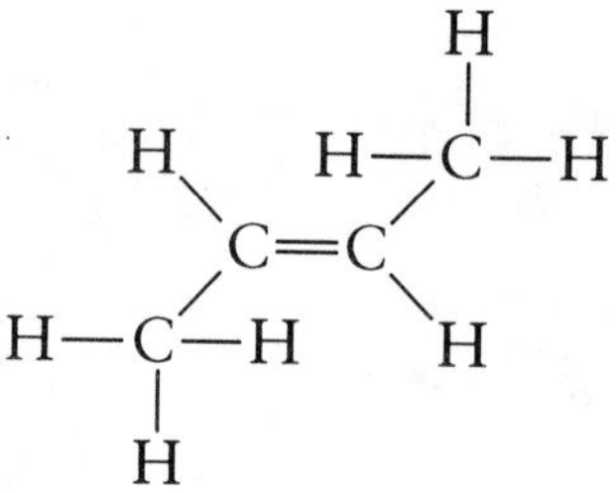

trans-2-Butene

(b) C_4H_8: 2 cycloalkane isomers

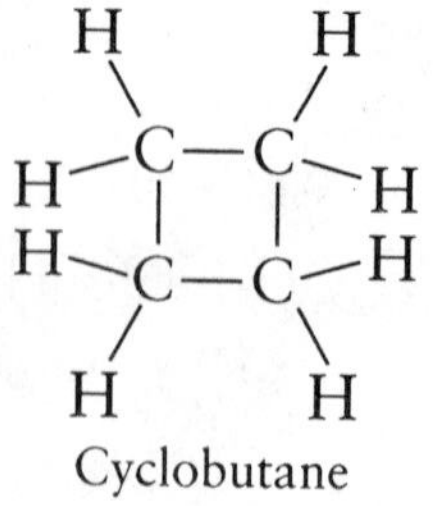

Cyclobutane

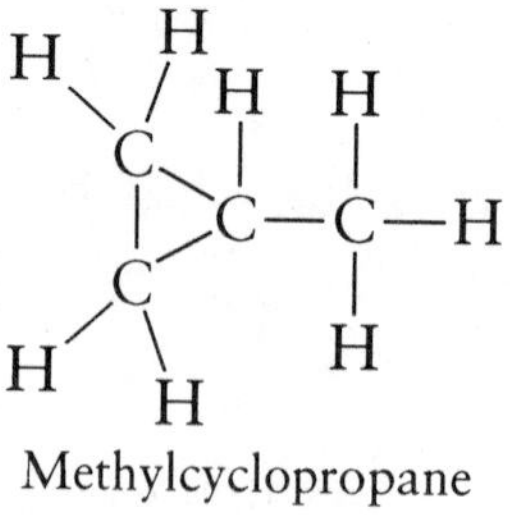

Methylcyclopropane

18.23 (a) Butane is C_4H_{10}, cyclobutane is C_4H_8. Because they have different formulas, they are not isomers.

(b) 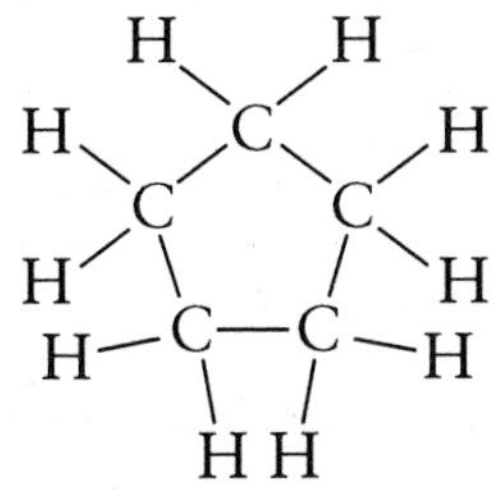 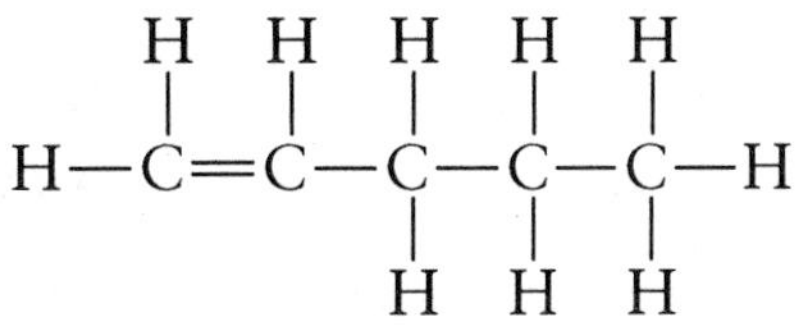

Cyclopentane (C_5H_{10}) Pentene (C_5H_{10})

Same formula, but different structures; therefore, they are structural isomers.

(c) Same formula (C_5H_{10}), same structure (bonding arrangement is the same), but different geometry; therefore, they are geometrical isomers.

(d) Not isomers, because only their positions in space are different and these positions can be interchanged. Same molecule.

18.25 If only two isomeric products are formed and they are both branched, then the only possibilities are

(a) 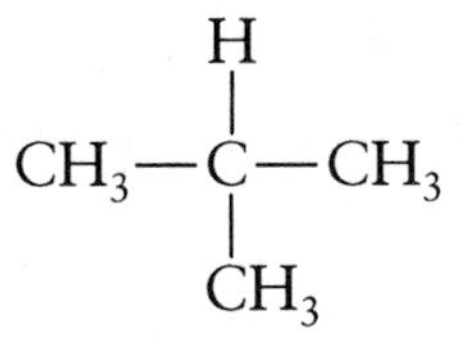 (b) 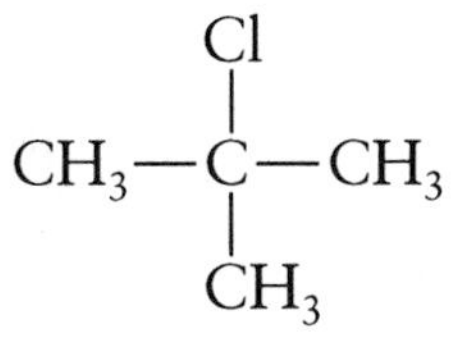 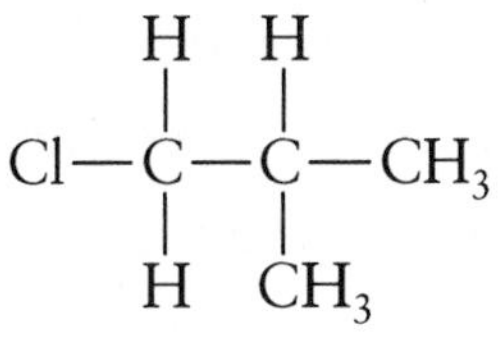

Note: All methyl groups are equivalent.

18.27 (a), (c), and (d) are optically active.

(a) 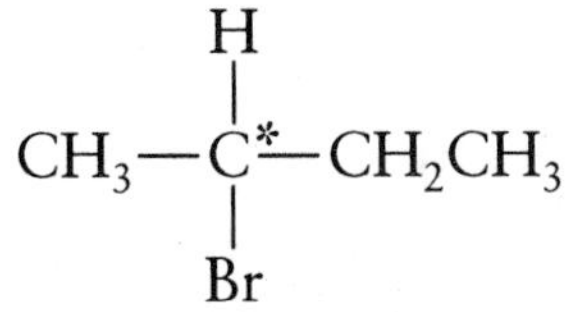 (c) 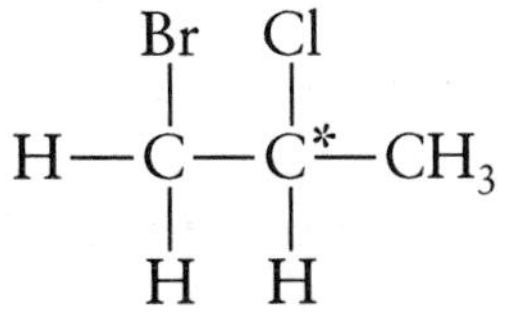

(d) 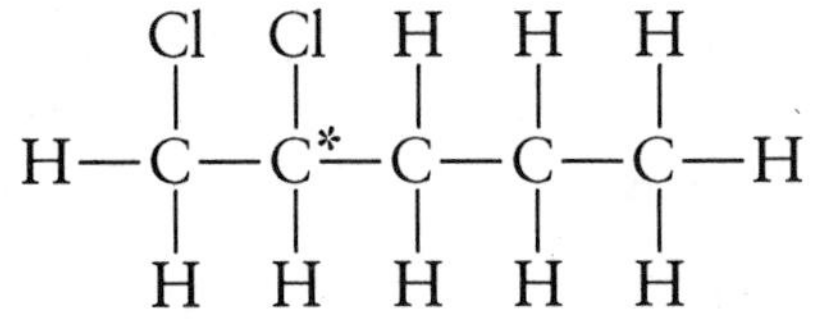

18.29 The difference can be traced to the weaker London forces that exist between branched molecules. Atoms in neighboring branched molecules cannot lie as close together as they do in the unbranched isomers. As a result of the molecules' irregular shape, the atoms in neighboring branched molecules are more effectively shielded from one another than they are in neighboring unbranched molecules.

---

* Indicates the chiral carbon atoms.

18.31 The balanced equations are

$C_3H_8(g) + 5\ O_2(g) \longrightarrow 3\ CO_2(g) + 4\ H_2O(l)$

$C_4H_{10}(g) + 13/2\ O_2(g) \longrightarrow 4\ CO_2(g) + 5\ H_2O(l)$

$C_5H_{12}(g) + 8\ O_2(g) \longrightarrow 5\ CO_2(g) + 6\ H_2O(l)$

The enthalpies of combustion that correspond to these reactions are listed in the Appendix:

Compound	Enthalpy of combustion $kJ \cdot mol^{-1}$	Heat released per g $kJ \cdot g^{-1}$
Propane	−2220	50.3
Butane	−2878	49.5
Pentane	−3537	49.0

The molar enthalpy of combustion increases with molar mass as might be expected, because the number of moles of CO_2 and H_2O formed will increase as the number of carbon and hydrogen atoms in the compounds increases. The heat released per gram of these hydrocarbons is essentially the same because the H to C ratio is similar in the three hydrocarbons.

18.33 There are nine possible products:

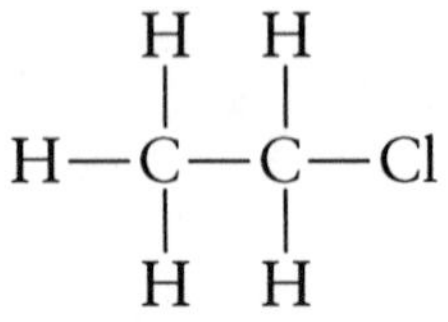

one monochloro compound

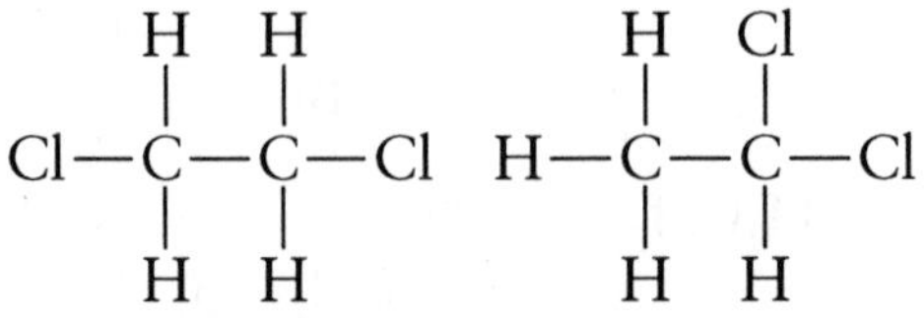

two dichloro compounds

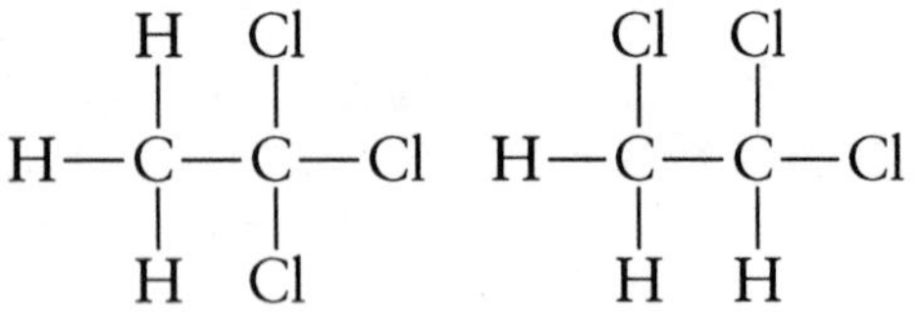

two trichloro compounds

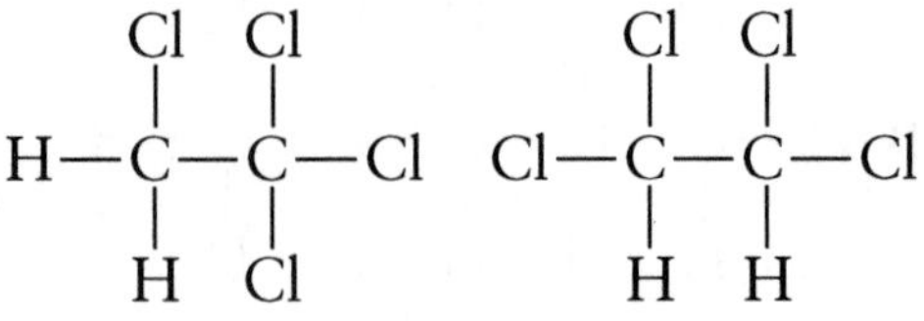

two tetrachloro compounds

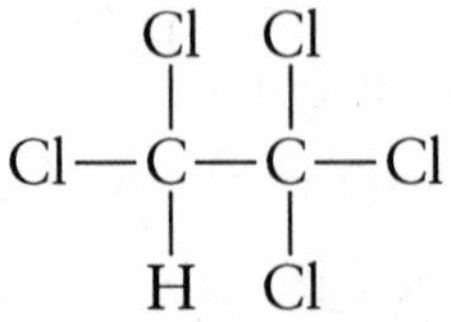

one pentachloro compound

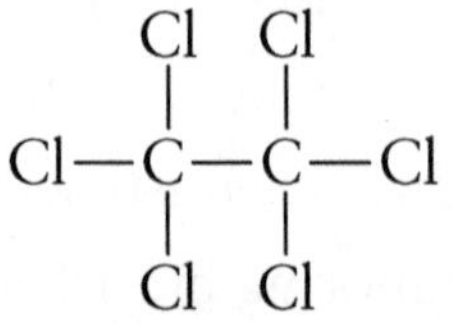

one hexachloro compound

None of these form optical isomers.

18.35

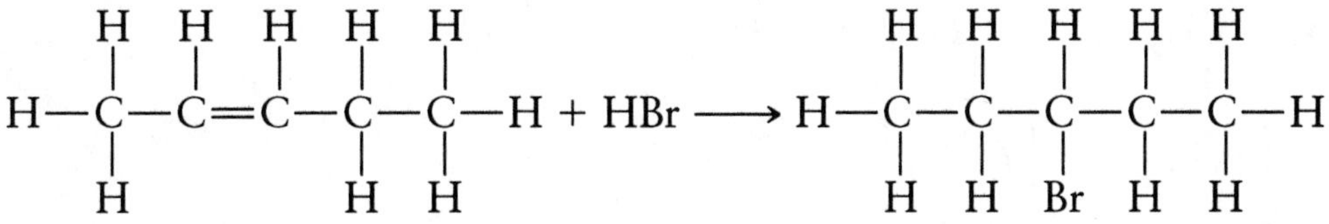

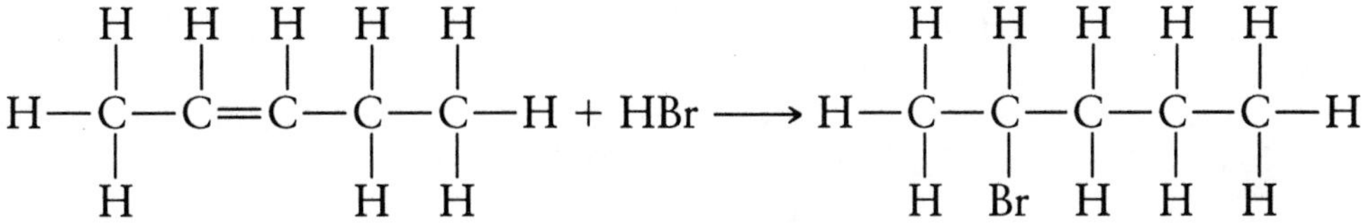

The first reaction produces 3-bromopentane. The second reaction produces 2-bromopentane. These are addition reactions.

18.37 $C_6H_{11}Br + NaOCH_2CH_3 \longrightarrow C_6H_{10} + NaBr + HOCH_2CH_3$

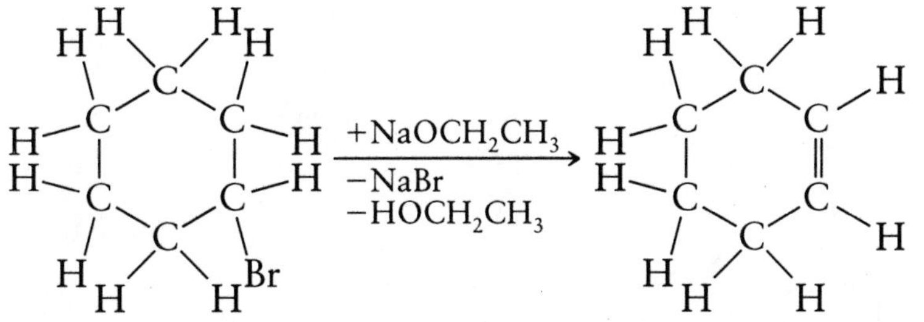

Elimination reaction

Elimination reaction

18.39 $C_2H_4 + X_2 \longrightarrow C_2H_4X_2$

We will break 1 X—X bond and form 2 C—X bonds.

Using bond enthalpies:

Halogen	Cl	Br	I
X—X bond breakage ($kJ \cdot mol^{-1}$)	+242	+193	+151
C—X bond formation ($kJ \cdot mol^{-1}$)	−2(338)	−2(276)	−2(238)
Total ($kJ \cdot mol^{-1}$)	−434	−359	−325

The reaction is less exothermic as the halogen becomes lighter. In general, the reactivity, and also the danger associated with use of the halogens in reactions, decreases as one descends the periodic table.

18.41 (a) 1-ethyl-3-methylbenzene (b) 1,2,3,4,5-pentamethylbenzene (or, because there is only one possible structure, pentamethylbenzene)

18.43 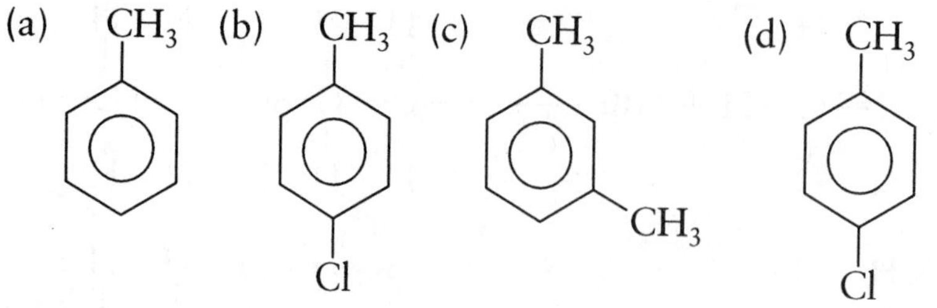

18.45 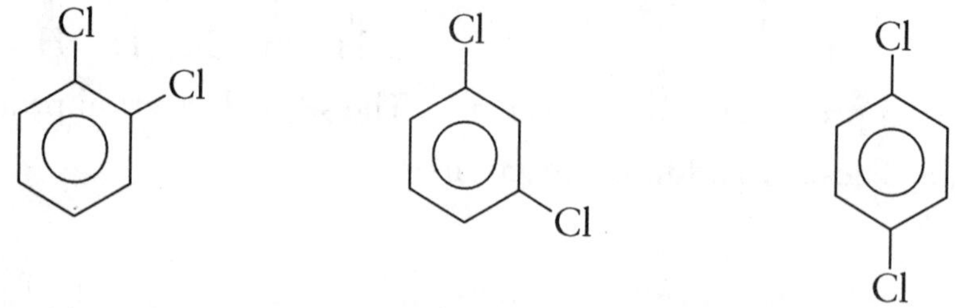

1,2-Dichlorobenzene or *ortho*-Dichlorobenzene (polar)	1,3-Dichlorobenzene or *meta*-Dichlorobenzene (polar)	1,4-Dichlorobenzene or *para*-Dichlorobenzene (nonpolar)

18.47 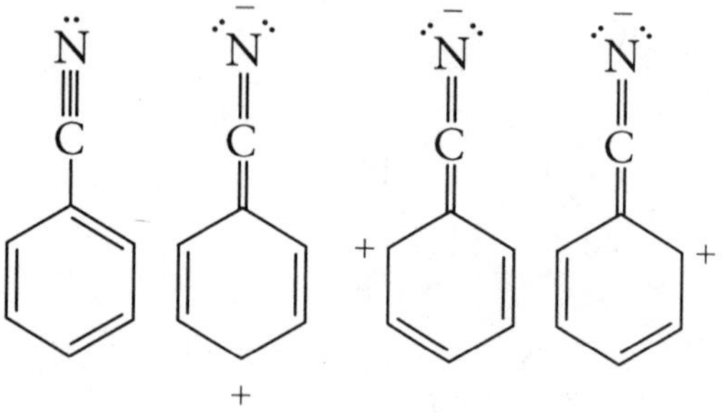

Electrophiles tend to avoid the ortho and para positions that develop slight + charges in the resonance forms.

18.49 Two:

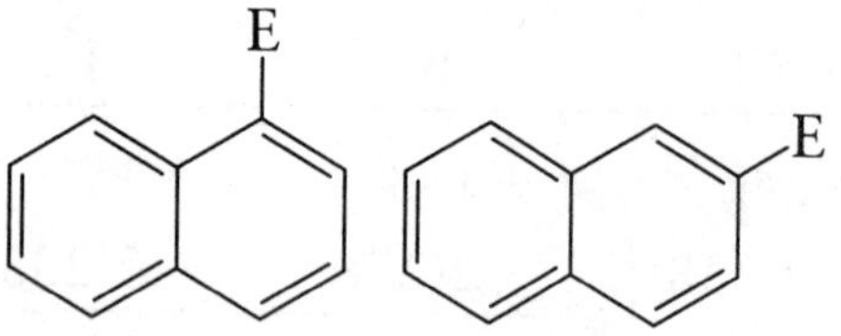

E = electrophile

18.51 (a) four σ-type single bonds (b) two σ-type single bonds and one double bond, consisting of one σ-bond and one π-bond (c) one σ-type single bond and one triple bond, consisting of one σ-bond and two π-bonds

18.53 (a) $CH_4 + Cl_2 \xrightarrow{\text{light}} CH_3Cl + HCl$, substitution

(b) $CH_2{=}CH_2 + Br_2 \longrightarrow CH_2Br{-}CH_2Br$, addition

18.55 (a) $CH_3CH_3 + 2\ Cl_2 \xrightarrow{\text{light}} CH_2ClCH_2Cl + 2\ HCl$, substitution ($CH_2ClCH_3$ may also be produced)

(b) $CH_2CH_2 + Cl_2 \longrightarrow CH_2ClCH_2Cl$, addition

(c) $HC{\equiv}CH + 2\ Cl_2 \longrightarrow CHCl_2CHCl_2$, addition

18.57 The double bond in alkenes makes them more rigid than alkanes. The atoms of alkene molecules tend to be locked into a planar arrangement by the π-bond and therefore they cannot roll up into a ball as compactly as alkanes. Thus, they do not pack together as compactly as alkanes, and so have lower boiling and melting points.

18.59 (a) propene (no geometrical isomers) (b) *cis*-2-hexene, *trans*-2-hexene
(c) 1-butyne (no geometrical isomers) (d) 2-butyne (no geometrical isomers)

18.61

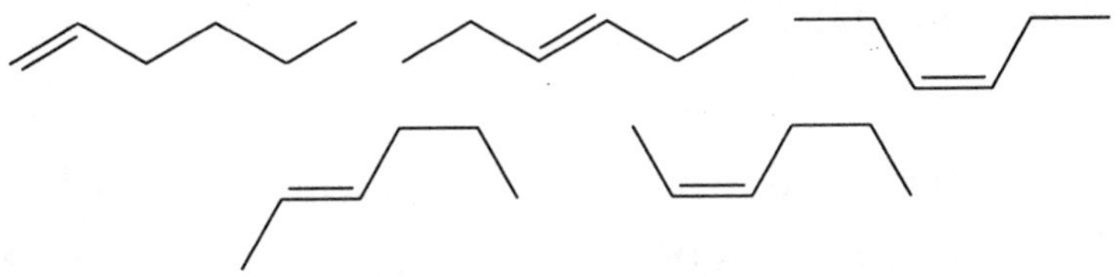

18.63 Aromatic carbon-carbon bonds are intermediate in length (139 pm), between a C—C single bond (154 pm) and a C=C double bond (134 pm). An aromatic ring is characterized by a delocalized π system, in contrast to the localized bonds present in aliphatic hydrocarbons.

18.65

$$\text{number of moles of H} = \left(\frac{3.32\text{ g } H_2O}{18.02\text{ g } H_2O/\text{mol } H_2O}\right)\left(\frac{2\text{ mol H}}{1\text{ mol } H_2O}\right) = 0.368\text{ mol H}$$

$$\text{number of moles of C} = \left(\frac{6.48\text{ g } CO_2}{44.01\text{ g } CO_2/\text{mol } CO_2}\right)\left(\frac{1\text{ mol C}}{1\text{ mol } CO_2}\right) = 0.147\text{ mol C}$$

$$\frac{0.368\text{ mol H}}{0.147\text{ mol C}} = 2.50\left(\frac{\text{mol H}}{\text{mol C}}\right) = \frac{5\text{ mol H}}{2\text{ mol C}}$$

Therefore, the empirical formula is C_2H_5. The molecular formula might be C_4H_{10}, which matches the general formula for alkanes, C_nH_{2n+2}. The compound cannot be an alkene or alkyne, because they all have mol H/mol C ratios less than 2.5.

18.67 (a) 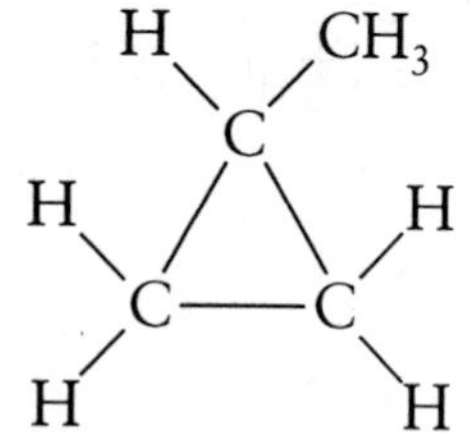 (b) 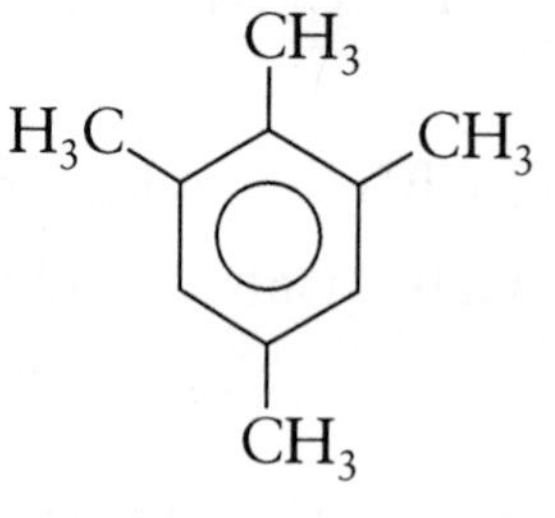 (c) 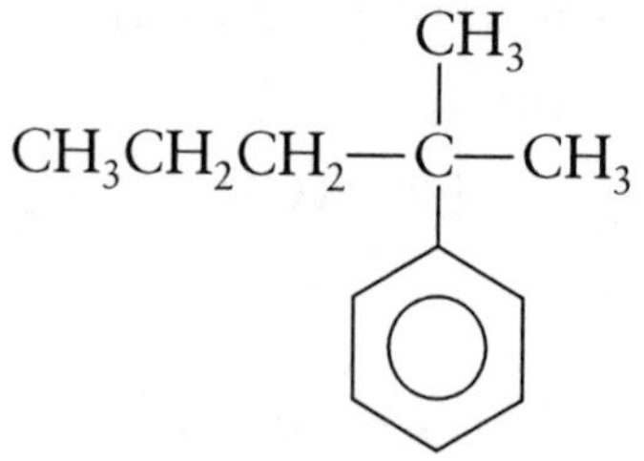

(d) 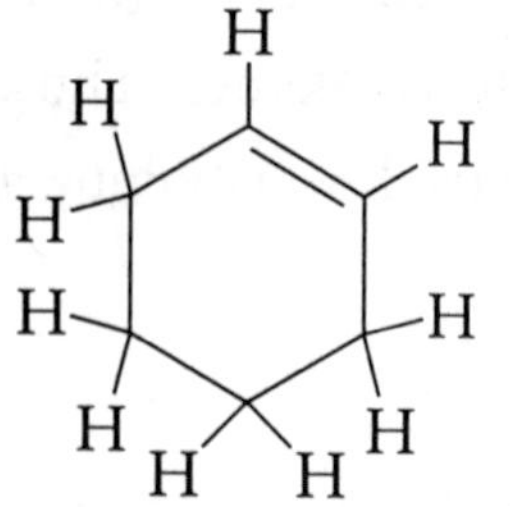 (e) 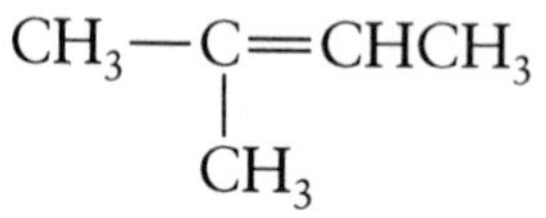 (f) 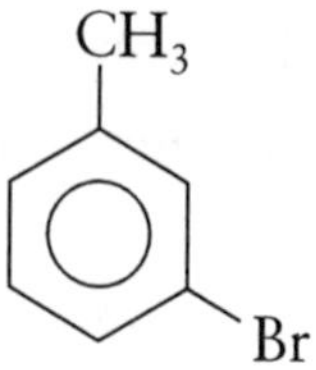

18.69 These hydrocarbons are too volatile (they are all gases at room temperature) and would not remain in the liquid state.

18.71 Cracking is the process of breaking down hydrocarbons with many carbon atoms into smaller units, whereas alkylation is the process of combining smaller hydrocarbons into larger units. Both processes are carried out catalytically and both are used to convert hydrocarbons into units having 6 to 10 carbon atoms, suitable for use in gasoline.

18.73 $CH_3(CH_2)_2CH(OH)CH_3 \xrightarrow[120°C]{H_2SO_4} H_2O(g) + CH_3(CH_2)_2CH{=}CH_2$

18.75 The NO_2 group is a meta-directing group and the Br atom is an ortho-, para-directing group. Because the position para to Br is already substituted with the NO_2 function, further bromination will not occur there. The resonance forms show that the bromine atom will activate the position ortho to it, as expected. The NO_2 group will *deactivate* the group ortho to itself, which in essence enhances the reactivity of the position meta to the NO_2 group. This position is also the position that is ortho to the Br atom, so the effects of the Br and NO_2 reinforce each other. Bromination is thus expected to occur as shown:

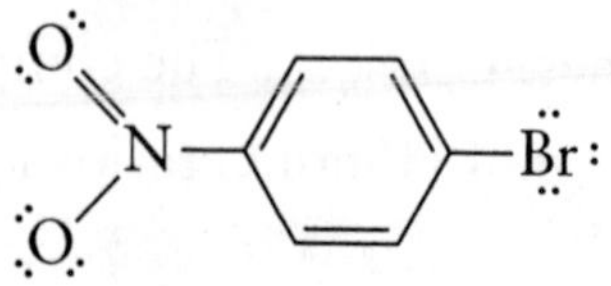

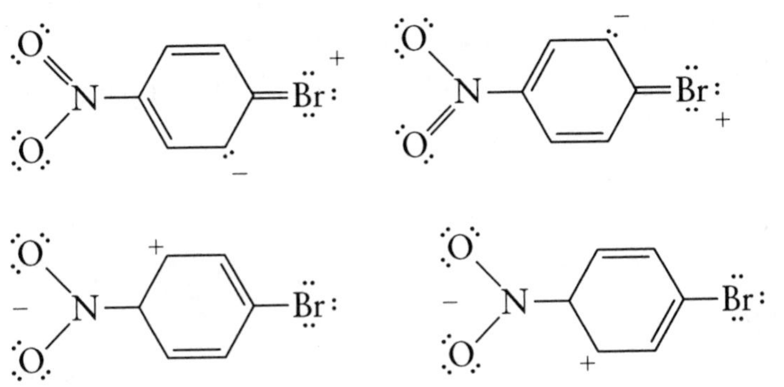

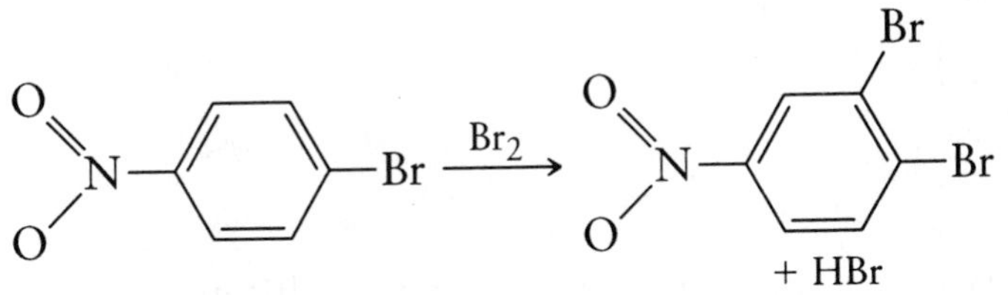

18.77 A mixture of ethane (0.500 bar) and hydrogen (0.300 bar) was placed over a metal catalyst and the equilibrium was established. The dominant reaction is the hydrogenation of ethane to methane. What will be the ratio of methane to ethane at equilibrium (to simplify the problem use values for 298 K)?

The balanced equation is

$C_2H_6(g) + H_2(g) \longrightarrow 2\ CH_4(g)$

The equilibrium constant for this reaction can be obtained at 500°C from the enthalpy and entropy changes of the reaction:

$$\Delta H°_r = 2\Delta H°_f(CH_4, g) - \Delta H°_f(C_2H_6, g)$$
$$= 2(-74.81\ kJ\cdot mol^{-1}) - (-32.82\ kJ\cdot mol^{-1})$$
$$= -116.80\ kJ\cdot mol^{-1}$$

$$\Delta S°_r = 2S°_m(CH_4, g) - [S°_m(C_2H_6, g) + S_m(H_2, g)]$$
$$= 2(186.26\ J\cdot K^{-1}\cdot mol^{-1})$$
$$- [229.60\ J\cdot K^{-1}\cdot mol^{-1} + 130.68\ J\cdot K^{-1}\cdot mol^{-1}]$$
$$= +12.24\ J\cdot K^{-1}\cdot mol^{-1}$$

$$\Delta G°_r = \Delta H°_r - T\Delta S°_r = -116.80\ kJ\cdot mol^{-1}$$
$$- (778\ K)(12.24\ J\cdot K^{-1}\cdot mol^{-1})/1000\ J\cdot kJ^{-1}$$
$$= -126.32\ kJ\cdot mol^{-1}$$

The equilibrium constant is obtained from

$$\Delta G°_r = -RT \ln K$$

$$(-126.32\ kJ\cdot mol^{-1})(1000\ J\cdot kJ^{-1}) = -(8.314\ J\cdot K^{-1}\cdot mol^{-1})(778\ K) \ln K$$

$$\ln K = 190$$

$$K = 3 \times 10^{82}$$

Because the equilibrium constant is so large, the reaction essentially goes to completion. At equilibrium then, there will be 0.300 bar of CH_4, 0.200 bar of C_2H_6 and essentially no H_2. The amount of H_2 can be calculated from the equilibrium expression:

$$\frac{{P_{CH_4}}^2}{P_{C_2H_6}P_{H_2}} = 3 \times 10^{82}$$

$$\frac{(0.300)^2}{(0.200)P_{H_2}} = 3 \times 10^{82}$$

$$P_{H_2} = \frac{(0.300)^2}{(0.200)(3 \times 10^{82})}$$

$$= 2 \times 10^{-83} \text{ bar}$$

18.79 If the molecule contains two carbon centers that have four different substituents attached but are arranged such that they are mirror images of one another *within* the molecule, the molecule will not be optically active. Such an example is shown below in general, for a 1,2—X_2—1,2—Y_2—1,2—Z_2 substituted ethane. Many other examples are possible—the only criterion being that the carbon atoms that have 4 substituents must have a mirror image carbon center within the molecule.

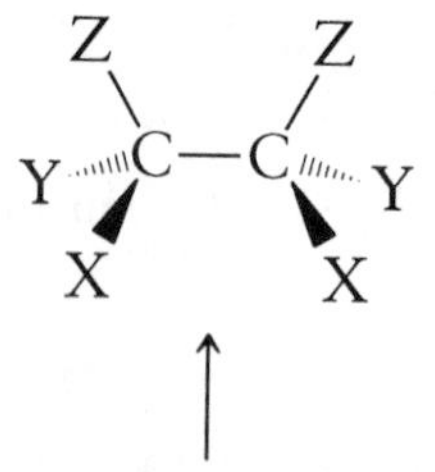

a mirror plane exists in the molecule

For comparison, the molecule

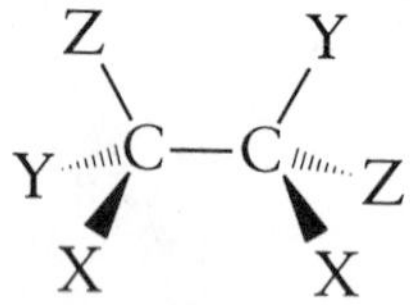

does have a nonsuperimposable mirror image and would exist as two members of an enantiomeric pair.

18.81 The presence of one bromine atom will produce in the ions that contain Br, companion peaks that are separated by 2 u. Any fragment that contains Br will show this "doublet" in which the peaks are nearly but not exactly equal in intensity. Thus, seeing a mass spectrum of a compound that is known to have Br or that was involved in a reaction in which Br could have been added or substituted with such doublets, is almost a sure sign that Br is present in the compound. It is also fairly easy to detect Br atoms in the mass spectrum at 79 and 81 u, confirming their presence. If more than one Br atom is present, then a more complicated pattern is observed for the presence of the two isotopes. The possible combinations for a molecule of unknown formula with two Br atoms is $^{79}Br^{79}Br$, $^{79}Br^{81}Br$, $^{81}Br^{79}Br$, and $^{81}Br^{81}Br$. Thus, a set of three peaks (the two possibilities $^{79}Br^{81}Br$ and $^{81}Br^{79}Br$ have identical masses) will be generated that differ in mass by two units. The center peak, which is produced by the $^{79}Br^{81}Br$ and $^{81}Br^{79}Br$ combinations, will have twice the intensity of the outer two peaks, because statistically there are twice as many combinations that produce this mass. All modern mass spectrometers have spectral simulation programs that can readily calculate and print out the relative isotopic distribution pattern expected for any compound formulation, so that it is possible to easily match the expected pattern for a particular ion with the experimental result.

CHAPTER 19
ORGANIC CHEMISTRY II: FUNCTIONAL GROUPS

19.1 (a) RNH_2, R_2NH, R_3N (b) ROH (c) 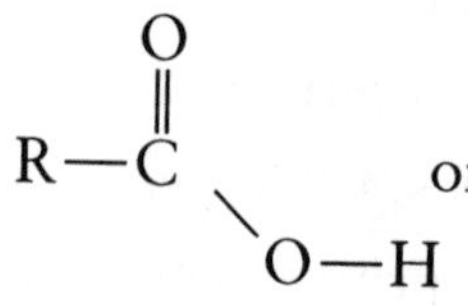 or RCOOH

(d) 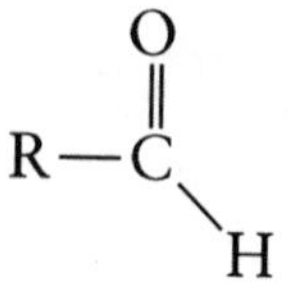 or RCHO

19.3 (a) ether (b) ketone (c) primary amine (d) ester

19.5 (a) 1,2-dichloropropane (b) 1,1-difluoroethene (c) 3-bromo-1-propyne (d) 1,1,1-trichloropropanone

19.7 (a) $CH_3CH_2CH_2CH_2OH$, primary alcohol (b) $CH_3CH_2CH(OH)CH_3$, secondary alcohol (c) 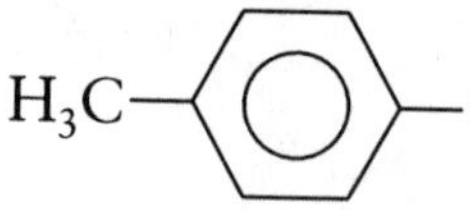, phenol

19.9 (a) diethyl ether (b) $CH_3OCH_2CH_3$

19.11 (a) aldehyde, ethanal
(b) ketone, propanone
(c) ketone, 3-pentanone

19.13 (a) 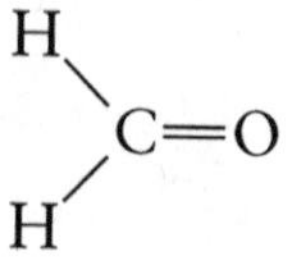 (b) 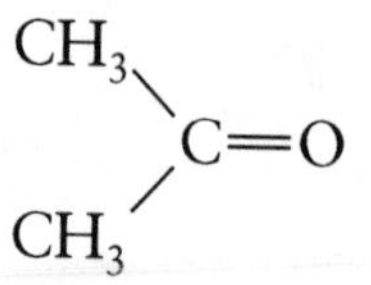 (c) 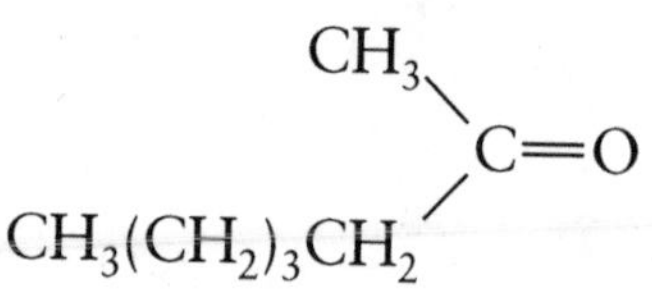

19.15 (a) ethanoic acid (b) butanoic acid (c) 2-aminoethanoic acid

19.17 (a) 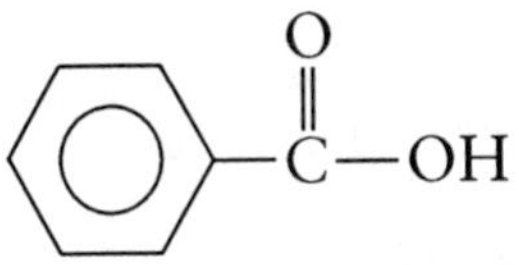 (b) 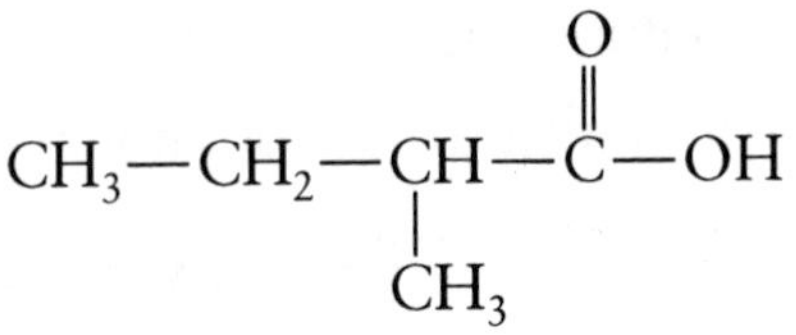

(c) 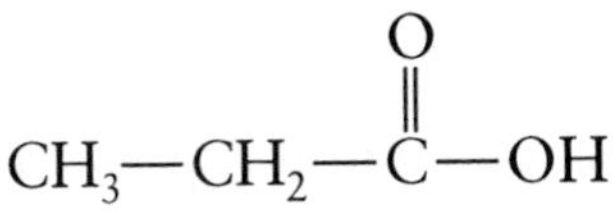

19.19 (a) methylamine (b) diethylamine (c) *o*-methylaniline or 2-methylaniline or *o*-methylphenylamine

19.21 (a) 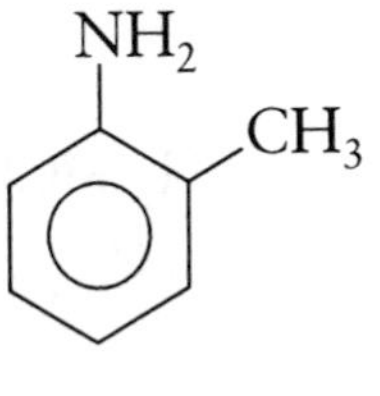 (b) 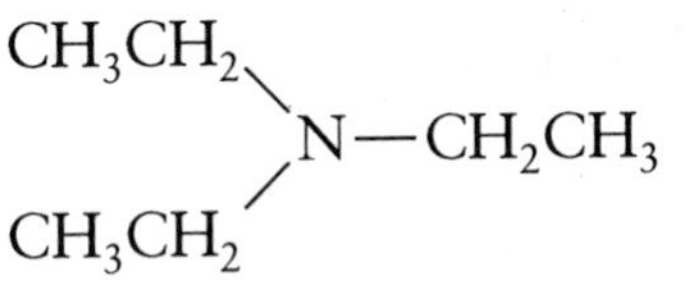 (c) 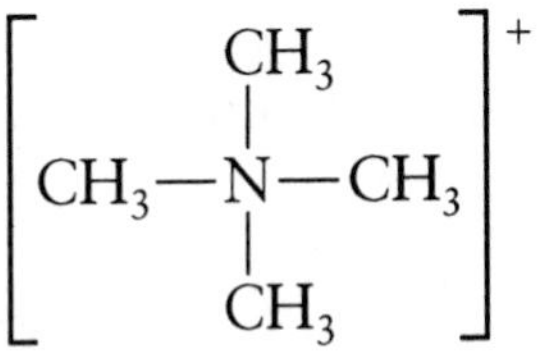

19.23 (a) The product is $C_6H_5(CH_3)CH(OH)$. (b) The mechanism must be S_N1, because an S_N2 mechanism would produce inversion of configuration; however, the product would still be optically active. The dissociation of the halide ion prior to addition of water allows the intermediate ion to form a racemic product.

19.25 Only (a) and (c) may function as nucleophiles, because they have lone pairs of electrons that will be attracted to a positively charged carbon center. CO_2 and SiH_4 have no lone pairs and cannot function as nucleophiles.

19.27 (a) ethanol (b) 2-octanol (c) 5-methyl-1-octanol These reactions can be accomplished with an oxidizing agent such as acidified sodium dichromate, $Na_2Cr_2O_7$.

19.29 (a) 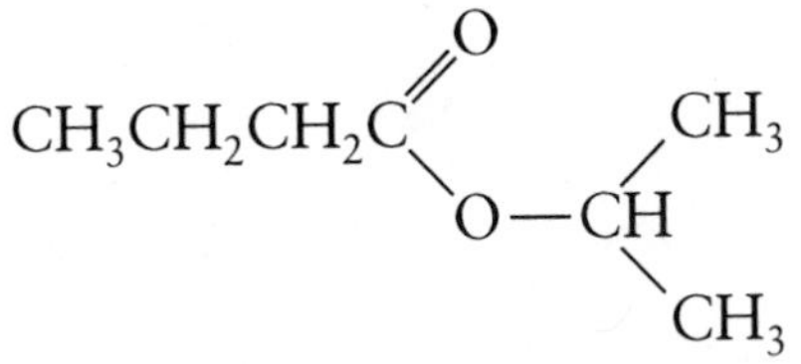 (b) 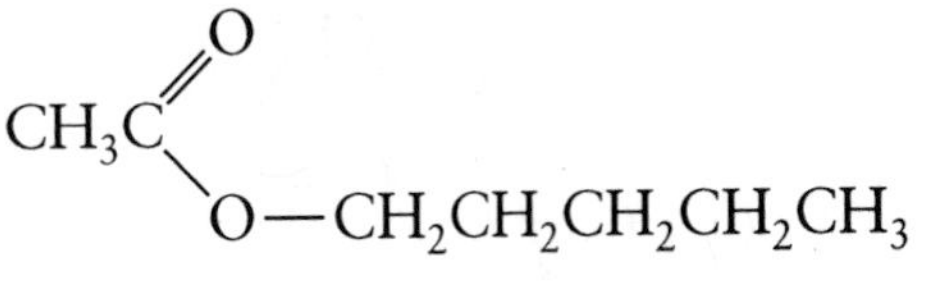

(c) 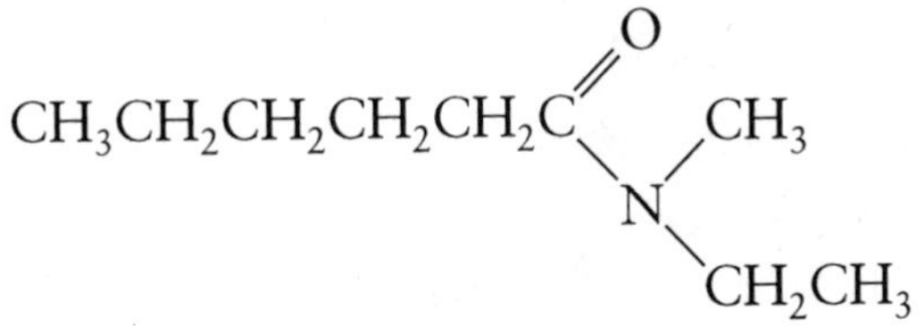 (d) 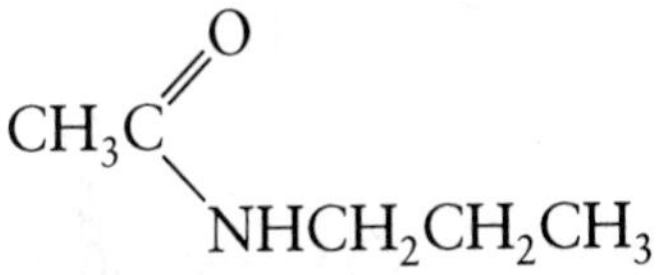

19.31 The following procedures may be used:

(1) Use an acid-base indicator and look for a color change.

(2) $CH_3CH_2CHO \xrightarrow{\text{Tollens reagent}} CH_3CH_2COOH + Ag(s)$

(3) $CH_3COCH_3 \xrightarrow{\text{Tollens reagent}}$ no reaction

Procedure (1) distinguishes ethanoic acid from propanal and 2-propanone. (2) and (3) distinguish propanal from 2-propanone.

19.33 (a) 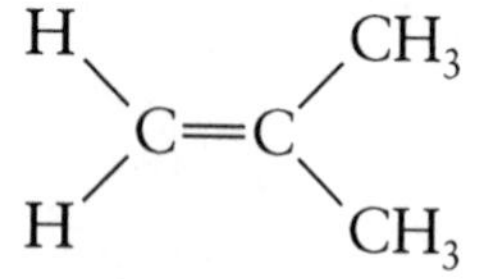 $—CH_2—C(CH_3)_2—CH_2—C(CH_3)_2—CH_2—C(CH_3)_2—$

(b) acrylonitrile: $CH_2{=}CH—CN$ 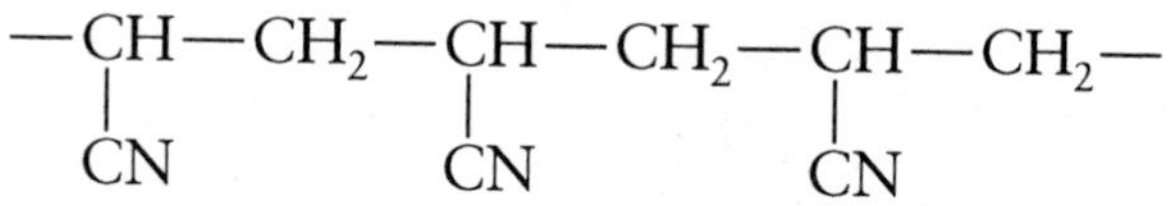

(c) 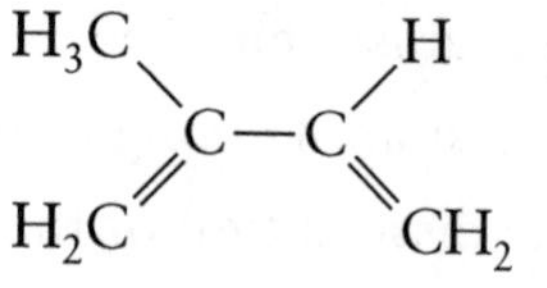

Isoprene

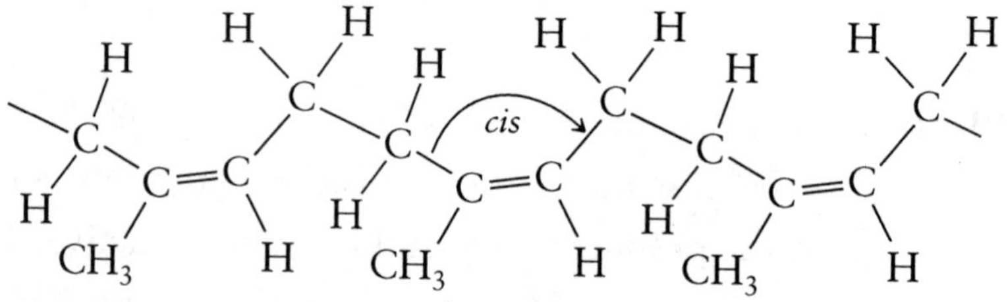

cis version

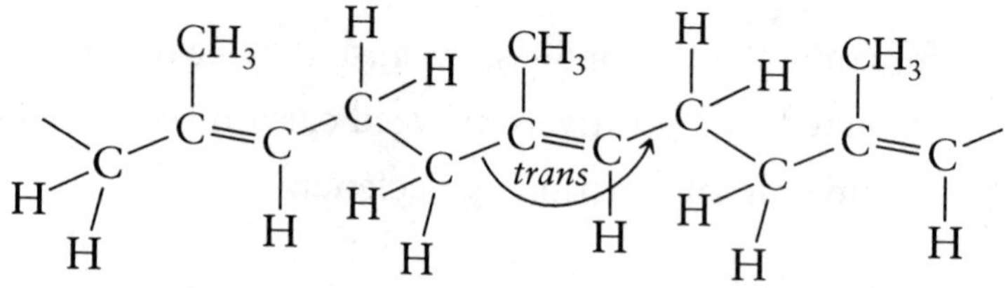

trans version

19.35 (a) 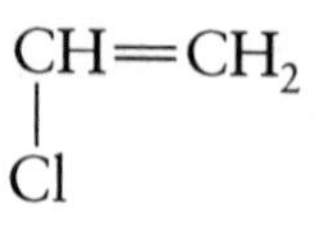 (b) 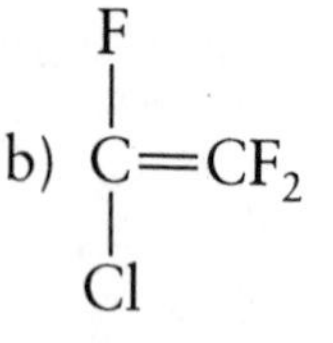

19.37 (a) 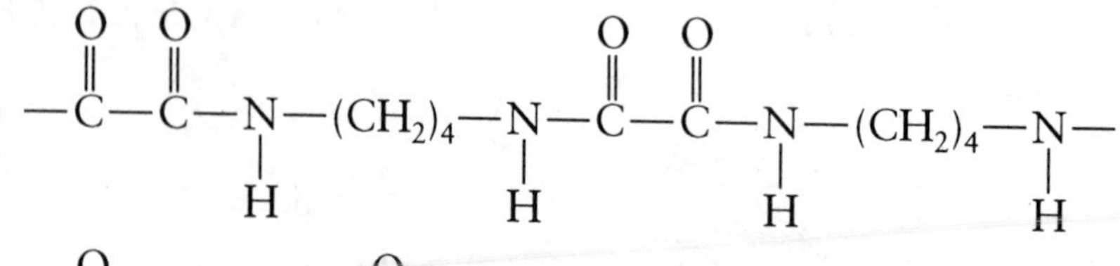

(b) 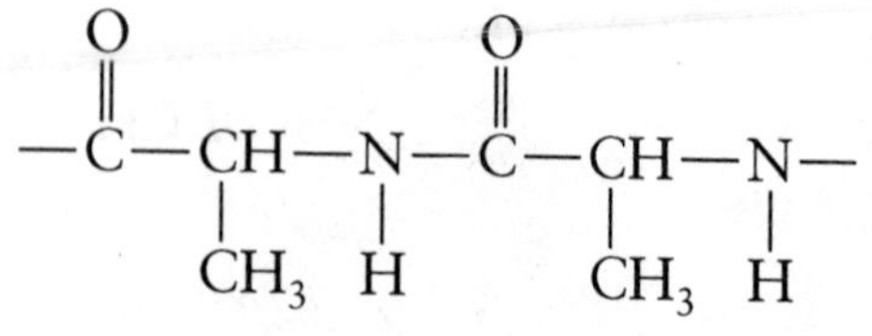

19.39 An isotactic polymer is a polymer in which the substituents are all on the same side of the chain.

A syndiotactic polymer is a polymer in which the substituent groups alternate, from one side of the chain to the other.

An atactic polymer is a polymer in which the groups are randomly attached, one side or the other, along the chain.

19.41 block copolymer

19.43 Larger average molar mass corresponds to longer average chain length. Longer chain length allows for greater intertwining of the chains, making them more difficult to pull apart. This twining results in (a) higher softening points, (b) greater viscosity, and (c) greater mechanical strength.

19.45 Highly linear, unbranched chains allow for maximum interaction between chains. The greater the intermolecular contact between chains, the stronger the forces between them, and the greater the strength of the material.

19.47 The Lewis structures of the molecules are:

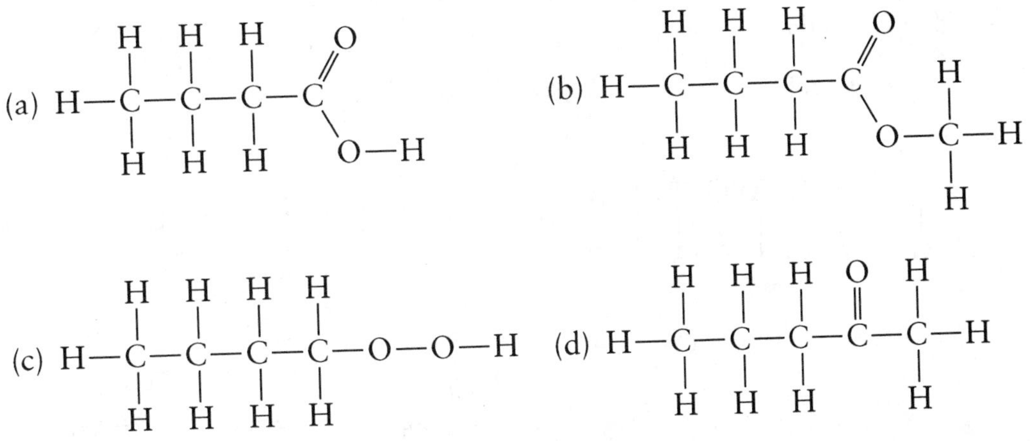

Of these, only (c) is peroxide. (a) is a carboxylic acid, (b) is an ester, and (d) is a ketone.

19.49 (a) 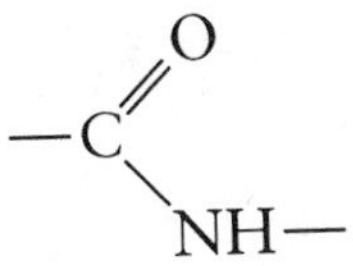 (b) amide (c) condensation

19.51 Side groups that contain hydroxyl, carbonyl, amino, and sulfide groups are all potentially capable of participating in hydrogen bonding that could contribute to the tertiary structure of the protein. Thus, serine, threonine, tyrosine, aspartic acid, glutamic acid, lysine, arginine, histidine, asparagine, and glutamine satisfy the criteria. Proline and tryptophan generally do not contribute through hydrogen bonding, because they are typically found in hydrophobic regions of proteins.

19.53 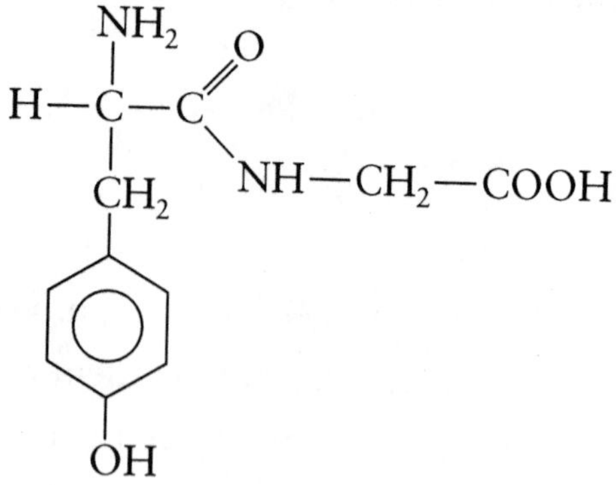

19.55 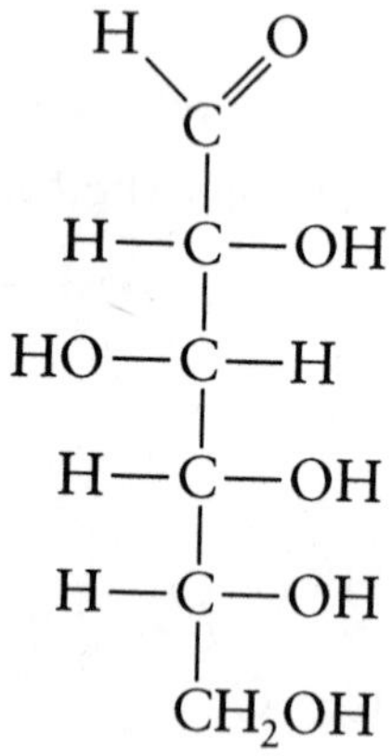

The functional groups are the alcohol, (—OH) and aldehyde 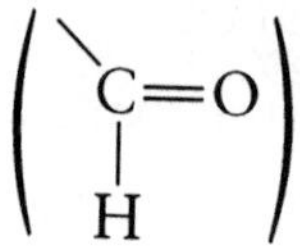 groups.

19.57 (a)
C A T G A G T T A
| | | | | | | | |
G T A C T C A A T

(b)
T G A A T T G C A
| | | | | | | | |
A C T T A A C G T

19.59 (a) 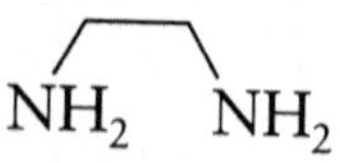 (b) 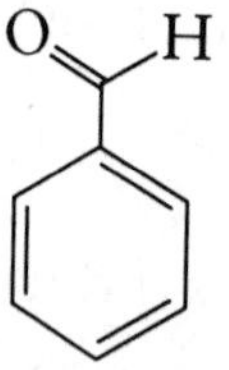 (c) 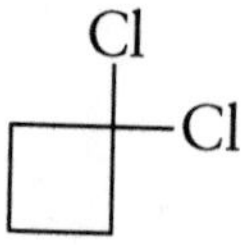 (d) 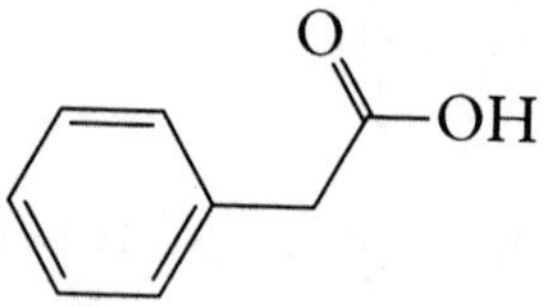

19.61 (a) $C_5H_5N_5O$ (b) $C_6H_{12}O_6$ (c) $C_3H_7NO_2$

19.63 (a) alcohol (—OH), ether (—OCH_3), aldehyde (—CHO)

(b) ketone 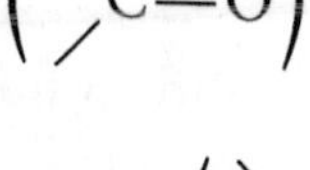

(c) tertiary amine 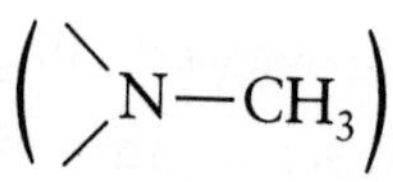, amide 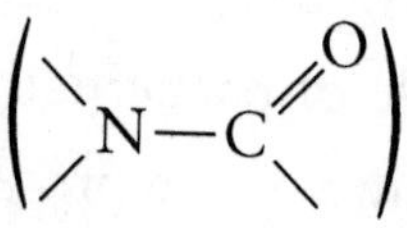

19.65 (a) 2-propanol (b) dimethyl ether (c) methanal or formaldehyde (d) 3-pentanone (e) methylamine

19.67 (a)

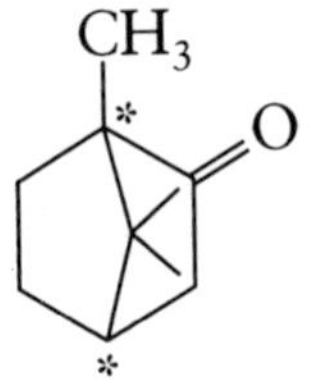

(b)

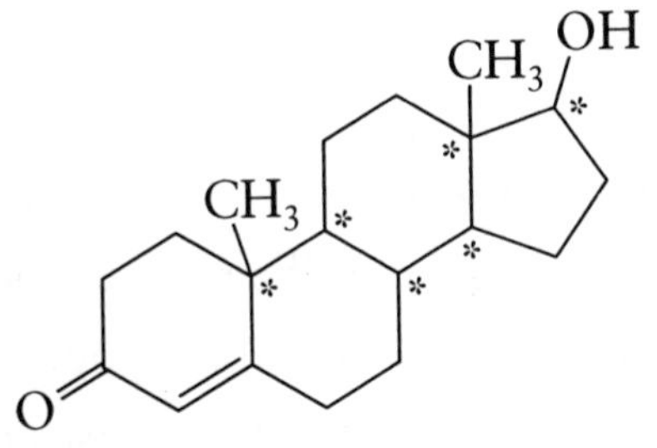

19.69 (a)

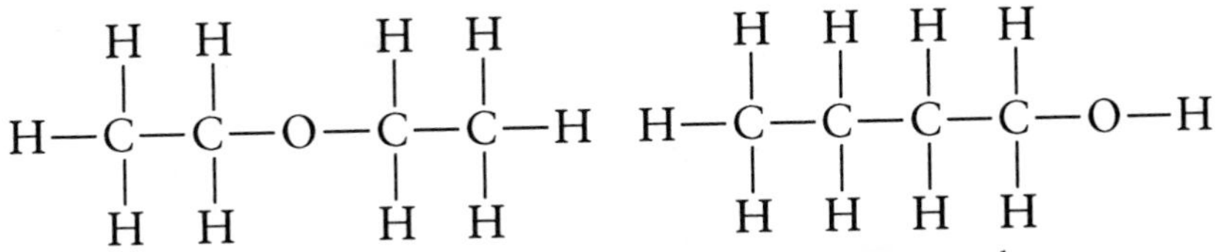

Diethyl ether 1-Butanol

(b) The principal forces between both of these compounds and water resulting in their solubility are London forces. Both molecules are likely to have very similar London forces with water because both contain the same atoms in a similar structural arrangement. However, 1-butanol can also undergo hydrogen bonding with itself, so the molecules are held together strongly in the liquid state, thereby resulting in a relatively high (117°C) boiling point.

19.71

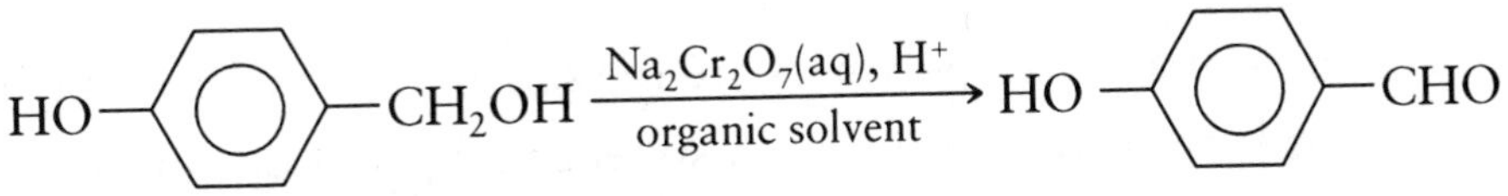

19.73 (a) radical (b) condensation (c) addition (d) radical (e) condensation

19.75 (a) oxidation (b) neither (c) reduction (d) oxidation (e) oxidation

19.77 (a) $HOCH_2CH_2OH + 2\ CH_3(CH_2)_{16}COOH \longrightarrow$

$$CH_3(CH_2)_{16}\overset{\overset{\large O}{\|}}{C}{-}OCH_2CH_2{-}O{-}\overset{\overset{\large O}{\|}}{C}(CH_2)_{16}CH_3 + 2\ H_2O$$

(b) $2\ CH_3CH_2OH + HOOCCOOH \longrightarrow$

$$CH_3CH_2{-}O{-}\overset{\overset{\large O}{\|}}{C}{-}\overset{\overset{\large O}{\|}}{C}{-}O{-}CH_2CH_3 + 2\ H_2O$$

* Indicates the chiral carbon atoms.

(c) $CH_3CH_2CH_2CH_2OH + CH_3CH_2COOH \xrightarrow{\Delta}$

$$CH_3CH_2\overset{\overset{\displaystyle O}{\|}}{C}OCH_2CH_2CH_2CH_3 + H_2O$$

19.79 Polyalkenes < polyesters < polyamides, due to the increasing strength of intermolecular forces between the chains. The three types of polymer have about the same London forces if their chains are about the same length. However, polyesters also have dipole forces contributing to the strength of intermolecular forces, and polyamides form very strong hydrogen bonds between their chains.

19.81 (a) n 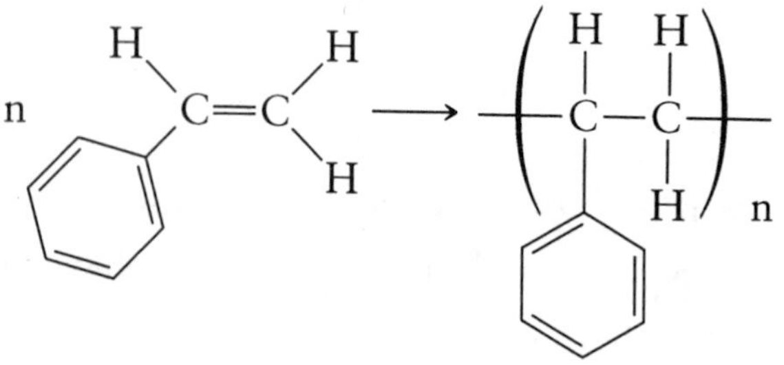

(b) n 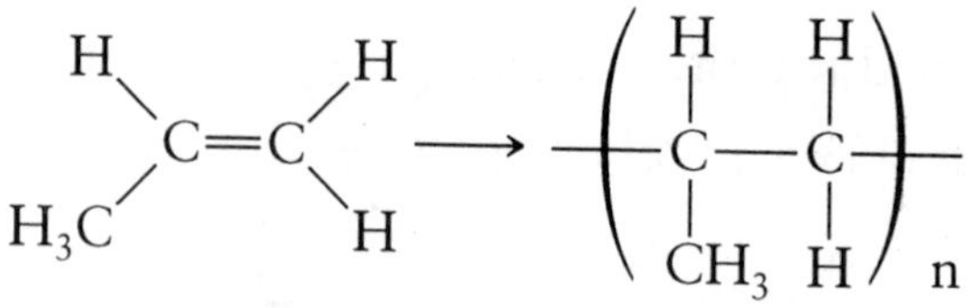

(c) The backbones of all three polymers are the same. They differ simply in what is attached to the long carbon chain. Both styrene and propylene produce polymers that have organic groups attached to the side of the polymer, in the first case a phenyl ring and in the other, a methyl group. These groups are known as *side* groups or *side chains.* For ethylene, there is only one way in which the pure monomer may be polymerized, but for propylene and styrene, the asymmetry in the monomer means that the final polymer will be more complex. This is illustrated in the more extended structures shown below, for a general side group R. Notice that the polymer may be very regular or somewhat irregular in the way the R groups become attached, with respect to each other. The presence of regular or irregular structures affects the properties of the polymers as does the nature of the side chains—not only due to their size but also due to their shape, polarity, etc.

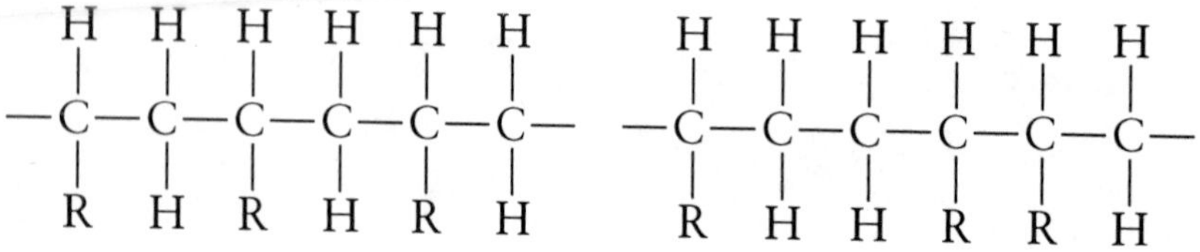

19.83 (a) Primary structure is the sequence of amino acids along a protein chain. Secondary structure is the conformation of the protein, or the manner in which the chain is coiled or layered, as a result of interactions between amide and carboxy groups. Tertiary structure is the shape into which sections of the proteins twist and intertwine, as a result of interactions between side groups of the amino acids in the protein. If the protein consists of several polypeptide units, then the manner in which the units stick together is the quaternary structure.
(b) The primary structure is held together by covalent bonds. Intermolecular forces provide the major stabilizing force of the secondary structure. The tertiary structure is maintained by a combination of London forces, hydrogen bonding, and sometimes ion-ion interactions. The same forces are responsible for the quaternary structure.

19.85 The basic difference between polyester fabrics and fabrics, such as wool, silk, and nylon, is that the latter group are all polyamides. The presence of the amide functional group, $-\underset{\text{O}}{\underset{\|}{\text{C}}}-\underset{\text{H}}{\underset{|}{\text{N}}}-$, allows for extensive cross-linking between polymer chains by hydrogen bonding with the amine group, —N—H, on one chain and the carbonyl group, >C=O, on another. Intermolecular attractions are thus greater between polyamide molecules than between polyester molecules. The less polar polyester molecules can more readily accept radon atoms among their chains.

19.87 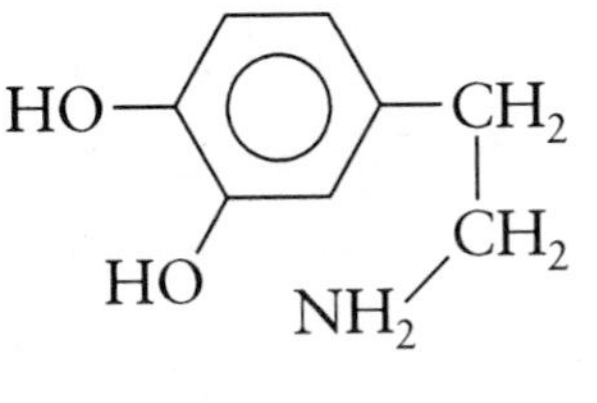

19.89 Condensation polymerization involves the loss of a small molecule, often water or HCl, when monomers are added together. 1,2-benzendicarboxylic acid reacts with ethylene glycol to yield

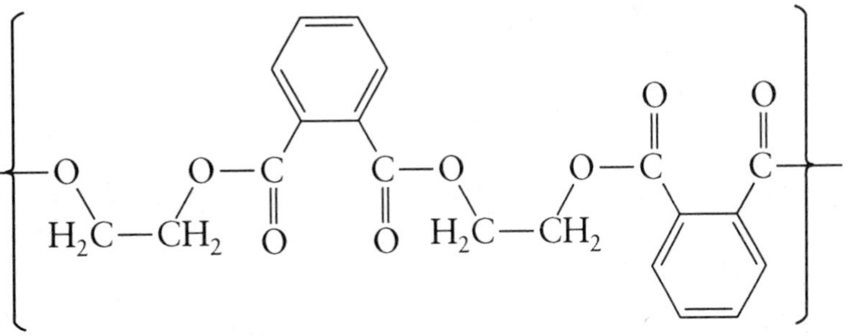

Dacron:

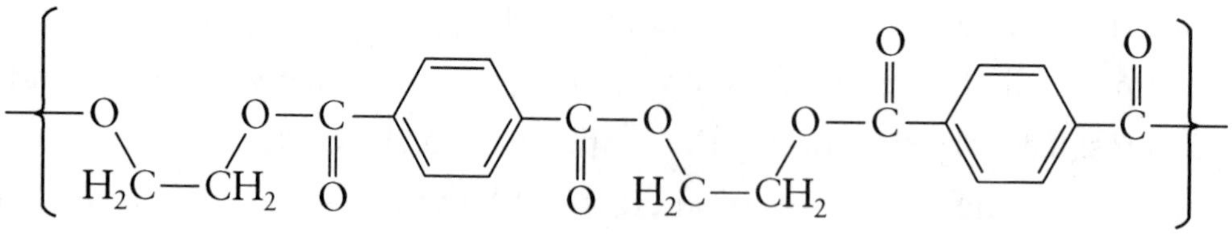

Dacron is more linear than the polymer obtained from benzene-1,2-dicarboxylic acid and ethylene glycol, so Dacron can be more readily spun into yarn.

19.91 Two peaks are observed with relative overall intensities 3 : 1. The larger peak is due to the three methyl protons and is split into two lines with equal intensities. The smaller peak is due to the proton on the carbonyl carbon atom and is split into four lines with relative intensities 1 : 3 : 3 : 1.

19.93 (a) ^{13}C

(b) 1.11%

(c) No. The reason is that the probability of finding two ^{13}C nuclei next to each other is very low. A ^{12}C nucleus next to a ^{13}C nucleus will not interact with the ^{13}C nucleus because the ^{12}C nucleus has no spin. Because the natural abundance of ^{13}C is 1.11%, the probability of finding two ^{13}C nuclei next to each other in an organic molecule will be 0.0111×0.0111 or 1.23×10^{-4}. Although such coupling is possible, it is generally not observed because the signal is so much weaker than the signal, due to the molecules with a single ^{13}C nucleus.

(d) Maybe. Because most of the carbon to which the protons are attached is ^{12}C, the bulk of the signal will not be split. The protons that are attached to the ^{13}C atoms will be split, but this amounts to only 1.11% of the sample, so the peaks are very small. Peaks that result from coupling to a small percentage of a magnetically-active isotope are referred to as satellites and may be observed if one has a very good spectrum.

(e) Yes. Although the splitting of protons by ^{13}C may not be observed because the amount of ^{13}C present is low, the opposite situation is not true. If a ^{13}C is attached to H atoms, the large majority of those H atoms will have a spin and so the ^{13}C will show fine structure due to splitting by the H atoms.